Jupiter Odyssey

The Story of NASA's Galileo Mission

Springer
London
Berlin
Heidelberg
New York
Barcelona
Hong Kong
Milan
Paris
Santa Clara
Singapore
Tokyo

David M. Harland

Jupiter Odyssey

The Story of NASA's Galileo Mission

David M. Harland
Formerly Visiting Professor
University of Strathclyde
Glasgow, UK

SPRINGER–PRAXIS BOOKS IN ASTRONOMY AND SPACE SCIENCES
SUBJECT *ADVISORY EDITOR*: John Mason B.Sc., Ph.D.

ISBN 1-85233-301-4 Springer-Verlag Berlin Heidelberg New York

British Library Cataloguing-in-Publication Data
Harland, David M.
Jupiter odyssey: the story of NASA's Galileo mission. –
(Springer-Praxis books in astronomy and space sciences)
1. Galileo Project 2. Space flight to Jupiter 3. Jupiter
(Planet) – Exploration 4. Astronautics – United States –
History
I. Title
629.4′3545′0973

ISBN 1852333014

Library of Congress Cataloging-in-Publication Data
Harland, David M.
Jupiter odyssey: the story of NASA's Galileo mission/David M. Harland.
p. cm. – (Springer-Praxis books in astronomy and space sciences)
Includes bibliographical references.
ISBN 1-85233-301-4 (alk. paper)
1. Galileo Project. 2. Jupiter (Planet)–Exploration. I. Title. II. Series.

QB661.H37 2000
629.45′55–dc21 00-055636

Printed by MPG Books Ltd, Bodmin, Cornwall, UK

Project Copy Editor: Alex Whyte
Cover design: Jim Wilkie
Typesetting: BookEns Ltd, Royston, Herts., UK

Printed on acid-free paper supplied by Precision Publishing Papers Ltd, UK

"I can't believe it, we're actually going to see a cometary impact – in our lifetimes!"

Gene Shoemaker

Carolyn and Gene Shoemaker, together with Melissa McGrath, Keith Noll and Forrest Hamilton of the Space Telescope Science Institute, await the first picture from the Hubble Space Telescope of the impact of comet Shoemaker Levy 9 with Jupiter in July 1994. (Picture courtesy of John Aiello, with thanks to Ray Villard)

Table of contents

List of illustrations

Cover shot: This artist's impression shows the Galileo spacecraft, with its fouled high-gain antenna, making an approach to Io, Jupiter's volcanic moon. (Courtesy, NASA; adapted from a painting by Michael Carroll)

Chapter 14: Fiery Io

Chapter 15: Passing the torch

Appendix

List of tables

Foreword

Project Galileo was born in an age of great ambitions and commitment. The Project was approved by the US Congress in October 1977 on the heels of the successful launches of Voyager 1 and Voyager 2 a few months earlier and the marvellously successful first landings on Mars by Vikings 1 and 2 in the summer of 1976. Galileo would be the first orbiter of an outer planet and place the first probe directly into an outer planet's atmosphere. Its target was the Jupiter System, which contains twice the mass of all the other planets combined. In many respects Jupiter offered the best laboratory to investigate the origin and evolution of our Solar System owing to its tremendous mass, horrendous magnetosphere and 'miniature solar system' of sixteen known natural satellites.

It was a time when it was well recognised that, regardless of the size and complexity of an interplanetary spacecraft, certain basic functions must be performed such as navigation and antenna pointing. So the paradigm was to carry as many scientific instruments to the planet as possible because the base costs of many irreducible functions are not significantly increased by adding instruments. Thus, Galileo was developed into a wonderful machine carrying eleven instruments on the Orbiter and another seven instruments on its Atmospheric Entry Probe. This is one of the most powerful, synergistic scientific payload suites ever devised. NASA selected the investigations with exactly that result in mind.

Much has been publicised about the many delays of Project Galileo. Originally it was to launch in January 1982 and arrive at Jupiter in July 1985. Ultimately, Galileo launched on 18 October 1989 and arrived at Jupiter on 7 December 1995. Due mainly to delays in the Space Shuttle Program and its various upper-stage candidates for Galileo, the Project was reprogrammed no less than five times. However, these delays resulted in the finding and fixing of many spacecraft and instrument problems. The Galileo spacecraft launched in 1989 was far more robust and capable than it would have been had it been launched earlier. In fact, two problems were discovered after the aborted 1986 launch campaign that were fixed during the post-Challenger stand-down, either of which would very likely have been catastrophic if not corrected. One of these involved upgrading the memory chips in the central computer with a newer, more reliable chip. This had the serendipitous result of allowing a doubling of the memory capacity within the available packaging volume. The additional memory was a Godsend when implementing the flight software workaround of the high-gain antenna deployment failure on the way to Jupiter.

It was indeed a tremendous tenacity that enabled the men and women of Project Galileo to get through the many technical and political challenges and ultimately launch a magnificent spacecraft on an awesome mission. The first year and a half of flight, including the absolutely essential gravity-assists from Venus and Earth-1, was smooth, with no major problems. Then, when the high-gain antenna (HGA) was commanded to deploy on 11 April 1991, it stuck. In the following years, everything imaginable and safe was done to try to open the HGA. Comprehensive analysis and testing of the spare HGA concluded that two or three of the umbrella-like antenna ribs were stuck to the central post because the lubrication on the heads of the pins that braced the ribs against the post during launch had been scrubbed off by vibration. There simply was no mechanism onboard for pulling the pins out, although a person could easily have done it with just one finger.

Innovation of the first magnitude came to the rescue when the Deep Space Network Advanced Technology Program showed that their research on data compression and coding would enable development and installation of new flight software on the Galileo spacecraft that would facilitate obtaining a full tape recorder load of images on every one of Galileo's orbits of Jupiter. The images would be recorded during the Jupiter and satellite encounters near perijove and then, in the balance of the one- to two-month orbit, the images would be read from the tape, small sections at a time, compressed in a spacecraft computer and, employing special encoding, transmitted to Earth over the low-gain antenna to upgraded and arrayed DSN antennas. When Galileo made the first-ever spacecraft encounters of asteroids – Gaspra in October 1991 and Ida in August 1993 – it had to do so without this new data return capability since it would not be available for uplinking to the flight computers until *after* it arrived at Jupiter. In the case of Gaspra, Galileo's subsequent return to Earth for the second and final gravity-assist enabled the data return at high rate close to Earth; for Ida, the prolonged interplanetary cruise and special editing techniques enabled the data return.

The last near-fatal event came two months before arrival at Jupiter. It appeared that the single tape recorder had broken. The workaround to the HGA failure crucially depended on this recorder. Again, some of the most remarkable detection work determined that, in fact, the tape recorder was not broken, but just stuck. After getting the recorder unstuck, and fully understanding its problems, very strict additional conditioning and usage rules were invoked to ensure that it could perform its now mission-critical function. In fact, the recorder problems were not fully understood until well after arrival at Jupiter. In my most painful decision as Project Manager, I directed that the recorder not be used in high-speed mode to record images of Io as had always been the plan. Though vindicated by subsequent diagnosis, I remain unforgiven by some of the scientists. I was most particularly gratified when Galileo made two close encounters of Io at the end of its first extended mission in 1999.

The Orbiter precisely delivered the Probe to its entry corridor and captured every bit of data the Probe transmitted. The Galileo Probe was an outstanding success, with all instruments providing excellent data and some data return as deep as 20 bars – twice the mission requirement. The Orbiter's Jupiter Orbit Insertion (JOI) was perfect. This marked the beginning of a phenomenally successful Orbital Mission. It had always been accepted that perhaps one of the ten satellite encounters over the two-year primary mission would be lost due to some anomaly. In fact, Galileo did ten for ten! It then went on to a two-year extended mission, including eight Europa encounters, and the long elusive Io was encountered at altitudes of 600 kilometres and 300 kilometres on 10 October and 25 November, respectively, of 1999. Galileo's bounty from Jupiter is absolutely stunning and we each have our favourites. For me and many others, the discovery of strong evidence of an extant liquid water ocean under Europa's ice crust and the implications for extraterrestrial life is the pinnacle. When Project Scientist Torrence Johnson and I had the fabulous occasion to show this 'ice rafts' imagery to Pope John Paul II at the Vatican in January 1997, he simply exclaimed, "Wow!".

With relatively minor, isolated temporary exceptions, all of the Galileo Orbiter's scientific instruments provided excellent data at every encounter and as appropriate in the orbital cruise periods between encounters. Galileo's scientific bounty easily achieved legendary proportions in the primary mission alone. Project Galileo has indeed justified its great namesake. The Project stands as living proof of the importance of large, capable, and – yes, complex – spacecraft in space exploration. The scientific Return on Investment (ROI) for these flagship missions has been outstanding. This is inappropriately obscured by the current enthusiasm for the smaller and the cheaper. The proof is readily available in the scientific journals and proceedings.

David Harland has written a most comprehensive, fast-paced story of Project Galileo. You will get a wonderful sense of the challenges and triumphs of this truly seminal Project.

In closing this foreword, I want to take this opportunity to express my tremendous pride in the Project Galileo Team, and I mean the entire team. Not only the people who operated Galileo so superbly for the eight-year primary mission and rebuilt its central nervous system inflight, but equally those who conceived, designed, built, and tested it through the 1970s and 1980s, and those who continued to operate it so expertly in the first – and now the second – mission extension.

Much is said about HUMAN presence in space. Galileo IS our HUMAN presence at Jupiter – that planetary system with twice the mass of all the other planets combined. It is our eyes, ears, touch, and taste. And while its voice is very weak, we gave it a new, very articulate language and the NASA/JPL Deep Space Network developed and installed the ultimate state-of-the-art hearing aids. Through this miraculous communication triumph we have succeeded in effectively connecting our magnificent robotic senses aboard Galileo with the rest of our bodies here on earth. So we have indeed, as never before, looked and listened and touched the

Jupiter System and learned extensively about the giant planet of our Solar System. It has not been some inanimate object doing this – it has been we humans operating the senses we built and sent to Jupiter nearly a billion kilometres away.

Project Galileo is a tremendous success of the human creativity and spirit. It will surely be recorded as one of the great feats of the twentieth century.

Bill O'Neil*
Galileo Project Manager
Primary Mission 1990–1998

*Mr. O'Neil participated in Project Galileo from its inception: As JPL's Mission Design Department Manager 1977–1980; as Project Galileo Science and Mission Design Manager 1980–1990; and as Project Manager 1990–1998.

Acknowledgements

I would like to express my thanks to a number of people at the Jet Propulsion Laboratory in Pasadena (managed jointly by NASA and the California Institute of Technology), and most particularly to: Robert Carlson for information on Galileo's near-infrared mapping spectrometer and for reading part of the draft; Glenn Orton for his diary entries and for putting me right on a few points; Steve Mikes for data on Galileo's power supply; Todd Barber for data concerning the performance of Galileo's propulsion system; Marc Rayman for his encouragement and for reading an early draft; and, of course, I am grateful to Bill O'Neil for contributing the Foreword.

At the Lunar and Planetary Laboratory in Arizona, I would like to thank: Alfred McEwen for sending information about Io; Laszlo Keszthelyi for insight into Io and for permission to use certain illustrations; Greg Hoppa for help with Europa and for permission to use certain illustrations; and Bill Sandel for help with the UVSTAR observations made during the Shuttle flight that went down in more popular history for the fact that it marked John Glenn's return to space.

I would also like to thank: Larry Travis of NASA's Goddard Institute for Space Studies for information on Galileo's integrated photopolarimeter and radiometer; Paul Schenk of the Lunar and Planetary Institute in Houston for help with maps of Io and for permission to use certain illustrations; Don Banfield of Cornell University for permission to use certain illustrations; Steve Miller of University College London for help with Jupiter's electro-jet winds; Jennifer Blue of the US Geological Survey (a consultant to the International Astronomical Union's Working Group for Planetary System Nomenclature) for helping to track down maps of asteroids Gaspra, Ida, and Dactyl; and Bill Harris of the University of California at Los Angeles and of the Planetary Data System for tracking down the time lines for Galileo's tour of the Jovian moons.

I must also acknowledge the help of Donald Savage of NASA Headquarters, Ray Villard of the Space Telescope Science Institute, and David Levy of the Lunar and Planetary Laboratory, in identifying individuals in the frontispiece. Roger Launius of the NASA History Office is to be thanked for supplying several very interesting books relating to space missions of yesteryear.

Finally, I would like to thank Patrick Moore for his early encouragement, David

Woods for commenting upon an early draft, Tom Maver of the University of Strathclyde for high-speed access to the Internet, and, of course, Clive Horwood of Praxis for his eternal enthusiasm.

Author's Preface

Jupiter Odyssey: The Story of NASA's Galileo Mission is not a political and financial history of the Galileo spacecraft's tortured gestation.* Nor does it dwell upon the Pioneers and Voyagers which blazed the trail to Jupiter, as these missions have already been reported in both official and popular histories. This book focuses specifically on Galileo's odyssey-like journey through the Solar System and relates, *in depth*, what it discovered along the way.

While *en route* to Jupiter, Galileo had opportunity to make observations of Venus; the Earth and its Moon; a pair of asteroids – one of which would turn out to have a companion; and the crash of a comet into Jupiter. In fact, from the mission's conception, the story spans a quarter of a century. It is a tale of dogged tenacity and of imaginative engineering solutions to some of the toughest technical and political problems ever faced by a mission for planetary exploration.

As this book was being written, Galileo was beginning its so-called 'Millennium Mission'. The joint observations it was to make with Cassini as that spacecraft passed through the Jovian system in December 2000 would symbolically hand the 'torch' of exploration from one spacecraft to the other, and mark the switch of focus from Jupiter to Saturn.

I have borrowed liberally from the Planetary Image Archive managed by the Jet Propulsion Laboratory in Pasadena. When possible, I have mosaicked related imagery to reconstruct wide views, and in many cases I have annotated the imagery to name features pertinent to the text. If you find the annotations irksome, feel free to download the original imagery, and consider this volume as a handy guide book.

David M Harland
Kelvinbridge, Glasgow
May 2000

* *History of the Galileo Project to Jupiter* by Michael Meltzer will be published in 2003 by the NASA History Office as SP-4236.

1

Early days

THE DISCOVERY

While living in Pisa, Italy, Galileo Galilei is said to have dropped differently sized masses from the 'leaning tower' in order to demonstrate that they would all fall at the same speed – a prediction at odds with the conventional Aristotelian view that they would fall at different rates, the heavier ones reaching the ground first. In overturning Aristotle, Galileo became the first experimental physicist. His demonstration was not welcomed by the Church, however, which favoured Aristotle's view. In 1592, Galileo moved to Padua, which fell under the authority of Venice, and whose administration was rather more open-minded.

In 1608 Hans Lipperschey, a Dutch spectacle-maker, invented a device utilising two lenses through which it was possible to make distant objects appear closer; he called this a 'looker'.[1]* In May of the following year, Galileo heard about the invention. After building one, he set out to improve the design[2] in order to be able to identify ships on the horizon a few hours earlier than they would normally become evident, which, to Venetian merchants, was a significant advantage. Later in the year, he indulged his own longstanding interest, and started to observe objects in the sky. On 7 January 1610, he focused his attention on Jupiter, and, in addition to resolving it as a disk, he saw three small star-like points arranged in a line nearby. Over successive nights, not only did he note that these three subsidiary objects changed their relative positions, but he spotted a fourth.[3] As he accumulated observations, he realised that these objects were circling around Jupiter. This was contrary to the accepted view, as taught by Ptolemy, that things moved around the Earth on the 'celestial sphere'. Galileo favoured the theory proposed by the Polish astronomer Nicolaus Copernicus in 1543[4] that only the Moon circled the Earth, with everything else moving around the Sun. While the existence of the Jovian satellites did not actually prove Copernicus's theory, it certainly contradicted Aristotle, and so Galileo once again found himself in serious strife with the Church.[5, 6]

It is possible that Galileo was not actually the first person to see what he referred

* For Notes and References, see pages 413–424.

to as "these marvellous things which have lain hidden for all ages past". Simon Marius, the German astronomer – a longstanding rival of Galileo who had also received a Dutch-built instrument – may have observed the objects a month earlier, but Galileo is credited with the discovery as he was the first to publish.[7] Galileo called them the 'Medicean Stars', in honour of Cosimo de Medici, a leading member of one of the merchant families that had ruled Florence and Tuscany for some two centuries. However, Johann Kepler, a German astronomer, referred to them as the Galilean satellites. In 1614 they were individually named by Marius, in order of increasing range from the planet, as Io, Europa, Ganymede and Callisto.[8]

THE NEXT FEW CENTURIES

In 1675, the Danish astronomer Olaus Roemer was pondering the reason for the inaccuracies in the tables that he had drawn up to predict the times that Jupiter's moons entered and emerged from Jupiter's shadow. In a moment of inspiration, he realised that the times were accurate but the light did not reach Earth instantaneously. Contrary to contemporary wisdom, light travelled at a finite speed. The discrepancies in the observed timings were related to the changing distances between the two planets as they independently moved around the Sun. This discovery provided a measure of the physical scale of the Solar System.[9] Modern measurements have shown that light travels at approximately 300,000 kilometres per second, and Jupiter's orbit is 5 AU from the Sun.[10]

Early dynamical studies revealed that the orbital periods of Io, Europa and Ganymede were related, with Europa's period being twice that of Io's, and Ganymede's being twice that of Europa's, but the extent to which these resonances would influence the physical structures of the moons was not clear. The popular expectation was that any moons that had formed out in the frozen waste of Jovian space would be icy, because the energy in sunlight of all wavelengths – referred to as insolation – is dissipated with the square of the distance, so at Jupiter it is just 4 percent of that at the Earth's distance from the Sun.

As more powerful telescopes were developed, it became feasible to resolve the moons into disks and albedo variations were then reported. A 'polar spot' was observed on Ganymede by W.R. Dawes in 1849, and maps were published by E.M. Antoniadi in Paris in 1927 and by E.J. Reese in America in 1951. Maps by Bernard Lyot of all four satellites were published a few years later. In terms of the distribution of the main albedo features, Lyot's Ganymede map was in overall agreement with that of Reese, but had little in common with Antoniadi's map. The maps published by A. Dollfus in 1961 were generally regarded as being the most accurate. Over the years, additional satellites were discovered, but these were small asteroidal bodies and even less was known about them.[11]

What of the planet itself?

Jupiter's radius is about 10 times that of the Earth. When astronomers interpreted the periods of Jupiter's satellites using the 'laws of gravity' formulated by Sir Isaac

Newton in 1687, they were able to deduce that although the planet is physically large, it has a very low overall density, a mere 25 percent of the Earth's, and this meant that it is composed predominantly of hydrogen – it is a 'gas giant'. As planets were discovered farther out from the Sun, it was found that, at 318 times the mass of the Earth, Jupiter is not only the biggest planet in the Solar System but is more massive than all the other planets combined.

It was also obvious that Jupiter's disk does not show a solid surface, but an atmospheric circulation system comprising latitudinal 'belts' and 'zones', decorated by a multitude of 'spots', 'ovals', 'barges' and 'plumes', some of which are long-lived. The axis of rotation is almost perpendicular to its orbital plane, so there is no 'seasonal' influence to its weather during its 12-year-long orbit. As individual features in Jupiter's upper atmosphere were tracked moving across the disk, it became clear that although the planet is physically enormous, it rotates very rapidly – in only 10 hours – and this makes its equator bulge to produce an oblate spheroid. Furthermore, the rotation rate on the equator is rather faster than at higher latitudes, and this has become known as 'super rotation'. The rotational period at the equatorial strip is some five minutes less. Individual atmospheric features (the spots, ovals, etc.) have their own rotation rates, so they progressively drift in longitude.[12]

What is a gas giant?

In the late 19th century, astronomer Robert Proctor posited, "Jupiter is still a glowing mass, fluid probably throughout, still bubbling and seething with the intensity of the primaeval fires, sending up continuous enormous masses of cloud, to be gathered into bands under the influence of the swift rotation of the giant planet." He was thinking of Jupiter in terms of a 'failed' star. The prevailing theory at that time was that the Sun was hot because it was still contracting and was transforming gravitational potential into heat. Nuclear energy had yet to be discovered. As Jupiter has only about one-thousandth of the Sun's mass, it is tiny compared to a star; and even if it was an order of magnitude larger it would barely qualify as a 'brown dwarf' – the smallest type of star currently believed to exist. In the 1920s, Sir Harold Jeffreys proved that Jupiter's visible surface could not be hot. He suggested that a very deep, but tenuous, gaseous envelope hid a rocky core englobed by a mantle of ice and solid carbon dioxide.

A key advance was the spectroscopic identification of ammonia and methane in the Jovian atmosphere, as well as the expected hydrogen and helium. Other compounds were only minor constituents, together contributing only 1 percent. In 1932, Rupert Wildt refined Jeffreys' idea by proposing that the rocky core was surrounded first by a thick layer of water ice and then by an ocean of condensed gases. The next real advance was the independent suggestion, in 1951, by W.R. Ramsey in England and W. DeMarcus in America, that the core was metallic hydrogen rather than rock, and that this was surrounded first by an ocean of liquid hydrogen and then by a hydrogen-rich gaseous envelope.

The significance of Jupiter's core being metallic was that it would readily conduct electrical currents, and these currents would in turn generate a magnetic field. The likelihood of Jupiter having a strong magnetic field was a prediction-in-waiting. In 1955, in America, B.F. Burke and K.L. Franklin realised that Jupiter emitted radio waves. It was a chance discovery because they were not actually studying the planet; in tracking down a source they realised that it came from the giant planet. Such emissions suggested that Jupiter possessed a strong magnetic field. As observations of the radio emission accrued, it became evident that there was a periodicity correlated with Jupiter's axial rotation because the magnetic field in the core was dragging the entire magnetosphere around with it; so this 9 hour 55 minute periodicity represented the rate of rotation of the core. It was eventually realised that the 'burst' emissions correlated with the motion of the innermost Galilean satellite, Io, which – in some way – was interacting with the Jovian magnetic field. The discovery in 1958 by Explorer 1, America's first satellite, that the Earth is encircled by intense belts of charged particles prompted speculation that similar belts might encircle Jupiter. The magnetic axis is inclined at 9.5 degrees to the planet's rotational axis, so Io, orbiting in the planet's equatorial plane, is periodically swept by an alternating magnetic field.

Observations of the motions of the moons around Jupiter enabled their bulk densities to be estimated. At 3.5 g/cm^3, Io's density is comparable to the Earth's Moon, which is clearly a rocky body. The other Galilean satellites have lower densities, which means that they contain significant fractions of water. It was noticed that Io brightens somewhat upon emerging from Jupiter's shadow. Evidently, a reflective material settles onto the surface in the darkness, and evaporates when illumination is restored. Further, multicolour photometry revealed that Io's spectrum is distinctly reddish – in fact, it is the reddest body in the Solar System, even redder than Mars, which is often called 'the red planet'.

There was only so much that could be discovered about Jupiter and its system of satellites by peering at them from afar, however. As the 'Space Age' matured, the role of Solar System study passed from telescopic astronomers to robotic exploring machines.

2

Reconnaissance

PIONEERING DEEP SPACE

After an abortive early attempt by the US Air Force to send a series of probes towards the Moon, NASA's Ames Research Center in Mountain View, California, picked up the 'Pioneer' name and assigned it to probes which were dispatched into solar orbit to report on the state of the solar wind. Some of these spinning 'particles and fields' spacecraft were stationed slightly within the Earth's orbit, and some were just beyond it. All were extremely successful. In fact, four were still operating in the 1990s.

When Ames set out to develop a spacecraft to venture further out into the Solar System, it selected Jupiter as the objective. With the agency's budget in a state of decline, it would probably be simpler to extend a successful programme than to initiate a new one, so Ames offered the proposal as an advanced Pioneer – it would be a particles and fields platform, after all – and the Jovian mission was authorised in February 1969.

The previous probes had been drum shaped, and had rotated both for stability and to be able to sweep their sensors in three dimensions in order to chart the electromagnetic fields in the immediate vicinity. The new spacecraft would make similar measurements, but a different configuration was imposed by the mission requirements. Although the earlier design had used solar cells around the drum for power, this would not be practicable in the outer Solar System and, therefore, a radio-isotopic power source was used. As a large antenna would be required to communicate from such a distance, the design took the form of a cluster of instruments set on the rear of the large parabolic dish, with the spacecraft rotating to hold this facing the Earth. Although the spacecraft weighed only 260 kg, an Atlas-Centaur would be needed to send it to Jupiter.

On 2 March 1972 and 5 April 1973, respectively, two identical spacecraft were built and launched as Pioneers 10 and 11. By launching them a year apart, Ames would not only provide complete redundancy, but would also be able to postpone the final commitment to Pioneer 11's fly-by trajectory until after it saw how its predecessor fared in the Jovian magnetosphere.

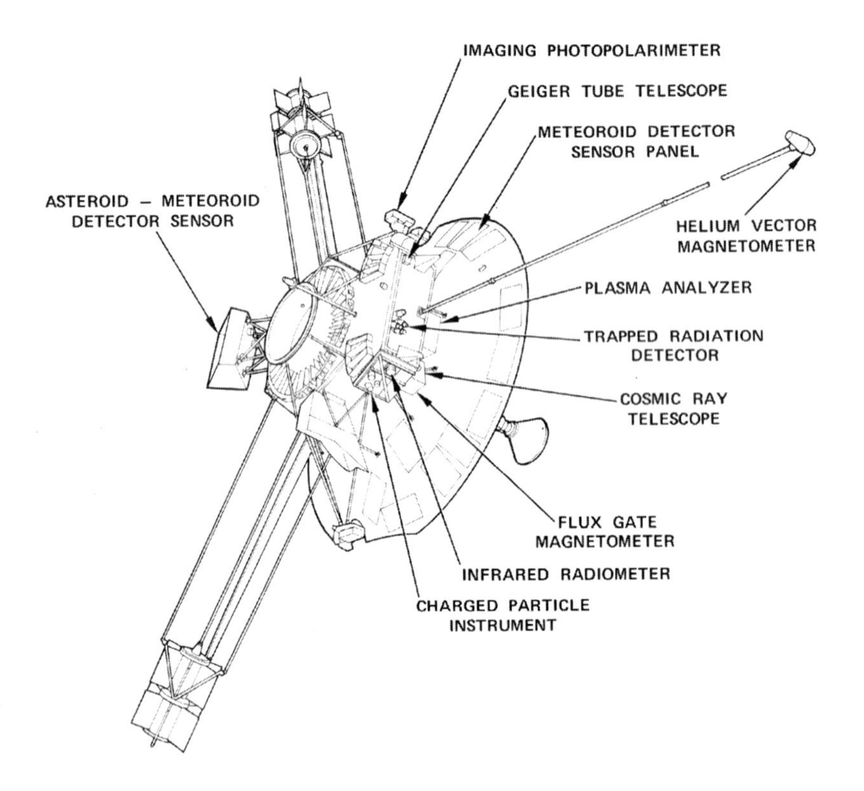

A diagram of the configuration of the Pioneer spacecraft dispatched to the outer Solar System. (Courtesy, NASA SP-349)

During its interplanetary cruise, Pioneer 10 reported on the state of the solar wind, its results complementing those from its predecessors in the inner Solar System.

Pioneer 10 lived up to its moniker by being the first spacecraft to attempt to fly through the asteroid belt. There had been fears that the spacecraft would be battered by micrometeoroids in this 'gap' between the orbits of Mars and Jupiter, and possibly even be destroyed by striking a larger chunk of debris. Its successful passage indicated that the risks had been overestimated. On 26 November 1973, when 110 Rj [1*] from Jupiter, its instruments detected the first sign of the planet's magnetosphere, which was much further out than expected.

Although the spacecraft did not have a television camera, it did have a photometer; and as this scanned Jupiter as the spacecraft rotated it was possible to assemble pictures by stacking these strips side by side. As this was a time-consuming process, the spacecraft was able to build up only two dozen images during the final day of its approach. Once processed back at Ames, these images revealed features in the Jovian atmosphere in much greater resolution than had ever been achieved by

* For Notes and References, see pages 413–424.

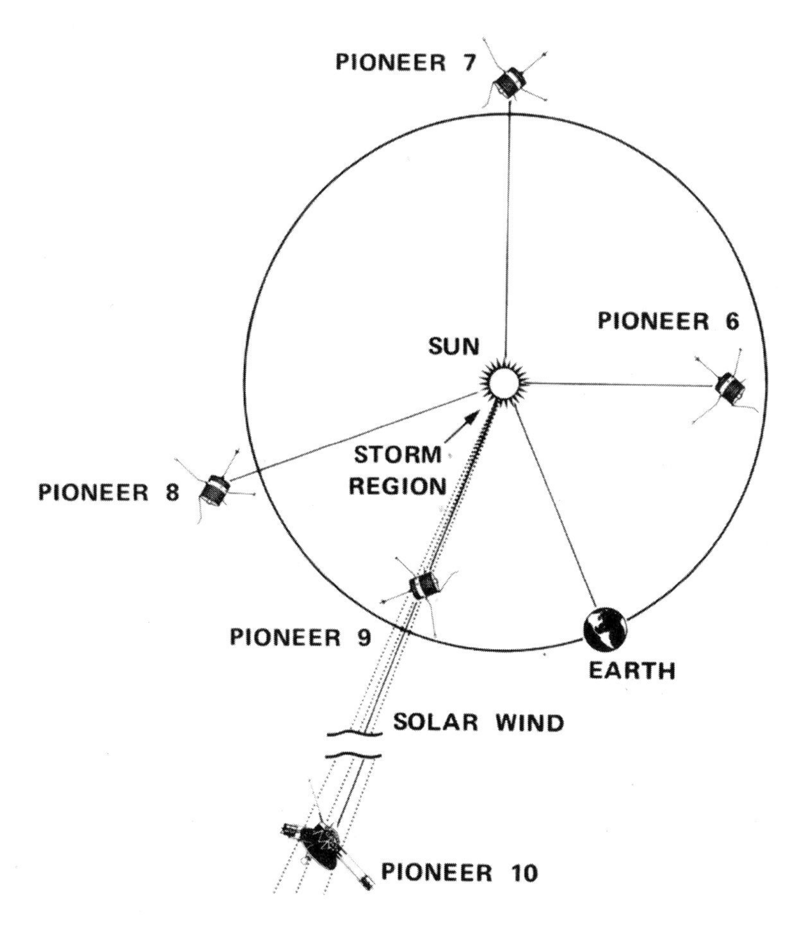

A diagram of how the Pioneers heading for Jupiter coordinated with their predecessors in the inner Solar System to study the solar wind. (Courtesy, NASA SP-349)

even the largest terrestrial telescope. Furthermore, information on the temperature and pressure within the atmosphere was secured by an infrared radiometer and by measuring how the radio signal was degraded as the spacecraft flew behind the planet's limb. The results revealed an unexpectedly intricate atmospheric structure of vast storm systems.

Pioneer 10's trajectory had been defined so that it would pass behind Io, in order that the spacecraft's radio signal could be used to 'sound' any atmosphere, and it identified a tenuous ionosphere at an altitude of 120 kilometres, making it the first planetary satellite to be found to possess an ionosphere.

The Pioneer photometer's imaging facility had also produced the first look at the Galilean satellites. A radiation glitch prevented Pioneer 10 from scanning Io,[2] but it was imaged by its successor – once, as viewed from the perspective of peering over the moon's north pole. Pioneer 10 secured a single image of Europa, but it was not

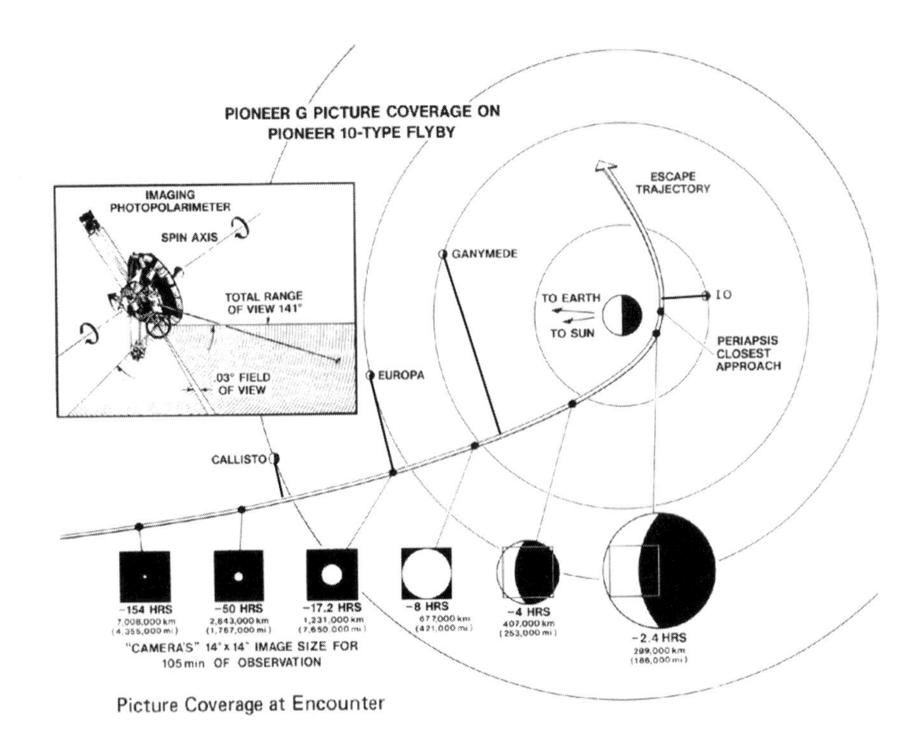

A diagram of Pioneer 10's trajectory through the Jovian system in December 1973. (Courtesy, NASA)

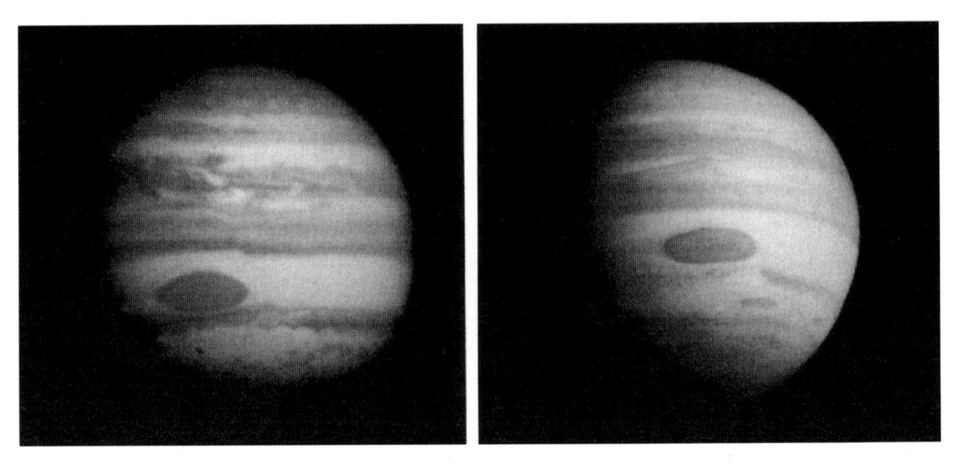

The Pioneers did not carry cameras, but by scanning a photometer across the disk of the planet, they were able to assemble pictures showing the structure of Jupiter's dynamic atmosphere in unprecedented detail. (Courtesy, NASA SP-349)

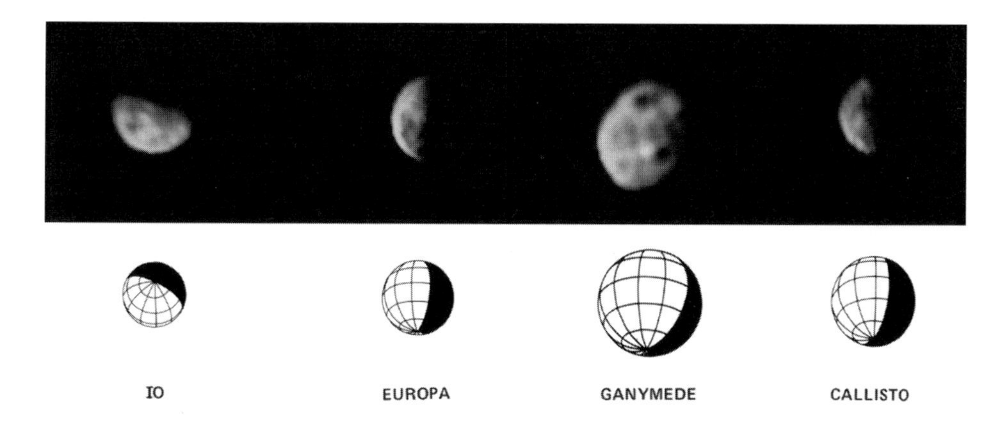

Although computer-enhanced, these Pioneer line-scan views of the Galilean moons of Jupiter show little detail. Nevertheless, they were very welcome because they provided broad hints at the surface morphology. (Courtesy, NASA SP-349)

close enough to reveal much surface detail on the extremely reflective surface. Ganymede and Callisto were both scanned several times. Ganymede's north polar region was confirmed to be very bright, and a number of large dark regions were identified, but the imaging technique was unable to resolve any feature with a width smaller than about 400 kilometres.

When Pioneer 10 made its closest approach to Jupiter on 4 December 1973, it was within one minute of its original schedule, which was amazing after a two-year voyage. After completing its planetary encounter, the spacecraft headed on out of the Solar System. Having survived the radiation 130,000 kilometres (1.86 Rj) above the giant planet's cloud tops, it had established that its successor would be able to penetrate more deeply into the magnetosphere. Pioneer 11's aim was adjusted to enable it to approach the planet three times closer than its predecessor so that the fly-by would deflect the spacecraft's trajectory and boost it out towards Saturn.[3]

Meanwhile, in September 1974, Charles Kowal, an astronomer at the California Institute of Technology, used the Palomar Observatory's Schmidt telescope to discover Jupiter's thirteenth satellite, Leda. This turned out to be the fourth member of a family of captured asteroids orbiting about 11 million kilometres from the planet.

Pioneer 11 was to approach Jupiter's south pole to avoid the very intense equatorial radiation belt on the way in and be accelerated through it on the far side, so that, despite flying much more deeply into the magnetosphere, its total radiation dose would be no more than that which Pioneer 10 had endured.

After its 43,000-kilometre encounter with Jupiter on 3 December 1974, Pioneer 11 became the first spacecraft to reach Saturn. It was, however, a long trip, and the spacecraft did not arrive until September 1979.

Both Pioneers are now heading out of the Solar System, but their instruments continue to report on the solar wind as they seek the 'heliopause' where the Sun's magnetic field yields to the interstellar medium.

VOYAGING FAR

John Naugle, NASA's Associate Administrator for Space Science, had emphasised that the Pioneers were 'precursor missions' whose function was to 'lay the groundwork for the outer planet exploration program'. They had certainly opened the door, but the next generation of spacecraft to venture out to Jupiter would not be advanced Pioneers.

Planetary imaging was the speciality of the Jet Propulsion Laboratory (JPL) in Pasadena, which was operated jointly by NASA and the California Institute of Technology. Its Mariners were three-axis stabilised platforms equipped with state-of-the-art cameras. Mariners had been first to reach Venus (1962), Mars (1965) and Mercury (1974). With early fly-bys and then orbiters, they had mapped Mars. In fact, the spacecraft which delivered the Langley Research Center's Viking landers into Martian orbit in 1976 were modified Mariners, so it was natural that JPL should follow up the Pioneers and explore the outer planets. Meanwhile, having decided that planets were interesting in their own right, Ames set out to design a probe to penetrate Venus's extremely dense atmosphere.[4]

In the early 1970s, JPL promoted a plan to dispatch a series of 'Mark 2' Mariners on what it referred to as the Grand Tour of all the outer planets. However, funding difficulties meant that the ambitious project had to be scaled back, and the Mariner design adapted for conditions in the outer reaches of the Solar System.[5]

The similarity of the mission resulted in a configuration like that of the Pioneers, but with a more comprehensive suite of instruments for ultraviolet spectroscopy, infrared spectrometry and radiometry, photopolarimetry, and high-resolution imaging, all mounted on an articulated scan platform, as well as 'particles and fields' instruments to follow up on the Pioneers' survey of the Jovian magnetosphere. A powerful X-Band system was installed to transmit all the data that the instruments would produce, and the Deep Space Network's 26-metre antennas were upgraded to 34 metres to enable them to receive the signal from such a distance. A trio of radio-isotopic power units were installed. It would require a Titan IIIE-Centaur – the most powerful rocket combination in the inventory – to dispatch the new 815-kg spacecraft to Jupiter.

In view of the tremendous distances involved, and the extended flight times – a decade as opposed to a few months for a target in the inner Solar System – this programme was named Voyager.

Following standard procedure, two identical spacecraft were built in case ill-fortune befell one of them. They were both to be launched in the same 'window' in 1977. A last-minute problem resulted in a switch, and Voyager 2 was launched first; however, Voyager 1 was then sent on a faster trajectory and soon overtook its mate.

The Voyagers revealed the highly dynamic Jovian atmosphere in unprecedented detail, and a series of pictures taken over a 10-hour period were linked to produce a 'movie' depicting the planet rotating on its axis.

The Galilean satellites were the 'stars', however. Voyager 1's passage through the Jovian system provided an opportunity to view the small inner moonlet Amalthea on the way in, and to image Io, Ganymede and Callisto on the way out. At its closest

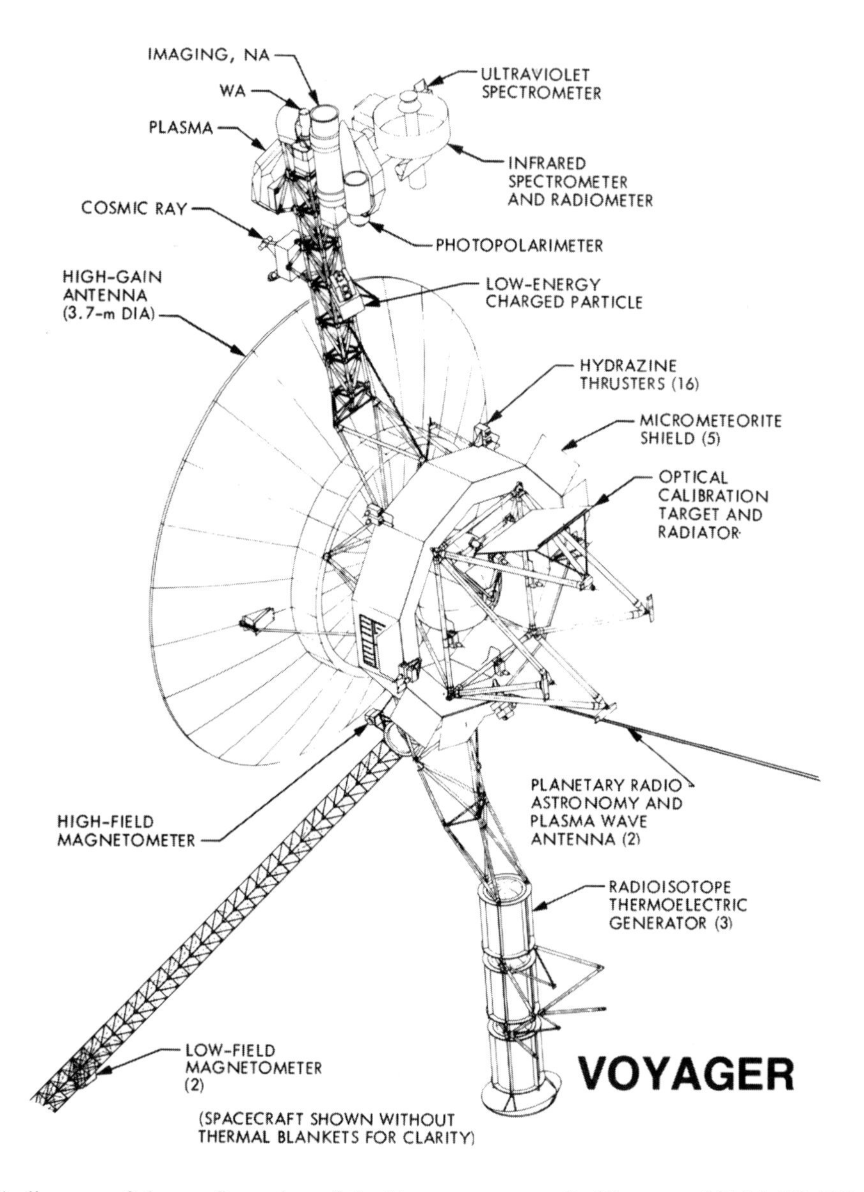

A diagram of the configuration of the Voyager spacecraft. (Courtesy, NASA SP-439)

point to Jupiter (4 Rj) the spacecraft was well within the orbit of Io. In fact, its trajectory took it within 20,000 kilometres of the moon's south pole, which facilitated imaging the side of Io which permanently faced the planet.

Although Voyager 1's trajectory did not offer an opportunity to study Europa, Voyager 2 made a close fly-by of this moon on the way in, just before the spacecraft's closest approach to Jupiter, but this time Io was inconveniently positioned on the other side of the planet.

Voyager 1 took this picture of Jupiter on 5 February 1979, a month out from the planet. The state of the atmosphere had changed considerably since the Pioneers. In particular, there was considerably more activity in the vicinity of the Great Red Spot (in fact, this area had been unusually quiescent in the mid-1970s).

Ganymede and Callisto received the best overall surface coverage because Voyager 2 was able to view the hemispheres that had been in shadow when viewed by its predecessor, but only parts of Europa and Io were mapped in detail.

Nevertheless – at long last – the "marvellous things" discovered by Galileo Galilei had been revealed to be miniature worlds, each with its own highly distinctive geological history.

Both spacecraft used their encounters with Jupiter to fly on to Saturn, but only

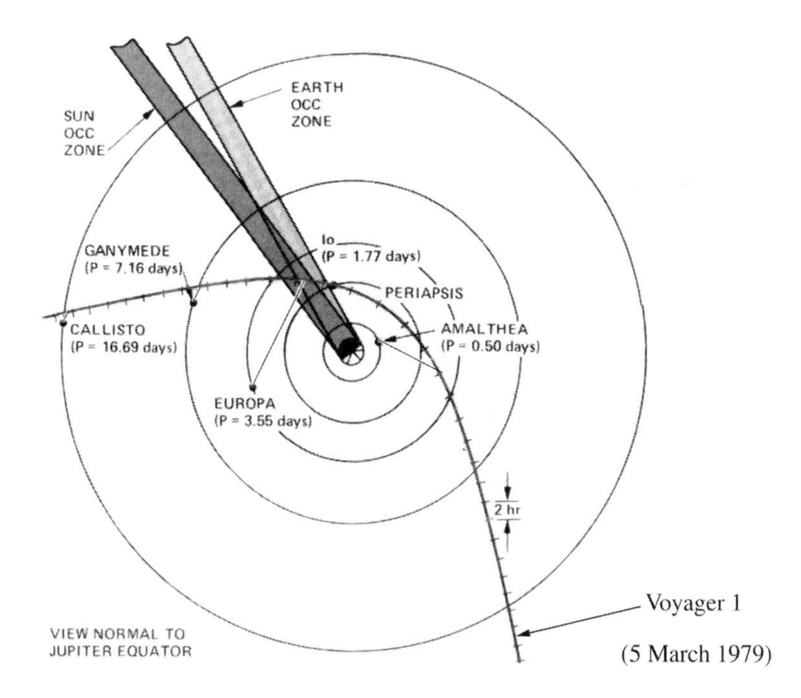

A diagram of Voyager 1's trajectory through the Jovian system. (Courtesy, NASA SP-439)

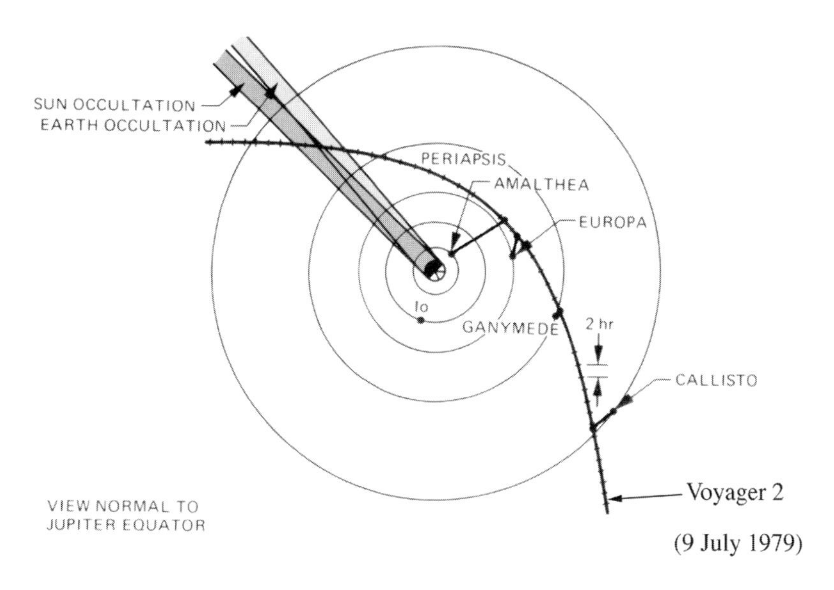

A diagram of Voyager 2's trajectory through the Jovian system. (Courtesy, NASA SP-439)

Voyager 2 made the long trek to Uranus and Neptune. Like their predecessors, both spacecraft are now heading out of the Solar System.

As for Jupiter, missions which flew through the system in a few days were able to provide only 'snap shots' of what was evidently a highly dynamic system. The next logical step was to place a spacecraft in orbit, to study the long-term behaviour of the planet's magnetosphere and atmosphere.

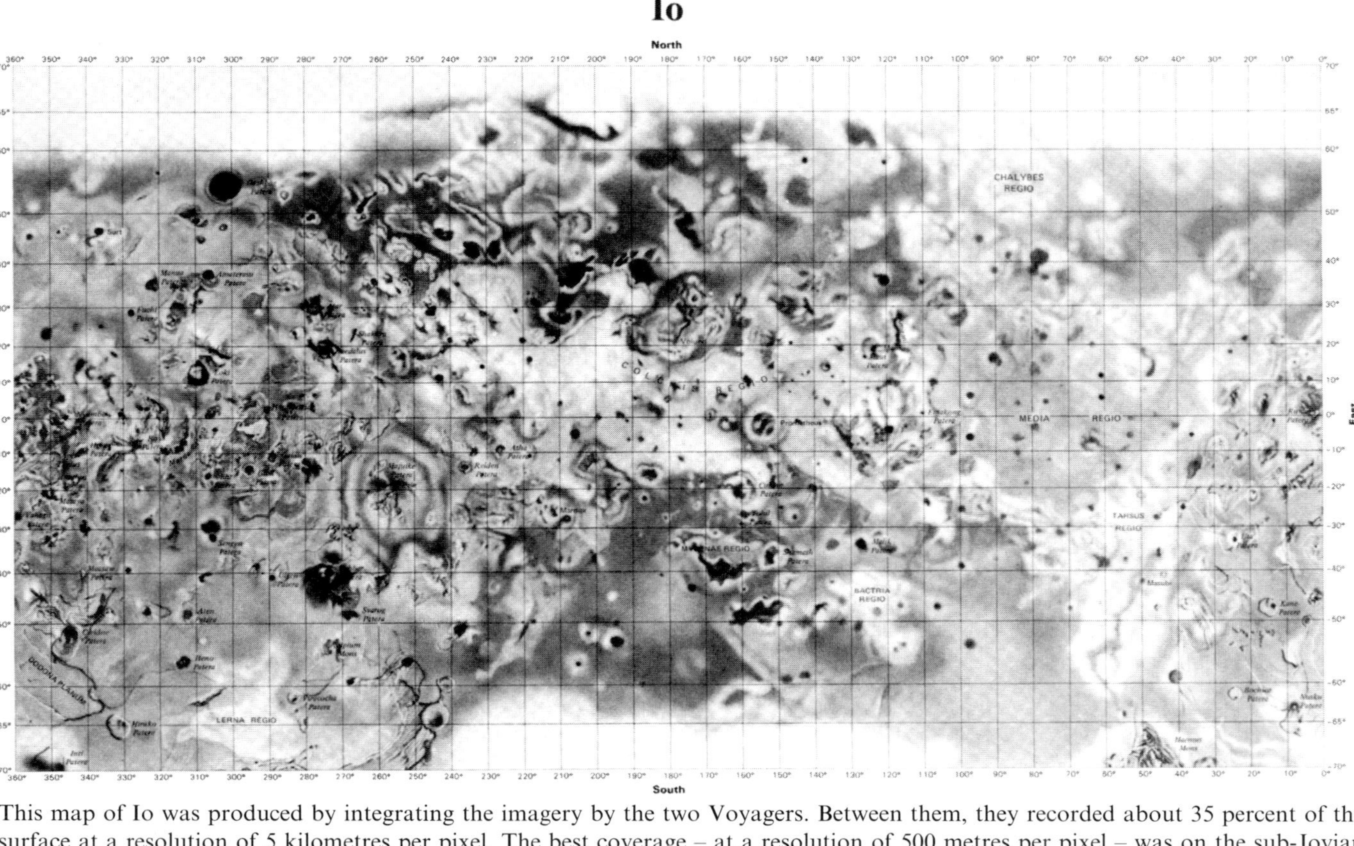

This map of Io was produced by integrating the imagery by the two Voyagers. Between them, they recorded about 35 percent of the surface at a resolution of 5 kilometres per pixel. The best coverage – at a resolution of 500 metres per pixel – was on the sub-Jovian hemisphere. Against expectation, Io turned out not to be cratered. It is the most volcanically active body in the Solar System. (Courtesy, USGS)

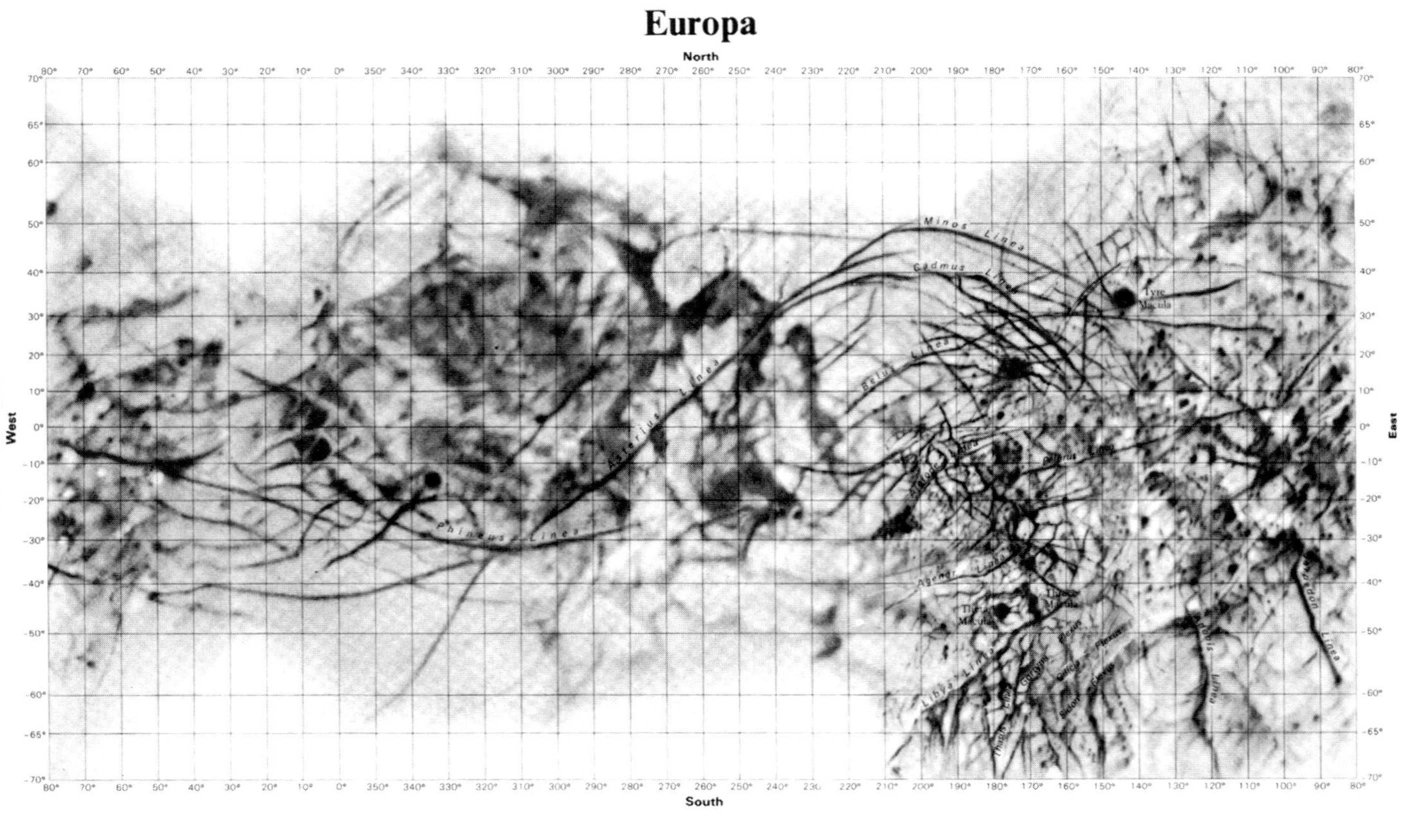

This map of Europa was produced by integrating the imagery by the two Voyagers. The best imagery was taken by Voyager 2. Although it provided a resolution of 2 kilometres per pixel, it was able to map only a small part of the surface. Completely enshrouded in ice, Europa is billiards-ball smooth. The lack of craters suggested that the ice froze 'recently'. (Courtesy, USGS)

Ganymede

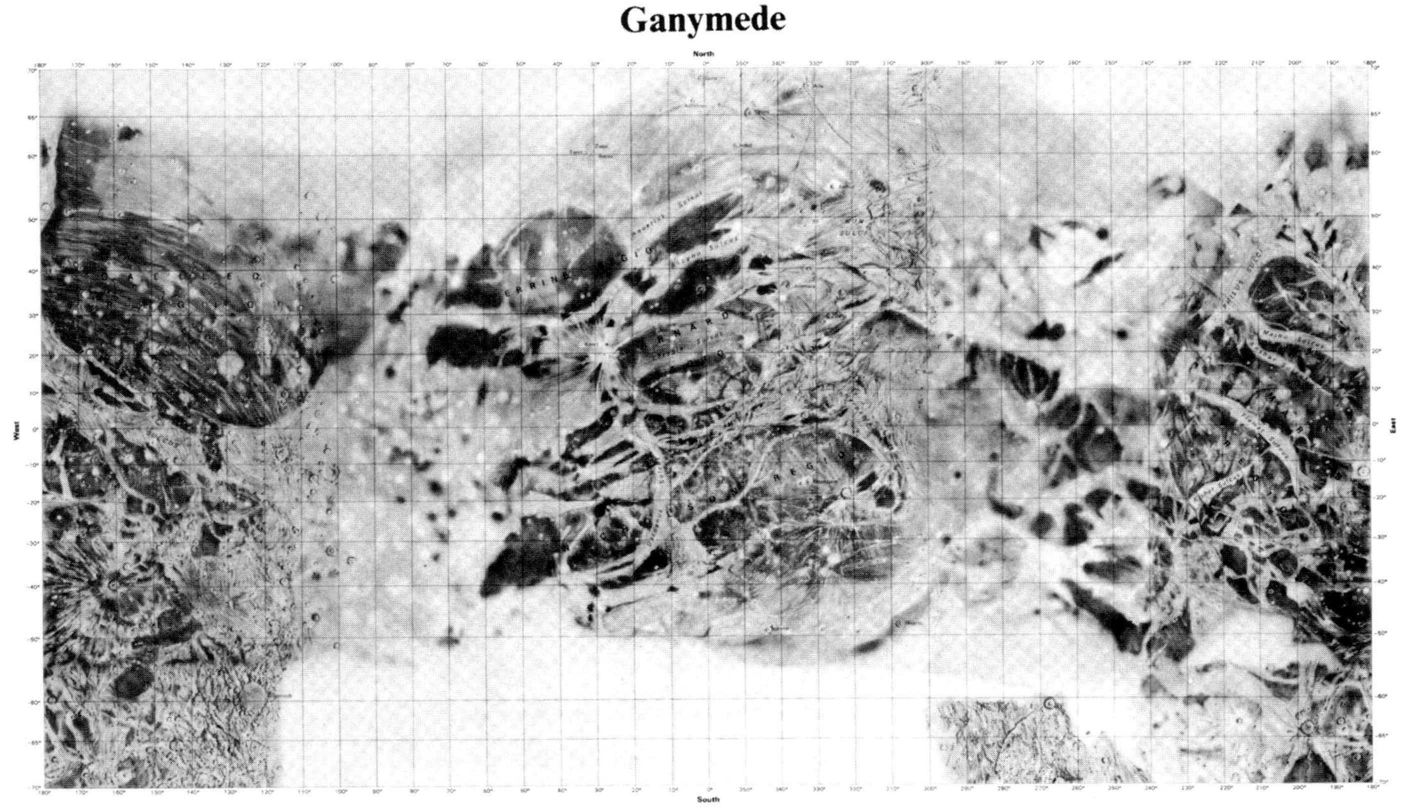

This map of Ganymede, the largest of the Jovian moons, was produced by integrating the imagery by the two Voyagers. Between them, they mapped 80 percent of the moon with a resolution of 5 kilometres or better per pixel. Dark blocks of the moon's surface have undergone extensive tectonic rifting and brigher material has risen to fill the gap. The bright polar areas have been almost completely resurfaced. (Courtesy, USGS)

Callisto

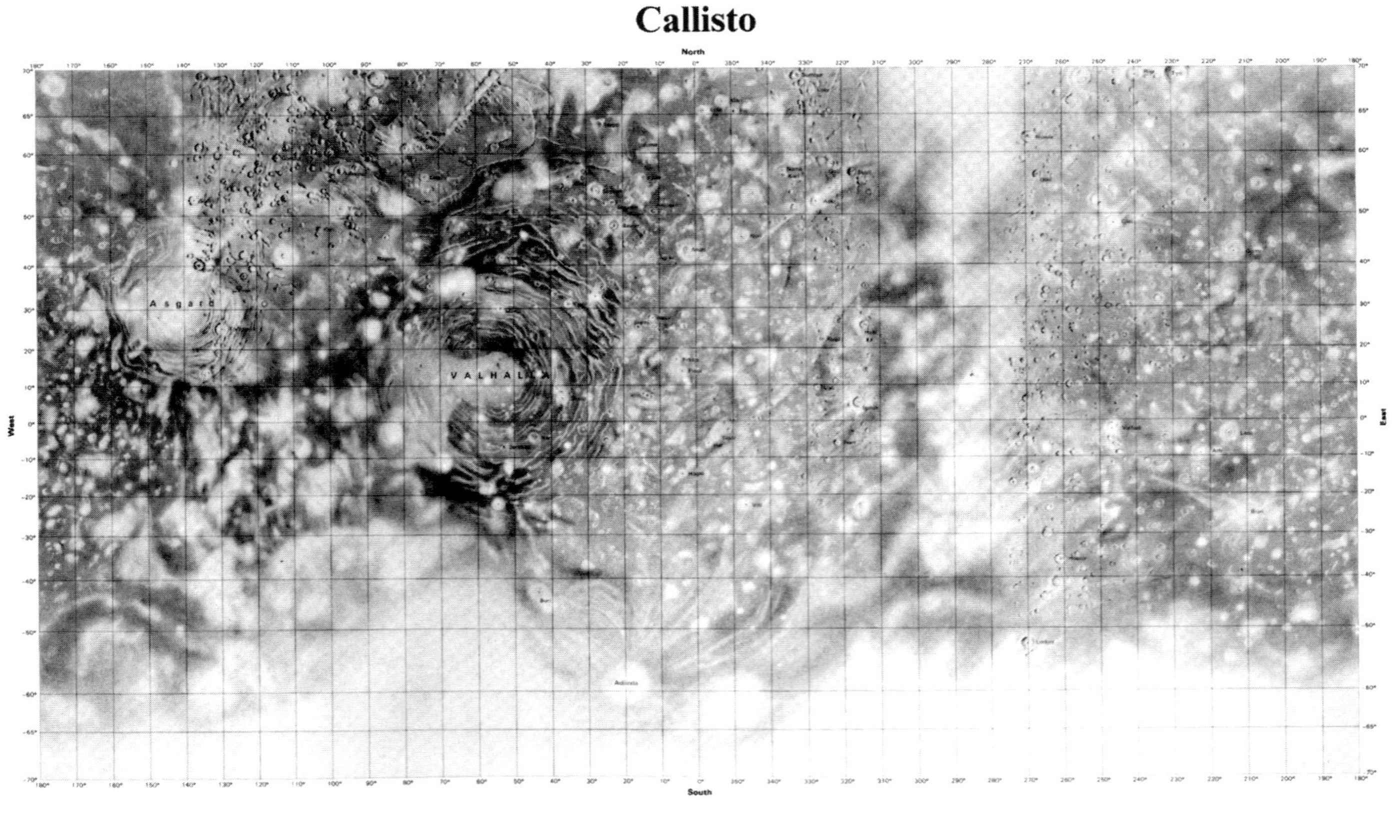

This map of Callisto was produced by integrating the imagery by the two Voyagers. Between them, they mapped 80 percent of the moon with a resolution of 5 kilometres or better per pixel. Most of its surface is a monotonous cratered terrain, broken only by the large multiple-ringed basin structures. The fact that it is peppered with craters right to the limit of the Voyager resolution indicates that it is a *very* old surface. (Courtesy, USGS)

3

Galileo's ordeal

TWO INTO ONE

In 1972, when NASA set out to develop the Shuttle as the 'National Space Transportation System', it had several interplanetary spacecraft at the planning stage. It was expected that the Shuttle would enter flight-test in 1978 and become operational within a year,[1]* at which time it would supersede conventional 'expendable' rockets, so all spacecraft scheduled for launch in the 1980s were to be designed to ride the Shuttle.[2]

The Ames Research Center's two Pioneers had just been launched towards Jupiter, and the logical follow-on mission seemed to be to put a spacecraft into orbit around the giant planet and 'drop' a probe into the atmosphere. In 1975, Ames was authorised to start developing this Pioneer Jupiter Orbiter and Probe, for launch in January 1982. In an administrative shake-out a few months later, however, NASA reassigned responsibility for all new planetary spacecraft to the Jet Propulsion Laboratory (JPL), and the project was transferred from Mountain View to Pasadena.

Although at that time the Voyagers were still being constructed, JPL had already started to plan a follow-on mission with a spacecraft entering Jovian orbit in order to explore the planet's atmosphere, its magnetosphere, and its family of satellites. Ames's spacecraft was to have been another 'spinner' configured for sensing particles and fields, with only a limited capability for imaging. The Voyagers were three-axis stabilised, so JPL's design was stabilised and carried high-resolution imaging instruments. As a compromise, a spacecraft combining spinning and non-spinning sections was devised. Despun antenna mounts were utilised by communications satellites, so this two-part vehicle involved low technological risk. However, the compromise resulted in a larger vehicle than either Center would have developed on its own. Although this transformed the spacecraft into a heavyweight, the Shuttle had been publicised as a space-faring truck, so this was not considered to be an issue.

* For Notes and References, see pages 413–424.

The integrated project was labelled Voyager Jupiter Orbiter and Probe in 1976, approved by Congress in July 1977, and renamed 'Galileo' the following year.

BIRTH PAINS

JPL accepted the 1982 'launch window' given to Ames because this would permit Galileo to make a close pass by Mars *en route*. This would make the cruise a little longer and slip the arrival at Jupiter from 1984 to 1985 but the gravitational 'slingshot' would enable Galileo to carry a larger propellant tank for its primary mission. However, this decision also took the overall mass of the spacecraft – and the rocket stage that would boost it out of Earth orbit – to the extreme limit of the Shuttle's predicted payload capacity.

As spacecraft development proceeded, it became apparent that the Shuttle would not fly in 1978, but NASA was confident that the Shuttle would fly in 1979 and JPL was assured that the manifest would be reordered to enable Galileo to launch on schedule. Unfortunately, the Shuttle's main engines and thermal protection suffered protracted development problems. In July 1979 the agency finally acknowledged that the Shuttle would be lucky to start flying by 1980 and informed JPL that it could not be certain of launching Galileo in 1982.

Slipping the launch would mean that Galileo would not be able to use a Mars fly-by to pick up energy. The spacecraft was to be sent on its way by a three-stage 'planetary' version of the Inertial Upper Stage (IUS). This was not powerful enough, on its own, to send Galileo all the way to Jupiter. The logical solution would have been to transfer Galileo to a Titan-Centaur, the most powerful expendable launcher in the inventory, and use the 1982 window. However, a cost–benefit analysis used to justify the development of the Shuttle had relied on a busy flight schedule to force down operational overheads. Off-loading a payload as prestigious as Galileo would set a poor precedent, and JPL was ordered to strip Galileo to suit the IUS's capabilities. By this point, however, it was also obvious that the initial configuration of the Shuttle would not meet its planned cargo capacity. This meant that JPL's task was not simply a matter of refitting the smaller propellant tanks, but scientific instruments would also have to be omitted. A few months later, NASA suggested that if the probe could be sectioned off and paired with a simple carrier of its own, so that each could be sent directly to Jupiter on its own three-stage IUS, then JPL could have *two* Shuttle launches in 1984.[3]

Other forces were at work, however, and by the end of 1979 NASA was under pressure to develop a more powerful planetary stage for the Shuttle.

The IUS had been developed for the Shuttle by the Department of Defense. It had specified a rocket which burned solid propellant so that it would be simple to operate. However, a solid propellant rocket engine is not as effective as one that burns high-energy liquids. The Shuttle's prime function was to ferry mass into low orbit, so it seemed sensible to make every kilogram count.

The high-performance Centaur had proved itself as an upper stage for the Atlas and Titan launchers. Surely, it should be possible to adapt the Centaur for carriage in the Shuttle's bay? The Centaur's performance derived from the fact that it burned hydrogen and oxygen. The use of cryogenic propellants would make the Centaur a much more difficult payload to service on the Shuttle, but it would be worth it. The three-stage IUS was cancelled in December 1980 and, one month later, Congress cancelled the two-stage variant of the IUS that had been intended to put heavy communications satellites into geostationary orbit. In January 1981, the agency was told to adapt the Centaur for use with the Shuttle. The Centaur would be able to send the complete Galileo spacecraft straight to Jupiter, but as it would take time to adapt the Centaur, Galileo's launch was slipped to 1985 by which time, on the original plan, the spacecraft should have been starting its exploration of the Jovian system.

Although JPL made excellent progress, its work was being undermined by NASA politics. The Marshall Space Flight Center in Huntsville, Alabama, had supervised NASA's interest in the IUS. The Lewis Research Center in Cleveland, Ohio, had developed the Centaur together with General Dynamics. Deleting the IUS in favour of the Centaur was a boon to Lewis, but it hurt Marshall, so lobbyists set to work. In November, Congress reversed itself – the Centaur was out, the IUS was reinstated, and Galileo was once again in trouble. But General Dynamics had lobbyists, California's economy was dominated by aerospace, and Congressional elections were due in 1982. The Centaur was therefore reinstated in July, but this time the IUS was not cancelled. Both stages would be built in order to provide operational flexibility. Galileo was finally saved, but the time wasted by the political manoeuvring had delayed the Centaur so much that the launch was slipped once again, this time to the May 1986 window (opportunities for a direct route to Jupiter occurred at 13-month intervals).

Galileo was paying a stiff price for its brief ride into orbit aboard the Shuttle. Nevertheless, it had fared better than several other projects that had been cancelled outright and their budgets devoured by the Shuttle's overruns. The policy of forcing all payloads onto the Shuttle was playing havoc with the space science programme. Instead of the Shuttle picking up the load in a vibrant programme, delays and cancellations were killing off its customers. In this phase of its development, the Shuttle was a predatory beast. It finally flew in April 1981. After its fourth test flight, it was declared operational in 1982.

Meanwhile, Galileo's development proceeded apace. Once committed to the Centaur, JPL allowed Galileo to grow to take full advantage of this stage's capacity. No longer constrained by a Mars slingshot, the navigators sought 'targets of opportunity' for its passage through the asteroid belt – a fly-by of 29 Amphitrite[4] was possible on 6 December 1986. No asteroid had yet been inspected close-up, so this *first* would be a welcome scientific bonus.

CHALLENGER

On 28 January 1986, Galileo was at the Kennedy Space Center, ready to be mated with its Centaur, preparatory for launch on 21 May.

Although the countdown for STS-51L Challenger was running on NASA TV, most of the JPL staffers were preoccupied by events much further away. Voyager 2 had finally reached the planet Uranus, and amazing images of its moons were streaming in. The Press was present in force, and the scientists were preparing a major presentation. Before this could start, however, the Shuttle 'malfunctioned', killing its crew. In the aftermath of the disaster JPL was instructed to prepare Galileo for the next Jovian launch window, in June 1987, on the assumption that by then the Shuttle would be back in service. The spacecraft was returned to JPL, stripped down, and placed into storage. Within months, however, NASA decided that it would no longer run the Shuttle's main engines at their maximum thrust, which posed a problem for Galileo as it meant that the Shuttle would not be able to lift a Centaur with Galileo mounted on its nose.

In the light of this, JPL decided to fly a Centaur with a *partial* propellant load, and to make up the energy deficit by a gravitational slingshot. Mars was no longer favourably placed, but a close fly-by of the Earth would be sufficient. In this revised scheme, the Centaur would boost Galileo out on a long ellipse that would produce a return to Earth, so that this encounter would extend the spacecraft's aphelion and send it all the way to Jupiter. The roundabout route – dubbed the Earth Gravity-Assist (EGA) – would see Galileo-as-built to Jupiter, but the initial loop to the asteroid belt would add *three years* to its flight time.

WHICH WAY?

After a thorough review of the Shuttle's ascent-phase abort modes, NASA decided in June 1986 that carrying a Centaur with cryogenic hydrogen and oxygen would pose a severe risk for a Shuttle attempting an emergency landing, so the Centaur was cancelled again – this time with no possibility of a reprieve. Unfortunately for Galileo, the three-stage IUS had long-since been cancelled, and the spacecraft had put on too much mass for the two-stage version to set it up for the EGA option.

With the straightforward EGA now insufficient, JPL's interplanetary navigators set out to investigate multiple-encounter options. Even making a second Earth fly-by would not provide sufficient energy to make up for the difference in capability between the IUS and the *partially* fuelled Centaur. The only option was to investigate the possibility of sending Galileo in the *other* direction, towards the Sun, and start the mission with a close pass by Venus. The Venus–Earth–Earth Gravity-Assist (VEEGA) trajectory was 'discovered' on 1 August 1986.[5] Although this would allow an IUS to send Galileo-as-built to Jupiter, this would be at a terrible penalty in flight time – the spacecraft would now be dispatched in October 1989; a Venus fly-by in February 1990 would ease its aphelion out beyond the Earth's orbit; an Earth fly-by in December would extend the aphelion out to the asteroid belt; and the second Earth encounter in December 1992 would sling it out to meet Jupiter in December 1995. The journey would take fully six years, and the spacecraft would arrive a decade later than originally planned! On the other hand, with opportunities for secondary scientific objectives at Venus and the Earth/Moon system, as well as 951

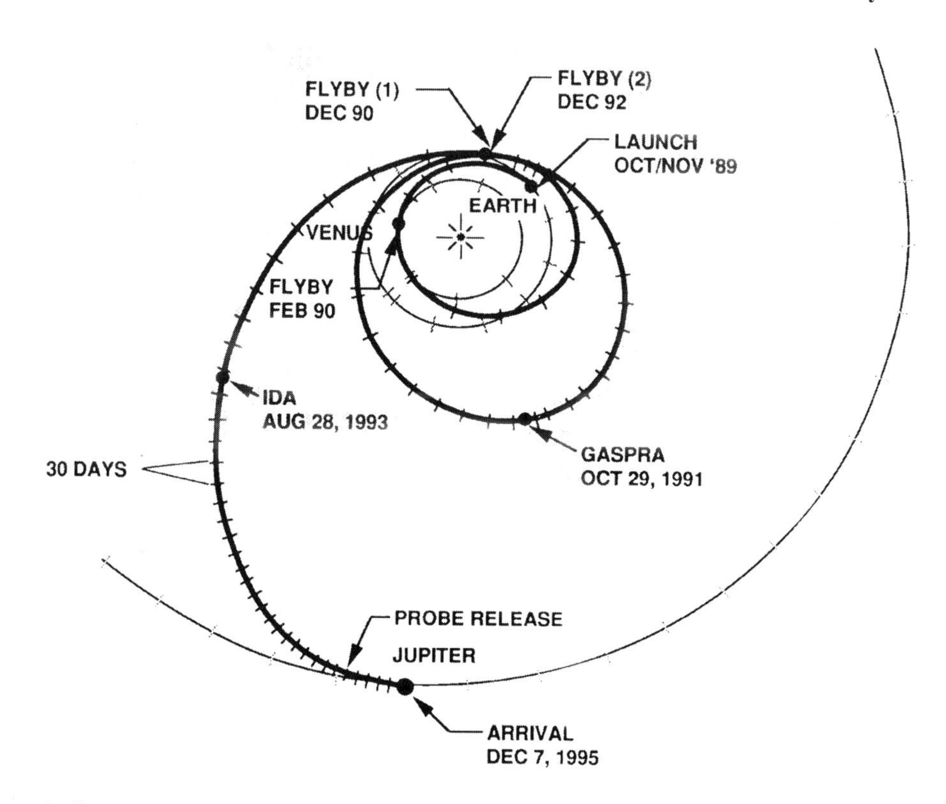

A diagram showing how the Galileo spacecraft was to reach Jupiter. The use of fly-bys of Venus in February 1990 and the Earth in December 1990 and 1992 for 'gravity assist' prompted the acronym VEEGA. (Courtesy, NASA)

Gaspra and 243 Ida in the asteroid belt in October 1991 and August 1993 respectively, this circuitous route had its compensations.

With no other alternative, JPL set out to modify Galileo to endure increased thermal stress, because insolation is twice as intense at Venus as it is at the Earth's distance from the Sun. The spacecraft had been designed to work at Jupiter, where the energy of sunlight is just 4 percent of that at the Earth. A change in procedure was also required: in previous circumstances, Galileo would have unfurled its umbrella-like high-gain antenna as soon as it was safely on its way, then orient itself to maintain its antenna facing the Earth. The antenna, however, was too delicate to be exposed to intense sunlight, so it would have to remain tightly furled around its axial tower until Galileo had performed its first pass by the Earth. The simplest way to protect the antenna was to mount a small disk-shaped sunshield on top of its tower, and then orient the spacecraft so that it maintained its axis towards the Sun. This meant that while it was in the inner part of the Solar System the spacecraft would have to rely upon a rear-mounted low-gain antenna. A wide disk was mounted across the top of the main body of the spacecraft to shade its primary systems. With these decision made, a sense of stability returned to the project.

This artist's depiction shows Galileo mounted on the nose of its Inertial Upper Stage, shortly after its deployment from the Space Shuttle Atlantis on 18 October 1989. Note that the spacecraft's various booms are in their stowed positions.

INTO SPACE

The window for Venus opened on 12 October 1989, and ran until 21 November. The first countdown was scrubbed long before the astronauts were due to board Atlantis as a fault had shown up in one of the engines. Bad weather intervened at the next launch attempt, on 17 October. An earthquake conspired to raise the tension overnight by shaking the Air Force's satellite control centre in Sunnyvale, California, which was to check out the IUS in orbit prior to deployment. The next day, however, 18 October, the countdown ran very smoothly and at 12:54 EDT,

STS-34 ascended into the Floridian sky on its pillar of flame. Many of the Galileo engineers, scientists, and managers were present to see their creation on its way. They greeted the lift-off with applause, grew tense, then cheered with delight when the Shuttle jettisoned its spent solid rocket boosters.

The payload was carried on an annular cradle at the rear of the payload bay. The 17-metre stack of the IUS and Galileo completely filled the cavernous bay. The first act in bringing the stack to life was to elevate the cradle to about 30 degrees. The IUS was heavily instrumented. Each of its stages had redundant control systems. Once Sunnyvale had verified the IUS, the power umbilical to the Shuttle was disconnected, committing the vehicle to its internal power supply, and the cradle was elevated to about 60 degrees.

At the appointed time, the ring-clamp holding the IUS in place was released by pyrotechnic bolts and a spring ejected the 18-tonne stack, which drifted lazily over the roof of the orbiter's cabin as it moved clear. "Galileo is on its way to another world," reported mission commander Donald Williams. "It's in the hands of the best flight controllers in this world – fly safely." It was another successful mission for the Shuttle programme.

4

An exploring machine

A HEAVYWEIGHT

At 3 tonnes, the Galileo spacecraft was much larger than its predecessors.[1]* However, most of this mass derived from the nature of its mission. Most of the atmospheric probe's 340 kg was accounted for by the heat shield, and the main spacecraft carried 935 kg of propellants for the rocket engine that was to put the spacecraft into Jovian orbit.

The scientific instrumentation accounted for only a small percentage of the overall mass of the spacecraft. Nevertheless, it was the most capable exploring machine ever developed. It had a suite of complementary instruments designed to explore the Jovian system *in depth*, pursuing three primary scientific themes: the composition and structure of the planet's atmosphere, the dynamics of its magnetosphere, and the nature and evolution of its system of rings and moons.

SPIN-BEARING ASSEMBLY

The two sections of the 'dual spin' spacecraft were mechanically connected by the spin-bearing assembly. The spacecraft's computer would continuously monitor the relative angular displacement between the two sections using an optical encoder and regulate the electrical motor that drove the bearing to maintain the rotation rate. The despun section was actually driven in the opposite direction by the motor to cancel the rotation. It could operate in several modes: 'inertial', 'cruise' and 'all-spin'. In 'cruise' mode the spinning section would turn at about 3 rpm, but when a higher degree of stabilisation was required the two sections would be locked together and spun up to 10 rpm in the 'all spin' mode. There was, however, a complication: the axially located main engine was at the 'base' of the despun section, but was rigidly attached to the spinning section, so all the propellant lines and engine cables were routed through the hollow centre of the spin-bearing.

* For Notes and References, see pages 413–424.

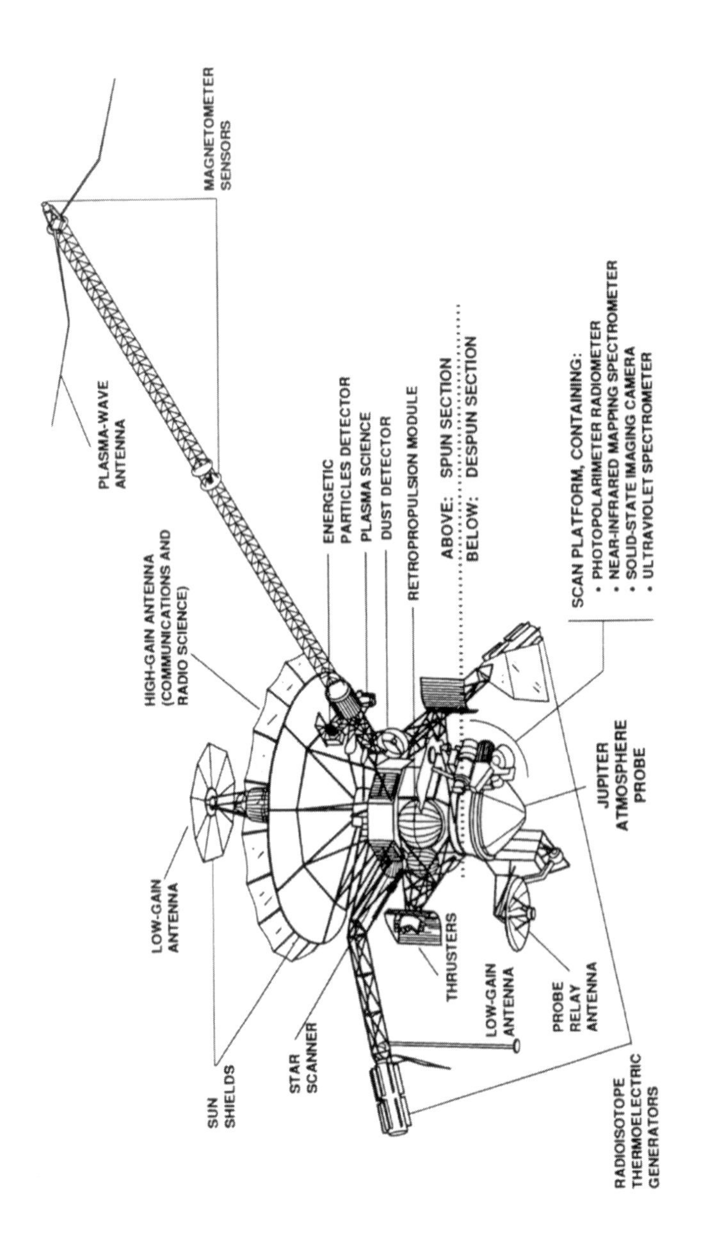

A diagram of the configuration of the Galileo spacecraft.

The bearing's design posed a challenge in transferring power and data between the parts of the spacecraft. With one part spinning with respect to the other, it was not practicable to utilise cables because they would become wrapped around the rotary mechanism. The solution was to transmit power and low-rate data signals via 48 annular 'slip rings', and high-rate data by way of 23 rotary transformers. In the slip rings, a flexible brush contact rode a large ring electrically connected to the spinning side. The rotary transformers comprised pairs of coiled wires around a ferrite core, with one coil electrically connected to the spun section, and the high-rate signals were transmitted without contact by means of the magnetic field generated by pulses of current in the coils.

COMPUTERS

Galileo's dual-spin design was a challenging departure from its three-axis stabilised and all-spin predecessors. It combined the sky-sweeping continuous rotation desirable for the particles and fields instruments with an inertially stable platform for precisely aiming optical instruments, so the design posed a unique set of control problems.

The resulting Attitude and Articulation Control Subsystem (AACS) was therefore the most sophisticated application of control system technology yet deployed on a spacecraft. Its heart was an ATAC-16MS 16-bits/word microprocessor. The primary attitude control reference was the star scanner on the spinning section. When the star scanner was not usable (such as during manoeuvres) data from two orthogonally mounted twin-axis gyroscopes on the scan platform was used.

Galileo would also have to be capable of operating autonomously for long periods, and of undertaking attitude changes on its own. In addition to redundant computers with a high level of fault protection, the AACS was able to detect failures and switch to redundant components in order to continue operations as long as possible in an anomalous situation.

The AACS communicated with the Command and Data Subsystem (CDS), which received commands and transmitted telemetry and scientific data to the Earth.

COMMUNICATIONS

Communications with Galileo was to be by way of NASA's Deep Space Network, which operated stations spaced at roughly 120-degree intervals in longitude (Goldstone in California, Canberra in Australia, and Madrid in Spain) for a continuous link during intensive periods.

To transmit the data from Galileo's imaging instruments (particularly the large-format CCD imaging system which would generate data at a prodigious rate) the high-gain X-Band antenna was to be capable of sustaining a rate of 134 kilobits per second. The deep space Pioneers and Voyagers had incorporated rigid dish-shaped antennas, but for Galileo it was decided to use the deployable dish designed for

NASA's geostationary Tracking and Data Relay Satellites, which was pinned to a central tower for launch and unfurled once the satellite was on station. It would form a 4.8-metre wide molybdenum mesh dish mounted axially on top of the spinning part of the spacecraft.

Galileo had two low-gain antennas to report telemetry. One was mounted on the end of the high-gain antenna's tower and the other projected from the rear of the spacecraft, for use when the main antenna was not aimed at the Earth. The transmission rate on this S-Band system would be typically 40 bits per second, but once Jupiter was at superior conjunction – on the far side of the Solar System from the Earth and the Sun degraded the signal – it would reduce to 10 bits per second.

Galileo had a 1.1-metre L-Band dish to receive the data from the probe as it descended into the Jovian atmosphere. This was articulated on the despun part of the spacecraft (opposite the scan platform) in order to track the probe. This radio relay was crucial to the probe's mission, as it could not transmit directly to the Earth. Although the task was simple in concept, it posed an engineering challenge because the acquisition of the probe's signal, tracking the probe, and processing its data had to occur autonomously. The receiver had to 'lock' on to the probe's signal within 50 seconds of the start of data transmission. A redundant design was selected to minimise the risk of equipment failure. It had two receivers and two ultrastable oscillators, and the antenna was designed to simultaneously receive two channels differentiated by frequency and polarisation with each receiver tuned to a specific channel and individually commanded by the spacecraft's CDS. The probe would transmit its data over both of these channels. The ultrastable oscillators would serve as independent references in measuring the doppler on each signal. Within 15 seconds of powering up, the receivers would search the entire 70-kilohertz bandwidth to locate the probe's signal. The relay system was developed by the Ames Research Center and fabricated by Hughes Aircraft Company.

DATA STORAGE

Although Galileo was to relay its probe's data to Earth, a magnetic tape recorder was added to store the data as a back-up in case a fault prevented real-time relay. The tape had a capacity of 900 megabits.

PROPULSION

Galileo's retro-propulsion module was developed by the German Space Agency (DARA) and donated as part of Germany's contribution to the mission. The bi-propellant engine burned monomethyl hydrazine in nitrogen tetroxide. It integrated an axial 400-newton primary engine, a dozen 10-newton thrusters, four propellant tanks and two tanks of helium pressurant.

The retro-propulsion module was mounted under the despun part of the spacecraft, but the two parts of the spacecraft would be synchronised (in the 'all spin' mode) for firing the main engine. However, as the probe was to be tucked against the nozzle for carriage, the engine would not be able to be fired until the probe had been released, which was not scheduled to occur until Galileo approached Jupiter.

The bi-propellant liquid rocket engine, which was built by Messerschmitt–Bolkow–Blohm (now Daimler-Benz Aerospace), represented a major achievement because it had to remain dormant in space for years, then start several times and perform flawlessly.

POWER

All spacecraft designed for the outer Solar System have to carry their own power supply because transducers for converting solar energy into electricity are impracticable, so, like the Pioneers and Voyagers which had preceded it, Galileo had Radio-isotope Thermal Generators (RTGs) which converted the heat released by 11 kg of plutonium-238 dioxide into electrical power. A pair of RTGs were used, and between them they provided about 500 watts. As the radio-isotope decayed, the power they produced declined, so this resource degraded at a few percent per year. Surplus power was shunted to heaters in the propellant tanks.

The RTGs were mounted on 5-metre booms which could be adjusted to refine the angular momentum of the spacecraft, and so eliminate 'wobble'.

ORBITER INSTRUMENTS

The Galileo orbiter carried 11 scientific instruments, but its radio transmitter also supported a number of radio science studies. The dual-spin design was to support the particles and fields community, who required a spinning spacecraft, and the imaging community, who required a stable platform. As a result, all the particles and fields instruments were carried on the spinning part to facilitate all-sky observations, and the imaging suite was mounted on the scan platform articulated from the despun part of the spacecraft. A list of the instrumentation is given in Table 4.1.

Imaging

The imaging suite comprised:

- solid-state imaging system
- near-infrared mapping spectrometer
- photopolarimeter and radiometer
- ultraviolet spectrometer

all of which were bore-sighted to facilitate complementary simultaneous observations. The spacecraft's AACS held the scan platform stable against any residual wobble affecting the remainder of the spacecraft.

Table 4.1: Orbiter experiments

Experiment	Principal investigator
Imaging	
Solid-state imaging system	Michael Belton, National Optical Astronomy Observatories, Kitt Peak, Arizona
Near-infrared mapping spectrometer	Robert Carlson, JPL
Photopolarimeter and radiometer	James Hansen and Larry Travis, NASA's Goddard Institute for Space Studies
Ultraviolet spectrometer	Charles Hord, Laboratory for Atmospheric and Space Physics, University of Colorado
Extreme-ultraviolet spectrometer	–
Particles and fields	
Magnetometer	Margaret Kivelson, Department of Earth and Space Sciences, University of California at Los Angeles
Plasma wave spectrometer	Don Gurnett, University of Iowa
Plasma detector	Lou Frank, University of Iowa
Dust detection system	Eberhard Grün, The Max Planck Institute for Astrophysics, Heidelberg, Germany
Energetic particle detector	Donald Williams, Applied Physics Laboratory, The Johns Hopkins University, Baltimore, Maryland
Heavy-ion counter	–
Radio science	
Celestial mechanics	John Anderson, JPL
Radio propagation	Taylor Howard, Stanford University

Solid-state imaging system

Galileo's solid-state imaging system was an 800 × 800 pixel Charge-Coupled Device (CCD) mated to a 1,500-mm, f/8.5 aperture narrow-angle Cassegrain telescope. The silicon technology was sensitive not only across the visible wavelength range (0.30 to 0.65 micron) but also the near-infrared (0.7 to 1.0 micron). The spectral range of Galileo's camera was three times that of Voyager's.

The CCD array also had a faster action than the videcon tube in the camera employed by the Voyagers and could be commanded to use one of 28 exposure times ranging between 0.004 and 51.2 seconds. Although the slow-scan videcon's calibration had varied with the intensity of the light source, the CCD's linear response facilitated photometric measurement. Furthermore, as the CCD was 100 times more sensitive than the videcon, it would be able to observe features on Jupiter's nightside and the moons within the planet's shadow.

An eight-position filter wheel could be stepped to obtain images in several different ranges, which could then be combined electronically on Earth to produce colour images. Three of the instrument's filters were in the near-infrared. Atmospheric gases and rock minerals impose absorption bands upon the solar reflection in the near-infrared, so this opened up the possibility of spectro-photometry to determine the composition of planetary atmospheres and the mineralogy of satellite surfaces.

However, the use of a solid-state camera also represented a risk, because it would be much more sensitive to interference from the intense radiation in the Jovian environment. Although a 1-cm-thick tantalum shield protected the camera, it was nevertheless expected to incur transient effects when the spacecraft was in the most intense part of the Jovian magnetosphere. Also, the CCD subsystem was chilled to 163 K in order to minimise sensitivity to the neutrons emitted by the RTG power units. It required the scan platform to remain stable for mosaicking – there had to be some overlap to assemble a mosaic; if the scan platform ruined the camera's aim, the elements would not match.

The instrument was designed and assembled at JPL using a virtual phase CCD supplied by Texas Instruments and RCA-1802 microprocessors. Although the telescope, shutter, and filters were inherited from Voyager, they had been improved to better reject off-axis scattered light to eliminate flare.

"The design of the solid-state imaging system was dictated by a combination of goals and constraints," explained Kenneth Klaasen, one of the science coordinators. "The need to study both atmospheric motion and geological formations dictated a high-resolution large-format camera, while the need to study the composition of satellite surfaces and the vertical structure of features in Jupiter's atmosphere dictated the use of several spectral filters within the range 0.4 to 1.1 microns. Accurate mapping and atmospheric velocity measurements required a camera with excellent geometric fidelity, while precise photometric requirements necessitated a linear detector, stable calibration, and adequate data encoding. Low lighting situations, such as observations of the auroras, lightning, and the Jovian ring system, required a detector of very high sensitivity and an optical system with low scattered light. Constraints on the design included limitations on the available telemetry rate, potential image smearing caused by residual motions in the scan platform, use of large amounts of shielding to protect the instrument from Jupiter's harsh radiation environment, limited electrical power and mass, and protection from contamination during launch and from propellant by-products in flight."

Near-infrared mapping spectrometer

The near-infrared mapping spectrometer was designed to measure the spectrum of reflected sunlight and use spectral absorption bands to identify the composition in the Jovian atmosphere or the minerals on the surface of one of its satellites.[2,3,4]

Performing simultaneous high spectral and spatial resolution infrared imaging in the Jovian environment – where insolation is only 4 percent of that at the Earth –

required an optical system with high light-gathering power and the most sensitive detector technology.

The instrument comprised an 800-mm focal length f/3.5 Ritchey–Chretien telescope which fed a spectrometer, which in turn illuminated a linear array of 17 indium antiminide and silicon photovoltaic diode detectors. One dimension of spatial scanning was provided by 'wobbling' a secondary mirror within the telescope to provide a line of contiguous pixels. On one half of the up/down mirror scan, the movable diffraction grating in the spectrometer was held still to place a given range of wavelengths across the 17 individual detectors. At the extremes of the mirror scan, the grating was stepped to the next setting. The size and number of grating steps could be set for specific encounter conditions and scientific objectives. At its highest spectral resolution, the instrument could develop a 408-wavelength spectrum for each pixel. By slowly slewing the spacecraft's scan platform, the instrument was able to assemble a two-dimensional image from a series of linear strips. Its 'push broom' mode of operation required the scan platform to hold the instrument stable. In effect, it was capable of making an image in each selected wavelength in the range 0.7 to 5.2 microns.

However, Galileo's infrared spectrometer's spatial resolution was not as fine as that of its solid-state imaging system, so the two instruments complemented one another. When it was conceived, it was considerably more advanced than the multispectral imaging systems utilised by NASA's contemporary Earth Resources Technology Satellite. In a real sense, it was Galileo's primary sensor.

Photopolarimeter and radiometer

In many respects, the integrated photopolarimeter and radiometer was three instruments in one: a photometer, a polarimeter and a radiometer. Combining three functions made a flexible and powerful instrument, but it required some compromises and a highly innovative design. It built up an image by scanning a single-element detector back and forth across its target. It had low resolution, however. Close to Jupiter, a single pixel would cover 2,000 kilometres of the Jovian atmosphere – which was a resolution some two orders of magnitude lower than that of the solid-state imaging system. However, because all the imaging instruments were co-aligned on the scan platform, the camera's high-resolution 'visual' view would provide the context to interpret what the other instruments were sensing.

The photopolarimeter's design derived almost directly from the cloud photopolarimeter on the Pioneer Venus Orbiter. Polarisation is the suppression of the vibration of 'light waves' in a given direction; and the way that a planetary atmosphere polarises reflected sunlight is determined primarily by cloud particles. Rayleigh scattering is scattering by molecules; it produces a very distinctive polarisation signature, and is most prominent at shorter wavelengths – in fact, this is the reason our sky appears blue. Measuring the relative contributions of Rayleigh scattering in the Jovian atmosphere would establish the amount of gas that was above the cloud tops. Most of the polarimetry was to be done at the shorter end of the instrument's range, which ran from 0.4 micron all the way out to 45 microns. The photometer used seven narrow wavelength bands in the visible and near-infrared to measure methane and ammonia absorption, and thus determine the vertical structure of clouds and haze. Because the radiometer sensed so far into the infrared, it would be able to measure thermal energy leaking from Jupiter's interior. Different wavelengths would correspond to emissions from different depths. It could integrate over the entire spectral range (in effect, using a 'clear' filter) to measure the total energy, including both thermal emission and reflected insolation, and it could also observe through a filter which passed only reflected insolation; the overall thermal radiation emitted could be derived by subtracting one measurement from the other.

The output from the integrated instrument's telescope was passed through a filter/retarder wheel that could be stepped across 32 positions. Polarimetry used both filters and 'half-wave' retarders. After passing through these, the light entered a prism that isolated the vertically and horizontally polarised components for simultaneous measurement of their intensities by a pair of silicon photodiode detectors. The source's polarisation could be determined by rotating the beam using the retarder. For photometry, the output from the selected filter was passed to the polarimetry unit's detectors to enable the intensity to be measured – thus avoiding the added complexity of a separate optical path and detector. For radiometry, a mirror assembly sent the output from the selected filter to a conical mirror which focused it onto a pyroelectric detector.

Of course, one instrument sharing time among three functions meant less time for any one of them, but the advantage in having the functions and wavelengths sampled with exactly the same field of view were deemed to be adequate compensation.

The integrated instrument was designed and built by the Santa Barbara Research Center in California, which had supplied radiometers for many deep-space missions and sensors for the Landsats which were the operational form of the Earth Resources Technology Satellite.

Ultraviolet spectrometer

The ultraviolet spectrometer's function was to measure spectral features in the wavelength range of 1,150 to 4,300 Ångströms in order to determine the composition and structure of the upper Jovian atmosphere – including auroral and air-glow phenomena. Many molecules that reach Jupiter's upper atmosphere are dissociated by exposure to ultraviolet insolation, and the scattered ultraviolet light carries their spectral signatures. In addition to detecting any volatiles outgassing from the moons, the instrument was able to detect molecules on the surfaces of the moons by their absorption features in reflected sunlight, and the scattering observations could distinguish between ice and frost and deduce the sizes of the crystal grains.

The ultraviolet spectrometer, developed and built by the Laboratory for Atmospheric and Space Physics at the University of Colorado, had a 250-mm Cassegrain telescope, a ruled grating, and a trio of photomultiplier detectors.

Extreme-ultraviolet spectrometer

The extreme-ultraviolet spectrometer was mounted on the 'underside' of the spacecraft's spinning section and observed a narrow ribbon of space perpendicular to the spin axis. Its main role was to measure the size and shape of Io's torus.

Particles and fields suite

The particles and fields instruments were mounted either on or at the base of the science boom which was deployed from the spinning part of the spacecraft:

- magnetometer
- plasma wave spectrometer
- plasma detector
- dust detection system
- energetic particle detector
- heavy-ion counter.

The mutually supportive nature of Galileo's particles and fields suite was emphasised by Jim Willett, one of the science coordinators: "each of the particles and fields instruments senses only a portion of the whole picture. By combining the results of all the instruments, we will be able to more exactly shape our model of the Jovian system."

Magnetometer

The role of the magnetometer was to measure the strength and direction of the magnetic fields in the spacecraft's immediate environment, as well as distortions in those fields due to interactions with plasma or magnetised bodies. If the field strength change was correlated with a rotation in the field vector, it could be inferred that the spacecraft was crossing the boundary of a magnetosphere. This was a much more sophisticated instrument than its Pioneer and Voyager precursors.

The magnetometer was mounted on a telescoping boom which extended from the end of the science boom. Two clusters of three sensors were employed – one cluster halfway along the boom and the other at its tip, some 11 metres from the spacecraft's axis – in order to minimise interference from the body of the spacecraft and its electronic systems. The measurements were analysed by the instrument's data processor, drawing upon data indicating the spacecraft's orientation to derive the spatial context. Magnetic effects due to spacecraft were measured and subtracted during on-board data processing.

The magnetometer was designed by the University of California at Los Angeles, and built by the Westinghouse Electric Company employing a 'ring-core sensor' supplied by the Naval Surface Weapons Center and a microprocessor supplied by the Radio Corporation of America.

During the long journey to Jupiter, the magnetometer would study the interplanetary field, including fast streams and interplanetary shock waves. The Pioneers and Voyagers had shown that the size, shape and internal structure of the Jovian magnetosphere all varied, but they had not been able to linger to study the dynamics in detail. Galileo was to measure magnetic fields from the outer fringe of the magnetosphere (where it interacts with the solar wind) to where it couples with the planet's upper atmosphere, and do so over an extended period.

Plasma wave spectrometer

Because it comprises low-density ionised gas, a plasma is a very good electrical conductor with properties which are strongly affected by electric and magnetic fields. Individual ions and electrons interact with one another by both emission and absorption of low-frequency 'waves'. Plasma waves occur both as electrostatic oscillations – which are similar to sound waves – and as electromagnetic waves. Such waves are induced by instabilities within the plasma. The localised interactions between the waves and the particles strongly control the dynamics of the entire structure. Plasma waves are difficult to study from afar, and *in situ* measurements are necessary. The plasma wave spectrometer was designed to measure the density of charged particles in the spacecraft's immediate environment, and the electromagnetic waves in the plasma, in order to determine the contents of the solar wind or of a magnetosphere.

The instrument was basically a sophisticated radio receiver, with an electric dipole antenna consisting of a hinged gull-wing antenna pair spanning 6.6 metres mounted at the tip of the magnetometer boom to measure the electric fields, and a pair of search coil magnetic antennas on the high-gain antenna's tower to measure the magnetic fields. Almost simultaneous measurements of the electric and magnetic field spectrum permit electrostatic waves to be distinguished from electromagnetic waves, and the spatial context could be derived from the spacecraft's orientation. Similar instruments had been carried by the Voyagers.

Plasma detector

This subsystem was designed to study low-energy plasmas to determine their temperatures, densities, bulk motions and composition.

The instrument used two electrostatic analysers to measure the energy per unit charge of the positive ions and the electrons, both separately and simultaneously. It used a set of slit-shaped apertures to build up a fan-shaped field of view which, as the spacecraft turned, determined the direction and velocity of the charged particles. A trio of mass spectrometers measured the mass per unit charge of the ions to identify the species. The plasma analyser was sensitive across the range 1 to 50,000 volts, which was considerably better than the 100 to 4,800 volt range of the Pioneer instrument and the 10 to 6,000 volt range of the Voyager instrument. Also, at 5 seconds integration time rather than 200 seconds, the temporal resolution of Galileo's instrument would enable it to identify short-term fluctuations. Furthermore, as neither of Galileo's predecessors had used mass spectrometers, this would provide the first direct measurements of the composition of the plasma.

Dust detection system

Galileo's dust detection system was a revised version of the impact plasma micrometeoroid detector flown on the Highly Eccentric Orbit Satellite launched in 1972 by the European Space Research Organisation to study the Earth's environment. Its function was to gather statistics on dust in the spacecraft's vicinity by measuring the velocity, mass, charge and direction of flow of submicron-sized particulates.

The instrument consisted of a multi-coincidence detector and a microprocessor to control its operation and to process the data. Ions entering the sensor were first detected by the charge that they induced as they flew through the entrance grid. This signal was only considered if the ion subsequently hit the impact plasma detector at the rear of the instrument, which detected plasma produced when the particle hit a gold target. After separation by an electrical field, the ions and electrons of the plasma were accumulated by charge-sensitive amplifiers to give two coinciding pulses of opposite polarity. The pulse height (corresponding to the total charge) was a function of the mass and the velocity of the particle, but the rise time of the pulses depended only on the particle's speed. It was therefore possible to derive both the mass and impact speed of the dust particle. The trajectory and true speed of the particle could be inferred from the event time and the spacecraft's trajectory and orientation. The system could resolve up to 100 impacts per second.

Energetic particle detector

The energetic particle detector really combined two silicon solid-state detector systems. The low-energy magnetic measurement subsystem was to measure ions and electrons, and the composition measurement subsystem was to measure ions ranging in mass from hydrogen to iron. The detector was to determine the composition, intensity, energy and angular distribution of the charged particles within the Jovian magnetosphere, in order to identify the processes which replenish the reservoir of energetic particles. In effect, it would create three-dimensional maps of the particles within the magnetosphere, and show how they varied with time.

Heavy-ion counter

This instrument was added to the particles and fields suite when the Voyagers revealed just how intense the 'radiation' in the Jovian magnetosphere really was. In addition to electrons and hydrogen and helium ions, it was seething with much heavier oxygen and sulphur ions that had been 'snatched' from the cloud that surrounds Io by the planet's magnetic field and accelerated to high energies. Although, these ions had bombarded the Voyagers, their electronics had been resistant. The solid-state systems of the more sophisticated Galileo spacecraft, however, were more delicate and were likely to suffer 'single-event upsets' when radiation randomly 'flipped' bits which might result in a chain reaction of erroneous commands, possibly with disastrous results.

In 1984, in order to monitor the threat from heavy ions, it was decided to add an instrument based on the Cosmic Ray Science instrument that had been carried by the Voyagers. "We repackaged the instrument, updated the electronics for Galileo's requirements, and redesigned the sensor system to optimise the detection of heavy ions," explained Ed Stone, the Voyager instrument's team leader. The remainder of the spacecraft was systematically 'hardened' to make its electronics more resistant to penetration by heavy ions. Ironically, being 'old technology', the heavy-ion counter was one of the safest instruments on Galileo!

The heavy-ion counter used five single-crystal silicon wafers of different thicknesses from 30 to 2,000 microns, each of which had arrays of gold and aluminium electrodes across its surface. A fast-moving ion may, for example, pass through the first and second silicon wafers and stop in the third, ionising the silicon through which it travelled. Once these ionisations had been collected as electrical signals and amplified, the charge and speed of the particle could be inferred – for ions travelling at the same velocity, the signal was proportional to the square of the charge. If there was no signal from the final wafer, it could be deduced that the heavy ion had been stopped in the preceding wafers, and all of its energy has been accounted for. It was possible to measure heavy ions having energies as low as 6 million and as high as 200 million electron-volts per nucleon, a range which included not only oxygen and sulphur, but all of the elements from carbon and nickel in terms of their atomic masses. This was "nearly three times the capabilities of Voyager instruments", Stone pointed out. "It will watch for changes over time", to provide a method of assessing the threat to the spacecraft.

In order to integrate the heavy-ion counter into the overall system architecture, it was designed to share a communications link with the extreme ultraviolet spectrometer, so that only one of these two instrument could observe at any given time.

Radio science

Galileo's radio signal facilitated a number of experiments. These exploited the ability of the Deep Space Network antennas to monitor the properties of the signal very precisely.

The radio propagation team measured the minute changes in frequency, power, time delay and polarisation of the spacecraft's radio signal as the spacecraft passed behind a body during occultations. As the spacecraft passed behind Jupiter's limb it would be possible to 'profile' the temperature, pressure and layering of the upper atmosphere. As it passed behind a moon, it would be able to measure the physical properties of an ionosphere. When the spacecraft was on the far side of the Sun, the manner in which its signal was degraded would provide insight into the physics of the solar corona. "One man's noise is another man's data," observed Jay Breidenthal, the science coordinator for the radio propagation team.

By measuring how the spacecraft's signal was modified by the doppler effect during a fly-by of a moon, the celestial mechanics team would be able to chart the gravitational field and, as a result, deduce the internal structure. In fact, during these measurements a two-way radio link would be established, with the spacecraft echoing a signal dispatched from Earth. The raw data for the analysis was the fine differences between the signals sent and received.

Also, while the spacecraft was *en route* to Jupiter, it would be possible to attempt to detect gravitational waves, the ripples in space which 'must' exist but have yet to be detected. They have such long wavelengths that a very large detector – billions of kilometres long – may well be required to detect them. A spacecraft far from Earth offered an opportunity to test the idea. Frank Estabrook, leading the gravity wave experiment, estimated that Galileo had a 40 percent chance of detecting something.

THE ATMOSPHERIC PROBE

Galileo's 340-kg atmospheric probe comprised an outer deceleration module and a descent module. The deceleration module's forebody aeroshell and aft cover completely enclosed the descent module. Its role was to protect the descent module during the entry phase, and slow it by aerodynamic braking so that a parachute could be deployed, at which point the deceleration module would be jettisoned. The heat shield, which was designed to withstand 15,500 °C, accounted for 220 kg of the probe's overall mass. The heat shielding was supplied by General Electric (now Lockheed Hypersonic Systems) of Philadelphia, Pennsylvania. The probe itself was integrated by Hughes Space and Communications Company in El Segundo, California.

The descent module – which accounted for just 35 percent of the entry mass – contained the scientific instruments, a power supply, a data-processing system, and a transmitter to send data to the main spacecraft.

The probe would draw power from the main spacecraft while it was being carried, but in independent flight it would utilise a trio of lithium-sulphur batteries with a total capacity of 18 amp-hours – sufficient to operate the probe's instruments during a descent phase lasting just over an hour.

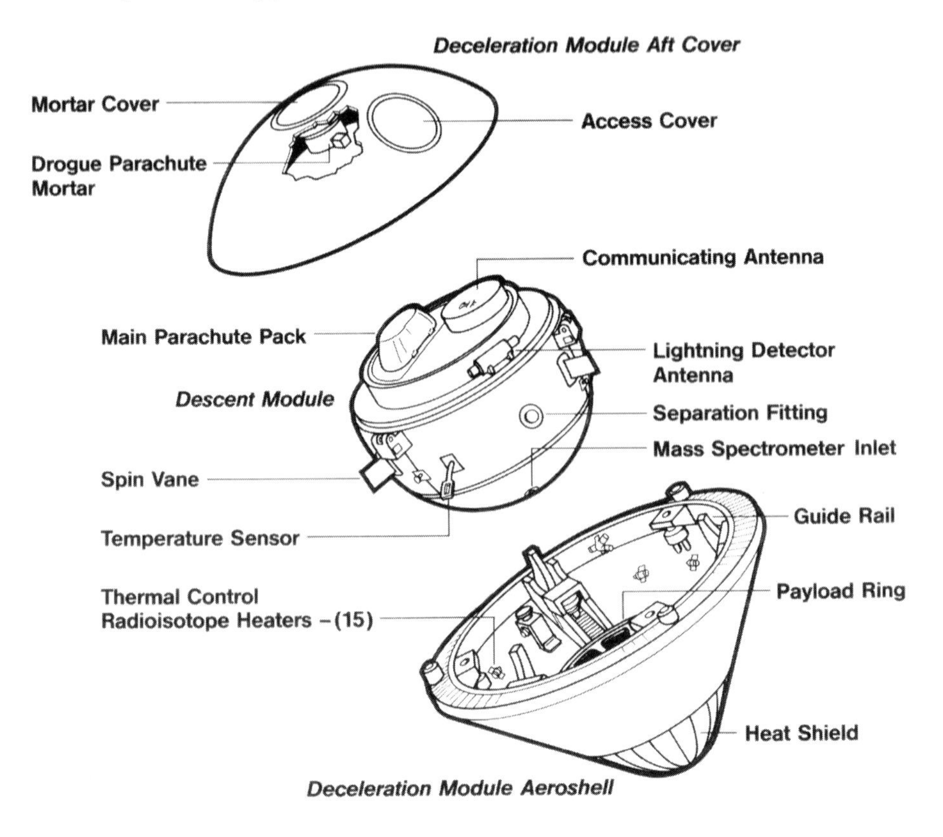

A diagram of the configuration of the Galileo spacecraft's atmospheric probe. (Courtesy, NASA Ames)

PROBE INSTRUMENTS

Of the probe's nine scientific experiments, two derived their data from monitoring the radio transmission to the main spacecraft. The probe's instruments suite was selected in 1977. Of the seven instruments carried, only six were designed to study the atmosphere:

- energetic particle instrument
- lightning/radio detector
- atmospheric structure instrument
- neutral mass spectrometer
- nephelometer
- net flux radiometer
- helium abundance detector

It was also intended that the probe should conduct a doppler wind experiment (see Table 4.2).

Magnetospheric studies

Energetic particle instrument

The energetic particle instrument was the only instrument aboard the probe which was not designed to study the atmosphere. It was to record the electromagnetic environment within the intense belts of radiation above the equator – the equivalent of the Earth's Van Allen belts. It was to complement the measurements which were to be made by the instruments on the main spacecraft as it passed overhead.

The two-element 'telescope' used silicon surface barrier detectors to make omnidirectional measurements of four species of particles – electrons, protons, helium nuclei, and heavy ions – at high counting rates. Since the heat shield would be in place during the entire experiment, any particles which reached the detector would have to have the energy to penetrate the shield. It was to be activated about three hours out from Jupiter, shortly before the probe crossed Io's orbit. The first data was to be taken while within Io's torus, at 5 Rj, with further samples being taken at 4 Rj and 3 Rj, and then it would run continuously from 2 Rj – the closest Pioneer 10 had approached – to the entry interface to establish the spatial and energy distributions of the energetic particles. To date, the only way to study Jupiter's inner magnetosphere had been by the synchrotron radiation produced by the relativistic charged particles in the radiation belts.

To save the probe's power supply for the descent phase this early data would be stored in solid-state memory for subsequent transmission. The experiment was developed by Ames, in collaboration with the University of Kiel.

Atmospheric studies

Lightning/radio detector

Prior to the Voyager fly-bys, there had been speculation that lightning bursts in the

Jovian atmosphere powered non-thermal radio emissions from the planet. Night-side imagery caught flashes of lightning, and the plasma wave experiment detected radio signals characteristic of an electrical discharge propagating in the magnetosphere. To follow up on these studies, Galileo's lightning/radio detector had a radio-frequency antenna and a pair of photodiodes located behind fish-eye lenses. The radio receiver was to measure magnetic signals in the 10-hertz to 100-kilohertz frequency range with a waveform analyser and a spectrum analyser. The instrument was funded and supplied by Germany, as part of that nation's collaboration in the mission.

Like the energetic particle instrument, the lightning/radio detector would be activated on the run in to Jupiter. It was to take measurements at 4 Rj, 3 Rj and 2 Rj. The measurements in the magnetosphere were to complement the data from the energetic particle instrument. During the descent phase, the lightning/radio detector would operate continuously. Since both instruments operated in space, they shared data-processing facilities.

Atmospheric structure instrument

The atmospheric structure instrument's function was to measure the temperature, density, pressure, and molecular mass of atmospheric gases as a function of altitude. During the entry phase, the instrument was to infer these values from the rate of deceleration, as measured by accelerometers, but it was to measure the temperature and pressure directly (the density was a computed parameter) throughout the descent phase.

Neutral mass spectrometer

The neutral mass spectrometer was directly and repeatedly to sample Jupiter's atmospheric gases during the descent phase to provide a detailed analysis of the atmosphere's composition.

The gas would enter the instrument through two inlet ports near the top of the probe. Once the gas had been ionised by an electron beam whose energy could be varied, the ions would be fed into a quadrupole analyser (four hyperbolically shaped rods) and a voltage applied to filter the ions by mass. Varying the voltage and radio frequency in the analyser could select ions of a specific mass and charge for the counter.

With its broad mass and sensitivity range, the instrument would be able to measure almost everything that entered it, and was therefore ideal for this exploratory mission. The *in situ* data would complement remote-sensing instruments on the main spacecraft, which would make a broader survey, but would be limited to observing the uppermost part of the atmosphere. The state-of-the-art instrument was built by the Goddard Space Flight Center at Greenbelt, Maryland.

Nephelometer

The nephelometer was to measure the physical structure of Jupiter's clouds in the pressure range 0.1 to 10 bar, and to determine the location of cloud layers, and the size, concentration and shape of individual cloud particles (the shape of a cloud particle would indicate whether it was in an ice or a liquid phase).

The instrument, built for Ames Research Center by the Martin-Marietta Corporation of Denver, Colorado, was similar to those used by the atmospheric probes of the Pioneer Venus mission, incorporating both a forward scatter unit and a backward scatter unit. As soon as the descent module was released, the nephelometer was to extend a short (13-cm) arm which had forward scatter-sensing mirrors located on its tip. Particles in the gap between the body of the probe and the mirrors on the arm would be illuminated by an infrared laser, and the size of the particles would be determined by the intensity of scattered light intercepted by mirrors set to detect light scattered at various angles from the main beam.

When integrated with the probe's rate of descent (as derived from the doppler on its radio transmission), the data from the nephelometer, atmospheric structure instrument, and neutral mass spectrometer would profile the atmosphere's composition, pressure, temperature and particulate properties as a function of altitude.

Net flux radiometer

Fly-by spacecraft had been limited to measuring the radiation leaving Jupiter's cloud tops, but the probe's net flux radiometer was to sample directly the local radiation flows within and below the Jovian clouds.

The instrument used a rotating optical head in which detectors viewed a 40-degree cone of the atmosphere through a diamond 'window', illuminating pyroelectric detectors and filters sensitive from the visible to infrared wavelengths – one to measure the solar energy, one to measure the integrated infrared energy flux, and three to help to identify selected atmospheric species. During the descent phase, it would rotate to derive the upward and downward radiant energy fluxes as a function of depth and thus profile the ratio of radiative heating and radiative cooling which induces the buoyancy differentials that drive atmospheric motions. The vertical profile of net radiation flux at specific wavelength bands would also identify regions where the atmosphere absorbed radiation relatively strongly. Molecular hydrogen is the major source of gaseous opacity in Jupiter's atmosphere. However, in some parts of the spectrum hydrogen is transparent. If it were not for absorption by minor constituents – such as methane, ammonia and water vapour – the atmosphere would radiate tremendous amounts of heat to space.[5] By profiling the net flux as a function of altitude at wavelengths transparent to hydrogen, it would be possible to estimate the abundances of such constituents. The net flux radiometer was built by Martin-Marietta Denver Aerospace. Its 'remote-sensing' data would complement the direct measurements of particulates by the nephelometer, and compositional data by the neutral mass spectrometer.

Helium abundance detector

Although the ratio of helium to hydrogen in Jupiter's atmosphere would be measured by the neutral mass spectrometer, this particular ratio was deemed to be sufficiently important to justify the inclusion of a separate instrument. This was adapted from a compact commercial product for measuring methane in a coal mine. It was a folded interferometer which operated by comparing the refractive index of a sample of Jupiter's atmosphere with a 'reference' gas mixture within the instrument. If it could make 30 measurements between the 2.5- and 10-bar pressure levels, it would be able to determine the ratio to an accuracy of 0.0015 – which would be an order of magnitude improvement on the value derived from Voyager data. It was built by Messerschmitt–Bolkow–Blohm using an interferometer supplied by Carl Zeiss Optics.

Doppler wind experiment

The doppler wind experiment was quite straightforward in concept, if not in application. It set out to infer Jupiter's wind speeds by monitoring the motion of the probe as it descended on its parachute deeper into the atmosphere. The raw data for this investigation would be the doppler on the probe's L-band transmission to the main spacecraft. Interpreting the results would be a complex process, drawing upon insight from the instruments making *in situ* measurements.

Table 4.2: Probe experiments

Experiment	Principal investigator
Magnetospheric studies	
Energetic particle instrument	Harald Fischer, University of Kiel, Germany
Atmospheric studies	
Lightning/radio detector	Lou Lanzerotti, AT&T Bell Laboratories
Atmospheric structure instrument	Al Seiff, Ames Research Centre
Neutral mass spectrometer	Hasso Nieman, Goddard Space Flight Centre
Nephelometer	Boris Ragent, Ames Research Centre
Helium abundance detector	Ulf von Zahn, Bonn University
Net flux radiometer	Larry Sromovsky, University of Wisconsin at Madison
Doppler wind experiment	David Atkinson, University of Idaho
Radio science	
Jovian atmosphere	Taylor Howard, Stanford University

5

The long haul

18 OCTOBER 1989

Torrence Johnson, one of the original Galileo proponents and the Project Scientist, described the VEEGA route as "innovative billiard shots". To make it feasible, the spacecraft would require to fly an extremely accurate trajectory. Although it could make 'corrections' using its own rocket engines, this would use propellant, and that would restrict the spacecraft's capability when it finally reached Jupiter, if indeed it was able to reach the planet at all. The first step was for the Air Force's IUS to place the interplanetary wanderer on an accurate course for Venus.

Ten minutes after it had been deployed, the IUS activated its flight control system. Half an hour later it took a series of star sightings to update its inertial platform. A rocket burning solid propellant tends to accrete combustion products in its nozzle, so the IUS spun itself up to even out any irregularities in its thrust. At 20:15 EDT, its first stage ignited. It delivered a thrust of 185,000 newtons for two and a half minutes, then released the second stage. The second stage ignited five minutes later, delivering 27,500 newtons of thrust for a minute and a half, by which time the vehicle was moving away from the Earth at 11.5 kilometres per second and was in solar orbit.

After the escape manoeuvre, the IUS stabilised itself and turned so that Galileo was facing the Sun. The spacecraft then deployed the booms carring its RTG power cells and some of its scientific instruments. Then the IUS spun up to about 3 rpm and released its payload. Galileo immediately reported its healthy status to the Deep Space Network over its low-gain antenna.

"Galileo is on its long, multi-year odyssey to Jupiter," proclaimed a highly satisfied acting associate administrator for space flight.

AGAINST THE WIND

Galileo's roundabout route to Jupiter made possible a valuable study of the interplanetary medium at a wide variety of locations and over an extended timescale.

The solar wind is a plasma. It comprises mostly protons and electrons, but there is also a percentage of heavy ions. The Sun blows off vast blobs of plasma, in the form of 'coronal mass ejections'. The solar wind is fairly 'gusty', varying in speed from 200 to 400 kilometres per second. Because plasma is electrically charged, the solar wind drags the Sun's magnetic field out with it. The first *in situ* investigation of the solar wind was undertaken in 1962 by Mariner 2 which, as it happened, was also on its way to Venus.

Galileo was to monitor the physical and chemical properties of the solar wind, the ambient electromagnetic and electric fields, the characteristics of the dust, and cosmic rays throughout its interplanetary cruise and report the data once a week.

The solar wind penetrates to the farthest reaches of the Solar System, and eventually yields to the interstellar medium. Many other spacecraft, including the Pioneers and Voyagers, which were now departing the Solar System,[1*] were also reporting on the solar wind. In addition, the Shuttle would soon launch Ulysses – a spacecraft that would take the direct route to Jupiter and use the planet's immense gravity to redirect it into an orbit which would enable it to sample the solar wind emerging from the Sun's 'polar regions'. The final decade of the twentieth century therefore heralded a renewed 'golden age' for deep space missions.

On 26 October 1989, Galileo's atmospheric probe, tucked in tightly against the nozzle of the spacecraft's main engine, was subjected to six hours of tests to verify its status, and was then powered down. It was to be given such a check annually.

Radio tracking by the Deep Space Network showed that the IUS had performed accurately, but it was decided to make two small corrections to refine the aim for Venus. The first would serve to calibrate the small 10-newton thrusters, and the second would eliminate any residual errors. In space, the earlier a correction is made, the less propellant it will take to achieve the required adjustment. The refinements – on 9 November and 22 December – were made by repeatedly pulsing the thrusters. Tracking established that the thrusters were performing at 102 percent of their rated thrust. The near-perfect trajectory imparted by the IUS was welcome news because the commitment to visit the 'targets of opportunity' in the asteroid belt would be dependent upon having a healthy propellant margin.

CLOUD-ENSHROUDED VENUS

A number of spacecraft, both American and Soviet, had investigated Venus. Magellan had been launched by Shuttle in May 1989. Although it was a few months ahead of Galileo, it was taking a 'slow' trajectory, and would not arrive until August 1990. It was to enter orbit and use a synthetic aperture radar to map the surface through what appeared to be a permanent cloud cover. The Pioneer Venus Orbiter had been monitoring the atmosphere and its interaction with the solar wind since 1978.[2]

* For Notes and References, see pages 413–424.

If Galileo was to be able to fulfil its mission with the minimum expenditure of propellant, it would have to navigate a narrow corridor. When it made its closest approach – flying 16,000 kilometres from the planet's centre – it was within 5 kilometres of the middle of its corridor. A gravitational 'slingshot' is a technique whereby some of a planet's orbital energy from its motion around the Sun is transferred to a spacecraft. In the process, the spacecraft's trajectory is deflected around the planet and its speed is increased around the Sun. The Venus fly-by on 10 February 1990 added 8 kilometres per second to the spacecraft's speed. Galileo observed the planet and its electromagnetic environment over an 8-day period, starting a day before closest approach, storing 81 images of the planet on tape for later transmission to Earth.[3]

The infrared spectrometer provided a new view of the planet's atmosphere, and produced the highest resolution imagery of cloud circulation ever achieved, enabling small-scale structures over mid-latitudes to be studied for the first time.

Although Venus rotates exceptionally slowly – it takes 243 terrestrial days to turn once on its axis – the upper atmosphere is dominated by a jetstream that circles the planet every four days. The jetstream is from east to west[4] at about 370 kilometres per hour. Through a violet filter, the view of Venus was dominated by the sulphuric acid clouds in the upper atmosphere. A comparison of the violet and near-infrared imagery confirmed that the 'super-rotation' of the atmosphere decreases with depth. In fact, the Soviet Venera probes which had landed on the surface had revealed that

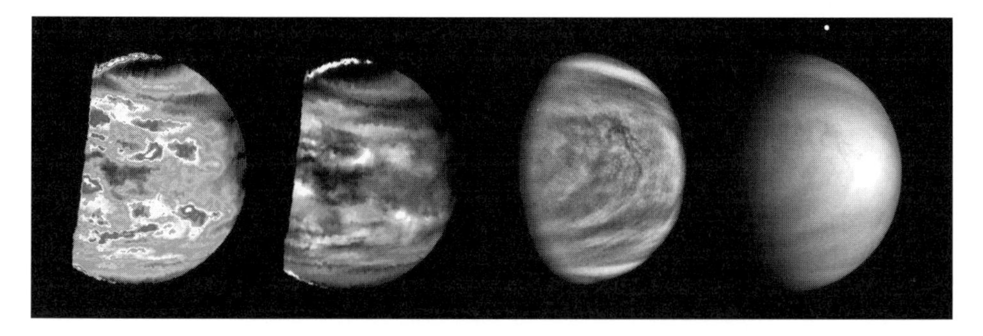

Four views of Venus taken by Galileo. The violet filter views (on the right) were taken six days after the encounter from a range of almost 3 million kilometres and show the illuminated hemisphere. They show details of the sulphuric acid clouds which form at the top of the atmosphere. The thin dark filaments visible in the right image correspond to details in the one on the left, which shows considerable convective activity spanning a wide area in the equatorial zone. Note the strong specular reflection from the planet at the sub-solar point. The 2.3-micron images (on the left) of the dark side of the planet were taken during Galileo's approach from a distance of 100,000 kilometres. At this wavelength the view is of radiant heat from the 'middle' atmosphere, located some 15 kilometres below the cloud tops and about 50 kilometres above the surface. The signal is attenuated by the sulphuric acid, so the brigher features correspond to thin patches of upper-level cloud. Note the thin slivers of the illuminated hemisphere poking in the polar zones in the night-time views.

the air at the surface is stagnant. Galileo's data put constraints on the depth to which the meridional circulation persists.

The main constituent of the atmosphere had long been known to be carbon dioxide, and the jetstream made the upper atmosphere uniform. The composition of the lower atmosphere remained obscured by the monotonous cloud cover until 1984,[5] when it was realised that there were two narrow spectral 'windows' in the near-infrared. At these wavelengths, it was possible to view the radiation emerging from below. In principle, these absorption features would enable the minority constituents of the atmosphere to be determined all the way down to the surface. Galileo's infrared spectrometer was able to make such 'soundings' of the composition of the atmosphere. At a wavelength of 2.3 microns, it saw the turbulent cloudy middle atmosphere, some 10 to 15 kilometres below the visible cloud tops, about 50 kilometres above the surface. As the radiant heat from the low atmosphere was attenuated passing through these patchy clouds, Galileo was able to 'see' their structure through the upper cloud deck. At equatorial latitudes, the clouds appeared to be fluffy and blocky. Farther north, they were stretched out into east–west filaments by winds of 250 kilometres per hour. The polar regions were capped by thick clouds at this altitude.

The spectrometer's spatial resolution was only 50 kilometres, but highland and lowland terrains could clearly be identified.[6] Surface detail could be enhanced by 'de-clouding' (as the process was dubbed) by subtracting the signal representing the middle atmosphere from the signal representing the hot surface and the atmosphere below an altitude of about 10 kilometres. The very steep (8 degrees per kilometre) thermal gradient implied that the surface temperature varied by 100 degrees from the top of the tallest mountain to the floor of the lowest plain – a vertical range of 13 kilometres.

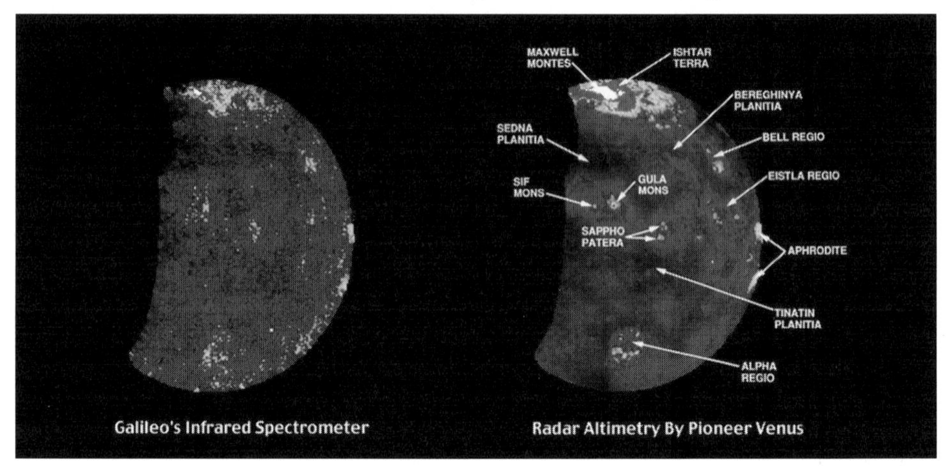

The temperature of Venus's surface is correlated with altitude, so Galileo's 1.8- and 2.3-micron data could be processed to 'subtract' the atmosphere, and infer the topography of the surface with a resolution of 50 kilometres per pixel. The 'infrared altimetry' map compares favourably with the low-resolution radar map produced by the Pioneer Venus Orbiter. (Courtesy Robert Carlson, JPL)

Earlier spacecraft had detected radio signals that appeared to be due to lightning in Venus's atmosphere. In addition to the visual flash, a lightning bolt gives rise to a broadband burst of radio waves.

Galileo's plasma wave spectrometer detected electromagnetic pulses from Venus, thereby confirming the presence of lightning within the dense atmosphere. The cause of such electrical discharges was a mystery. Terrestrial lightning occurs when strong updrafts cause moist air to condense out droplets of rain. The rapid air motions prompt the build-up of electrical charges, which discharge to the ground and from one cloud to another. Water vapour is conspicuous by its absence on Venus, however. On Earth, lightning can also occur in the plume of gas and dust emitted by an explosive volcano, because the particles in the roiling cloud are electrically charged. Perhaps the lightning on Venus indicated volcanic activity? However, the pressure of the dense atmosphere – almost 100 bar at the surface – would tend to inhibit plume activity.

For the next few weeks, Galileo flew within Venus's orbit. Throughout this time, it kept its delicate systems hidden behind the sunshades. After its 0.7 AU perihelion on 25 February 1990, the spacecraft headed back out. Trajectory corrections on 10 April and 11 May refined the aim for its first Earth fly-by.

Charged particles have an effect on radio waves, much like the heat from a fire has on the air above it, making objects beyond appear to shimmer. This is called scintillation. Although it was a nuisance to the communications engineers, the radio interference which occurred when the spacecraft was aligned close to the Sun in the sky offered an opportunity to make observations of the solar corona. The energy of the radio signal was monitored either side of a superior conjunction in order to study the region in which the solar wind was generated.

In addition to continuing to monitor the interplanetary medium, Galileo made observations of Comet Austin in May and Comet Levy in August. The extreme-ultraviolet spectrometer was able to detect the vast hydrogen comas which surround comets.

After another small trajectory correction on 17 July, the mission planners knew that they would have enough propellant to take a close look at Gaspra, so this was confirmed. The 1.28 AU aphelion on 23 August was barely outside the Earth's orbit. To reach Gaspra, orbiting a distance of 2.2 AU, Galileo would have to make an extremely accurate slingshot on its first Earth fly-by. In late November, as the Earth loomed, Galileo transmitted the data it had taped during the 8-day encounter with Venus – this had been delayed until the spacecraft had returned to the Earth's vicinity in order to be able to use the low-gain antenna at a higher data rate.

HELLO EARTH

Galileo's second planetary encounter, on 8 December 1990, marked a historic first – this was the first time that a spacecraft had made an approach to the Earth as if it were a 'target' for investigation.

The Earth has a magnetic field. This is believed to be generated by the 'dynamo effect', as vast electrical currents circulate within its metallic core. The magnetic field extends into space, where it interacts with the solar wind. The space that is dominated by the Earth's field is called the magnetosphere.

Directly up-Sun, the Earth's magnetic field actually halts the solar wind, and eases it aside. Plasma continues to impact, however, and forms a shock wave. Although the Earth is moving around the Sun in its orbit, its 'bow shock' is directly in line with the Sun, not in the direction it is travelling. Down-Sun, the magnetosphere is drawn out into a tail, variously known as the magnetotail, or geotail, which extends for at least 1,000 Earth radii. The interface between the magnetosphere and the solar wind is the magnetosheath (the particles and fields community that investigates these things has a penchant for long names). The inner magnetosphere rotates with the Earth on a daily basis, so the plasma contained within it is highly compressed up-Sun and rarefied down-Sun, and this rotational cycle gives rise to some intriguing electromagnetic effects. The distended magnetotail does not rotate, so there is a complex region between these two sections – rotating and not rotating.

Plasma from the solar wind can leak into the Earth's magnetosphere around the magnetic poles. The electrically charged plasma flows back and forth along the magnetic field lines, and when it interacts with the upper atmosphere it stimulates auroral displays. Radiation – mainly electrons and protons – becomes trapped in the inner magnetosphere and is concentrated above the equator in the dense toroidal Van Allen belts. Particles down the magnetotail escape to the solar wind. If it were not for this leakage, the magnetosphere would long since have become saturated.

The Earth's magnetosphere had been intensively studied by a variety of spacecraft over the three decades of the 'Space Age', but Galileo's particles and fields instruments were well-suited to the task. As the spacecraft flew in from its modest aphelion, its trajectory brought it in over the night-side of the Earth, and it crossed the magnetosheath at a range of 560,000 kilometres. As the magnetosphere was fairly active, the spacecraft was able to observe several magnetic 'sub-storms' as it worked its way up the magnetotail. The data from Galileo's 16-hour run up the tail was to be correlated with that from other satellites and from ground stations to form a three-dimensional snapshot of the state of the magnetosphere.

Although the Moon had been extensively mapped in the mid-1960s by the Lunar Orbiters, and the final Apollo missions had conducted mineralogical scans of narrow swaths of terrain close to the equator, Galileo's infrared spectrometer was more advanced. The fly-by provided an opportunity to verify its calibration and capabilities.

At this time, the Moon was several days beyond its 'full' phase, so all but the trailing limb was illuminated. As it flew up the Earth's magnetotail, Galileo turned its scan platform to look at the Moon and caught it as a crescent displaying Mare Imbrium and Oceanus Procellarum on the limb. This region included some of the Apollo landing sites, so the multispectral data from the infrared spectrometer could be calibrated against the 'ground truth' of the rocks taken from these sites. Generally speaking, lunar terrain is classified as being either 'highland' or 'mare' material. The highlands are the pulverised remains of the original crust. The dark maria are smooth lava flows, many of which are contained within large impact basins. Galileo showed that the 90-kilometre wide crater Copernicus had punched through the Procellarum lava flow and excavated the underlying highland crust, which explained why this crater and its rays of ejecta are so bright. Reflectance spectroscopy had been done from Earth, but a telescopic study was capable of 2 kilometres resolution at best and was obviously restricted in scope to the near side of the Moon. Galileo caught up with the Earth from behind, and the Moon was leading the Earth, so the spacecraft approached no closer to the Moon than the Earth was. Nevertheless, its best imaging resolution of 400 metres per pixel represented a major improvement on telescopic studies.

At its closest point of approach to the Earth, at 20:35 UT, Galileo was within half a second of its scheduled arrival time and its altitude was 960 kilometres – as against the intended 952 kilometres. As Project Manager Bill O'Neil proudly explained: "Delivery accuracy was better than 99 percent in the aim point and time of closest approach for both of the last two trajectory correction manoeuvres." Even so, the anti-'nuke' activists protested the encounter, seemingly under the mistaken impression that if Galileo somehow 'spun out of control' it would re-enter the Earth's atmosphere and distribute its 'nuclear fuel' far and wide.

As Galileo flew 1,600 kilometres above Africa, the particles and fields instruments noted its flight through the inner Van Allen belt. The plasma wave spectrometer detected lightning. As the signal travelled through the electrons flowing in the Earth's magnetosphere, the lower frequencies of the broadband radio burst were retarded by 'dispersion', which arose from an interaction between the electrons and the electric field component of the electromagnetic wave. When such a signal was transformed to the audio range in order to be heard, the pitch started high and decreased, so such signals had been dubbed 'whistlers'. The rate at which the pitch fell off provided information on the environment through which the signal had to pass.

Some of Galileo's instruments were more sophisticated than those carried by the satellites whose role was to monitor the Earth's environment from orbit. The infrared spectrometer saw very high level clouds over Antarctica. This offered insight into the development of the ozone hole over the south pole. The atmosphere is stratified, and each layer has its own temperature profile. In the troposphere, which extends from sea level to the base of the stratosphere, there is a decrease in temperature with altitude. In the stratosphere, thanks to the presence of ozone, the profile reverses. But above the stratosphere, in the mesosphere, the temperature decreases again, with the coldest part of the atmosphere being a mere 130 K. In the rarefied atomic oxygen above the mesosphere, in the thermosphere, the temperature soars again due to solar radiation.

Galileo's observation of high altitude clouds over Antarctica was significant because the computer models used to predict ozone depletion were in need of data on the amount of water in the mesosphere. The extreme cold produced clouds of water-ice crystals, rather than water droplets, and the ice enabled chemicals to come together and react in ways that would not otherwise have been practicable. Although such high-level clouds had been seen as late as October, they were considered very rare in December.

A few hours after closest approach, Galileo again turned its attention to the Moon, and this time saw it as 'half' phase. The view was similar to what it would have been from Earth, with Mare Imbrium and Oceanus Procellarum on view, but Galileo's vantage point enabled it to see a little way around the leading limb and it was able to view Mare Orientale. By an unfortunate bureaucratic quirk, despite the fact that 'Orientale' is Latin for 'Eastern', this feature lies mostly just beyond the western limb! This is because, in the age of telescopic astronomy, the Moon was viewed upside-down, with south at the top, and the leading limb was defined as being the eastern one because it faced the observer's eastern horizon. However, once NASA started to plan to send people to the Moon, the International Astronomical Union redefined the eastern horizon to be that over which an observer on the lunar surface would see the Sun rise.[7]

Late on 9 December, as it withdrew, Galileo was presented with a 'full moon', but it was not the disk that we observe under this illumination. Mare Imbrium and Oceanus Procellarum were truncated by the limb and Mare Orientale was in the centre, resembling a vast 'bullseye', and ideally illuminated for the infrared spectrometer to survey it mineralogically by measuring its solar reflection spectrum.

One of NASA's Lunar Orbiters had captured Orientale magnificently. It is the classic example of a large multiple-ringed impact structure. The infrared spectrometer found that its blanket of ejecta was mostly plagioclase-rich highland rock. It had been thought that such a vast impact would have excavated pyroxene-rich rock from deep within the crust, so this surprising discovery was a major bonus for lunar scientists. Galileo's imagery also found evidence of mare material underlying the ejecta. This was a significant discovery, because most of the dark lava flows on the Moon post-date the formation of the basins in a period known as the Great Bombardment. Orientale's well preserved ring structure implied that it was one of the last such structures to be excavated, as the bombardment tailed off some 3.8 billion years ago.

Galileo also resolved another lunar mystery. Various lines of evidence suggested that there was an enormous basin on the far side, spanning the area between Aitken Crater and the south pole. In 1962, Bill Hartmann and Gerard Kuiper had suggested that the Mountains of Leibnitz which poke over the southern limb during favourable 'librations' were the rim of a vast crater, and so it turned out. Galileo's multispectral imagery strongly supported the case for it being an impact basin. At 2,500 kilometres wide, this Aitken Basin is the largest such structure yet seen in the Solar System. It would have taken a rock about 160 kilometres in diameter to make such a vast crater. Despite its size, it has been so heavily 'modified' by subsequent impacts that it is very difficult to see its outline visually.

When the Clementine spacecraft conducted a global mineralogical survey a few years later, it confirmed Galileo's result, and the altimeter on the later Lunar Prospector mapped this basin as a 10-kilometre deep cavity. So, whereas Orientale was one of the youngest basin structures, Aitken was one of the oldest.

Carl Sagan of Cornell University, one of the original proponents for the Galileo mission, exploited the fly-by to test whether spacecraft observations could provide unambiguous proof that life existed on Earth. At 1 kilometre resolution, the best

The Orientale basin on the Moon is the classic example of a multiple-ringed impact structure. It formed near the end of the Great Bombardment 3.8 billion years ago. The impact excavated a cavity and laid down an annular blanket of ejecta several kilometres thick. There are small dark lava flows in the centre and between the rings of mountains, whose 6 kilometre tall scarps face inwards. The outermost ring is 900 kilometres wide. On a favourable libration, the outer ring can be seen poking over the limb by telescopic observers, and it had been named the Cordillera Range. The true extent of the structure was revealed by Lunar Orbiter 4 in 1967.

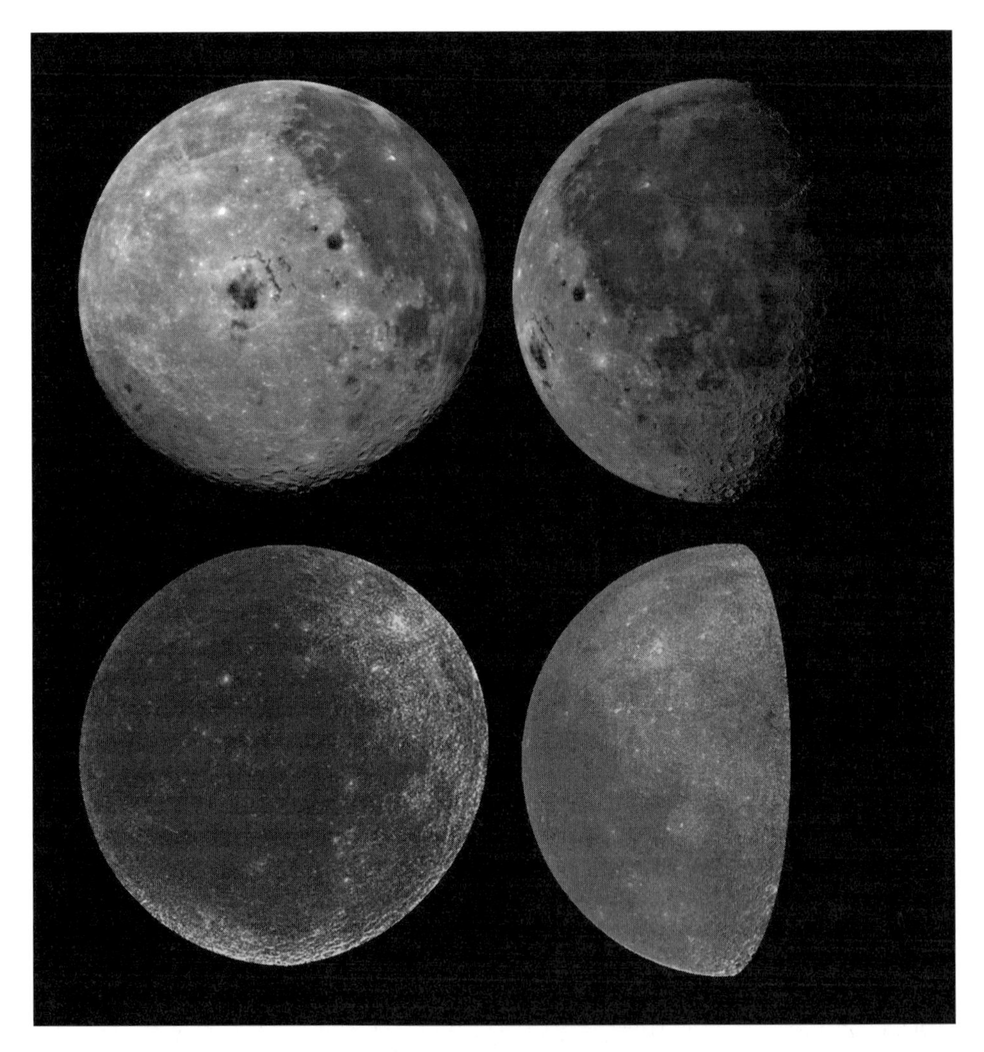

These two pairs of Galileo images show the Moon in visible (upper) and near-infrared (lower) wavelengths. To an observer on the Earth, the Moon would have been in 'last half' phase displaying the hemisphere dominated by the dark Oceanus Procellarum. From its vantage point, Galileo was able to see over the western limb, and its full-disk view was centred on the Orientale basin. The infrared imagery (which is best viewed in 'false colour' form) depicts the surface mineralogy. Although the Oceanus Procellarum area shows rich detail, most of the far side, including the Orientale basin, is 'highland' rock. The massive Aitken basin, which dominates the southern part of the far side and extends down to the South Pole, is difficult to distinguish.

imagery was insufficient to show any structures such as cities, highways, or fields. Nor did it see any artificial lighting at night. The atmosphere's high oxygen content, considered together with the barest trace of carbon dioxide, was highly suggestive of biological activity. So too were the levels of methane and of nitrous oxide. Methane, a by-product of biological activity,[8] was observed in far higher concentration than the equilibrium level for the oxygen present. Water was consistent with life, but it did not require or imply it. Some areas glowed in the near-infrared, but were not distinguished in the red, which hinted at organic activity.[9] The only form of life directly observed was Australia's Great Barrier Reef, which was recorded by the infrared spectrometer but there was no way of knowing that it was alive.[10] The only indications of *intelligent* life were the coherent emissions in the radio spectrum.[11] In addition to auroral and ionospheric emissions, Galileo's plasma wave spectrometer detected narrow-band emissions which grew in strength as the spacecraft closed in on the Earth's nightside. It became clear, however, that the signals originated on the surface as they rapidly diminished when the spacecraft crossed the terminator into daylight – enhanced on the day-side, the ionoshere was trapping the signals. The modulation of the signals was clearly relaying information, which implied intelligent technological life on the surface.

Starting on 11 December, when some 2 million kilometres away and rapidly withdrawing, Galileo took a picture of the Earth every minute for a 25-hour period. Up to this time, only geostationary satellites had been able to provide hemispherical views, but because they remain stationary they cannot observe the Earth's rotation on its axis. Furthermore, because Galileo's path was far to the south, it saw Antarctica turning at the pole, so Galileo's time-lapse 'movie' was unprecedented.

"We had seven days of detailed observations of the Earth and Moon," said a clearly happy Torrence Johnson. "This was just a small preview of what Galileo will do on every one of its orbits of Jupiter."

On 11/12 December 1990 (as it withdrew after its first fly-by) Galileo took a series of images to create a time-lapse 'movie' which showed the Earth rotating. The first image (left) shows South America with Africa on the limb. In the next image South America has gone, and western Australia is presented. The final image shows the Pacific Ocean a day later. The viewpoint was far to the south, so Antarctica can be seen rotating at the bottom of each image.

The spacecraft had performed flawlessly, both during its interplanetary cruise and during the encounters. The Earth fly-by in particular had served an engineering function, by enabling the engineers to develop a 'feel' for how to fine-control the spacecraft's scan platform. "Every time we turned on an instrument, we were really learning how to operate it," reflected Clayne Yeates, the Science and Mission Design Manager. "It's been invaluable for us, to collect this data."

The slingshot was sufficient to ease Galileo's aphelion out beyond the orbit of Mars, to the inner edge of the asteroid belt. On 19 December 1990, the spacecraft refined its trajectory in order to encounter Gaspra in October 1991.

FOULED ANTENNA

On 11 April 1991, with Galileo now safely in the relatively benign environment outside the Earth's orbit for which the spacecraft had been designed, the command was sent to deploy the high-gain antenna. The unfurling of this 4.8-metre diameter dish was eagerly awaited because it would enable the spacecraft to transmit at a rate thousands of times greater than was possible using its low-gain antenna. Unfortunately, something went awry and the structure jammed. A Deployment Anomaly Team was promptly formed, drawing in mechanical, electrical, thermal and materials specialists, as well as staff from the Harris Corporation, the contractor who had supplied the antenna structure.

The high-gain antenna was based on the umbrella configuration of NASA's Tracking and Data Relay Satellite (TDRS) system. The deployment sequence called for catches to release the pins which held 18 graphite-epoxy ribs tight against the central support tower. An electric motor was then to splay the ribs out to stretch a thin molybdenum mesh into a parabolic dish. The pin release was immediately beneath the small circular 'tip shield' which had been mounted on the end of the tower to protect the antenna's structure from the Sun while the spacecraft was within the Earth's orbit.

It was clear that the command had been received, and that the deployment mechanism had been activated because the motor had drawn an electrical current. However, telemetry showed that the motor had operated at a power level that was higher than expected. From this it was concluded that it had encountered an unexpected frictional force. Another anomaly was the microswitch that was to have deactivated the motor after a few seconds – by which time the pins ought to have released – did not trigger. Evidently at least some of the pins had failed to release. The next line of evidence was less direct: after the deployment attempt, the spacecraft had assumed a slight wobble, which confirmed that some of the pins had released and that the antenna was partially deployed and was asymmetrical. Finally, in certain orientations, a Sun sensor's field of view was masked by one of the ribs of the antenna. To some extent this provided a way of verifying the asymmetrical configuration. On 30 April, having factored in all the evidence and consulted widely, it was formally concluded that "one or more ribs are probably restrained in the 'stow' position, resulting in an asymmetrical partial deployment".

The failure of the high-gain antenna to deploy was a potential 'show-stopper', as it was to have been able to sustain X-Band transmissions at a rate of 134 kilobits per second. If it could not be coaxed out, then the spacecraft would be unable to report its findings when it reached its final destination! The spacecraft would have to rely upon its low-gain antenna, which, from Jupiter, was capable of transmitting a mere 10 bits per second. To convey a sense of how catastrophic this was to the mission, just imagine trying to access the internet via a 10 bps modem! But the general public's response was typical: "Give them a few days, and JPL will work its magic."

Determined not to lose the primary mission, JPL studied the release mechanism's design to try to devise a recovery strategy. Perhaps the pins had been insufficiently lubricated? Perhaps during the long time spent in the shade some of the pins had 'cold-bonded' to the mechanism. Could this bond be broken? If the tower could be flexed it might crack the bonds, and then the motor might pull the pins free and deploy the dish.

On 20 May, Galileo was turned 40 degrees from its normal Sun-pointing attitude to remove most of the antenna tower from the tip shield's shadow. It was hoped that thermal expansion of the tower would release the pins; but it didn't. On 10 July, with the spacecraft now 1.84 AU from the Sun, it was turned almost all the way around, putting the tower in deep shadow for 32 hours to induce thermal contraction. The spacecraft was then spun-up to apply centrifugal force to the ribs in the hope that this would release the pins; but it didn't. Galileo was returned to Sun-pointing attitude and the engineers considered their options. "We have to get colder," Bill O'Neil announced optimistically.

On 13 August 1991, with Galileo now 1.98 AU from the Sun, the antenna was again placed into shadow – this time for 50 hours – but even that didn't release the pins. There was no time for another attempt, however, because Gaspra was looming and JPL had now to prepare for the encounter.

951 GASPRA

When NASA announced that Galileo would make a close fly-by of asteroid 951 Gaspra, it fell to telescopic observers to refine the 'elements' of the asteroid's orbit so that the spacecraft could aim for it. However, even as Galileo approached, Gaspra's location was known only to within an ellipsoid about 200 kilometres across, which was not good enough. In September, Galileo took a series of distant pictures of Gaspra set against the stars to enable the navigators to locate the asteroid within a 50-kilometre ellipsoid – which was barely adequate.

It is important to realise that when Galileo undertook a fly-by, it did not sense and track its target. The spacecraft's computer had to be told where to aim the camera, and the time exposure to use. Mission planners had therefore to develop an observational sequence which took into account the position of the target with respect to the spacecraft, the orientation of the spacecraft, and how to orient the scan platform to aim the instruments. If they made a mistake, the spacecraft would 'miss' its target and a unique observation would be wasted. When taking long-shots in 'telephoto' mode, a misalignment would be all too easy. Even if the instruments were aimed correctly, a wrong exposure would make the result worthless. The exposure had to be precalculated, based on the range to the target, its albedo, and how well it was lit by the Sun. Galileo was no tourist, shooting pictures of whatever took its interest. In fact, even if an alien spaceship flew alongside it, Galileo would stick to its programmed imaging sequence.

Because this would probably be humanity's one and only look at Gaspra, Galileo had been instructed to make as many observations as possible. "The team has been extremely ambitious about going for all they can with the Gaspra encounter," emphasised Bill O'Neil.

When still 30 minutes out, at a range of 16,000 kilometres, Galileo took a mosaic of nine images, each through three filters to form a colour image with a resolution of 160 metres. The spacecraft's trajectory was so accurate that it was virtually certain that Gaspra would be in the camera's field of view. Over the following 15 minutes, as the asteroid loomed larger, Galileo took a sequence of 49 monochrome images. What would turn out to be the highest resolution mosaic was composed of a pair of images taken at a distance of 5,300 kilometres, 10 minutes before closest approach. The final picture, with 50-metre resolution, was taken a few minutes later but captured only a partial view. The scan platform's fastest rotation rate of 1 degree per second was insufficient to 'pan' as Galileo flew by Gaspra at 8 kilometres per second, so no imaging was scheduled for the 1,600-kilometre point of closest approach, which was at 22:37 UT on 29 October 1991.

It turned out that Galileo was within 5 kilometres of the intended position and 1.5 seconds of its schedule. "The key word of this encounter was precision," observed Michael Belton of the National Optical Astronomical Observatories, Tucson, Arizona, the leader of the imaging team.

If the high-gain antenna had been available, it would have been possible to transmit an 800×800 pixel image in under a minute. On the low-gain antenna running at 40 bits per second – which was the highest rate that that antenna could operate with Galileo so far from the Earth – it would take fully 72 hours to transmit an image. Transmission rate is a trade-off with 'received power', which is in turn a function of the antenna used to receive the signal. The energy that a signal carries is inversely proportional to its transmission rate. The transmission cannot exceed the rate at which the signal's content is overwhelmed by background 'noise'. Also, interference is increased if the signal passes close to the Sun. It was clearly impracticable to transmit all of the Gaspra data at 40 bits per second, so it had been decided to wait until either the high-gain antenna could be coaxed out, or until the spacecraft was heading back towards the Earth in the autumn of 1992, when its proximity would enable the low-gain antenna's transmission rate to be increased. A single image was, however, played back immediately on the low-gain antenna just to whet the appetite. The selected image was the central element of the early nine-element mosaic, on the basis that if the fly-by had gone to plan, there was a 95 percent chance that the asteroid would be within the field of view. "We were overjoyed when we saw the first dozen lines of the image, and realised that we'd got the asteroid right on target," Belton recalled.

"The early images were better than expected," said Torrence Johnson. "We had hoped to see the asteroid's shape, and maybe a couple of craters."

For this first *in situ* look at a main belt asteroid, the science objectives were to characterise the asteroid in terms of its size and shape; to establish the nature of its surface, both in terms of composition and the degree of cratering; and to study its interaction with the solar wind. Up to this time, the only small Solar System objects to have been resolved in detail were Phobos and Deimos, which are most likely asteroids which Mars had captured. Galileo found Gaspra to be an irregular $18 \times 10 \times 9$-kilometre potato-like body – a mountain tumbling through space.

"While we know quite a bit about this object statistically from ground-based observations," Johnson pointed out, "there is still a lot of controversy about details."

Gaspra was discovered in 1916 by Grigori Neujmin of the Simeis Observatory in the Ukraine. He named it after his favourite retreat on the Crimean Peninsula. Gaspra orbits the Sun at a distance of 2.2 AU. The main belt's mean range is 2.8 AU, so Gaspra orbits near its inner edge.

Photometric observations of Gaspra's 'light curve' had established that it rotates in approximately 20 hours. In terms of its spectral classification, it is 'S' type, meaning that it is believed to be similar to the 'stony' meteorites[12] and is composed of both metals (iron and nickel) and rocky minerals (pyroxene and olivine), which suggests that it has undergone a high degree of chemical differentiation. Yet it is only a few kilometres wide. Even 1 Ceres, the largest of the main belt asteroids, is a mere 800 kilometres across, which is much too small to have induced differentiation into a metallic core and a rocky mantle. However, the spectral characteristics of 4 Vesta are basaltic, implying that its surface is either exposed gabbro or is a fragment of a lava flow that formed on the surface of a now-shattered progenitor.[13] How could such small bodies come to be differentiated?

Many asteroids share orbital characteristics that suggest that they are parts of a larger body that was broken up by a major impact. Gaspra is a member of the 8 Flora family. An analysis of its shape suggested that the asteroid really might be a composite of two distinct objects. The infrared spectrometer noted a compositional difference between its two ends. Gaspra's overall shape is irregular, and possibly even 'faceted'. If it is a fragment of a progenitor body, then its gross configuration may reflect collisional spallation of the parent.

At the 21st General Assembly of the International Astronomical Union, held shortly before Galileo's fly-by, astronomer Tom van Flandern of the US Naval Observatory in Washington DC suggested that Gaspra would be surrounded by a cloud of debris "rather like a faint comet halo", but the dust detector did not report any increase in events while passing the asteroid. Thermal data from the infrared spectrometer showed that Gaspra's surface is not 'bare' rock; nor is it a 'deep' regolith. The thermal inertia – which is twice that of the lunar surface – is consistent with a 'modest' regolith that may be a thin dust layer, a combination of rocky outcrops and dusty patches, or a mix of coarse and fine material forming a layer of rubble.[14] Actually, the fact that it had a regolith at all was a surprise, because it had been thought unlikely that such a small body would be able to retain impact ejecta.

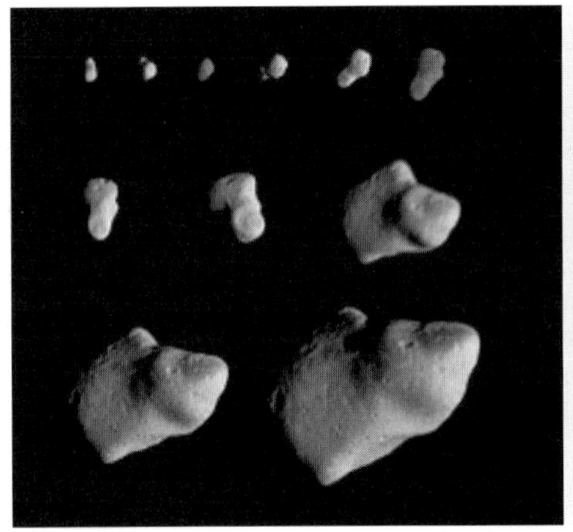

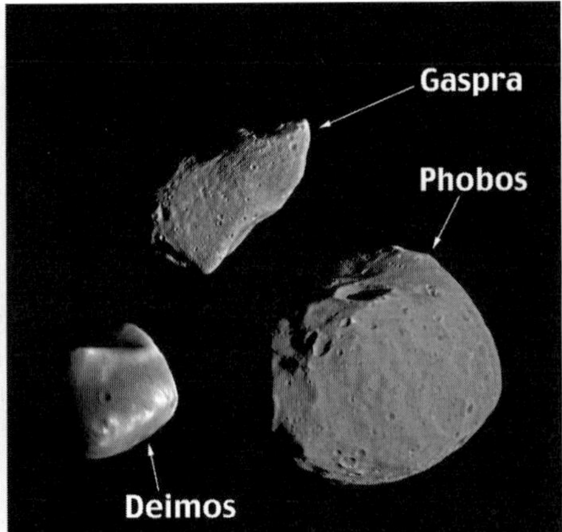

Galileo took a sequence of pictures as it approached asteroid 951 Gaspra on 29 October 1991, documenting almost a full 7-hour rotation of the $18 \times 10 \times 9$-kilometre irregular body, which is effectively a mountain tumbling through space. It was our first close-up look of a 'main belt' asteroid. Mars's two small moons, Phobos and Deimos, are believed to be captured asteroids (these had been imaged by the Viking Orbiters in 1977). There are striking differences between the three.

Being so small, Gaspra produces only a weak gravitational field – only 0.0005 as strong as the Earth's. The escape velocity is a mere 10 metres per second. It had been expected that the craters would have sharp rims as the ejecta would have been blasted into space, so it was a surprise to find the craters softened by a blanket of regolith that might be of the order of a metre deep – and even if all the ejecta from all the observed craters had been retained, this layer would be no more than 10 metres thick. It is unlikely that the process of meteoritic 'gardening' can have had time to reduce the outermost material to a powdery constitution so, in contrast to its ancient lunar counterpart, Gaspra's regolith is not 'mature'.

Gaspra has major depressions, indentations, ridges and craters – the largest of which is about 1.5 kilometres in diameter. Although it appears to be almost uniformly grey, Gaspra's surface displays subtle variations in colour which correlate with topographical features. The spectra of the ridges and the crater rims seem to be slightly bluish and the low-lying areas are rather reddish, indicating a process of 'weathering' in which exposure to sunlight modifies the reflectance. A similar, but less potent process 'ages' the Moon's surface. Such a process may well explain some puzzling discrepancies relating to the populations of the different 'types' of asteroid.[15]

An analysis of more than 600 craters in the 'high phase' imagery by Clark Chapman,[16] a planetary scientist at Southwest Research Institute, Boulder, Colorado, and a member of the imaging team, concluded that although the population is dominated by 'fresh' craters several hundred meters wide superposed on a landscape which appears to have been smoothed on a vertical scale of hundreds of metres, there are few overlapping craters, and degraded craters are under-abundant. There are some subdued depressions greater than 500 metres across, some of which appear to be associated with the linear grooves and so may well be pre-existing impact craters that have been deeply blanketed or otherwise degraded. This study concluded that the crater population differed from that of Phobos, but bore some comparison with those of the Moon and Mars at that scale. The distribution of crater sizes serves as a measure of the length of time that a surface has been exposed. Gaspra has many small craters, but relatively few large ones. The crater counts indicated that the surface is at least 200 million years old.

There are two large dents running over a substantial fraction of the surface that appear to be scars from ancient impacts on the body off which Gaspra was spalled. Striations similar to those on Phobos suggested fractures resulting from the shock of large impacts, but these were just beyond the 160-metre resolution of the early imagery. "We can't tell if these are ridges, or grooves caused by an impact that didn't quite break up the asteroid," explained Joe Veverka of Cornell University, Ithaca, New York, and a member of the imaging team. The low angle of illumination on the highest resolution imagery offered ideal conditions to inspect such subtle topography, but the geologists had to wait until late 1992 for these images to be transmitted.

When the full data was replayed, the particles and fields scientists received a real surprise. A minute before closest approach, the magnetometer detected a 'mild' shock in the solar wind, and the field vector rotated towards Gaspra. Three minutes later, as the spacecraft drew away, the field reverted to the prevailing solar wind. There were two possible explanations: either Galileo had passed through a small

This close-up view of asteroid 951 Gaspra is a mosaic of two images taken by Galileo from a range of 5,300 kilometres, 10 minutes before closest approach. The resolution is about 54 metres per pixel. The abundance of small craters was a surprise. The prominent linear features are believed to be related to fractures, which is consistent with this being a fragment of a larger body that was shattered by a massive impact.

eddy in the solar wind, as it did from time to time, and by chance this occurred while near the asteroid; or Gaspra could have an intrinsic magnetic field which was strong enough to deflect the solar wind. The implications of Gaspra having a magnetic field were profound. "If Gaspra has a permanent magnetic field, this finding would have a bearing on its thermal history and have implications for the history of the magnetic field of the early Solar System," explained Margaret Kivelson, the magnetometer team leader who was based at the University of California at Los Angeles.

One theory posited that asteroids with a significant iron content may have been magnetised early in the formation of the Solar System. When the Sun 'switched on', it is believed to have blown away what remained of the solar nebula which, in effect, would have been a dense solar wind. (Stars of the T Tauri type are in the process of doing this.) If this early solar wind carried an intense magnetic field, this could have induced electrical currents within small asteroids and 'cold-magnetised' them.

However, the fact that the infrared spectrometer had found Gaspra's surface to be rich in metallic silicates[17] suggested that the asteroid had undergone chemical differentiation as a result of being heated. Gravitational compression stresses rock, heating it. If a body is sufficiently massive to form a nickel–iron core, circulating electrical currents will generate a magnetic field that will be 'fossilised' into crystallising iron-rich rock. The combination of the chemical composition and the intrinsic magnetic field suggested that Gaspra was part of a rocky mantle from a larger differentiated body. This was consistent with the theory which held that asteroids are the left-over debris of fragmented planetoids. So, too, was the fact that there seemed to be families of asteroids. By this theory, the 'M' type asteroids were parts of a metallic iron core. Some meteorites, evidently derived from iron asteroids, had also been found to be magnetised. Asteroid 4 Vesta's basaltic surface was further support for this evolved progenitor theory.

Galileo's fly-by of Gaspra was a milestone in the study of asteroids, whose name meant 'star-like'. Most of our knowledge was derived from spectroscopy, and Galileo provided the first 'ground truth' by which to assess previous observations and theories. "It's been a shot in the arm," reflected Bill O'Neil.

The asteroid specialists were eager to know whether Gaspra was typical, and an encounter with another asteroid was programmed for late 1993, but the decision to pursue this would not be made until mid-1992. While Galileo was at its 2.3 AU aphelion in December – the farthest it had yet ventured from the Sun – it cold soaked the high-gain antenna tower to try to release the stuck deployment pins, but again to no effect.

ULYSSES

The European Space Agency's Ulysses spacecraft had been launched by NASA's Shuttle on 6 October 1990. Compared to Galileo, it was a lightweight, and had been able to fly the direct route to Jupiter using the two-stage IUS augmented by a small solid-propellant Payload Assist Module. The spacecraft was not capable of making major course corrections, so it was crucial that it be accurately dispatched.

Fortunately, it was placed 'in the groove', on track for the 160 kilometres wide fly-by corridor at Jupiter.

In addition to instruments to study the solar wind, Ulysses carried a dust detector identical to that aboard Galileo. As the spacecraft moved out beyond 3 AU from the Sun – beyond the asteroid belt – it encountered a series of dust streams.

On a clear night at an observatory above most of the troposphere – such as on the summit of the 14,000 foot extinct volcano Mauna Kea, on the Big Island of Hawaii – it is possible to see, with the naked eye, interplanetary dust reflecting sunlight.

In fact, there are two distinct dust populations. The faint glow resembling an inverted cone poking up over the horizon about half an hour after sunset is believed to be dust blown off by comets making their perihelion passages. The dusty glow directly opposite the Sun in the sky, visible at midnight, is believed to be concentrated in the zone between Mars and Jupiter and to be derived from collisions between asteroids. This dust re-radiates in the infrared the light that it absorbs from insolation, so it had been well observed by the Infrared Astronomical Satellite (IRAS) in 1983. Most of the dust in the inner Solar System orbits in prograde fashion (which is consistent with its having been shed by short-period comets) but the dust streams that Ulysses was passing through were following hyperbolic trajectories at speeds of up to 200 kilometres per second. At first it was suspected that the dust that Ulysses had detected was of interstellar origin and was 'streaming' through the Solar System, but as the spacecraft neared Jupiter it became clear that the dust was of Jovian origin.

When Ulysses reached Jupiter in February 1992, the plan was not to enter into orbit, but to make a very close fly-by so that the slingshot would swing it back into the inner Solar System. Why bother? By making the fly-by over the planet's north

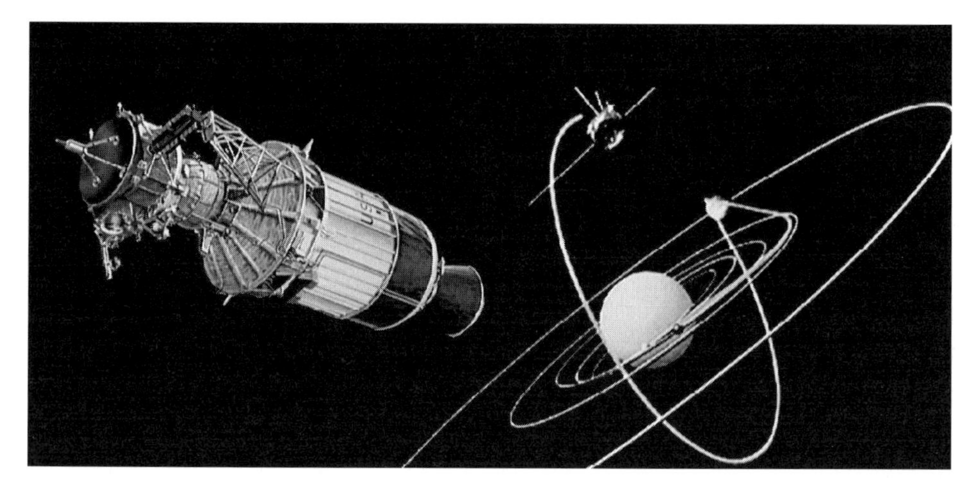

The Ulysses spacecraft was dispatched by an Inertial Upper Stage (left) on a trajectory to Jupiter. It used Jupiter's gravity to swing back into the inner Solar System, but on an orbit highly inclined to the plane in which the planets orbit the Sun, so that it would be able to observe the solar wind at high solar latitude.

pole, the plane of the exit trajectory would be nearly perpendicular to the ecliptic. The new orbit would provide Ulysses with an unprecedented view of the Sun's south pole in 1994 and then of the north pole a year later.

For Ulysses, the Jupiter fly-by was simply a means to an end, but its passage through the Jovian system offered a welcome opportunity to check the state of Jupiter's magnetosphere in advance of Galileo's arrival, to establish whether there had been any major changes since the Voyagers had passed through in 1979.[18] Although Ulysses was a particles and fields platform, most of its instruments were deactivated because they were not shielded to survive the intense radiation circulating close to Jupiter, but the magnetometer and the radiation detectors took data through the encounter. It noted the spacecraft's passage through the bow shock at 17:33 UT on 2 February as a sharp rise in the intensity of the magnetic field. At this point, Ulysses was still 113 Rj from the planet. The crucial gravitational slingshot occurred on 8 February.

At 12:02 UT, when it was 448,000 kilometres (6.3 Rj) from Jupiter's centre, Ulysses was deep within the magnetosphere. However, its route was perpendicular to the most intense belt of radiation in Io's plasma torus and Jupiter's gravity had accelerated the spacecraft. An hour after its closest approach it shot straight through the torus at 98,000 kilometres per hour. "We have survived a difficult encounter," reported a relieved Derek Eaton, ESA's Ulysses Project Manager.

The results were rather surprising. The Jovian magnetic field was evidently more complex than had been believed. Its outer part did not rotate with the rest, which was locked into the 10-hour rotation of the planet's metallic core. The most intense radiation belt seemed to be rather more oblate than the Earth's, no doubt due to the rapid rotation.

Telescopic observers were accruing a database on the distribution and temporal behaviour of volcanic activity on Io in order to provide a basis for interpreting the observations that Galileo would make. Ulysses found Io's torus to be less intense than expected, which suggested that the level of volcanic activity had diminished since 1979, when almost a dozen eruptions were pumping dust and gas into space. A reanalysis of the Voyager data had identified at least 20 hot spots.[19]

High-resolution ultraviolet observations by the Hubble Space Telescope reported by Paola Sartorelli of the University of Padua in Italy and Francesco Paresce, an ESA representative at the Space Telescope Science Institute in Baltimore, Maryland, had revealed that Io's volcanic atmosphere was not only less dense than it had been during the Voyager fly-bys, but was rather 'patchy', with some areas being 1,000 times as dense as others.[20]

For Ulysses, of course, this was only a step towards its primary mission. "We're ready to explore the poles of the Sun," confirmed JPL Director Ed Stone. (Although the spacecraft had been supplied by ESA, the project was operated jointly with NASA.) As Ulysses dipped below the ecliptic in June 1992, it crossed the transition zone marking the boundary of the turbulent disk of plasma flowing out from the Sun's equatorial zone and penetrated the far simpler flow at mid-latitudes. At 800 kilometres per second, the solar wind in this region was twice as fast, less dense, and smoother, but the tilt of the Sun's magnetic axis meant that the transition zone was

On 3 November 1973, as Mariner 10 set off for Venus (and then Mercury) it tested its camera by returning an unprecedented view of the Moon's north polar region (left). Galileo recorded the same area (right) as it made its second approach to the Earth on 7 December 1992, in this case to provide context for a mineralogical survey.

ragged, so for the next 13 months the spacecraft flew through a succession of whorls and eddies before finally leaving the turbulence behind.

EARTH AGAIN

The scientific 'take' from Gaspra was so impressive that on 25 June 1992 JPL authorised Galileo to make a second asteroid fly-by. First, however, the spacecraft would have to pick up some more energy from the Earth. A 20 metre per second correction on 4 August and a series of tweeks in November refined the trajectory. As it approached at a million kilometres per day, the spacecraft transmitted the data that it had recorded during the Gaspra encounter.

This time, Galileo flew by the Moon on the way in, flying over its north pole at a range of only 110,000 kilometres at 03:58 UT on 8 December. This region had been photographed in 1973 by Mariner 10 as it set off for Mercury, but it still held many puzzles. With a resolution of 1.1 kilometres per pixel, Galileo's best imagery considerably improved our knowledge of the topography, and the infrared spectrometer surveyed the composition of the surface in terms of the predominant minerals.

The spectrometer data revealed that the polar zone is rather more mineralogically diverse than had been believed. The dark maria are basalt enriched with pyroxene and olivine which originated in the mantle. Some of the 'light plains' in the north polar region had compositions indicative of a volcanic origin, but this lava was not as dark as that of the maria. Furthermore, it had evidently extruded *during* the Great Bombardment that excavated the basins which the darker lava subsequently flooded.

The name 'crypto-mare' was coined for these mare-like areas, which are partially masked by crustal rock ejected from subsequent basin-forming events. "These flows occurred early in lunar history," assured Ron Greeley, a geologist at Arizona State University in Tempe, and a member of the imaging team. This indicated that the Moon had been volcanically active earlier than had been thought. Mapping how the crypto-mare have been masked by ejecta would shed light on the process of cratering.[21]

"We did not know how to characterise the region near the north pole," said Robert Carlson, the leader of the infrared spectrometer team. "We now know that some of the smooth areas in the polar region are like the maria."

The infrared spectrometry of Mare Serenitatis clearly distinguished the various lava flows, and highlighted the patches that had been 'mantled' by dark titanium-rich pyroclastic material blasted out from explosive vents – one of which had been sampled by the final Apollo mission in 1972.

Intriguingly, the mountainous rim of the 650-kilometre diameter basin that would later be flooded by the upwelling lava of Mare Humboldtianum, showed a distinctly polygonal shape. This hinted that the impact may have struck terrain which was already deeply criss-crossed by faults. As seen from Earth, Humboldtianum is close to the limb, so the perspective makes its morphology difficult to study in detail.

Twelve hours after passing the Moon, Galileo flew above the South Atlantic at an altitude of 305 kilometres. It was within 700 metres of the centre of its corridor, and just 0.1 second early. Its trajectory took it through an 'anomalous' region where the inner Van Allen belt dips down several hundred kilometres over the coast of Argentina. A spacecraft flying through this 'South Atlantic Anomaly' receives a greater dose of radiation in a few minutes than it accrues during the rest of its orbit. Galileo, however, was moving considerably faster than an orbiting satellite. A series of images were taken during the encounter to assess the extent to which its solid-state CCD technology was affected by the radiation. Although the imagery was degraded at the highest sensitivities, the camera performed adequately.

As during the first fly-by, the infrared spectrometer found sheets of ice-crystal cloud high above Antarctica. "These clouds may be a common phenomenon over Antarctica," observed Robert Carlson in retrospect.

The ultraviolet spectrometer made significant observations by imaging the geocorona, the outermost part of the atmosphere consisting of atomic hydrogen which has 'evaporated' from the mesosphere and is leaking to space. The corona is 'swept' down-Sun and concentrated in the magnetotail by a process of absorption and re-emission of ultraviolet insolation. It was first photographed by John Young in 1972, from the lunar surface on Apollo 16. It was known to extend at least 95,000 kilometres down the magnetotail, but Galileo established that it actually ran for 400,000 kilometres, and so could actually impinge upon the Moon at times when this passed through the magnetotail.

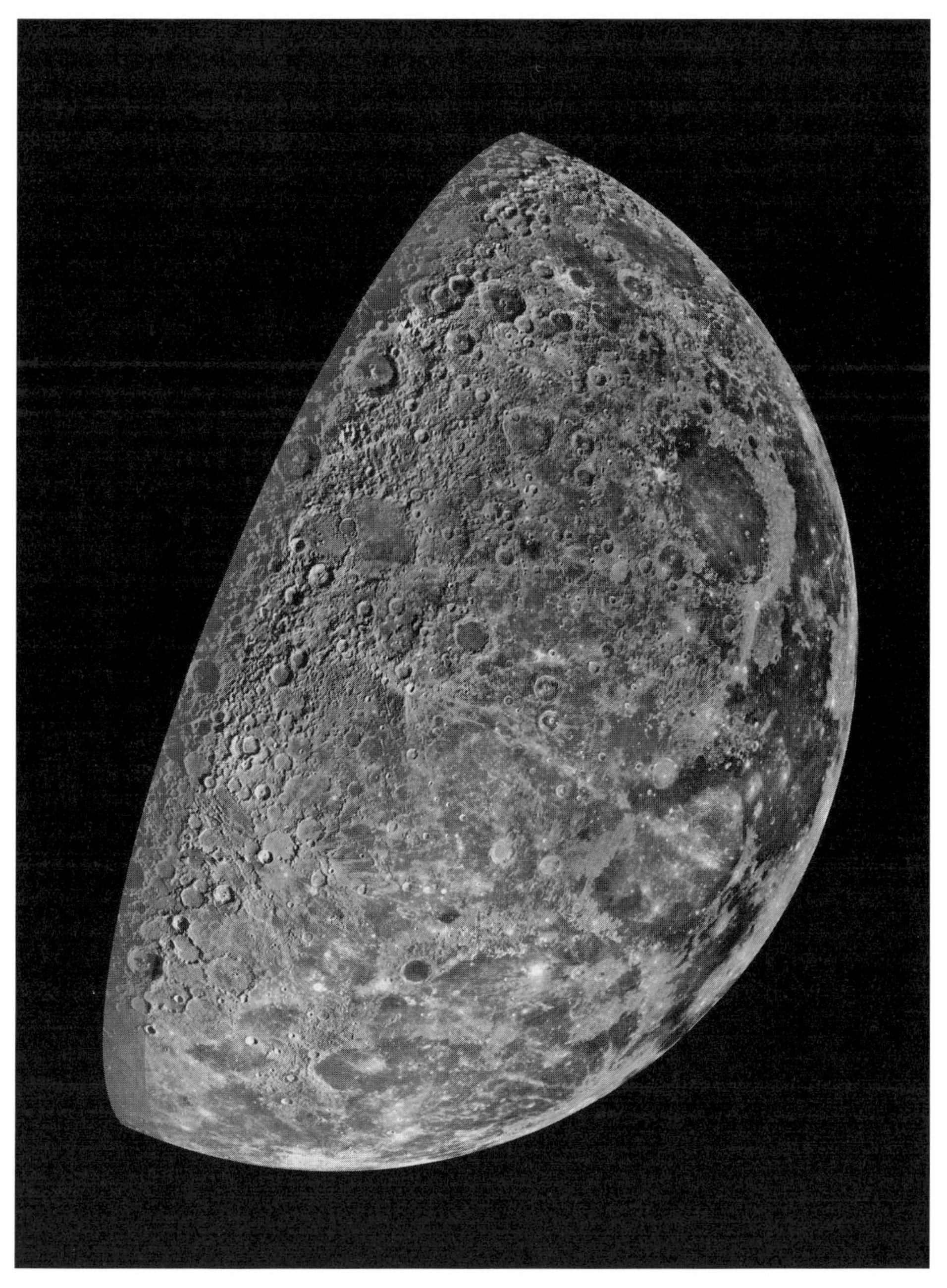

As Galileo made its second approach to the Earth on 7 December 1992, its near-infrared mapping spectrometer made an unprecedented mineralogical survey of the Moon's north polar region, providing a number of surprises. A false-colour representation is required to fully appreciate the detail in this survey – it is 'PIA00131' in JPL's Planetary Image Archive, so download a copy.

Galileo's near-infrared spectrometer made this mineralogical survey of the hemisphere of the Moon which we see from the Earth. Although the maria appear to be dark in the visual view, this multispectral image demonstrates that their lava flows have different compositions. The prominent ray crater in the southern hemisphere is Tycho.

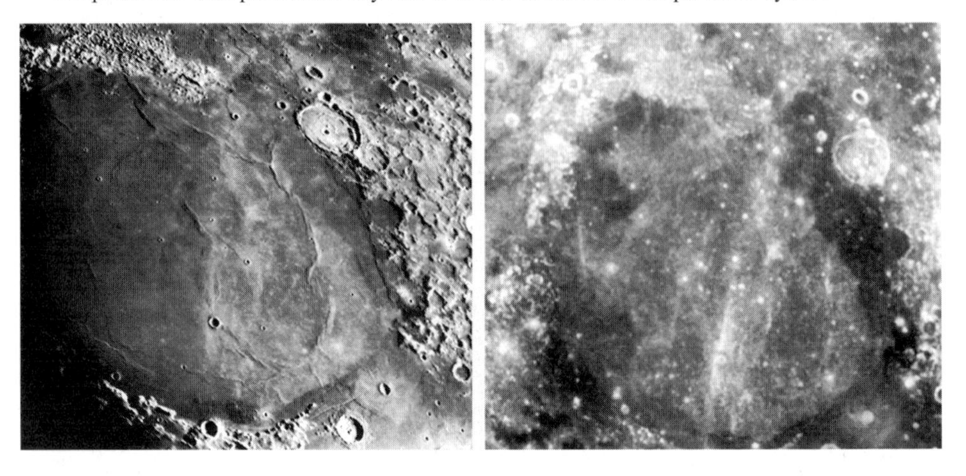

The telescopic view (left) shows the same general area as the enlargement of a Galileo near-infrared mapping spectrometer image (right) of the Serenitatis basin on the Moon. The chemically distinct mare lava flows around the edge of the basin are evident in the multispectral image, as are the patches of 'dark mantle' pyroclastic from ancient 'fire fountains' (the patch in the mountains forming the basin's eastern rim was sampled by the Apollo 17 astronauts in December 1972).

FINAL SLINGSHOT

When the IUS had boosted Galileo out of Earth orbit in August 1989, it had actually acted to reduce by 3.1 kilometres per second the heliocentric velocity that it would otherwise have inherited from the Earth's orbital motion, so that it would 'fall' towards Venus. The slingshot at Venus had added 2.3 kilometres per second in order to send the spacecraft back out towards the Earth, and the first fly-by had added 5.2 kilometres per second and sent the spacecraft on a two-year ellipse with its aphelion in the asteroid belt. This second Earth fly-by increased the spacecraft's speed by

On 8 December 1992, as it drew away from the Earth for the final time, Galileo took a family portrait showing the Moon close alongside the Earth's limb.

another 3.7 kilometres per second and eased its aphelion out to Jupiter's orbit. The overall effect had been to raise the spacecraft's heliocentric speed to 39 kilometres per second and set up a trajectory which was far beyond the capability of the IUS which Galileo had been obliged to utilise. In effect, this circuitous preliminary had created a three-year Hohmann minimum-energy transfer orbit.[22] Furthermore, passing over the Moon's north pole had also tilted the plane of Galileo's orbit, to enable it to reach its next target – the asteroid 243 Ida.

On 16 December 1992, as Galileo departed from the Earth for the final time, it returned a spectacular image depicting the Moon close to the Earth's limb.

"How sweet it is!" admitted a delighted Bill O'Neil, now that Galileo was finally heading for Jupiter. Torrence Johnson agreed. It was good to be leaving the inner Solar System, "not that we didn't like the local neighbourhood".

The accumulated Galileo fly-by results were the 'hot' news at the 24th Annual Lunar and Planetary Science Conference (LPSC) hosted in March 1993 by the Lunar and Planetary Institute and the Johnson Space Center co-located in Houston, Texas. By any measure, the spacecraft had already achieved a great deal and the prevailing mood was one of mounting expectation.

REVISED PLANS

On 29 December 1992, exploiting the fact that Galileo would never again be so close to the Sun, it turned to bake the high-gain antenna for 20 hours to induce thermal expansion and the deployment motor was pulsed 2,000 times to try to 'hammer' the stuck pins free. Over the following month, the motor was pulsed a total of 13,320 times, but to no avail. Whatever had fouled the antenna, it had evidently established a very secure bond. The engineers were still optimistic, however. On 10 March 1993, as part of a test of procedures to be employed during the deployment of the atmospheric probe, the spacecraft was 'all spun' to 10.3 rpm. Doing so enabled the 'wobble' resulting from the asymmetric configuration of the high-gain antenna to be measured in this regime, so that a plan could be developed to counter it during the probe's release. While the spacecraft was spinning, the antenna motor was pulsed again, in the hope that the increased centrifugal force on the ribs would release the stuck pins, but to no avail yet again.

Having tried every trick in the book, the project managers finally accepted that the primary mission would not be able to proceed as planned. It had been intended to return *50,000* images during two years of exploration of the Jovian system, but this would be impracticable over the low-gain antenna. To put the magnitude of the task into context, consider that after its historic Mars fly-by in 1965 Mariner 4 had taken 8 hours to transmit two dozen 200 × 200 pixel images at 8 bits per second. Each of Galileo's images contained *sixteen times* as many pixels, so how could it possibly explore the Jovian system – a vast planet with a highly dynamic atmosphere and magnetosphere, four large moons, some small moons, and a ring system – using an antenna that would be able to operate at a rate no higher than 10 bits per second? Apart from the high-gain antenna, the spacecraft was healthy. The engineers were in the process of developing a plan to perform the primary mission involving a reduced imaging programme that could be supported by the low-gain antenna. Nevertheless, it would take two weeks to return one image together with the associated engineering and navigational data. At this rate, it would take three years to replay a tape full of data from a *single* encounter with a Jovian moon, and even that figure was optimistic because it assumed that the Deep Space Network would be able to be devoted to the task 24 hours a day. "It was a bleak picture," admitted Talbot ('Tal') Brady, the Control and Data Subsystem Manager.

Les Deutsch, of the Telecommunications and Missions Operations Directorate, thought that something could be done by reprogramming the onboard computer to do image compression, and also by enhancing downlink telemetry encoding. He led a feasibility check with the Deep Space Network engineers. An engineering team then spent some three months fleshing out the details of what could be done on the ground and aboard the spacecraft. Implementation started in March 1993 under the joint leadership of Jim Marr (Project Galileo) and Joe Statman (Deep Space Network). Although the spacecraft would not approach Jupiter until late in 1995, it was all too obvious that preparing the new software would be a race against the clock.[23]

The data processor aboard Galileo, the Command and Data Subsystem (CDS), employed six 8-bit microprocessors of 1970s' vintage with a total memory of 384 kilobytes. The Attitude and Articulation Control Subsystem (AACS) used a faster 16-bit computer, but it had only 64 kilobytes of memory, nearly all of which needed to remain as it was. Most of the instruments were controlled by single 8-bit microprocessors with very limited memory. The constraints of memory, processing time, and data bus traffic were excruciating. The vast volume of data that the science instruments and engineering systems would collect would somehow have to be fed through the narrow 'data pipe' provided by the low-gain antenna.

Leading the implementation effort, Tal Brady was at the centre of a whirlwind, negotiating with the science teams to satisfy their requirements as far as possible, and supervising the writing and testing of new computer code. One early obstacle was locating programmers still fluent in the archaic assembly language of the CDS operating system.

"A science virtual machine was created and integrated into the existing operating system to support editing, compression, and 'packetisation' of the science data," explained Erik Nilsen, Deputy Team Chief of the Orbiter Engineering Team. A multi-use buffer was formed by using 80 kilobytes of memory within the science virtual machine. The data awaiting downlinking to Earth would be loaded into this buffer and prepared for transmission.

The science data would be collected and processed at high speeds, relative to the speed of the data pipe. The data would not necessarily be transmitted in the order that it was taken, so the new software had to prefix 'data packets' with headers identifying the data type, its size, and its collection time. Before imaging data could be inserted into playback packets, it would have to be compressed by ratios of up to 15 to 1, compared to its format on the tape. The particles and fields data would have to be compressed by a factor of 100, because this type of data was to be placed into real-time packets. To enable the Deep Space Network to detect and to rectify transmission errors – which would be critical when using data compression – the packets would have to be encoded. Every innovation was used to refine, compress and package the data so that only the most valuable information entered that slim pipeline to Earth, or, as the mantra claimed, "to ensure that every bit carried the maximum amount of data".

While improving the way the CDS would control and refine the data, the team looked for compression technology expertise. JPL researcher Kar-Ming Cheung responded by defining and developing the compression. When "constricting boundaries don't exist", he reflected, it was remarkable how "responsive" the entire organisation was. His implementation of integer cosine transform algorithms successfully compressed the camera's images and low-rate data from the plasma wave spectrometer. David Breda then developed a way to squeeze Cheung's algorithm into the computer.

While this was being done, seven of the instrument teams rewrote the control software for their instruments; the other instruments either lacked the memory capacity to be revised or could not be reprogrammed.

Bob Gershman, Deputy Manager of the Science and Sequence Office, managed to integrate all the various different strands of activity and liaised between scientists and design engineers to ensure that the spacecraft resources were not oversubscribed.

The key to success was the tape recorder, which had been installed to provide a back-up to the high-gain antenna during the time that the atmospheric probe was relaying data to the main spacecraft. If the real-time transmission was marred by interference, then it would be possible to replay the taped data. The tape recorder had facilitated the observations of Venus before the high-gain antenna was scheduled to be opened, and, after the antenna fouled, the tape recorder had saved the day at Gaspra. The primary mission would be reliant upon taping data during an encounter, for subsequent replay. Similar tape recorders had been used on other spacecraft, so confidence was running high.

With data editing and compression by the spacecraft, and upgraded Deep Space Network systems, the effective 'information rate' on the low-gain antenna would be able to be boosted by a factor of about 100, which would be sufficient to address all of the science objectives. It would be possible to return at most 10 percent of the planned 50,000 images, so the scientific objectives would have to be pursued by a strategy of combining medium-resolution mapping and very selected high-resolution imaging of representative features, rather than the originally intended comprehensive mapping.

243 IDA

After taking a series of long-range images to refine the position of its next target, Galileo made a slight trajectory correction on 26 August 1993. Two days later, with asteroid 243 Ida just 4 hours away and with everything set, Galileo mysteriously

Galileo took a sequence of pictures as it approached asteroid 243 Ida on 28 August 1993, documenting a full 4.6-hour rotation viewed from above the north pole.

stowed its scan platform and resumed its cruise configuration. The engineers interrogated the spacecraft's computer to determine why it had abandoned its programme,[24] and then set about restoring it. The recovery process was aggravated by the fact that it took almost an hour for a command sent by the Deep Space Network to reach the spacecraft and the result to return to Earth. This frantic effort was successful, however, and the instruments were fully operational at 16:52 UT when Galileo flew by the asteroid at a range of 2,410 kilometres.

As at Gaspra, a colour mosaic was taken early in the encounter to secure a complete view of one side of the asteroid at a resolution of 35 metres. The highest resolution mosaic was taken from a range of 10,500 kilometres. The final frame was a partial monochromic view with 24 metres per pixel resolution.

This time there could be no return to the vicinity of the Earth to enable the low-gain antenna to operate at a high rate, so the taped data would have to be replayed slowly. At the time of the encounter, Galileo was 550 million kilometres from the Earth, and the Earth was opening the range as it proceeded around its orbit. With Ida close to the Sun, as viewed from Earth, there was only just time in September to transmit the early mosaic. At 40 bits per second, it took *30 hours* to send *each* of the five frames. The rest of the tape was transmitted from mid-February through to late June 1994, by which time the Earth had completed half an orbit, and restored a clear line of sight.

Ida was discovered in 1884 by the accomplished asteroid hunter Johann Palisa, of Vienna in Austria. He named it after the mountain on Crete where Zeus once lived. Like Gaspra, its spectroscopic characteristics were of type 'S'. Its orbital elements suggested that it was a member of the 158 Koronis family of asteroids.

At $56 \times 24 \times 21$ kilometres, Ida turned out to be a little larger, and more heavily cratered than expected. It was rotating on its shorter axis with a 4.6-hour period. Most of Ida's craters are simple concavities, but some have flattened floors, and a

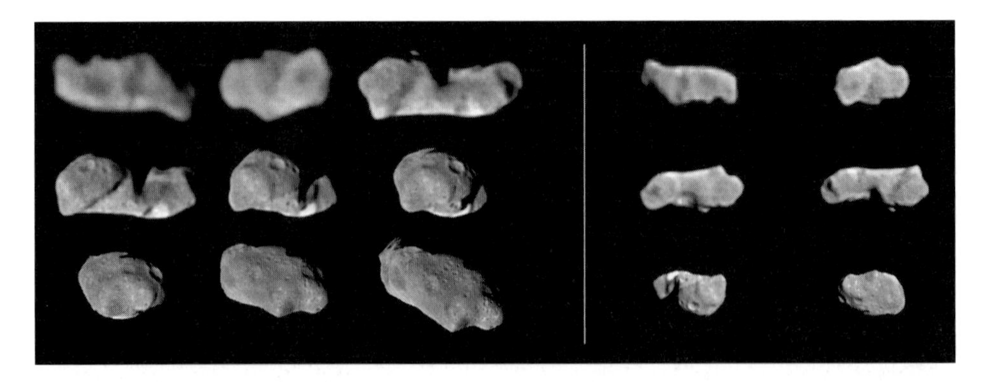

At 56 × 24 × 21 kilometres, asteroid 243 Ida turned out to be a little larger than expected, with ridges suggesting that it is a composite body. It was necessary to analyse a series of images to determine its complex shape – one end appears to be a rectangular block while the other end is irregular and dominated by a single depression 28 kilometres in diameter. A few linear features could represent either deep fractures or 'sutures' where distinct blocks had at some time settled against one another.

few have central mounds. There are even several intriguing chains of craters. The smooth crater rims suggested the presence of a fairly thick regolith, as did the presence of a few 'ray' craters. Since Ida was a member of a well-defined family which was thought to be the result of a recent fragmentation, this had prompted the belief that Ida's surface would be strikingly young, so its battered form was a considerable surprise.

A study[25] found that the population density of craters is high; that they show a wide range of degradational morphologies, at both large and small sizes; that the craters ought to have produced an appreciable regolith, typically 150 metres deep; and that craters greater than 2 kilometres wide (and especially those greater than 5 kilometres wide) are located preferentially on one region of Ida. The infrared spectrometer suggested that Ida is a composite of several objects. One end of it appears to be a rectangular block, and the other end is irregular, and is dominated by a single depression about 28 kilometres in diameter. A few linear features could represent either deep fractures or 'sutures' where distinct blocks settled against one another in the distant past.

The fact that Ida is more heavily cratered than Gaspra implies that its surface is older. Clark Chapman argued that if Gaspra and Ida have similar structural strengths, and have been exposed to the same population of projectiles, then Ida is 10 times older than Gaspra – 2 billion years old. However, if Ida suffered anomalously heavy bombardment in the immediate aftermath of the break-up of its Koronis progenitor, it could well be younger than its cratering record would suggest.

The infrared spectrometer accumulated observations spanning a complete revolution of the asteroid at a variety of spatial and spectral resolutions. The best data was taken about 4 minutes prior to closest approach, when the spacecraft was just 3,600 kilometres away. This measured 17 wavelengths at the excellent surface resolution – for this instrument – of 1.8 kilometres. The thermal inertia indicated the

This close-up view of Ida is a mosaic taken by Galileo on 28 August 1993 from a range of about 3,500 kilometres.

presence of a thick regolith which acted as an insulator,[26] and this was consistent with the visual evidence of topographic softening. Ida displayed 'weathering' similar to that found on Gaspra, believed to be progressive reddening of the reflectance spectra of material exposed to insolation.

The fact that Galileo's magnetometer reported a field rotation at Ida just as it had at Gaspra banished all remaining doubts regarding the interpretation of the Gaspra fly-by magnetometer data. Evidently the 'S' type asteroids possess intrinsic magnetic fields that are strong enough to ward off the solar wind.

However, the truly amazing revelation was that Ida has a companion. This was discovered by Ann Harch as she inspected the first day's downloaded imagery on 17 February 1994. As further imagery came in showing the companion in different positions, an analysis of the relative lines of sight established that the object was

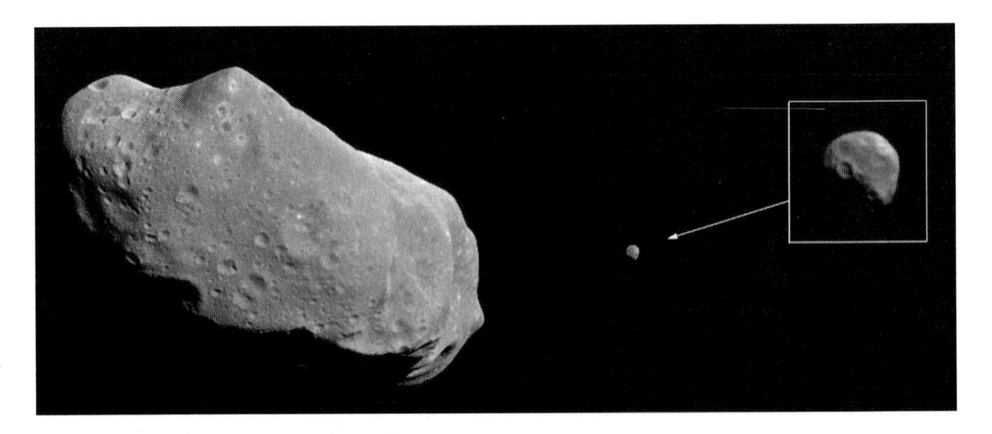

Galileo discovered that asteroid 243 Ida has a small companion, which has been named Dactyl. Although only 1.6 × 1.4 × 1.2 kilometres wide, Dactyl is surprisingly spheroidal. The insert shows Dactyl's cratered surface with a resolution of 40 metres per pixel. The perspective in this image is deceptive, because Dactyl is actually in the foreground, not 'next' to Ida.

about 100 kilometres away (for a sense of scale, recall that Ida's primary axis is 50 kilometres long). Was it just passing by? The calculated speed of the companion was 10 metres per second, which was sufficiently low to suggest that it was gravitationally bound to Ida, and was orbiting it. The likelihood of a random asteroid of such a size passing Ida at that speed just when Galileo was present was computed to be about one in a thousand. Although only 1.6 × 1.4 × 1.2 kilometres, the companion is surprisingly spheroidal.

The satellite was initially designated '1993 (243) 1', but subsequently named Dactyl, after the mythical bird which is reputed to have been seen by Zeus while atop Mt Ida.

The images from the early approach viewed Ida and Dactyl more or less in Ida's equatorial plane, so the spacecraft's perspective remained fairly constant. However, when it took its final image the spacecraft looked down on Dactyl's orbital plane. Although Dactyl was captured in 47 images, its orbit around Ida was difficult to determine because a range of possible orbits differed only in the view along the line of sight and it was recorded only for a small fraction of its orbit. Nevertheless, Dactyl's presence offered an opportunity to estimate Ida's bulk density. It would be the first accurate determination of an S-type asteroid's density, so the navigational team made a comprehensive analysis.

Had Dactyl formed independently and become gravitationally bound to Ida as a result of a low-energy encounter? This was ruled out because Ida's escape velocity is so low that it was unlikely to have captured a body as large as Dactyl. They evidently have a common origin and have always been bound together.

Dactyl was present in three of the infrared spectrometer's images taken 15 minutes before closest approach, at which time Galileo was some 12,000 kilometres out. Dactyl's spectral characteristics were shown to be similar, but not identical to

Ida's. Its surface was olivine, orthopyroxene, and clinopyroxene in roughly equal proportions. Ida's surface was predominantly olivine, with a little orthopyroxene.

The cratering density was consistent with Ida and Dactyl having a common history. Either Dactyl was chipped off Ida, or both asteroids are fragments of the Koronis progenitor. The possibility that Dactyl was chipped off Ida could be ruled out by Ida's low escape velocity. If the Koronis progenitor was of the order of 100 kilometres across, Ida must represent a major fragment. If the two asteroids established their association long ago, it is remarkable that their rather weak gravitational bond had proved so stable. Although tiny, Dactyl has a few craters 300 metres across, indicating that it has suffered major collisions, but none has disrupted its orbit around Ida.

Dynamical studies showed that orbits with a periapsis of less than 75 kilometres from Ida were unstable because the satellite would either strike or escape from irregular Ida during close encounters. This constraint placed a maximum limit of Ida's density of about 2.9 g/cm^3. And at the other extreme, highly elliptical or hyperbolic orbits were ruled out, because if Dactyl was in such an orbit it should have been individually resolved by the Hubble Space Telescope when it looked at Ida on 26 April 1994. These two conditions provided a preliminary value of 2.1 to 2.9 g/cm^3. To allow for uncertainty in the volume of Ida employed in the computation, this range was increased to 1.9 to 3.2 g/cm^3.

Ida's bulk density was of great interest because it could indicate whether the asteroid was made of rock which had been thermally processed deep within a larger body that had been fragmented by collisions. However, this surprisingly well-constrained density range indicated that Ida is actually fairly porous and/or is made of fairly light rocks. This conflicted with the classes of dense igneous rock indicated by the infrared spectrometer's mineralogical survey.

Photometric telescopic studies of asteroidal light curves to determine their rotational periods had hinted that some of these bodies may have companions,[27] but Dactyl was definitive proof. Intriguingly, Steven Ostro of JPL had employed the Arecibo radio telescope in Puerto Rico to bounce radar off asteroids passing close to the Earth. In 1989 he had found that 4769 Castalia consisted of two lobes in contact, each several kilometres diameter. When 4179 Toutatis came close in December 1992, the Goldstone Deep Space Network antenna, acting as a radar, found this too was a 'contact binary'. In this case, one of the lobes was larger than the other. Perhaps they were once gravitationally bound, had nudged one another in a low-energy collision, and had become locked together.[28] Although these bodies are not actually satellite systems – at least not today – asteroidal companions may turn out to be fairly commonplace.

In early October 1993, with Ida far behind, Galileo began a five-day period of pulsing its lateral thrusters – 2,000 pulses per day, for a grand total of 10,000 firings – to impart the required 38.6 metre per second change in velocity to realign its trajectory for Jupiter. The Ida encounter had been sanctioned only after it had been verified that the spacecraft would have enough propellant for this manoeuvre, otherwise the option of inspecting Ida as a target of opportunity would have been forsaken.

Although Jupiter was still far off, course corrections were most propellant-efficient when performed early, so in mid-February 1994 Galileo made a 0.1-metre per second manoeuvre to refine its trajectory into a collision course with the giant planet as a preliminary to releasing its probe.

GRAVITY WAVES

From 28 April to 11 June 1994, Galileo took part in a novel experiment to attempt to detect gravity waves. Albert Einstein's 1916 Theory of General Relativity posited that the Universe is pervaded by 'gravitational waves', just as it is swept by electromagnetic waves.

Many experiments had been devised to verify this prediction but none had proved positive. Attempts to measure such a subtle effect on the Earth were complicated by 'background noise'. The final Apollo crew had taken an innovative detector to be set up on the lunar surface, but it had malfunctioned. The rationale had been that if two instruments on different planets detected simultaneous signals, this could only be due to a gravitational wave passing through the Solar System. Another strategy was to use spacecraft on interplanetary missions. The Deep Space Network measured the doppler on Galileo's signal very accurately in search of a characteristic triple-pulse as low-frequency "ripples in the curvature of space–time" buffeted the spacecraft and the Earth in a predictable manner.

Although it was thought that gravitational waves would be rare at the level of sensitivity of such an experiment, a positive detection would undoubtedly lead to a Nobel Prize. A negative result would also be useful, because it would place constraints on models of the phenomenon.

This experiment had been left until Galileo had finally departed the inner Solar System. It was scheduled for when the spacecraft was close to opposition in order to minimise interference to the radio signal from the plasma of the solar wind, so it was able to be performed twice, first in 1994 and again in 1995. Frustratingly, no 'signal' was detected.

SHOEMAKER LEVY 9

In March 1993, Gene and Carolyn Shoemaker and David Levy were utilising the Palomar Observatory's 18-inch Schmidt camera to look for comets. The night of 23 March was cloudy, pre-empting the planned observations, but they nevertheless managed to take a pair of plates of a region in the constellation of Virgo, not far from Jupiter in the sky, which turned out to have historic significance.

After doing pioneering work on impact craters, during which he conclusively proved that 'Meteor Crater' in Arizona is aptly named, Gene Shoemaker had founded the US Geological Survey's Branch of Astrogeology at Flagstaff, Arizona. He played a leading role in planning the Ranger and Surveyor lunar programmes and in training the Apollo astronauts in the gentle art of field geology. Having expanded his interest in craters to the bodies that excavated them, he was now searching for comets. Carolyn tackled the demanding job of analysing the plates, employing a stereo-viewer to identify all the objects that had moved in the time between the two exposures. David Levy was a literary scholar and an inveterate comet seeker. He had been yearning to look for comets ever since reading *Starlight Nights*, the autobiography of renowned comet-hunter Leslie Peltier. By March 1993, Levy had discovered 21 comets, 8 of them jointly with the Shoemakers.

When she examined the plates two days later, Carolyn Shoemaker found a peculiar linear feature with a fan of light on one side – it looked just like "a squashed comet".

Alerted to the presence of the strange object, Jim Scotti utilised the 36-inch telescope at the Kitt Peak National Observatory, Tucson, Arizona. It was equipped with a CCD camera, so he was not only able to confirm the discovery but also to add detail. "I've been trying to pick my jaw up off the floor," he acknowledged upon returning the Shoemakers' call. It was actually a comet that had broken up, and its fragments were now strung out in line.

Once sufficient observations had been made to derive the orbit of Shoemaker Levy 9 – as it was designated – it was determined that it had broken apart while passing 2,000 kilometres above Jupiter's cloud tops in 1992. This was well within the Roche limit, the distance within which the stresses of the planet's differential gravity will rip a solid object apart. Furthermore, although the fragments were now climbing away from Jupiter, they did not have the energy to escape and had been captured. Perhaps this was the way that Jupiter acquired its small outer moons.

On 22 May, Brian Marsden of the Central Bureau for Astronomical Telegrams, which was located at the Smithsonian Astrophysical Observatory, announced that Shoemaker Levy 9 was likely to hit Jupiter in July 1994.

"I can't believe it," exclaimed a delighted Gene Shoemaker upon hearing the news, "we're actually going to see a cometary impact – in our lifetimes!"

Over a period of six days in late July the string of fragments would plunge into Jupiter's atmosphere at 200,000 kilometres per hour. This averaged out to one fragment every 6 hours. The planet rotated in 10 hours. All the comet fragments would strike at the same latitude in the southern hemisphere. "Comparing the mass of Shoemaker Levy 9 fragments to Jupiter is a bit like comparing a gnat with an elephant," Clark Chapman pointed out. Even so, "these impacts should profoundly affect the planet's atmosphere". Unfortunately for the astronomers the impacts would occur just before local dawn, so they would take place just beyond the limb. The rapid rotation would carry the impact sites into view within minutes, however. Still, it was frustrating that the impacts themselves would not be able to be seen.

If, as expected, the comet fragments penetrated several hundred kilometres down into the Jovian atmosphere before exploding, this would probably blast out hot gas from far below the visible cloud tops. What we know of the atmosphere is derived from observing the reflection spectrum of the uppermost clouds and from the infrared emission which leaks out from between the clouds, so the cometary impacts offered an unprecedented opportunity to study the deeper atmosphere. Even if plumes of hot gas were not ejected, it was likely that every impact would create a

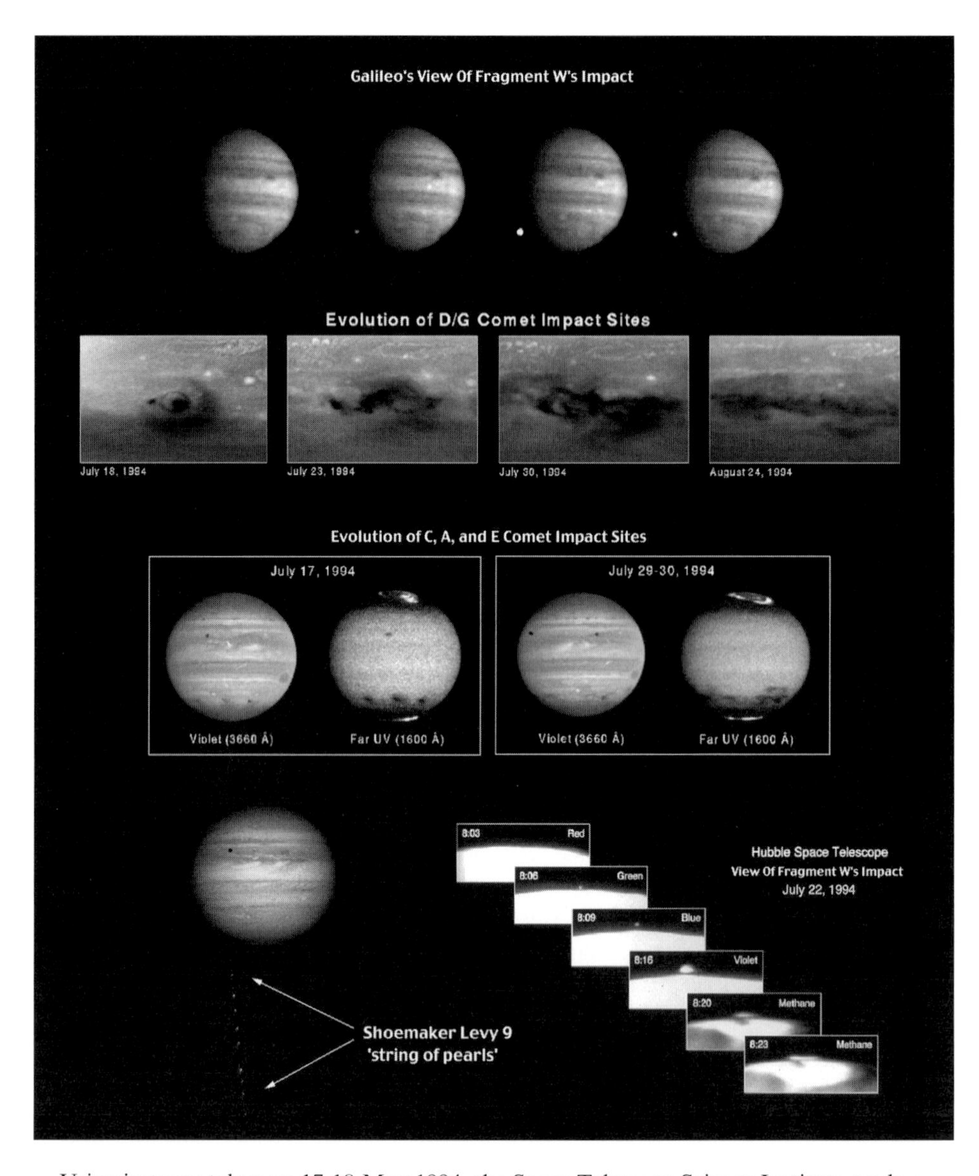

Using imagery taken on 17-18 May 1994, the Space Telescope Science Institute made a mosaic showing Shoemaker Levy 9's 1 million kilometre long 'string of pearls' falling towards Jupiter.

Although 238 million kilometres from Jupiter, Galileo was uniquely located to observe the impacts – all of which occurred at a latitude of about 44 degrees south, just beyond the dawn terminator. A four-image sequence taken over a 7-second period recorded the impact of fragment 'W' on 22 July. A few minutes later, the Hubble Space Telescope saw the plume poke over the limb.

The Hubble Space Telescope documented the evolution of the impact sites for fragments 'D' (17 July) and 'G' (18 July). Most of the ring structure in the 18 July image▶

disturbance in the visible clouds. There was even a prediction that so many impacts in such a short period would disrupt the stable banded structure of the southern atmosphere!

Voyager 2 would have a direct view of the impacts, but it was 6 billion kilometres away, which was much too far for its camera to see anything of use – the planet would barely cover two pixels. However, its ultraviolet spectrometer would be able to determine precisely when each fragment of the comet impacted. Galileo would be 238 million kilometres from Jupiter, and it too would have a direct view. Jupiter would span 66 pixels in its camera. On a planet 140,000 kilometres in diameter, even a resolution of 2,000 kilometres per pixel was adequate for serious observations.

The planning of Galileo's observations was complicated by the fact that it had to follow a preprogrammed sequence, but there was an uncertainty of several minutes in the times of the impacts. Galileo could not simply watch Jupiter and then snap a picture whenever something interesting happened. It was decided to take a series of observations bracketing the predicted time, and an innovative procedure was devised to manage the vast amount of data this would generate. At Galileo's range from Jupiter, the planet would occupy only a small portion of the pixels in the camera's CCD array. It would therefore be possible to make multiple exposures on a single image by slewing the camera a little between each of a series of exposures taken a few seconds apart. This 'on-chip mosaicking' would enable Galileo to take hundreds of images without overloading its ability to transmit the data via the low-gain antenna. The new software was uploaded in early July, just in time. A sample of the data was transmitted in mid-August, but as the spacecraft was approaching conjunction the full replay was postponed until early in 1995.

Although Galileo secured unique data, and ground-based telescopes made observations, it was the Hubble Space Telescope that stole the show with a remarkable series of images of the aftermath of the impacts.

Analysis revealed that the impacts involved four distinct phases. Firstly, as a fragment dug into the atmosphere it heated the gas by friction, leaving a meteor-like tail. This was the 'flash' phase. Once the fragment had penetrated several hundred kilometres, it exploded and created a bubble of hot gas derived from its own vaporised material and the local atmosphere. This was the 'fireball' phase. Galileo had been able to study the fireballs of glowing gas, viewed against the dark side of the terminator. As the hot gas expanded, it emerged from the fragment's entry

was due to 'G', although the small dark spot to the left was what remained of 'D'. By 23 July, the strong Jovian winds had greatly dispersed the fall-back debris, and this disruption progressed through the following month. An image taken several hours after the impact of 'E' on 17 July showed (from left to right) the 'C', 'A' and 'E' sites. The sites were darker in the far-ultraviolet because the debris from the fireballs strongly absorbed this part of the spectrum. A fortnight later, the debris was in an advanced state of dispersal. In addition to the dark impact sites, the ultraviolet views recorded the auroral displays near the poles. The fact that the planet's magnetic poles are offset by 10 degrees from the rotational axis meant that the full circle of the northern aurora was presented. (HST imagery is courtesy STScI).

corridor and mushroomed far above the visible cloud tops. This was the 'plume' phase. Fortunately, the Hubble Space Telescope and terrestrial observers were able to view the tops of some of the plumes, poking over the limb. The most spectacular views of the plumes were in the infrared, and some were so bright that they momentarily outshone the planet itself. Finally, as the gas cooled, it fell back onto the cloud tops and produced intense thermal emission upon impact. This was the 'splash' phase. The result was a dark splotch that persisted for months. These were best observed by the Hubble Space Telescope as it was capable of the best spatial resolution, but ground-based telescopes, being less in demand, provided longer-term documentation.

Calculations showed that the main comet fragments penetrated to a depth of 100 kilometres before exploding, and that their plumes rose 3,000 kilometres above the cloud tops – this was further into space than the comet's trajectory on its original fly-by.

The cometary fragments had previously been labelled by letters of the alphabet in the order that they would arrive. Some fragments subsequently split ('P' did so twice) and others ('J', 'M' and 'T') faded away.

Galileo made observations using its camera, its ultraviolet spectrometer, its plasma wave spectrometer, its infrared spectrometer, and its integrated photo-polarimeter and radiometer. It took images of fragments 'D', 'E', 'K', 'N', 'V' and 'W', and infrared spectrometry of 'C', 'F', 'G' and 'R'. The data downloaded in August included a sequence of four images of the impact of fragment 'W', a photometric light curve for fragment 'K', and infrared, ultraviolet and photometric data for fragment 'G', which had turned out to be one of the largest impacts. Astronomers calculated the duration, size and temperature of each fireball in order to gauge the energy that had been released. Galileo's data put strong constraints on the flash phase. The solid-state imaging system provided brightness profiles for fragments 'K', 'N' and 'W'. This characterised the flash and fireball phases. The brightness peaked within 10 seconds of initial flash detection and then faded for some 20 seconds until it passed below the limit of detection. Significantly, the intensity of the peak brightness for all three fragments was within a factor of two. It was puzzling, therefore, that the long-term effects of these three impacts showed such diversity. At its peak brightness, the emission from 'N' was about half as bright as that from 'K', which implied that the emission was from the bolide's penetration of the atmosphere rather than the subsequent fireball, which appeared to be rather weak.

The correlated infrared, ultraviolet and photometric observations enabled the effect of the impact of fragment 'G' to be characterised in detail. When first spotted by Galileo, the fireball was apparently about 7 kilometres in diameter, and had a temperature of at least 8,000 K, which is hotter than the surface of the Sun. The infrared spectrometer detected it 5 seconds later and recorded the fireball's expansion – after a minute and a half, the fireball was several hundred kilometres in diameter and it had cooled to 400 K.

The infrared spectrometer's observations of fragments 'G' and 'R' showed that they both created super-hot fireballs which persisted for a minute or so, before cooling. Plume material began to fall back onto the cloud tops about 6 minutes later, growing progressively brighter as the fall-back progressed. However, the instrument monitored the splashes for only 3 minutes. Fortunately, ground-based data showed that, in each case, the fall-back took 10 minutes, with the peak brightness occurring about midway through. The striking similarity in the timelines for these two impacts was puzzling, because the 'G' fireball was four times brighter than that of 'R' and the splash from 'G' was twice as intense as that from 'R'. It seemed as if the characteristics of the splash derived from the fireball and plume physics, rather than from the mass of the comet fragment.

The emission spectra from the glowing fireballs indicated the presence of many molecules such as methane, hydrogen sulphide, carbon monoxide and some molecules that had not been detected on Jupiter before, but some of these may represent the cometary material. All of this material fell back, to be absorbed into the atmosphere. Cometary impacts have chemically enriched Jupiter over the years. The spectroscopic observations of the splash also showed the presence of hydroxyl (OH) and methyl (CH) free radicals in the plume material. Although the hydroxyl ions were probably dissociated water molecules, it was not possible to say whether this was from the water clouds believed to exist 100 kilometres beneath the visible surface, or from the icy fragment. Similarly, the methyl ions could have been from Jupiter's atmosphere, or from cometary hydrocarbons.[29]

The large dark splotches that marked the fall-back sites may well have been concentrations of micron-sized carbon particles derived from methane, that remained suspended in the upper atmosphere in much the same way that dust from volcanically generated plumes persists in the Earth's stratosphere.

Several features are notable in the image of the 'G' impact that the Hubble Space Telescope took immediately after it rounded the limb. There is a dark thin ring, a dark streak inside this ring, a broad horseshoe-shaped feature to the south, and a small spot trailing behind (this is actually the impact site of 'D', which hit the day before 'G'). The thin ring may be a shock wave in the atmosphere, moving outward from where 'G' exploded below the cloud tops. At the time of the image, the ring was about one half of an Earth diameter. If it was a shock wave, based on the time of impact it must have been radiating at about Mach 1.7 (about 580 metres per second at that pressure). The dark streak inside the ring is almost certainly the path of the fragment, with the entry point being on the southern end. This ends near the centre of the thin ring. The broad horseshoe-shaped feature is the debris which has splashed back from the fireball.

Concentric rings were observed to radiate out from the dark cores of the impact scars. An analysis suggested that these were waves propagating through the atmosphere, akin to ripples when a stone is dropped into a pond. The speed and structure of the waves suggested that the atmosphere at the depth to which the fragments penetrated was very rich in water, far more so in fact than had been deduced from the Voyagers' remote-sensing data of the cloud tops.

One mystery was that the Hubble Space Telescope, and some Earth-based telescopes, saw some of the impacts start just as soon as Galileo did – as if looking 'through' Jupiter's upper atmosphere. "In effect," reflected Andrew Ingersoll, a

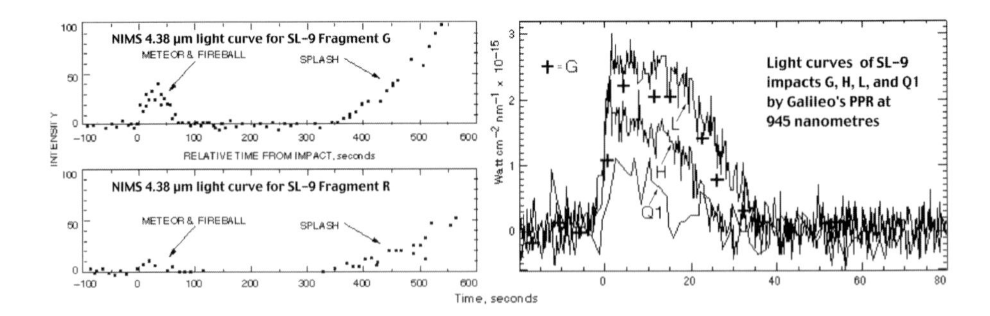

In addition to its camera, Galileo observed the crash of Shoemaker Levy 9 using some of its other instruments to determine the 'light curves' of the events. The near-infrared mapping spectrometer observed fragments 'G' and 'R', clearly showing the similarity of the various phases of the impacts. The integrated photopolarimeter and radiometer observed fragments 'G', 'H', 'L', and 'Q_1' showing the dramatic increase in radiation from the fireball phase of the impacts.

Galileo interdisciplinary scientist, "we are apparently seeing something we didn't think we had any right to see."

Torrence Johnson had a possible explanation. "The Hubble observations of fragments 'G' and 'W' could conceivably be due to scattering of light from the Galileo events off comet dust or other material at very high altitude. There may have been earlier, smaller impacts going on that were too faint for Galileo to detect."

One side effect of the impact of Shoemaker Levy 9 with Jupiter was the public awareness that such events really did take place and that the chance of a comet or an asteroid striking the Earth was a genuine threat. Such events are believed to have given rise to mass extinctions. It is believed that the dinosaurs were wiped out by an impact some 65 million years ago. One of the most prominent ray craters on the Moon is Tycho. It was formed around 100 million years ago and serves as a stunning illustration of the degree of damage which the Earth must bear from time to time.

As a sad footnote to the Shoemaker Levy 9 story, Gene Shoemaker was killed on 18 July 1997 in a car crash near Alice Springs in Australia, while on his way to investigate a possible impact crater.

PROBING THE SUN

As Galileo passed through conjunction on 1 December 1994, the degradation of its signal was used to probe the charged particle environment of the inner solar corona, where the solar wind forms. This was a particularly favourable conjunction for the solar scientists because the spacecraft did not actually pass behind the Sun, it skimmed the solar disk, so the Deep Space Network was able to develop a continuous record for a month on either side of the 0.2-degree angular separation at the moment of closest approach.

The emerging picture was one in which the solar corona is permeated by a wide variety of ray-like structures which are organised by the overall solar magnetic field. The degree of solar magnetic activity varies over an 11-year timescale. (Actually, the field polarity 'reverses' every 11 years making it a 22-year cycle.) By this point, Galileo had been in space for over five years. While in the inner Solar System, it had repeatedly passed behind the Sun as viewed from the Earth – superior conjunction – and had provided useful data as solar activity declined from its peak down to its minimum.

Ulysses, once again close to the Sun, and at high solar latitude, was reporting on the low-density plasma that rapidly moves clear of the Sun in the polar regions. On 26 June 1994, its path took it higher than 70 degrees solar latitude, and it officially began its first polar passage. On 13 September, at 80 degrees, Ulysses was at its most southerly solar latitude and some 2.3 AU from the Sun. It concluded its first polar passage on 6 November. It crossed the ecliptic on the far side of the Sun on 5 March 1995. For a month on either side of this crossing, Ulysses was within the turbulent disk. The 1.34 AU perihelion passage occurred on 12 March, just north of the ecliptic. The second polar passage took place between 19 June and 30 September 1995, with the most northerly latitude attained on 31 July. With its primary mission now complete, Ulysses' trajectory returned it to Jupiter's orbit, but when it got there, the planet was not present so the spacecraft looped back to begin its second orbit.

STORMY WEATHER

Throughout the interplanetary cruise, Galileo's magnetometer, dust detector and extreme-ultraviolet spectrometer had reported in near-real time on the spacecraft's environment. In late June 1994, the dust detector registered an increase in the rate of dust impacts, which indicated that the spacecraft had penetrated a stream of dust.

As Ulysses had moved out beyond 3 AU, heading for Jupiter, it had recorded 11 dust streams which had been narrow and had followed parallel paths (that is, the dust was in collimated beams). The fact that Galileo found the dust streams in the same region of space was an indication that it was not a transient phenomenon. The data confirmed that the dust was moving at 50 to 100 kilometres per second, which was sufficient to escape from the Solar System. In July 1994, the dust detector was reprogrammed, as Carol Polanskey, JPL's team chief for this instrument, explained, "to take advantage of the knowledge gained from the Ulysses experiment".

The 'normal' event rate was one strike every few days, on average. By the end of the year, Galileo had passed through seven such streams of dust, each more intense and wider than its predecessor – in fact, the most recent stream had taken three weeks to cross and had resulted in 1,000 to 2,000 per day. The dust was more intense than had been observed by Ulysses, so whatever the source was within the Jovian system, it seemed to have increased its output. As some of the individual atoms in the dust particles had been ionised, the particles had acquired an overall electrical charge of up to 5 volts. Evidently Jupiter's magnetic field had accelerated these electrically charged particles sufficiently for them to escape the giant planet's gravity. The sheer amount of dust suggested that the source was Io's volcanoes. But what could make such collimated beams? Remarkably, considering the collimation of the dust streams, Galileo was still 160 million kilometres from Jupiter.

On 28 July 1995, by which time Galileo was only 80 million kilometres from Jupiter, the spacecraft penetrated what became the most intense storm of interplanetary dust ever detected. At its peak, the microscopic dust particles were hitting the detector at a rate of 20,000 per day. The instrument's normal reporting rate for the interplanetary cruise was twice per week but as the strikes rose the reporting was increased to thrice daily in order to more closely follow the structure of the stream. By early October, the rate had decreased to only a few hundred strikes per day. For insight into the fine structure of the stream, the instrument was switched to high time resolution to monitor the stream over a 10-hour period during which Jupiter – together with its co-rotating magnetosphere – would make a complete revolution. By the time Galileo reached Jupiter, it was once more flying through clear space.

THE VEEGA BONUS

The scientific targets of opportunity "had exceeded all expectations", enthused Bill O'Neil, referring to the secondary objectives facilitated by the VEEGA path to Jupiter ... and Shoemaker Levy 9 had been the icing on the cake. Galileo had not yet reached its primary target, but it had already achieved a great deal.

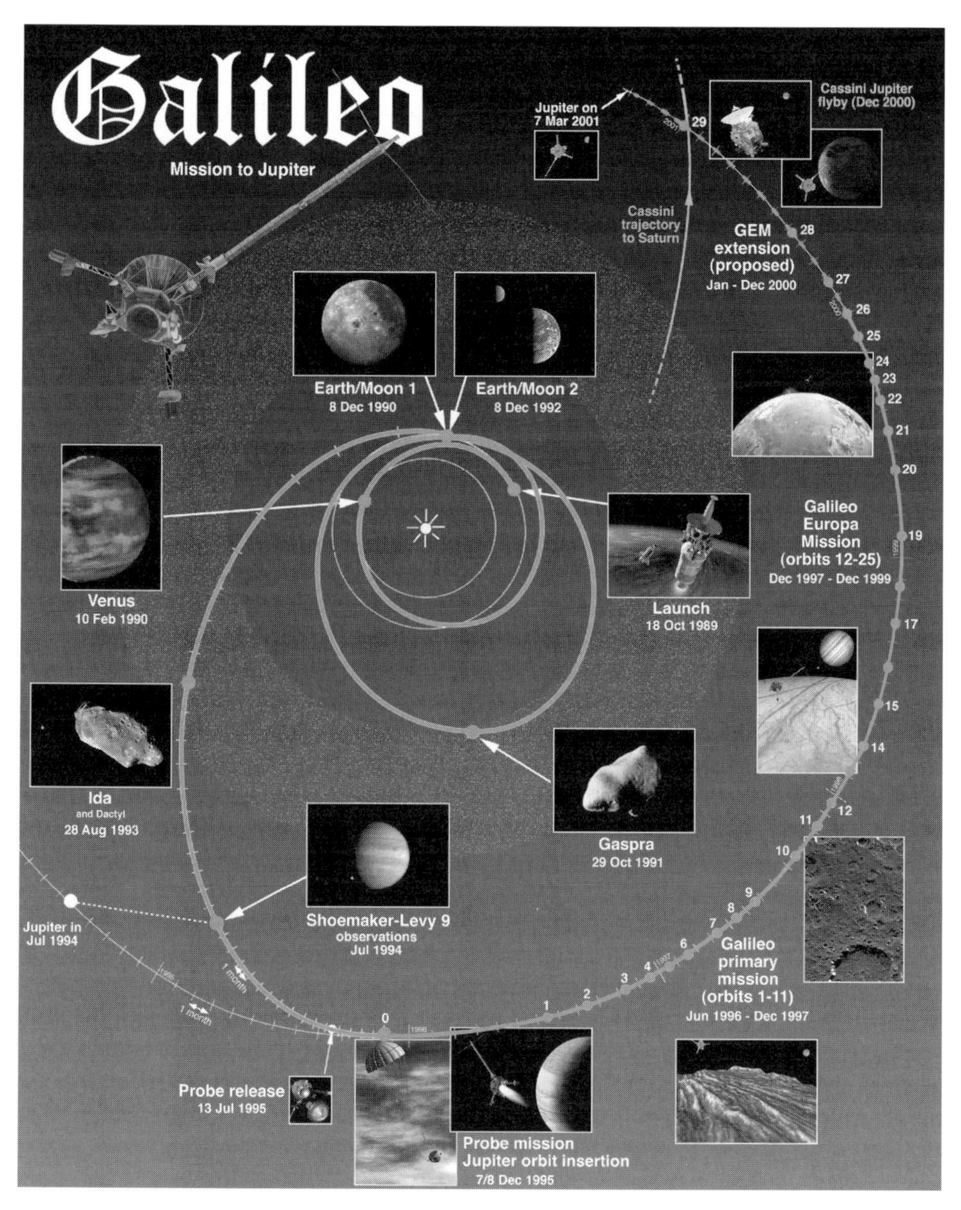

A chart documenting the way-points during Galileo's odyssey to Jupiter. (Courtesy, NASA)

6

Target in sight

PROBE AWAY!

On 29 January 1995, Galileo finished replaying its Shoemaker Levy 9 data. The next day, the complex task of reprogramming its various computers began.

The new software was to see the spacecraft through the approach to Jupiter using only the low-gain antenna. It would handle the science acquisition on the final run in, store the atmospheric probe's data on the tape recorder, and perform the engine burn to insert the spacecraft into Jovian orbit. The spacecraft started running the new software on 24 February.

"This was the first time a planetary spacecraft's computer had been completely reloaded in flight," pointed out Bill O'Neil, the Galileo Project Manager. "Because of the great complexity and criticality of this process we allowed six weeks for it, and we were surprised, and elated, when the whole thing was done without a significant problem within one day of the nominal 25-day schedule."

Of course, in a way, the approach phase was the easy part of the primary mission. A great deal of work remained to be done to write the software that would handle the encounters with the Jovian moons; but the mission was a series of tasks, each of which was a potential show-stopper until it had been successfully achieved, and in the early part of 1995 the priority was to release the atmospheric probe. This had been tested soon after launch in 1989, and again in 1990 and 1992, and had been found to be in perfect condition each time. Although it had been designed for a three-year cruise to Jupiter, the probe had already been in space for longer than this. A thorough check of the probe's status, completed on 15 March, confirmed that the battery was in peak condition and the timer that would awaken the probe six hours prior to its arrival at Jupiter was working. The accelerometers that would enable the probe's atmospheric structure instrument to gather data during the entry phase were also calibrated. In addition, the check-out data was stored in Galileo's solid-state memory and then processed by the software which was to handle the probe's data relay in order to confirm that this functioned properly.

"The Galileo atmospheric probe's systems are functioning as designed," reported Marcie Smith, the probe's manager at Ames.

The trajectory correction made after the Ida encounter had set Galileo on a collision course with Jupiter, but if the probe was to survive its penetration of the giant planet's atmosphere it was crucial that it should approach along a very narrow corridor. On 12 April, a series of 64 pulses of the thrusters produced an 8 centimetres per second refinement so that the probe, once released, would hit the centre of its entry corridor.

On 18 April, after a wide-ranging review of the probe's status, Ames gave the formal go-ahead for the probe to undertake its full observational programme. In the event that the battery had deteriorated, the contingency plan had been to omit the pre-entry observations of Jupiter's inner magnetosphere, to safeguard the atmospheric investigation. A comprehensive review of Galileo's status on 21 May confirmed the spacecraft's readiness to release the probe.

The probe was powered up on 5 July. While being carried, of course, it had drawn power from the main vehicle. At 05:32 UT on 11 July, the umbilical was severed by an explosively driven guillotine. The probe was essentially passive; only its cruise timer was active. Then at 08:37 UT on 12 July, Galileo reoriented itself and spun up to 10.5 rpm in 'all spin' mode. To bestow the greatest degree of gyroscopic stability upon the probe, one of the RTG booms was finely adjusted to minimise the wobble resulting from the asymmetrically deployed high-gain antenna.[1]* At 05:30 UT on 13 July, three explosive bolts fired to release the probe. Small springs pushed the two vehicles apart at 0.3 metre per second – of course, the final impulse had been taken into account when plotting its subsequent trajectory. From now on, Sir Isaac Newton would be in the probe's driving seat. The release attitude had been carefully arranged so that the probe's primary axis would be aligned on its velocity vector when it penetrated the Jovian atmosphere. The spin would ensure that it remained stable both during the five months of its 82 million kilometre 'ballistic fall' to Jupiter and as it penetrated the atmosphere.

The successful-release signal from the main spacecraft was received by the Canberra Deep Space Communications Complex 37 minutes after the event. It actually took a few minutes to process the signal and pass it to JPL and Ames, but the doppler on the signal resulting from the push-off springs alerted the Canberra engineers to the fact that the probe had indeed been released.

It was "standing room only", reflected Bill O'Neil, in the von Kármán Auditorium for the post-release Press Conference. "We're delighted to have successfully released the probe on its Jupiter atmospheric mission, after having carried it for almost six years."

"We're very excited to have the probe mission underway," confirmed a delighted Marcie Smith.

The successful release of the probe was a relief, because it had covered the main engine's nozzle. If the probe had not released, the engine would not have been able to insert Galileo into orbit around Jupiter, in which case the mission would have defaulted to a frustrating fly-by.

* For Notes and References, see pages 413–424.

In order to test the engine, it was to be used to move the main spacecraft off the collision course it had adopted for the probe's release. First, however, the engineers wanted to make a brief firing to 'clear its throat'; it had, after all, been dormant for six years. A two-second burn on 24 July verified the engine's health.

The engine was used to perform the Orbiter Deflection Manoeuvre on 27 July. It began at 07:00 UT. At the nominal engine rating, the burn was scheduled to last 306 seconds. In fact, the engine underperformed its nominal 'steady-state thrust' by some 3 percent – perhaps because the propellant feed system pumped propellant at a lower rate than intended. This was not a problem, however, because the computer kept the engine firing until an accelerometer showed that it had achieved the desired 61 metre per second change in velocity. The engineers had stipulated 6 percent as the acceptance criterion. The deflection manoeuvre could have been made by the small thrusters, but it was an excellent opportunity to calibrate the main engine in preparation for the mission-critical orbital insertion burn. "This is a joyous morning," assured Bill O'Neil at the post-burn Press Conference. However, the centre of attention was held by the German engineers who had designed and built the engine.

Although Galileo had achieved the desired trajectory adjustment, there were nevertheless a few worrying anomalies. In preparing the engine, the engineers had been puzzled by telemetry indicating a difference in pressure between the tanks containing its fuel and oxidiser. It burned mono-methyl-hydrazine in nitrogen tetroxide, and the tanks were pressurised by helium. After this burn, telemetry suggested that a 'check valve' – which is intended to facilitate one-way flow – had not closed. If the helium valve to the oxidiser tank was still open, then there was a possibility that the oxidiser would feed back through the pressurisation system.

"If the oxidiser check valve is stuck open, large temperature changes in the propellant tanks could cause oxidiser vapour to migrate to the fuel plumbing," explained Todd Barber, the JPL propulsion engineer on the engine team, "which would cause undesirable chemical reactions."

This was a masterful understatement, because the reactants for the innovative bi-propellant engine were hypergolic – that is, they would ignite if they came into contact. A programme of thermal management was therefore instigated, using small electrical heaters to regulate the tank pressures to prevent feedback.

The faint hope that the shock of igniting the main engine might shake loose the high-gain antenna's ribs proved to be wishful thinking. For those with an optimistic outlook, there was still a chance that it would be shaken free during the orbital insertion burn.

Because Jupiter would be close to conjunction when Galileo arrived, the Canberra Deep Space Communications Complex was upgraded in mid-1995 so that it would be better able to pick up the spacecraft's signal. The received power of the signal would be barely 1/10,000th of that which the high-gain antenna would have been able to provide.

On 23 August, the L-Band antenna which was to receive the atmospheric probe's data was deployed and slewed through its full range of action to verify that it was functional.

The deflection manoeuvre which had moved Galileo from the probe's collision course had been designed to aim it towards Io. In fact, the burn had put the spacecraft on a collision course with Io, so a 1 metre per second correction was made on 28 August to allow it to pass 1,000 kilometre in front of this moon, as called for in the flight plan.

As Galileo approached Jupiter, one by one its instruments were switched on and checked, but when the extreme-ultraviolet spectrometer was activated on 2 October, it malfunctioned. An anomaly team then tried to determine what was wrong, and how to fix it. The spectrometer was reprogrammed on 8 October and it restarted properly – just in time to begin observations scheduled for the approach to Jupiter. In fact, the next day, Galileo formally finished the interplanetary cruise phase of its mission and initiated 'Jupiter approach' activities.

THE ENTRY SITE

The Hubble Space Telescope had been observing Jupiter ever since it had been disfigured by Shoemaker Levy 9 in order to monitor the rate at which the impact scars 'healed'. This also provided a long-term study of the atmosphere to help to select the best location for the probe to sample. Detailed records of the prevailing weather systems would also assist in interpreting the probe's results.

On 4 October 1995, the Hubble Space Telescope started taking a sequence of pictures of Jupiter to determine the wind speeds in the vicinity of the atmospheric probe's point of entry. Although the probe's trajectory was known, predicting the entry site was not straightforward. It was known that the probe would enter about 6 degrees north, on the boundary between the Equatorial Zone and the North Equatorial Belt. As a matter of fact, it was also known that the longitude of the entry site would be about 5 degrees west. So what was the mystery over the entry site? As a gas giant, Jupiter's atmospheric features migrate in longitude within the belts and zones. Features at different latitudes migrate at different rates, and even within a specific band the various features tend to migrate independently. Jupiter's zero of longitude had been defined arbitrarily in terms of the rotation of the planet's core – which could be inferred from the magnetic field. The task was to predict which feature would be at the entry site's longitude when the probe arrived.

This was the Hubble Space Telescope's last opportunity to observe Jupiter in 1995, because the planet was approaching conjunction and the telescope could not be pointed near the Sun in case it momentarily lost its attitude control and allowed the Sun to shine into its tube, irretrievably damaging its instruments. Until early 1996, the task of monitoring Jupiter's atmosphere would therefore fall to terrestrial telescopes, particularly the infrared telescopes, as these could observe during the day by 'subtracting' the sky from the signal, and hence follow Jupiter through conjunction. On 5 October, NASA's Infrared Telescope Facility located atop Mauna Kea joined in the Jupiter survey. Its 4.8-micron imagery sensed heat radiated to space rather than solar reflection, so the bright features corresponded to gaps in the clouds, through which heat was leaking from below.

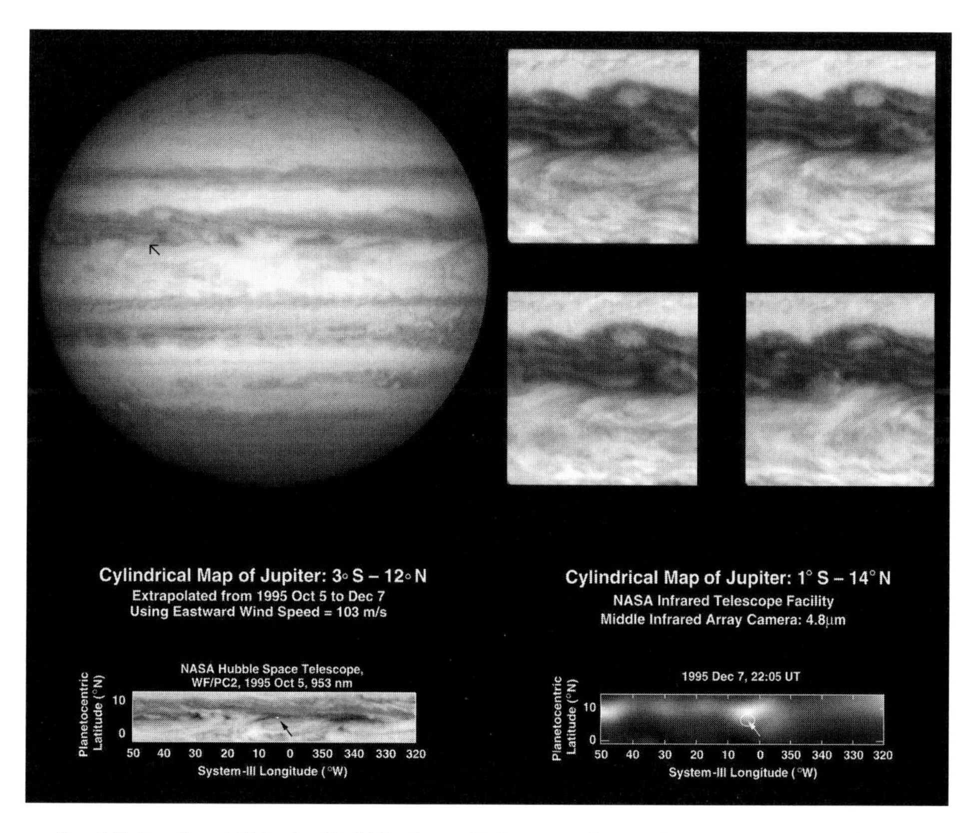

On 4/5 October 1995, the Hubble Space Telescope observed the site on the boundary between the bright Equatorial Zone and the dark North Equatorial Belt where Galileo's probe was predicted to enter the atmosphere. At 0.95 micron, the bright features were ammonia clouds reflecting sunlight. The site is marked by an arrow in the full disk. An analysis of the four detailed images (which span 30 degrees in longitude) permitted the wind speeds to be estimated so that the motions of the features could be extrapolated to 7 December, when the probe would arrive. It was predicted that the probe would enter the fringe of an 'infrared hot spot'. This proved to be so, as the image taken at that time by NASA's Infrared Telescope Facility indicates. The bright features in the 4.8-micron image indicate heat leaking through holes in the ammonia clouds. (Courtesy STScI and IRTF)

Characterising the nature of the entry site would be important in interpreting the data from the probe because conclusions pertaining to the whole planet would be inferred from a single direct sample. If the sample site seemed typical, then the results could probably be considered to be representative. But if the entry site was atypical, this would be manifested in the results from the probe. In this case, it would be important to be able to trace the history of the feature into which the probe had descended.

Tracking of individual clouds on Jupiter established that the winds at the latitude at which the probe was to penetrate ranged from 400 to 540 kilometres per hour, travelling to the east. One of the probe's key objectives was to profile the winds as a function of depth. If Jupiter's winds were driven by insolation, and sunlight did not penetrate far below the cloud tops, then the winds would decline at depth. If the winds persisted far below the depth to which sunlight could penetrate, this would show that the weather system was driven by heat from the planet's interior.

The telescopic observations indicated that the probe would enter near a so-called 'infrared hot spot', which meant that it would penetrate a particularly clear zone in an otherwise cloud-dominated atmosphere.

DISASTER?

By 11 October, Galileo was 36 million kilometres from its target, and the approach phase required the spacecraft to take a number of images of Jupiter, both for navigational purposes and also to help to document the probe's entry site. The first item on the schedule was to be a colour image. With Jupiter near conjunction, it would take almost a month to replay this over the low-gain antenna. The data compression software was not yet available (this would not be uploaded until early in 1996) so it had been decided that this would be the only science image of Jupiter taken during the approach.[2] Immediately after storing the trio of component images (each taken with a different filter) the tape recorder was ordered to rewind to replay the stored data. However, it malfunctioned. The telemetry suggested that the tape had failed to stop after the command for it to rewind.

A similar tape recorder had been used by Magellan, which had mapped Venus with no problems. In the case of the Cassini spacecraft, however, which was then in final development, a high-capacity solid-state memory was to be used rather than a reel-to-reel recorder.

By a spooky coincidence, a few hours later, while testing an identical tape recorder on the ground, the tape was ripped from its reel. If this was what had occurred to Galileo's machine, then the mission was in serious trouble. The failure of the ground test unit turned out to be an unfortunate red herring. For a week or so it focused the troubleshooters on how *it* had failed, but then various failure modes were studied based on the scheduled sequence of activities and the resulting telemetry. On 20 October the tape recorder was commanded to advance, which it did. Despite the acceptance of the rewind command, the tape was still at the end of the third recorded image. Evidently the mechanism had been active, but the tape had not moved. It was decided that the tape had probably been worn when the capstans had turned, so the machine was ordered to advance a little and redefine the 'start' point. The image of Jupiter, which was now on the wrong side of the new marker, had to be abandoned.

Once the tape recorder was confirmed still to be usable, Bill O'Neil "made the most painful decision of my career. Until the probe's data was returned to Earth, we would use the recorder only to record and playback the probe's data." He also decided not to run the tape recorder at high speed until after the probe's data had been secured. This would impose a more "benign" load on the tape.

The tape recorder had been operating at high speed when it had malfunctioned. Since being retrieved, it had been used only at its slower speed. With Jupiter imminent, the engineers were reluctant to test it at high speed. Low speed was sufficient for recording the probe's data. The high time-resolution particles and fields data that was to be taken while passing through Io's torus could also be recorded "at no significant risk".

One day out, the plan had been for all of the imaging instruments (the camera, the infrared spectrometer, the ultraviolet spectrometer, and the integrated photopolarimeter and radiometer) to make observations of Jupiter, Europa, Io and the small inner moonlets Thebe and Adrastea. Unfortunately, this decision ruled out the imaging observations on the final approach. "All imaging and other high-rate data had to be eliminated from the arrival sequence," O'Neil lamented. These instruments yield data in such bulk that the tape recorder would have to run at high speed. Given "the paramount importance of the probe", this was deemed an unacceptable risk.

A domestic two-track tape recorder has one forward track and one reverse track. Galileo's recorder had four tracks, two forward and two backward. It had been intended to use three of the four tracks to record science data during the final five days of the approach, right up to half an hour after the Io encounter. The fourth track was to record particles and fields data through to an hour and a half after orbital insertion, and the data from the atmospheric probe. The data from the probe now had absolute priority, but it would still be possible to record the high time-resolution particles and field data for 50 minutes while passing through the torus, a region that the spacecraft was unlikely to venture into again.

This serious tape recorder problem only two months from Jupiter prompted three parallel activities. One was to troubleshoot the tape fault. Another was to figure out how to maximise Galileo's orbital mission in the event that the tape proved to be irretrievable (on 13 October, it was determined that it would be possible to route data from the imaging instruments directly into the solid-state memory of the spacecraft's main computer,[3] by-passing the tape recorder). The third activity was to write the probe's data into solid-state memory in order to back-up the tape recorder in the event that it was available, and to provide a fall-back if it was not. As the tape recorder had been installed to back up the high-gain antenna during this critical phase of the mission, this was significant reprogramming so late in the approach. The main obstacle to using the solid-state memory for probe data was that it was to have been used by other instruments during the very busy flight around Jupiter. However, the cancellation of most of the in-bound programme released some memory for storing the probe's data.

October was therefore a hectic time for the Galileo teams, with the mood fluctuating on a daily basis as one problem was overcome only to expose another. There was a rather subdued celebration of the sixth anniversary of Galileo's launch, on 18 October.

"While not yet panic stricken," admitted David Atkinson, the University of Idaho scientist who headed the probe's doppler wind experiment, upon hearing about the faulty tape recorder, "I have to admit that I will have some trouble sleeping for a while." If the tape recorder could not be recovered it was likely that the data for his experiment would be sacrificed.

Clark Chapman, a planetary scientist at Southwest Research Institute, Boulder, Colorado, and a member of the imaging team, was philosophical upon hearing the news. "I can think of twenty times during this mission when I've felt the way that I do now – and most them meant nothing. The good news, is that [our systems] are very resilient."

"Life can be a roller coaster," said Glenn Orton upon first being told that the tape recorder had been recovered, and then that the imaging of the probe's entry site had been cancelled. He wanted to correlate what the spacecraft's high-resolution remote sensing could 'see' within the atmosphere with *in situ* 'ground truth' from the probe's net flux radiometer. He would have to rely upon telescopic observations, which could not resolve the atmospheric structures in such fine detail.

"It's like watching one of those old movies – 'The Perils of Pauline'," reflected Torrence Johnson. "We've had practically everything happen to this poor spacecraft along the way!"

O'Neil was adamant that even if the tape recorder failed at some point and Galileo turned out to be denied both the high-gain antenna and its mass storage medium, it would still not be crippled. "There is a whole lot more we can do than we imagined, before we looked at this in detail." With the new data compression software, data would be able to be stored in the solid-state memory and later replayed by the low-gain antenna, but this would severely constrain its mission's scientific capacity. On the other hand, apart from these ailments, the spacecraft was supremely healthy and it was the only game in town.

INTO JOVIAN SPACE

As Galileo closed within 20 million kilometres of Jupiter, it finally emerged from the dust streams. The magnetometer was now reporting in near-real time, seeking the first contact with the Jovian magnetosphere. On 6 November, the extreme-ultraviolet spectrometer started long-range observations of Io's plasma torus to determine its size and shape. Meanwhile, engineers continued the process of powering up and testing the instruments. Although the tape recorder's fault had meant cancelling its observations, the infrared spectrometer was put through its paces and passed with flying colours.

And what of the atmospheric probe? It was not possible to know, because it was to remain silent during its 'fall' to Jupiter. When asked on 9 November to comment, O'Neil simply said that the probe was "believed to be in excellent health".

The magnetometer detected the magnetosphere on 16 November, when a lull in the solar wind permitted the planet's magnetosphere to balloon suddenly. The bow shock washed back and forth over the spacecraft several times before Galileo finally entered the relatively peaceful magnetosphere on 26 November, at which time the planet was 9 million kilometres off. "With the spacecraft now in the magnetosphere," said a delighted O'Neil, "we begin our first direct measurements of the Jovian system." Jupiter's magnetosphere is the largest discrete structure in the Solar System. It is millions of kilometres across and tens of millions of kilometres long. In fact, if it could be seen by the naked eye, it would be larger than the Moon in our night sky – it would span 1.5 degrees, compared to 0.5 degree for the Moon.

At about this same time, the Deep Space Network started monitoring the degradation of the spacecraft's signal for the imminent conjunction to secure further scintillation data on the inner solar corona. These observations were to continue through to mid-January 1996.

On 17 November, Galileo started executing the 'critical event software' that was designed to see it through the probe's data relay and orbital insertion. It would therefore fly the final 21 days of the approach entirely autonomously. As O'Neil explained, the new software had been made extremely resilient so that the spacecraft would continue "no matter what". In this 'gung ho' spirit, the computer would ignore most of the anomalous conditions that would normally prompt it to 'safe' itself, and it would cascade down a stack of back-up procedures.

By early December, Galileo was closing on Jupiter at 70,000 kilometres per hour, and the enormously deep 'gravitational well' was drawing it in ever faster. "Basically, we're ready," confirmed Jim Erickson, the deputy manager of the science sequencing office, on 6 December.

ARRIVAL DAY

For Galileo, 7 December 1995 – 'Arrival Day' – was destined to be the busiest day of the entire mission, with a schedule including:

- Europa fly-by
- Io fly-by
- penetrating the torus for the close approach to Jupiter
- probe data relay
- orbital insertion
- Jupiter occultation.

EUROPA

Galileo's in-bound trajectory took it close to Europa, and the angle presented an excellent opportunity to view the south pole, an area which would not be so readily observed during the much closer fly-bys of the primary mission. However, when the spacecraft flew some 32,500 kilometres over the moon's icy surface at 13:08 UT,[4] its scan platform remained stowed.

Once inside Europa's orbit, the particles and fields instruments began continuous reporting at a low time-resolution. They would do so until one hour after orbital insertion in order to make a 14-hour sample of the inner magnetosphere.

AMAZING IO

As the Voyagers approached the Jovian system in 1979, very little was known for sure about Io. However, Io orbited very close to Jupiter and the planet's gravitational field drew in any material which strayed too close and accelerated it, so the expectation was that Io would be a heavily cratered body, rather like the Moon, but with a peculiar red colour.

When Voyager 1's early low-resolution imagery showed vaguely circular albedo features, these were interpreted as craters. But as the spacecraft closed in and its resolution improved it became evident that the dark features were not impact craters. Furthermore, the surface was so garishly mottled with orange-red patches that it was promptly dubbed 'the pizza moon'.

Astonishingly, a detailed survey revealed Io to have *no* impact craters. Since no moon can completely escape impacts, the absence of craters indicated that Io's crust was active and some resurfacing process had 'removed' the craters. Io was evidently highly volcanic. Nevertheless, nobody seriously expected to catch any volcanoes in the process of erupting.

However, on 8 March, as Voyager 1 flew beyond Jupiter, it turned back to take an image to record the position of Io's night-time hemisphere against the stars to assess its trajectory for Saturn. The extent to which the spacecraft's trajectory had been deflected by passing Io would provide a measure of the moon's moment of inertia, and an indication of its internal structure.

The image captured Io as a crescent, with its night hemisphere lit by 'Jupiter shine'. When Linda Morabito, a Voyager navigation engineer, enhanced the image to look for faint stars close to the moon's limb she spotted a faint plume projecting 280 kilometres from the bright limb. It was an active volcano! There was also an anomalous glow near the terminator. A second volcano! This vent was in darkness, but its plume of gas and dust was tall enough to catch the Sun. The Voyagers had infrared spectrometers which were able to identify 'hot spots' on the surface in the vicinity of what looked to be volcanic features. There were evidently many hot but inactive vents too.[5]

Just before Voyager 1 ventured into the Jovian system, a paper by Stanton Peale, Patrick Cassen and Ray Reynolds[6] in the journal *Science* reported the results of modelling the tidal stresses acting on Io. Europa's orbital period is twice that of Io and Ganymede's is twice that of Europa, and these resonances have made Io's orbit slightly elliptical – this eccentricity is only 0.0041, but this has proved to have a significant effect. The overall gravitational force acting on Io varies cyclically. All of the Galilean satellites rotate synchronously, and so maintain one hemisphere facing Jupiter. The 'tidal bulge' on Io rises and relaxes in response to the varying gravitational attraction and all this mechanical stress is converted to heat. Peale's team predicted that the tidal energy would be two to three *orders of magnitude* greater than that likely to arise from radiogenic sources. In addition to concluding that Io was highly thermally differentiated, they had tentatively predicted that it might even be volcanically active.

Io was not only active, but turned out to be the most volcanically active object in

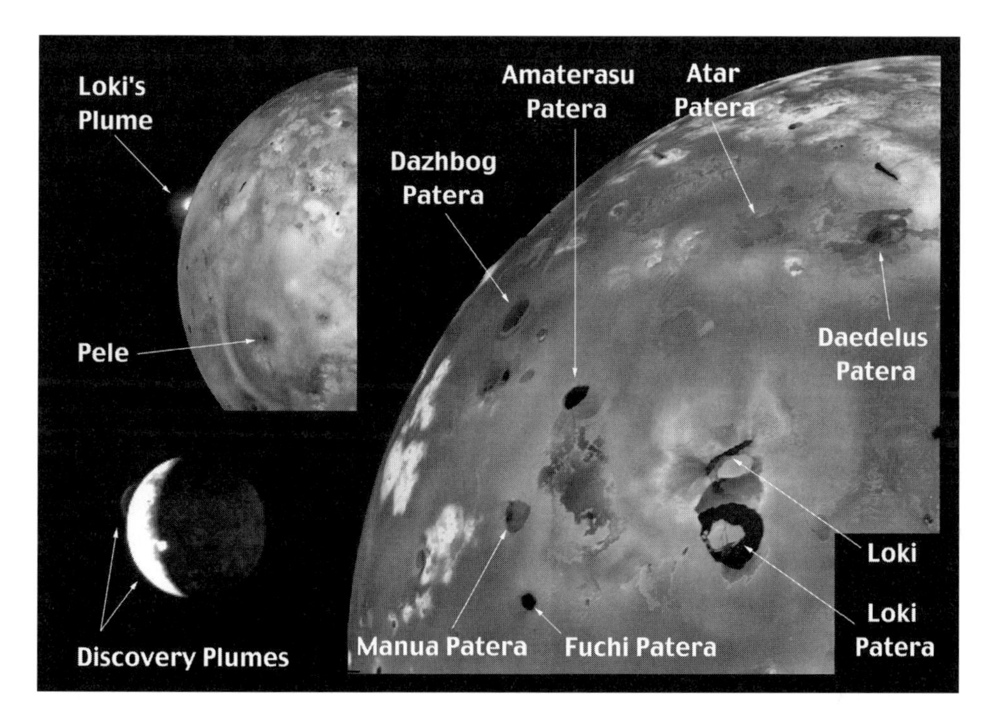

When Linda Morabito examined the 'navigational' image taken when Voyager 1 looked back at Io (lower left) she saw two anomalous glows, one on the limb and the other on the terminator. These turned out to be volcanic plumes. Voyager 2 caught a plume over Loki. The intensely volcanic surface of Io is littered with calderas and Loki, one of the largest, is 200 kilometres across.

the Solar System. In all, nine vents were detected with plumes rising from 70 to 300 kilometres above the surface. The umbrella shape at the 100-kilometre level indicated that material was fanning out and falling back to form a halo around the vent. Other halos indicated the sites of formerly active vents. When Voyager 2 flew by four months later, eight of these vents were still active. The exception was a plume that Linda Morabito had spotted, and which was later named Pele. As a result of the Voyagers, therefore, Io was transformed from a crater-scarred frozen hulk into an intensely active moon.

The trajectory refinement that Galileo had performed in late August had set up its approach so accurately that no additional manoeuvres had been needed. "That's like a hole-in-one," said Steven Tyler, the leader of the Orbiter Operations Group.

It had been hoped to take three long-ranged pictures of Io for navigational purposes, but the tape recorder fault had ruled this out. As long as the spacecraft flew within 1,000 kilometres of the moon, all would be well as far as achieving Jupiter orbit was concerned. Nevertheless, the precise fly-by distance would significantly influence the shape of the resulting 'capture' orbit. Without spacecraft imaging, the navigators had to rely upon the Deep Space Network's radio tracking. The early estimate was for a 940-kilometre fly-by, but as the spacecraft closed in this was progressively reduced. As Io loomed, it became evident that Galileo would fly about 890 kilometres ahead of the moon's leading hemisphere. Assuming a perfect orbital insertion burn, this slight undershoot would yield a capture orbit with a period about a week shorter than the planned 205 days. Although it would have been feasible to make a minor burn to correct this,[7] it was decided to accept the discrepancy because Ganymede, the first target of the orbital tour, orbited Jupiter in 7.2 days, and that encounter could be advanced by one orbit, from 4 July to 27 June 1996.

There was another benefit from a closer-than-planned fly-by. The retardation arising from the inverted slingshot would be increased. The fly-by had been a feature of the plan ever since the VEEGA route had been conceived, because its retardation of the spacecraft meant that less propellant would be needed for the insertion burn. The margin had nevertheless been tight. A closer-than-planned Io fly-by would therefore save propellant that would be able to be utilised to extend the primary mission's tour of the moons. In fact, the Io fly-by slowed the spacecraft by about 162 metres per second, which was about 20 percent of the overall change in velocity required for the spacecraft to be captured by Jupiter, which translated into a reduction of about 95 kg in propellant.[8]

At 17:46 UT, Galileo passed 892 kilometres above Io's volcanic surface but, as at Europa five hours earlier, the scan platform remained stowed. The only consolation was that particles and fields data was being collected and doppler tracking would permit the celestial dynamics team to derive its internal structure. One valuable observation, which unfortunately was ruled out by the need for a retardation effect, was an occultation. If the spacecraft had been able to pass behind Io, the refraction of the signal would have enabled the radio science team to probe the moon's ionosphere.

Rosaly Lopes-Gautier, a native of Rio de Janeiro, Brazil, had been starting her doctorate in planetary geology when Voyager discovered that Io was volcanically active. She had joined the infrared spectrometer team in 1991. After planning the observations for Galileo's 1992 Earth–Moon fly-by, she had developed the infrared plan for Galileo's one-and-only close look at Io.

So what had been lost? The in-bound trajectory had presented opportunities to observe the day-time hemisphere. Because Io's rotation was locked to its motion around Jupiter, Galileo's imagers would have been able to document the sunlit anti-Jovian hemisphere. Prometheus and Volund, two volcanoes in Colchis Regio in the centre of the anti-Jovian hemisphere, had been scheduled for observation. The point of closest approach was just beyond the terminator, and the low illumination and best resolution of 40 metres would have been ideal for observing surface relief. For half an hour after closest approach, the spacecraft was to have looked back 'over its shoulder' and the infrared spectrometer was to have tried to observe the plume rising from Loki, which was known to be active. Thermal and polarimetry scans were to have been made of the dark hemisphere by the integrated photopolarimeter and radiometer to identify the nature of the surface material and how it cooled after sunset.

In July, the Hubble Space Telescope had discovered "a dramatic change" on Io, with the appearance of a 350-kilometre wide yellowish-white patch that had been just a small white spot in March 1994.[9] "The new spot surrounds the volcano Ra Patera," said John Spencer of the Lowell Observatory in Flagstaff, Arizona, and a co-investigator for the radiometer. As it would have been on the dark hemisphere, Galileo's radiometer was to have observed this still-cooling volcanic centre.

The cancellation of the Io observations was a major scientific loss. "I'd be disappointed if I were a scientist," admitted Don Ketterer, the Galileo Programme Manager in Washington, "but I'd look to what I was going to get in the future."

"There's still the hope that we will go back to Io for a close look," insisted Rosaly Lopes-Gautier, undeterred, "maybe at the end of the mission, or with another mission – in fact, I'm already involved in the planning of a possible mission to Io!"

TORUS

In December 1973, when Pioneer 10 was occulted by Io, the refraction of the radio signal at the limb-crossing indicated that the moon had an ionosphere. Astronomers had noticed that Io tended to brighten slightly immediately after emerging from Jupiter's shadow, and this was taken to mean that a gas was settling onto the surface during the chilly darkness and was then sublimating to gas upon the return of sunlight. Telescopic spectroscopy found a halo around Io, and the yellow line-emission indicated that this was from neutral sodium. A significant fraction of the sodium cloud is piled up in front of the satellite, and the rest trails behind. Further observations established that the halo extends all the way around Io's orbit, forming a torus of ionised gas – a plasma.

So, even before the Voyagers arrived, it was known that Io had a tenuous atmosphere of neutral sodium, potassium and magnesium. However, it was concluded that Io's surface was probably salty, and that charged particles circulating in Jupiter's magnetosphere – which was presumed to be bathing the moon's surface – were 'sputtering' neutral atoms from the crust. The real process by which the torus was sustained became evident only when Linda Morabito spotted the volcanic plume on Io's limb. The Voyagers saw ultraviolet emission from ionised oxygen, and from both singly and doubly ionised sulphur in the torus. The plasma density of the torus was evidently related to Io's volcanic activity, because although Voyager 1 measured a plasma temperature of 100,000 K, this had decreased to 60,000 K by the time Voyager 2 passed through.

The volcanic eruptions on Io spew tonnes of material into space. At 1 kilometre per second, Io's escape velocity is so low that the plumes can rise several hundred kilometres. Some material condenses, and 'rains out' onto the surface, but much of it escapes. Given a positive or a negative electrical charge, these particles are susceptible to Jupiter's magnetic field, which attempts to 'snatch' them. But the magnetic field and the newly erupted particles are moving at different speeds. The particles are moving with Io around its orbit once every 42 hours. Jupiter's magnetic field, on the other hand, is rotating with Jupiter once in approximately 10 hours. "Jupiter's magnetic field is moving at about 8 kilometres per second faster than the particles ejected from Io," Duane Bindschadler of Galileo's science planning and operations team, pointed out. "The result, is a lot of interesting scientific phenomena."

In fact, the situation is very complex, because it is not simply a matter of the magnetic field and the recently ejected particles. The particles ejected previously will have been progressively accelerated, and so will be moving along with Jupiter's magnetic field. All these particles, fast and slow, form the doughnut-shaped torus. As Lou Frank of the University of Iowa leader of the plasma wave spectrometer team summed up, the torus is "the beating heart of the Jovian magnetosphere".

The particles and fields data that Galileo took while passing through Io's torus was stored on tape, but the scientists were in for a long wait because it could not be transmitted until the data compression software was uploaded in early June 1996.

PERIJOVE

Galileo's trajectory took it well within Io's orbit, and deep within the intense radiation belt of the inner magnetosphere. Despite its shielding, there was a possibility that some of its solid-state electronics would be disabled, and possibly even permanently damaged.

In fact, during its swing around Jupiter, the spacecraft was expected to receive a dose close to 50,000 rads, or one-third of the total dose that it would accumulate during its entire two-year primary mission. "When we were building Galileo," said Matt Landano, the Deputy Mission Director, "we put in a lot of shielding to offset the expected effects of Jupiter's environment, but we won't know how well a job we did until we fly through it."

Galileo's closest approach to Jupiter occurred at 21:54 UT, some four hours after Io. This was at 4.0 Rj as measured from the planet's centre. The spacecraft passed 214,570 kilometres above the cloud tops. To a human pilot, the planet would have seemed the size of a basketball held at arm's length. However, 1,000 rads would be lethal to a human being. Galileo was not only going 'where no man has gone before', as the saying has it, it was going into a region of space where no human is ever likely to venture. Observations had been planned, but these too had been forsaken because of the problem with the tape recorder.

PROBE MISSION

Another surprising discovery by Voyager 1 was that Jupiter possesses a ring system. The rings showed up in an image that was taken just as the spacecraft crossed the equatorial plane to search for small inner moons. Indeed, three small inner moons were spotted. The ring was viewed virtually edge-on and, although it was inherently dark, it was visibly bright because it was forward-scattering sunlight. Primed by this discovery, Voyager 2 took a series of images which revealed the extent and structure of the ring system.

The rings comprise several narrow components, with the main one centred about

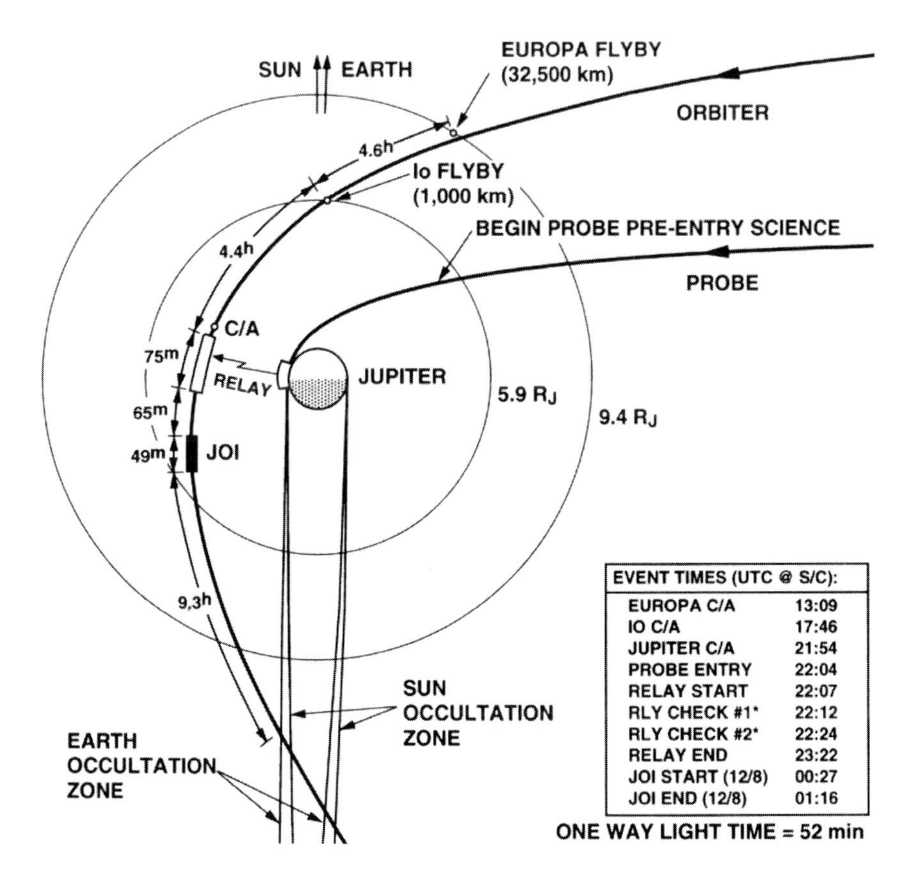

A diagram of the trajectories of the main Galileo spacecraft and its atmospheric probe, showing the sequencing of events upon arrival at Jupiter on 7 December 1995.

1.75 Rj from the planet. The light-scattering properties indicated that the rings are composed of small particles of dust – typically 2 microns across. Saturn's rings in contrast are composed of far larger chunks of ice, which is why they are so reflective. Although the main Galileo spacecraft would not fly close enough to Jupiter to encounter the ring system, the probe's trajectory had been selected to ensure that it would not pass through the dusty ring material on its way down to the planet.

The cruise timer awakened the probe's sequencer when Jupiter was still six hours distant, thus when Galileo crossed Io's orbit the probe was already active. The probe activated its energetic particle instrument three hours later to measure the fluxes of electrons, protons, helium nuclei and heavy ions in the electromagnetic environment down to 8,000 kilometres above the cloud tops, much closer to the planet than Galileo's particles and fields instruments would ever be able to sample directly. To save the probe's battery for the penetration of the atmosphere, this early data was stored in solid-state memory for subsequent transmission.

Ten minutes after Galileo made its perijove passage, the probe hit the planet's atmosphere. The probe's depth within the atmosphere would be measured in terms of the ambient pressure. Its 'entry interface' was defined to be 450 kilometres above the atmospheric level at which the gas pressure was 1 bar, which corresponded to 'sea level' on Earth.

A number of terrestrial telescopes had monitored the atmospheric features in the vicinity of the entry site during the final days of the probe's approach to Jupiter. The 3-metre reflector of NASA's Infrared Telescope Facility was able to observe longward of 5 microns when Jupiter was near conjunction, but it could not see anything worthwhile shortward of this wavelength. Combined with observations made by the 1-metre telescope at Pic-du-Midi, half a world away in the French Pyrenees, it was possible to form a near-continuous history of the probe's entry site.

Although the probe was 'falling' on a collision course, its trajectory would actually take it close around Jupiter's trailing limb, where the planet's gravity would whip it around so that it penetrated the atmosphere at a shallow angle. The mission's requirement for the probe's entry site was that it be between 1.0 and 6.6 degrees in latitude. It couldn't aim directly for the equator, because this would require the probe to pass through the dusty ring. Although it did not matter whether the entry site was north or south of the equator, it had long since been decided to aim for the most northerly permitted zone. A location on the dusk terminator was required so that the probe would follow the planet's rotation, because the heatshield needed to enter against the rotation would have comprised 90 percent, rather than 50 percent, of the probe's mass.

As predicted, the probe entered at the edge of an 'infrared hot spot' at 6.57 degrees north latitude and 4.94 degrees west longitude, close to the boundary between the bright Equatorial Zone and the darker North Equatorial Belt. It came in from the west, heading towards the dusk terminator, and made contact with the atmosphere at 22:04 UT. It had to penetrate in daylight to allow one of its instruments to measure how the light level varied with depth. It had to hit at 8.6 degrees. If it was 1.4 degrees shallower, it would bounce off the atmosphere and go back into space. If it was 1.4 degrees steeper, no heat shield would be able to save it and it would burn up. In fact, Galileo had established the probe's trajectory so accurately that it used only about 15 percent of the margin for error in its entry corridor. The attitude that the spacecraft had adopted for the probe's release had been to align it so that it would hit the atmosphere face on. When it hit the tenuous upper atmosphere at 170,000 kilometres per hour, the probe was the fastest-ever human artefact (at that speed, a flight from San Francisco to New York would take only a minute and a half).

A 340-kg blunt cone about 1.25 metres across and 0.86 metre high, the probe comprised the outer deceleration module and the inner descent module. The protective outer section of the probe accounted for half of the probe's mass prior to entering the atmosphere. The forebody of the heat shield was made of carbon-phenolic, and the afterbody was phenolic nylon. Phenolic heatshields were used extensively on re-entry vehicles for the Earth's atmosphere, but a return from Earth orbit involved entering the atmosphere at 8 kilometres per second. Galileo's probe would penetrate Jupiter's atmosphere almost six times faster than that. It sustained a punishing 230g deceleration. The shock wave which developed at the 'stagnation point' in front of the probe formed a 14,300 K plasma. The phenolic material glowed white hot, the surficial layer progressively vaporised, and the ablation carried away the heat.

Ames had constructed special high-speed arc-jet and laser facilities in which to simulate the conditions that the heat shield would have to endure, and the data had been fed into a computer model that 'ran' the entry so as to determine whether the heat shield would survive the passage. A 40 percent margin had then been added, just to be sure. It had been expected that 88 kg of the 152-kg forebody would be ablated. Sensors embedded within the phenolic revealed that it actually suffered considerably greater ablation, but its conservative design saw it through. All throughout the entry phase, the atmospheric structure instrument inferred the temperature and pressure in the near-probe environment using accelerometers which measured the deceleration forces.

After two minutes[10] of aerodynamic breaking – now some 400 kilometres below the entry interface, and with the probe having slowed to the speed of sound – a mortar was programmed to deploy the small drogue into the slipstream, and once the drogue had slowed the probe to 430 kilometres per hour the afterbody shield was to be released so that the 2.5-metre wide dacron main parachute could be deployed. Nine seconds later, what remained of the forebody shield was to be jettisoned and the descent module was to start *in situ* sampling of its environment, transmitting this data to the main spacecraft in real time.

In fact, although the plan had been to initiate sampling 50 kilometres above the 1-bar level, the deployment sequence began 53 seconds late, "apparently due to a wiring problem with the *g*-switches which told the probe when to deploy the chute," noted Bill O'Neil later, once the fault had been analysed.

Galileo was passing west to east, 215,000 kilometres above the probe, aiming its L-Band antenna at the probe's entry site.[11] As it dropped towards the probe's horizon, the spacecraft slewed this small dish to maintain contact with the probe. When it detected the probe's signal, Galileo sent a confirmation signal which took 52 minutes to reach Earth. Because the parachute deployment sequence was delayed, this signal came in about a minute later than expected – and this was a very long extra minute!

At JPL, it seemed as if the entire world's media had come to witness the unfolding of the day's events. Cherished replicas of previous robotic planetary explorers had been eased aside to accommodate the 1,000 invited guests. One of these was Donald Williams, the commander of the Shuttle mission that had given Galileo its ride into orbit. Administrator Dan Goldin was in the von Kármán Auditorium, which was standing-room only for the assembled engineers, scientists and media representatives.

"I got in and noticed that everybody was dressed up, but me!" recalled Claudia Alexander, who was one of many staff members assigned to explain to the reporters what was happening. Some of the reporters were unfamiliar with the mission concept. Some did not appreciate that the probe had been released in July; they were

The bell-shaped descent module of Galileo's atmospheric probe (top left) was encased within the conical deceleration module (top right) for flight. The deceleration module's task was to survive the aerodynamic braking of the entry phase. Once it had slowed below the speed of sound, the parachute was deployed (lower right) and the forebody heat shield was jettisoned (lower left). (Courtesy, Ames)

evidently waiting for confirmation that Galileo had 'dropped' the probe as it skimmed low over Jupiter's cloud tops, so that this would 'fall' into the atmosphere immediately below.

The mood in the Auditorium was "tense, but filled with anticipation", said Robert Gounley in an interview with the BBC. A deputy team chief of the Galileo Orbiter Engineering Team, he had devoted almost one-third of his life to the project.

As the time neared for the confirmation that the orbiter had received the probe's signal, the Galileo engineers and scientists lined up for the TV cameras in order that the world could share in their reactions to the signal, or to the absence of a signal. The radio wave carrying the news of Galileo's success was making its way Earthward at a paltry 300,000 kilometres per second.

The moment came – and went! Did this delay signify that something had gone wrong and that there would be no signal? Had the probe malfunctioned, and failed to transmit? Or had the probe's transmissions been missed by the main spacecraft? Or could it simply be a fault in the Deep Space Network? The engineers with the deepest appreciation of the task which had faced their automated spacecraft began to feel a little nervous.

As the clock ticked the seconds beyond the expected time, the tension in the von Kármán Auditorium rose palpably. Then a TV technician monitoring an external circuit shouted that the signal was in, and the room erupted in applause. It was a moment of triumph for the engineers at the Hughes plant in El Segundo in California, where the probe had been built.

"Of course," joked mission planning chief Jan Ludwinski, referring to the signal that the spacecraft had reported receiving, "we presume it was from the probe!"

Nobody knew why the signal had come in 1 minute late. For the moment, it didn't matter. What was important was that the probe had survived the atmospheric entry, was taking data, and Galileo was receiving it. The radio transmission ceased 61.4 minutes after entry interface. This included 57.6 minutes of atmospheric data.

The probe did not carry any imaging instruments, but six instruments were to measure the atmosphere in terms of pressure, temperature, cloud structure, lightning activity, composition, insolation from above, and thermal energy from below. The doppler on the probe's signal was measured by the main spacecraft to infer the probe's motion within the atmosphere. The rate of descent on the parachute was expected to decline from an initial 750 metres per second to about 27 metres per second in the deeper denser air.

The probe's 'baseline' mission was to sample for about 38 minutes, ranging from the 0.1-bar level to the 10-bar level, corresponding to a column running from 50 kilometres above, to 100 kilometres below the 1-bar level. The temperature was expected to be 110 K near the top of this column, and 335 K at the base. If the probe lasted 60 minutes, it would attain the 20-bar level at 134 kilometres, where the temperature would be 415 K.

When all the probe's instruments were functioning, its battery was sufficient for 75 minutes' operation – but there were several constraints. Firstly, just as radio waves cannot penetrate the deep oceans, the probe's signal would become increasingly attenuated as it descended into denser air. Also, as the main spacecraft flew towards the probe's horizon the low angle increased the 'air mass' through which the signal had to travel, further attenuating the signal. The key factor, however, was that Galileo would be able to receive data for only 75 minutes, because it would then have to turn away to prepare for the orbital insertion burn. In fact, the transmitter succumbed to the 425 K temperature at the 23-bar level, so it more than lived up to its requirement. If it had survived to its maximum 78 minutes it would have attained the 30-bar level at 163 kilometres, where the temperature would have been 465 K.

Bill O'Neil was "absolutely ecstatic" that the tremendously ambitious probing of an outer planet's atmosphere had been performed so successfully. "It's especially gratifying because so many of us have worked so hard for nearly two decades to get this first true 'taste' of Jupiter's atmosphere."

The inert probe continued to descend into the atmosphere. Its shell was not pressurised, so it could not implode. It undoubtedly succumbed to the rise in temperature that accompanied the increasing pressure. Its dacron parachute would have melted some 30 minutes after the end of the transmission, at a temperature of 530 K. The probe would then have fallen at terminal velocity and about 40 minutes later, at a temperature of 950 K and a pressure of 280 bars, the aluminium inner structure would have melted and leaked out from the titanium shell, which would itself have vaporised at a pressure of around 2,000 bars, where the temperature would have been some 2,000 K. After 9 hours, therefore, the probe would have become part of the atmosphere, slightly enriching it with heavy metals.

Galileo's probe is likely to remain a one-of-a-kind mission for the foreseeable future, since there are unlikely to be any such probes assigned to the atmospheres of the other outer planets.[12]

ORBITAL INSERTION

The probe's data having been received, Galileo now turned its attention to achieving orbit around Jupiter. It reoriented itself to aim the engine for a retrograde manoeuvre, then spun up to 10.5 rpm in order to even out any asymmetry in the thrust. When the engine had been fired in July, its performance had been calibrated. To attain the desired capture orbit, it would need to slow the spacecraft by 645 metres per second. The 400-newton engine would have to fire smoothly for almost 50 minutes.

If the engine failed to start, Galileo would sail on past Jupiter, as had the Voyagers, but in this case the problem with the tape recorder would mean that it would do so without taking any images! It would be able to replay the probe's transmission, but its own orbital mission would have been lost. If the engine malfunctioned, and Galileo was disabled,[13] then even this priceless data would be lost and the probe's fiery dive would have been for nothing.

In terms of the spacecraft's time, the 49-minute burn began at 00:27 UT on 8 December, but it was 17:19 PST on the momentous 7 December, Arrival Day, when the newly upgraded Canberra Deep Space Communications Complex received the signal confirming the ignition.

Galileo's star scanner had lost 'lock' on one of its guide stars for half an hour while passing through Io's torus, when it had been swamped with radiation. The star scanner had been operating as a back-up to the gyroscopic inertial reference, however, and the 'gung ho' computer had pushed on. For the insertion burn, it would keep the engine firing even if it underperformed, until the desired velocity had been achieved. In the event, the burn was perfect. It was within one-tenth of a percent of nominal. The trajectory was so good, in fact, that 'trim burn' options scheduled for 9 December and 2 January 1996 were not exploited.

At long last, Galileo's 3.7 billion kilometre flight through interplanetary space was over – it was now in orbit around its primary target. Jim Erickson was "ecstatic". After 16 years of work to follow up the Voyager fly-bys with a Jovian orbiter, this had been a "perfect" day. The only possible disappointment was that the thrust from the engine had not shaken loose the pins that had fouled the high-gain antenna. What a tremendous bonus it would have been if the big dish had finally unfurled!

JUPITER OCCULTATION

Less than an hour after entering orbit, Galileo crossed back out beyond Io's orbit. Shortly after that, it ceased sampling particles and fields data, having made a continuous record from one side of the inner magnetosphere to the other.

Nine hours after orbital insertion (at about the time that the last vestiges of the probe were blending into the Jovian atmosphere) Galileo passed behind Jupiter's trailing limb, as viewed from Earth. An hour earlier, it had boosted the strength of its radio signal. The Madrid Deep Space Communications Complex sampled the signal strength thousands of times per second to measure how it faded. Three and a half hours later, Galileo emerged from the leading limb and another set of data was secured. "It was beautiful," reported Randy Herrera, the coordinator for the radio science team. "The spacecraft signal just disappeared as predicted."

Knowing the spacecraft's trajectory in space, it would be possible to derive a 'refractivity profile' for the occultation. Different mixtures of gases have different refractive effects on the radio signal, so the observed profile could be used to refine the model of chemical composition, temperature, and pressure as functions of depth for the part of the atmosphere through which the signal passed.

BINGO!

In excess of 2,000 people packed into the von Kármán Auditorium for the wrap-up Press Conference. The Galileo 'home page' on the Internet had been read at least 250,000 times during the 'day' which had run from 06:00 PST on 7 December with the Europa fly-by, through 25 hours to the re-emergence from the occultation after orbital insertion. The most complex and the riskiest part of the entire mission was over.

Overnight, after months of criticism in the Press for its problems, Galileo was "the little spacecraft that could". A jubilant Dan Goldin roamed JPL's corridors exchanging 'high fives' with all and sundry. "Is this a great day, or what!?"

"Every Galileo achievement that we have shared with the world is the result of teamwork," said Neal Ausman, the Mission Director. "The team has time and time again had to overcome problems thought to be catastrophic."

In a sense, though, even before it could start, Galileo's orbital mission was fully 10 years late because if it had been launched in 1982 as initially planned it would have arrived in 1985 after a three-year flight involving a Mars fly-by and its mission would have been finished long before it was actually launched. The remarkable thing was that so many of the original Galileo team members were still players.

"Over the past 18 years," observed Lou D'Amario, the deputy chief of the navigational team, "I have designed and evaluated a tremendous number of interplanetary trajectories for Galileo. There were launch dates as early as 1982 and as late as 1989. The Jupiter arrival dates ranged from 1985 through 1995, and beyond. So many trajectories! Now it's over, the interplanetary phase of the mission is completed."

Arrival Day had been "stressful", admitted Ralph Reichert, the engineering office manager. "I've been working on this project for 17 years. Everything I've done has been in anticipation of this day; this critical event. I'd like to just bask in the glory for a while, but we have lots of work still to do."

"I'm only beginning to realise the magnitude of this success," admitted Greg Harrison, the technical group leader for power management, as people began to relax. But he acknowledged that while the engineering activities would continue without let up, the time was coming when the engineers would have to "pass the torch" to the science teams.

"Galileo's success today is truly a triumph of the human spirit and creativity," summed up Bill O'Neil. "Eighteen years of dogged tenacity, and imaginative engineering solution to some of the toughest technical and political problems ever faced by a project, finally paid off."

A large wall display in the corner of JPL's main plaza was regularly updated to indicate the positions of the robotic planetary spacecraft. The track for Galileo finally made contact with the circle representing Jupiter's orbit. By the end of the day, this intersection point had been adorned by an extra sign that said simply, "BINGO!".

7

Atmospheric probe

THE DATA

Only the first 40 minutes of the atmospheric probe's 57-minute transmission had been able to be stored in the limited amount of solid-state memory available as a back-up measure in case the tape recorder malfunctioned again.

Despite the fact that Jupiter was just two degrees from the Sun, as viewed from the Earth, Galileo immediately transmitted the data stored in its memory. Using the low-gain antenna at a rate of 10 bits per second, the replay was not completed until 14 December. Only when the data began to be received were the engineers able to confirm that all of the probe's instruments had actually functioned and reported data. Frustratingly, some 'frames' of the transmission were corrupted by passing through the solar corona. This would complicate the task of interpreting the data stream, which had been highly encoded by the compression process. Once the data had been transmitted, the spacecraft went into hibernation for the fortnight that it was to be out of contact. Conjunction also gave Galileo's hardworking engineers a welcome break.

For the Ames probe scientists, however, the successful receipt of their data promised lots of work ahead. In fact, while the data was in transmission many of the scientists attended the winter conference of the American Geophysical Union, which was held in San Francisco. On 18 December, they organised an informal meeting of the Project Science Group to preview the overall data, before setting off to process the results of their own individual instruments. Everyone was ecstatic with the quality of the data. For the principal investigators who had been waiting for two decades, this was a memorable day. As soon as Jupiter emerged from its conjunction, Galileo retransmitted the solid-state memory data, both to recover those parts which had been corrupted by solar interference and to verify the fidelity of the rest.

Carl Sagan had played a major role in steering Galileo through its protracted gestation. The probe's data was safely in, but he was looking forward to the tour of the moons. "It's sort of like your birthday – your parents have permitted you to see the pile of presents, but you can't open them yet; you don't know what's in them yet, but you know it's going to be great."

The experiment teams had agreed that they would not report their results until a joint Press Conference could be held. The so-called 'quick look' findings were presented at Ames on 22 January 1996.

"The quality of the Galileo probe data exceeds all of our most optimistic predictions," said Wesley Huntress, who was Associate Administrator of Space Science at NASA Headquarters. "It will allow the scientific community to develop valuable new insights into the formation and evolution of our Solar System, and the origins of life within it."

Nevertheless, the scientists warned that their analyses should be considered to be tentative until the tape-recorded data had been recovered and the engineering parameters it supplied had confirmed the calibration of the instruments. The tape was replayed between late January and mid-April (there were several lengthy breaks for engineering activities). The final results were announced in mid-March, and published on 10 May in the journal *Science*.

WITHIN THE RINGS

When the probe was three hours out from Jupiter, the energetic particle instrument – the only instrument that was not intended for atmospheric sampling – switched on to measure the fluxes of electrons, protons, helium nuclei, and heavy ions in the electromagnetic environment all the way down to within about 8,000 kilometres above the cloud tops, considerably deeper into the magnetosphere than even Pioneer 11 had ventured.

The gap between the inner edge of Jupiter's ring and the atmosphere had been expected to be fairly quiescent, but at a distance of 50,000 kilometres, the probe discovered a belt of intense radiation with a particle density ten times stronger than that of the Earth's Van Allen belts. Another surprise was the detection of energetic helium ions within this belt. This inner radiation belt might help to explain the high-frequency radio emissions which emanate from Jupiter.

THE ENTRY PHASE

Although the pressure at the entry interface was only 50 billionths of a bar, the probe was travelling at about 48 kilometres per second, so it experienced extreme aerodynamic braking.

Throughout the entry phase, the atmospheric structure instrument used the build up of the deceleration to infer the ambient density, pressure and temperature. Atmospheric density was proportional to the deceleration. As the pressure at a particular level in the atmosphere was equal to the accumulated material in the air column above, the atmospheric pressure could be derived from the vertical profile of the density. Later, after the other instruments had provided the chemical composition of the atmosphere, it would be possible to derive temperatures from the densities and pressures.

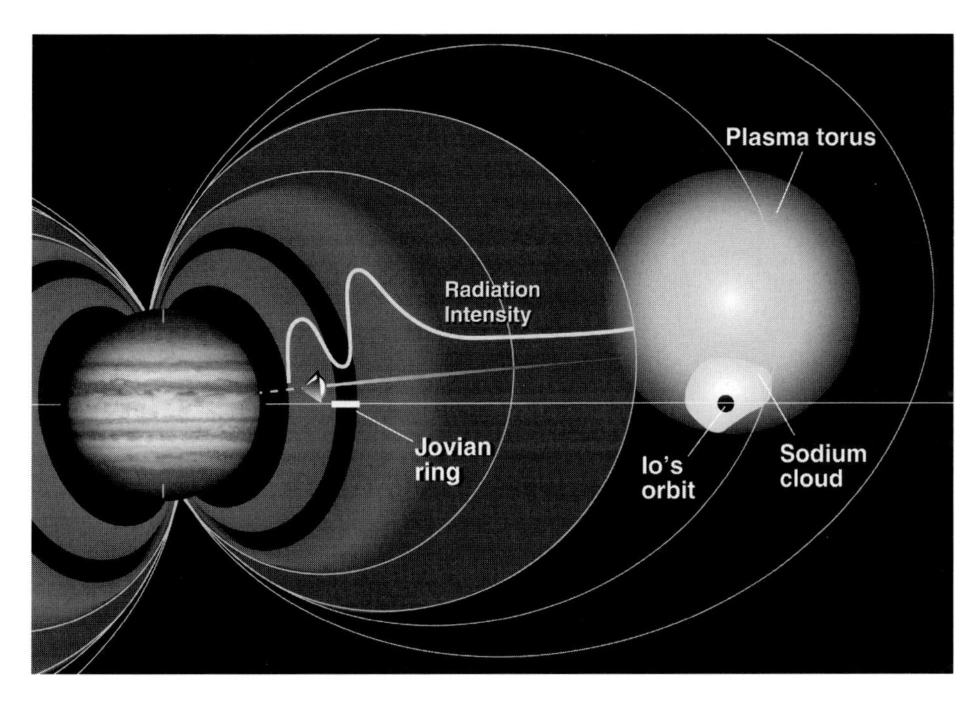

Once Galileo's atmospheric probe was within Io's orbit, it started its energetic particles investigation, measuring the radiation in the inner regions of Jupiter's magnetosphere. In this illustration, the diameter of the planet and the distances and sizes of the features are depicted in proportion and the 9.5-degree obliquity of the magnetic field axis with respect to the planet's rotation axis is shown. The measured radiation intensity clearly indicated the presence of two radiation belts (regions of trapped radiation), the inner of which had not been previously known. It contains energetic helium ions of unknown origin, and the radiation within it is 10 times stronger than the Earth's Van Allen belts. (Courtesy, Ames)

It turned out that 100 kilometres beneath the entry interface, the density of the atmosphere was 100 times greater than expected, and at 500 K it was significantly warmer than could be accounted for by insolation alone. Undulations in the temperature profile suggested that the heating process might involve 'atmospheric waves'.

BULK COMPOSITION

Apart from the Sun, Jupiter is the most significant body in the Solar System because it contains more mass than all the other planets combined. A key question that the probe was hopefully to resolve was how close was Jupiter's constitution to the solar nebula from which the planets formed? In other words, was the composition 'primordial', or had it been significantly 'enriched'?

"In some senses," noted Richard Young, one of Ames's leading probe scientists,

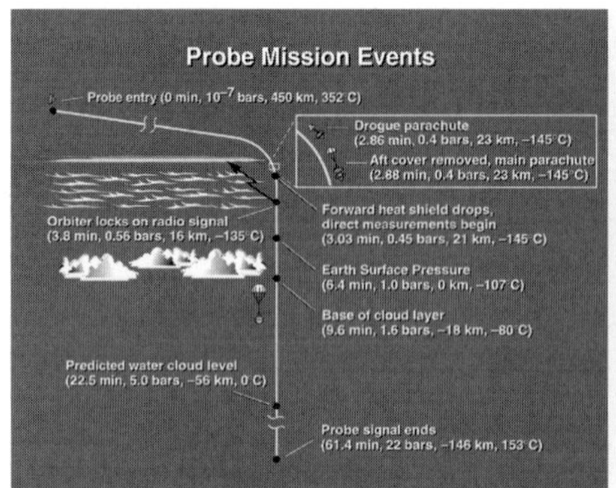

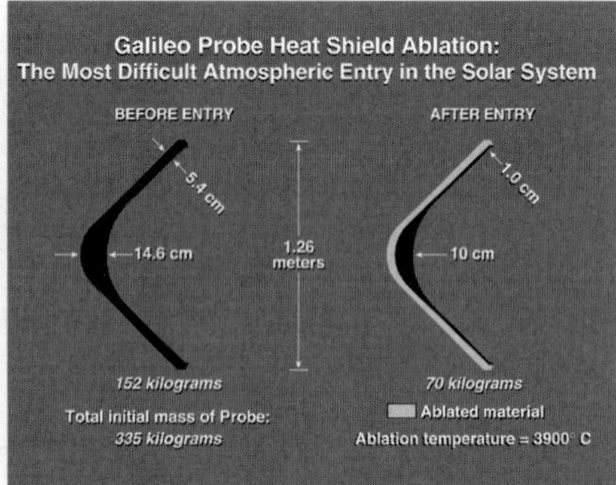

A diagram of the times, atmospheric pressures, altitudes, and temperatures at which the key events of the probe's mission took place. Altitudes are relative to the 1 bar pressure level. Times are relative to crossing the 'entry interface' 450 kilometres above the 1-bar reference, at 22:04:44 UT on 7 December 1995. The deceleration module's carbon-phenolic forebody heat shield suffered considerably greater ablation than expected, but it had been conservatively designed and it was able to protect the descent module during the 2-minute entry phase, entering the atmosphere at 170,700 kilometres per hour, with the frictional heating rising to a temperature of some 14,000 °C. (Courtesy, Ames)

Jupiter "is the Rosetta Stone of early Solar System development." The hydrogen-to-helium ratio was key to the theories of planetary evolution.[1]*

Jupiter is predominantly composed of hydrogen – as indeed is the Universe itself – with helium forming the second most abundant element. This was well known from spectroscopic remote sensing, but the probe's *in situ* measurements were to determine the precise fraction of helium. To do so, it had two instruments, the helium abundance detector and the neutral mass spectrometer, which measured the relative abundances of oxygen, carbon, neon and nitrogen, which were, in order, the next most abundant elements in the Universe. Considered together, the ratios of these elements would characterise Jupiter's formation and evolution. On the assumption that the Sun represents the nebula from which the planets formed, solar abundances were used as the points of reference.

Rupert Wildt had identified ammonia and methane in Jupiter's atmosphere in 1932. A few other species were detected in later years, and remote-sensing by the Voyagers had compiled a comprehensive survey. Although Jupiter's 'bulk' composition was expected to represent the primitive nebula at the location at which the planet formed, the Voyagers had established that methane was present in three times the solar abundance, for example – as a carbon–hydrogen compound, this represented carbon enrichment. But all of these studies had been restricted to the top of the atmosphere. Galileo's probe would add the third dimension to the sample.[2]

* For Notes and References, see pages 413–424.

In the Voyager era, it used to be thought that Jupiter formed by condensing directly from the solar nebula. However, the 'nucleation and collapse' models developed afterwards posited that a solid core formed first as a result of accretion of icy and rocky planetesimals, and once this had acquired sufficient mass it drew in gas from the solar nebula to form a large gaseous envelope. As the core converted gravitational potential to heat, its volatiles would have 'boiled out' and enriched the envelope. Additional species would have been infused by later infall of planetesimals and comets. Jupiter was therefore expected to be enriched over solar abundance in a very specific manner.[3] Hydrogen, helium and neon were all expected to match their solar abundances because they would have been derived from the nebula during the collapse phase. Nitrogen should be slightly enriched, and carbon, sulphur and oxygen should be significantly enriched from planetesimals. The heavy noble gases argon, krypton and xenon were expected to be insignificant, because they would not have condensed out of the solar nebula where the planet currently orbits.

In the 'quick look' data, the helium abundance detector implied that the helium abundance was 14 percent. Because this was much less than the expected value, theorists proposed that Jupiter had undergone more extensive internal evolution than had been thought. Specifically, they proposed that helium had gravitationally settled towards the interior – a process referred to as 'fractionisation'. In effect, the helium would have formed droplets and 'rained out'. The problem with introducing this process for Jupiter was that it conflicted with the clear evidence that Jupiter's core was hot, because the intense convective mixing would have inhibited such settling out. Saturn seems not to have such a strong internal heat source, there does appear to have been fractionisation, and its helium abundance is 6 percent. However, as Saturn has barely one-third of Jupiter's mass, its interior is not so hot and is not so convective. If Jupiter had undergone fractionisation, the models of planet formation would have to be substantially revised. When the calibrated data became available, however, the 'helium anomaly' vanished. The Jovian abundance had been expected to be be slightly less than the 25 percent solar value, so the final value of 24 percent was in excellent agreement with theory. The calibrated neutral mass spectrometer data confirmed the near-solar helium abundance.

"The revised helium abundance also indicates that gravitational settling of helium towards the interior of Jupiter has not occurred nearly as fast as it apparently has on Saturn," explained Young. "It may force us to revise the projection for the size of the rocky core believed to exist in the deep centre of Jupiter."

Jupiter is sufficiently massive for its gravitational field to have retained light elements (and their compounds) acquired from the solar nebula. The composition of less massive planets has been modified as these 'volatiles' boiled off, both during the planet's formation and during its subsequent evolution.

A consequence of increasing the helium ratio was a raising of other minority constituents, and the removal of several other surprises in the 'quick look' data. This was the case because 'correcting' the helium abundance had the effect of increasing the ratio of methane (measuring carbon), ammonia (nitrogen) and hydrogen sulphide (sulphur). As expected, all these 'heavy' elements exceeded their solar abundances. "This implies that the influx of small bodies such as comets into Jupiter over the aeons since its formation has played an important role in how the planet has evolved," confirmed Young.

The neutral mass spectrometer revealed the noble gases argon, krypton and xenon all to be present in much greater abundance than solar values, implying some form of enrichment. This was a major surprise, because it takes extremely chilly temperatures – far colder than should have been extant in the solar nebula at the point that Jupiter orbits – to condense these highly unreactive gases. Their isotopic ratios were near-solar, however. It was simply that there was an excess of the elements. An "agonising spectroscopic analysis", as Thomas Donahue of the University of Michigan at Ann Arbor reported,[4] confirmed that argon, krypton and xenon are present at two to three times the abundances that would expected if Jupiter had formed solely from the solar nebula. Furthermore, there was three times as much nitrogen as expected under prevailing models for how the Solar System formed.

"This new information might shake up our views of how the Solar System formed," noted Tobias Owen of the Institute for Astronomy at the University of Hawaii. "To catch such gases, Jupiter had to trap them physically, either by condensation or by freezing."

Most comets now populating the Oort Cloud are believed to have formed out about 30 AU from the Sun – the space now occupied by Uranus and Neptune – where the temperature of the solar nebula should have been in the range 45 to 65 K.[5] However, even this is 'warm'. The condensation of heavy noble gases can only take place in an extremely cold environment, at about 30 K. So, if Jupiter accreted planetesimals which formed at 30 AU, their internal temperature would have been too warm to have built up such a concentration of heavy gases.

Planetesimals formed in the Kuiper Belt – out beyond 40 AU – might have incorporated such noble gases, but to have been accreted by Jupiter they would necessarily have been in elliptical orbits, and they would have been heated as they strayed inwards the Sun and vented these volatiles. "This raises some intriguing possibilities," Owen added. "One theory, is that Jupiter was formed out in the region of the Kuiper Belt and was dragged to its present position." So, how could Jupiter have formed further out from the Sun than we see it today?

A spate of recent detections of extra-solar planets has provided a clue. As the planets formed by accretion from the solar nebula, the nebula would have acted like a fluid, and hence imposed a 'drag' that would have made the planets spiral in towards the Sun. The positions of the planets today correspond to where they were when the Sun 'blew off' its residual nebula, thereby removing the drag force.

"If Jupiter had migrated inward," reported Sushil Atreya, the Director of the University of Michigan's Planetary Science Laboratory, "it would have to have come from way out there, at 40 or 50 AU." For the first time, theorists began to study the consequences of Jupiter forming far from the young Sun. "It's either that, or the solar nebula was much, much cooler than the models have estimated."

Another intriguing possibility was that the planetesimals began to develop in the progenitor interstellar cloud of cold gas and dust even before it collapsed to form the solar nebula. "That would make these icy materials older and more 'primitive' than we expected," Owen explained.

"We certainly would like to know," Donahue pointed out, "whether this same thing is true of Saturn, Uranus and Neptune?"

As Richard Young of Ames put it, determining the composition of Jupiter's atmosphere – which needed *in situ* sampling – "was a primary scientific objective of the probe, because we knew it could change our understanding of Jupiter's formation and evolution." Indeed it had. Also, as Torrence Johnson had predicted, "The most exciting discoveries to come from Galileo will be those we have not thought of."

ATMOSPHERIC CIRCULATION

The latitudinal banding of the atmospheric circulation system is associated with jetstream winds. Long-term observations by the Hubble Space Telescope had established that these winds were very stable, both in terms of their positions and strengths, despite the turbulence in the shear zones and the presence of drifting structures such as Great Red Spot. How deep did these winds penetrate?

The doppler wind experiment used the probe's radio transmission (in conjunction with the other instruments) to infer wind as a function of depth. The probe's motion during the entry phase had to be inferred from knowledge of the trajectory. It could not be measured, because the probe did not start transmitting until it had deployed its parachute. In any case, the ionised plasma sheathing the probe during the entry phase would have inhibited a radio signal. It was necessary to establish the location and velocity of the orbiter during the relay, the location and velocity of the probe during the entry phase, and the doppler on the probe's signal. Allowance had also to be made for Jupiter's rotation. The atmospheric structure instrument's data during the entry phase was crucial to the doppler wind experiment. The temperatures, pressures, and densities deduced from accelerometers were later fed into a computer model based on the gas law and the law of hydrostatic equilibrium to derive the distance that the probe travelled from the point of entry to where it transitioned to a vertical descent and switched on its transmitter.

The objective was to determine the depth at which the strong surface winds diminished. It turned out that – at 540 kilometres per hour – the winds below the cloud tops were not only somewhat stronger than expected, but they persisted through the probe's descent, and *increased* slightly towards the end. When the calibrated data was replayed from the tape recorder, the measurement for the deep wind speed was increased to 640 kilometres per hour. Furthermore, the deep atmosphere was intensely turbulent. It seems that the Jovian circulation comprises an extremely deep jetstream system.[6]

The Earth's weather system is driven by differential solar insolation and by the latent heat that is liberated by the condensation of water vapour – that is, because the Sun is higher in the sky at lower latitudes, more solar energy is absorbed near the equator than at higher latitudes. The Earth has a system of high altitude jetstreams, but the tropospheric weather is dominated by cyclonic systems – swirling storms.

The fact that the winds remained strong below the depth to which sunlight could penetrate indicated that the Jovian weather system is driven by heat from the planet's interior. That is, either heat left over from the planet's formation is leaking from the interior, or perhaps Jupiter is still contracting and is transforming gravitational energy into heat. By taking our view of the Jovian atmosphere into the third dimension, the probe provided context to interpret the surface phenomena that we have only been able to observe from above for hundreds of years.

Since all of the giant planets of the outer Solar System display strong winds in their outer atmospheres, the revelation that Jupiter's weather system is driven by internal heat will assist comparative planetary meteorologists to refine models of Jupiter's smaller and colder relatives. Furthermore, the marked differences between the Jovian and terrestrial weather systems may also provide insight into our own planet's atmosphere.

Atmospheric stratification is determined by the vertical temperature profile, and it provides an important measure of how easily an atmosphere can be mixed, vertically. In the case of the Earth, there is a well-defined interface between the troposphere and the stratosphere, called the tropopause, and there is little mixing over this boundary, which is why aerosols derived from dust and chemicals injected into the stratosphere by volcanoes are so persistent.

The atmospheric structure instrument found that Jupiter's deep atmosphere is easily mixed. That is, because Jupiter's atmosphere is 'neutrally stratified' the heat escapes from the interior principally by convection, with the vertical cycling of the gas carrying the heat to the surface, where it is radiated to space. The tropopause can be thought of as being the level at which the temperature reaches a minimum. For Jupiter this is near the 0.1-bar pressure level. Above this level, in the stratosphere, convection ceases.

Furthermore, methane absorption of sunlight in the zone above the cloud tops warms the stratosphere, so that the temperature actually rises with altitude above the tropopause. This is referred to as an 'inversion'. The Earth's stratosphere shows a similar thermal profile, where ozone absorbs sunlight and reverses the decline in temperature with altitude that characterises the troposphere. The inversion in the Jovian atmosphere is far more pronounced, however.

"Jupiter is more like the Earth than we thought," reflected Andrew Ingersoll, "but Jupiter is also so much more! From its broiling, roiling, bottomless depths; through multiple cloud decks to frigid aurora-lit heights; through endless cloud canyons, searing dry voids, and centuries-long storms, Jupiter is an ideal meteorological laboratory."

CLOUD STRUCTURE

Not even the high-resolution Voyager imagery had been able to see through the outermost layer of cloud to discern what lay below, but it was believed that there would be distinct layers of clouds with compositions appropriate to the increasing temperature and pressure with depth.

The atmospheric scientists were hoping to confirm some basic assumptions, and

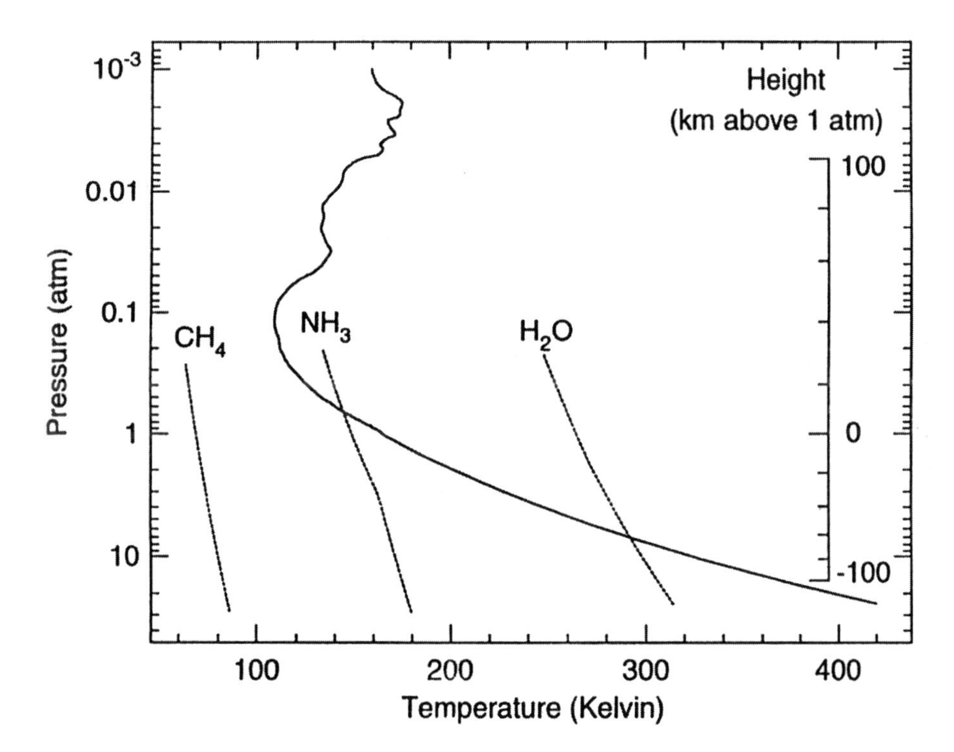

This thermal profile of the Jovian atmosphere maps temperature as a function of pressure. The tropopause at 0.1 bar marks the transition from a convective troposphere to a radiative stratosphere. A height scale is indicated, with the zero of height arbitrarily placed at the level where the pressure is 1 bar. For pressures of less than 1 bar, the temperatures are from data by the Voyager radio occultation measurements. For pressures greater than 1 bar, the temperatures are from Galileo's probe. Clouds are expected to form at levels corresponding to where the freezing lines intersect the temperature profile. Water and ammonia are shown (but ammonium hydrosulphide is not). The Jovian atmosphere is not cold enough to condense methane. (Courtesy Don Banfield, Cornell University)

to use the probe's data to select between alternative theories where there was debate. It was expected that the probe would initially descend through a frigid brown hydrocarbon aerosol haze just above the top of the white ammonia ice cloud. Once through the ammonia cloud, it was expected to drop through a thin layer of ammonium hydrosulphide cloud, then through a layer of water-ice crystals leading to a billowy water-rich cloud. After about 30 minutes (by which time the pressure would have risen to 8 bar and the probe would be at a depth of 80 kilometres) it was expected to drop out of this cloud deck into clear weather beneath.

The parachute deployment sequence triggered 53 seconds later than planned, so the probe began directly sampling its environment at the 0.35-bar level instead of at the intended 0.1-bar level. The atmospheric structure instrument promptly switched to direct sampling, but the late deployment meant that its profile of the ambient temperature and density began 25 kilometres lower than intended, at 25 kilometres above the 1-bar reference.

The nephelometer deployed a small arm and shone a laser across the 'sample gap', and used an array of mirrors to measure the scattering properties of the particles passing between the probe's body and the end of the arm. It was to establish the vertical structure of the clouds in terms of opacity and particle characteristics. The opacity was lower than predicted; that is, the visibility at the probe's entry site was better than expected. However, this could be explained by the fact that infrared telescopic observations had indicated that the probe would hit the fringe of one of the least cloudy parts of the atmosphere.

The net flux radiometer was so-called because it measured the difference between upward and downward energy flows (that is, the 'net flux') of insolation and thermal 'leakage' from the planet's interior. Its measurement of insolation as the probe swung and rotated beneath its parachute established the general cloud structure in terms of their optical thickness and the vertical separation of layers of cloud. This was in contrast to the nephelometer, which sampled the probe's immediate environment. It detected breaking through one expansive cloud deck. The Sun was very low on the horizon, and above the 0.6-bar pressure level the sky was brightest towards the Sun. The initially substantial variations in brightness with azimuth abruptly ceased when the probe passed through 0.6 bar. Within this cloud layer, the sky brightness became uniform. The 0.4- to 0.6-bar range corresponded to a temperature range of 125 to 145 K, some 15 to 20 kilometres above the 1-bar reference, which suggested that the probe had penetrated the well-observed topmost layer of ammonia cloud.

However, the nephelometer did not detect any cloud particles when the net flux radiometer reported the abrupt change in lighting. In fact, the nephelometer did not detect any thick, dense clouds at all, only very small concentrations of cloud and haze. The fact that the nephelometer did not see anything specific when the net flux radiometer noted the change in lighting implied that although the probe had descended beneath the ammonia cloud, it had not actually passed through it. Evidently, the ammonia cloud layer was patchy, and the probe had fallen through a gap. The nephelometer's only unambiguous observation of a tenuous cloud was at 1.6 bars, about 45 to 50 kilometres deeper. The 200 K temperature implied that this feature was ammonium hydrosulphide. The fact that the net flux radiometer did not see any change in lighting after the nephelometer had detected the probe's passage through this cloud implied that the ammonium hydrosulphide layer did not darken the zone below, because it was very thin. The nephelometer detected traces of cloud particles at a wide range of altitudes, but nothing dense enough to be classified as a cloud. The fact that the net flux radiometer reported a smooth variation in the solar flux below the 0.6-bar cloud deck suggested that solar heating penetrated to considerable depth – but then, it was a particularly clear site.

The net flux radiometer's data on thermal flux as a function of depth provided two crucial pieces of evidence pertaining to atmospheric composition. Firstly, the higher than expected net flux above the 1-bar level suggested that ammonia gas, which would have absorbed emergent infrared, was not as concentrated as expected, and this reinforced the conclusion that the layer of ammonia cloud was thin and patchy. Secondly, the thermal flux deeper down was greater than expected, indicating that water vapour, which would also have absorbed infrared, was conspicuous by its absence. Any water present would be 60 to 80 kilometres below the 1-bar reference, in the 5- to 8-bar range. Although Jupiter is a gas giant, and its envelope extends for thousands of kilometres, in terms of the probe's descent this pressure range can conveniently be considered to be the 'deep' atmosphere. This was consistent with the atmospheric structure instrument's temperature profile, which indicated that the atmosphere was remarkably free of condensation in the 6- to 15-bar pressure range, corresponding to 100 to 150 kilometres below the reference level. Initially, the absence of dense water vapour cloud was viewed as a major contradiction of the anticipated triple-decked cloud structure, but the site turned out to be anomalously dry by virtue of being an 'infrared hot spot'.

"The probe has sparked a lively worldwide scientific debate about the theories of planetary formation, and about the internal mechanisms in the huge Jovian atmosphere," Richard Young observed at the American Geophysical Union meeting in Baltimore in late May 1996.

The surprisingly arid nature of the entry site prompted discussion of whether this location was typical. The issue of water depletion could be resolved by saying that the sample site was representative and that – against expectation – the Jovian atmosphere was arid. This, in turn, suggested either that Jupiter had not accreted as many icy planetesimals as was believed or that the infall had taken place but the water liberated was not circulating in the atmosphere. Theorists revel in 'what if' to explain unusual observations, and so a variety of theories were soon offered for debate.

One theory reconsidered how Jupiter formed, and posited that the water had been trapped in the deep interior at an early stage. As gravitational energy was liberated, it was argued, and the core boiled off volatiles such as ammonia and hydrogen sulphide – as required to explain the observed enrichment of carbon, sulphur and nitrogen – the water was somehow retained. "There are problems with this view," wryly pointed out Tobias Owen, "the primary one being, why the ice remained in the hot core while the carbon-containing gases escaped."

Such theories notwithstanding, however, it was far more likely that the sampled site was a meteorological anomaly, and the surrounding atmosphere was as wet as the models predicted.

"We need more probes!" proclaimed Andrew Ingersoll lightheartedly. In fact, the only way to resolve this issue was for the Galileo orbiter to seek convincing evidence of water within the deep atmosphere using its remote-sensing instruments.

To peek ahead with this aspect of the story for a moment, as Galileo made successive perijove passes it became clear that Jupiter did indeed possess a richly complex and dynamic weather system. Cued by telescopic observers, Galileo studied a number of 'infrared hot spots' similar to the one the probe had entered. Water vapour strongly absorbs infrared (which is why infrared telescopes are located on high mountains above most of the troposphere). The infrared spectrometer could detect this absorption feature, and yield maps of water distribution in the Jovian atmosphere. The hot spots really were arid, but there were also vast tracts of the deep atmosphere which were saturated with water vapour. The infrared spectrometer eventually proved that water abundance varies by two to three orders of magnitude from location to location.

The oxygen in Jupiter's atmosphere was presumed to be present mainly as water vapour. The Voyagers had suggested an oxygen abundance of twice the solar value. The 'atmospheric waves' which the impacting fragments of Shoemaker Levy 9 had generated had suggested an abundance as much as ten times the solar value. The calibrated data from the probe's neutral mass spectrometer determined the oxygen abundance to no more than 15 percent of the solar value, but this was clearly not representative of the atmosphere as a whole. The probe's entry site was indeed a meteorological anomaly.

Prior to Galileo, the leading theory was that, like the Earth's weather system, atmospheric circulation on Jupiter was driven by differential insolation and the latent heat of condensation, and so it would be confined to a shallow zone in the upper atmosphere. The probe established that the winds extended to a depth of at least 150 kilometres and that the deep atmosphere was highly convective. A rising column of heated moist air cools with expansion and clouds of rain condense out. On Earth, this condensate is water. But on Jupiter the rising air also condenses out ammonium hydrosulphide as a higher, coloured, 'whispier' cloud, and then a layer of ammonia at the top. By the time the rising air emerges from the ammonia cloud layer, all the volatiles will have already condensed. Galileo's remote sensing confirmed the air above these updrafts to be very cold, very clear, and very dry. As this air flow 'turns over' north or south it is deflected by the Coriolis effect, sustaining the alternating latitudinal atmospheric circulation. As the air starts to settle it is compressed, and is heated. The descending cloud-free air appears visually dark, but because it is arid it is transparent at 5 microns, so downdrafts appear bright in the near-infrared and offer a 'window' into the deeper hotter region.

How deep did the downdrafts penetrate? "Considering that Jupiter emits more heat from its interior than it receives from the Sun," pointed out Andrew Ingersoll, "this upflowing interior should block a downflow of dry air." The infrared observations indicated that the downdraft at the probe's entry site 'soaked up' volatiles upon reaching a 'hot mixing zone' several hundred kilometres down, and was recycled into an updraft. However, it was possible that the system was even more dynamic. The strength of the winds in the deep atmosphere "may be evidence that Jupiter has high-speed wind currents extending thousands of kilometres deep into its hot, dense, atmosphere," Ingersoll speculated. By penetrating the region below the level to which sunlight can reach, the probe sampled the upper part of what may be a fairly uniform 'interior atmosphere' in which one continuous circulation pattern extends from the visible surface to a depth of about 16,000 kilometres. This convection would be driven by heat leaking from the interior. By the time it halted, underwent mixing, and was recycled, such a downdraft would have penetrated far beneath the level observable at 5 microns. A planetary convection cycle on this scale may yield alternating zones of rising air super-saturated with water, and descending arid air.

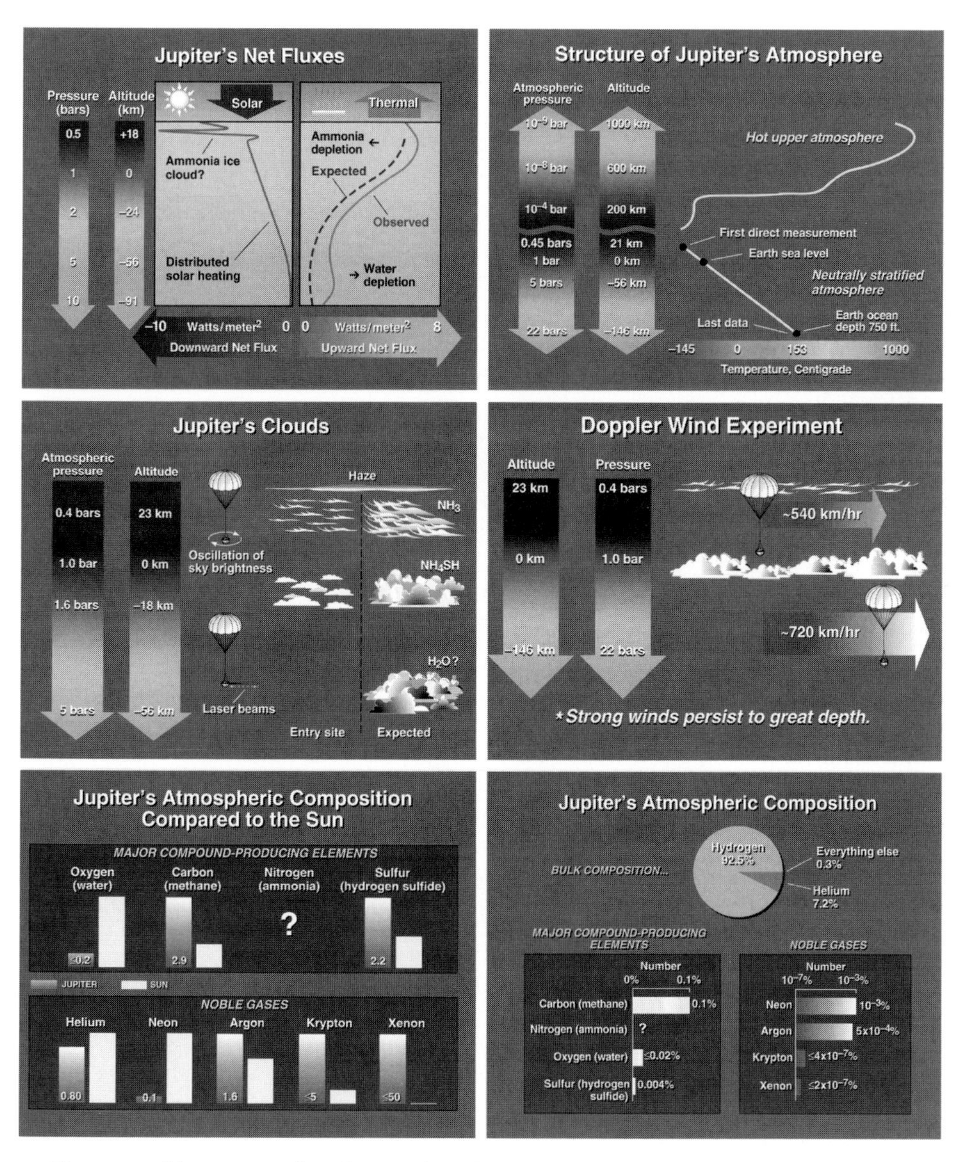

These graphics summarise the results of the probe's various atmospheric studies. (Courtesy, Ames)

"Perhaps Jupiter's interior heat comes out only in certain regions, where ascending currents bring up hot material from the planet's interior," offered Owen, "like the heat escaping from the Earth's interior" – which surfaces at volcanoes and at mid-ocean spreading ridges. "More of Jupiter's interior heat is also emitted at high latitudes," Young reflected. "Unfortunately, at the moment, we cannot put all of this into a [single] mechanism."

DISTANT LIGHTNING

The Voyagers observed brilliant flashes of lightning in Jupiter's atmosphere – much brighter flashes than occur on Earth – and the activity seemed to be continuous. Although when averaged out over time terrestrial lighting is evenly distributed geographically, Jupiter's appeared to be confined to high latitudes.

Galileo's probe carried a lightning and radio emission detector. This was a dual instrument designed both to register optical flashes in the vicinity of the probe and bursts of radio energy from more remote events. It was expected that radio signals reflecting off the underside of the ionosphere – an extension of the upper atmosphere in which free electrons circulate – would be able to propagate for tens of thousands of kilometres.

On Earth, lightning indicates thunderstorm activity in which there is water precipitation in strong updrafts zones. We live on the ground, so we tend to think of lightning in terms of the 'earthing' ground strike, but discharges between clouds are far more common.[7] Jupiter has no solid surface, so lightning there would be from one cloud to another.

Because there had been expected to be a thick deck of water cloud in the deep atmosphere, Jovian lightning was expected to be extensive but there was "a veritable absence of lightning", as Richard Young put it. The probe did not 'see' any flashes, but it 'heard' the radio bursts – some 50,000 in fact. The radio characteristics suggested that the lightning was about 12,000 kilometres away – a full Earth-diameter distant.

Although the average Jovian lightning bolt was far stronger (with 10 times more electrical current) than its terrestrial counterpart, the radio bursts implied that the number of bolts over a given time interval per unit area was about an order of magnitude less than the terrestrial value.

The absence of lightning in the probe's immediate vicinity was consistent with its entry site being arid and free of cloud.

ABSENT ORGANICS

The neutral mass spectrometer, which measured the chemical composition of the gas in the vicinity of the probe, found little evidence of organic molecules.

This was a severe blow to those who had hoped that the lightning would have replicated a classic laboratory experiment conducted in 1952 by biochemist Stanley Miller, a graduate student of Harold Urey, in which a sample rich in hydrogen and ammonia was subjected to prolonged electrical discharges and gave rise to advanced organic molecules, including amino acids.

Although the 'quick look' compositional ratios were revised when the calibrated data from the tape recorder became available, the abundance of organics remained "minimal". Evidently, the complex carbon-hydrogen compounds used by life-chemistry are rare. Some hydrocarbon molecules were present, but these were the 'short-chain' variety incorporating a few carbon atoms. The data placed extremely tight 'upper limits' on the abundance of 'long-chain' hydrocarbons.

The significance of the fact that the winds penetrated to such great depth was that even if lightning prompted the formation of complex organic molecules they would be destroyed once they were drawn down into the 'hot mixing zone'. Evidently, intense atmospheric mixing meant that it was not practicable to argue that the probe's entry site was anomalously poor in this respect.

So much for the creatures which Carl Sagan – in his 'Cosmos' TV spectacular – had imagined flying in and out of the clouds of the Jovian atmosphere.

MISSION ACCOMPLISHED

As planned, the helium abundance detector switched off 40 minutes after the probe started transmitting data, and the net flux radiometer and the nephelometer reduced to 'noise' because below the 14-bar level they had nothing to measure. Only the atmospheric structure instrument produced useful data throughout the descent. This was to have started reporting 50 kilometres above the 1-bar reference, but the delayed parachute deployment meant that it was half-way to the reference level before direct sampling started. It nevertheless determined the characteristics of a 165-kilometre deep column extending from about 25 kilometres above to 140 kilometres below the reference level.

The transmission ceased after 57.6 minutes when the communications system succumbed to the 425 K temperature at the 23-bar level. "The descent module's internal temperature was much closer to the outside temperature during the descent than we expected," Bill O'Neil explained. By this time, however, the probe had considerably exceeded its baseline mission.

8

The capture orbit

THE LONG CLIMB

Now safely in orbit, Galileo began the long climb towards the apojove of the highly elliptic capture orbit. The Deep Space Network's tracking indicated that its trajectory was so good that no corrections would be needed.

Once Galileo had transmitted the compressed data from the atmospheric probe which it had stored in its solid-state memory, it was instructed to replay the full data set on its tape recorder. Although the recorder had functioned flawlessly on 'Arrival Day', the tape fouled as its status was being tested on 18 January 1996. Intriguingly, it jammed transitioning from fast forward to fast reverse, just as previously. "This looks consistent with the theoretical model of the tape sticking to the guide head," reflected Bill O'Neil, "except that we were surprised it stuck when it had been stopped for just a few seconds." After two days of study, the tape was instructed to advance, which it did, recovering its functionality. The replay of the tape was finally started on 25 January, and allowing for several lengthy breaks for engineering activities, it ran through to mid-April.

The fact that the recorder's fault was recurrent didn't bode well for the forthcoming orbital tour. An exhaustive analysis confirmed that the tape was sticking to a 'dummy' erase head that was used to guide the tape. This led to a loss of tension, with the result that when the recorder tried to reverse, the tape slipped on the capstans. A consultative workshop of industry experts held by JPL in mid-March concurred with this conclusion. The pragmatic solution was simply to schedule a slight advance in order to impart tension prior to commanding the tape to reverse. To protect against the fault in the ground test unit, when the tape had detached from its reel, it was decided that the recorder's built-in 'functions' that were likely to produce this fault should be done step by step by the spacecraft's computer. Furthermore, the new software that was soon to be uploaded to execute the orbital tour was to monitor the tape's condition and to recover when it stuck. The new software (which would supersede that uploaded in early 1995 to run through orbital insertion) would be capable of data compression and a variety of telemetry speeds in order to boost the 'information rate' of the low-gain antenna by at least a factor of

10. The effective rate depended on the extent to which a given data set could be compressed. It would be possible to compress an image of Jupiter's atmosphere 20-fold, but high-resolution views of the very textured surfaces of the moons were not expected to compress by very much, if at all.

By March 1996, Galileo was nearing the capture orbit's 20 million kilometre apojove. On 13 March, it adopted the appropriate orientation for a prograde burn and spun up to 10.5 rpm for enhanced stability. The next day, it made the 24-minute 'perijove raise' manoeuvre, which added 377 metres per second, in effect doubling its velocity, to lift its perijove from 4 to 11 Rj, a little outside Europa's orbit.[1]*

The new perijove had to be well above Io's orbit to remain clear of the most intense radiation in the inner magnetosphere. If this burn had not been performed, Galileo would have succumbed to radiation damage after only a few low passes. Actually, the situation was rather more complex. The capture orbit was so long that when the spacecraft was out near apojove it was susceptible to perturbation by the Sun's gravity. Such perturbations would rob Galileo of orbital energy, lowering its perijove, with the result that it would have followed its own probe into the atmosphere after only a few passes.

During the primary mission, the perijove would vary between 9 and 11 Rj, as appropriate to the orbits to facilitate the series of Europa, Ganymede and Callisto encounters. To revisit Io, at 5.9 Rj, it would be necessary to lower the perijove. This would be permissable only after the primary mission, however.

To protect the engine against a suspected helium leak, the helium supply was isolated after the burn in order to 'trap' the oxidiser pressure below the fuel pressure and thereby eliminate the possibility of the oxidiser vapour migrating into the fuel system. Unfortunately, the fuel check valve malfunctioned during the burn and temporarily re-established the danger by dropping the fuel pressure below that of the the oxidiser! Afterwards, with the propellant tanks now almost empty, the risk of a leak was determined to be minimal.

The 'gravitational braking' derived from the inbound Io fly-by had effectively saved 95 kg of propellant. Of course this did not mean that Galileo was now 'fat' with propellant. Actually the 10 percent of propellant remaining would be barely enough for the 'trim' burns that would be required to establish close encounters with the moons during the orbital tour. Making the Io fly-by had meant that the spacecraft had been able to be launched with a lighter propellant load.

As propulsion engineer Todd Barber recalled: "Since the tanks were so near to being full, I'm not even sure 95 kg more propellant could have been loaded!" There was 925 kg of propellant at launch, so the 10 percent left after the perijove raise burn equated to about the amount saved by the Io fly-by. "Without the Io fly-by," pointed out Barber, "Galileo would have been out of propellant before the tour even started! That's a nice metric, I think." In effect, the duration of the mission would be determined by the rate of propellant usage – once the spacecraft ran out of propellant to set up encounters, its tour would end.

* For Notes and References, see pages 413–424.

On 15 March, before Galileo resumed its 3 rpm dual-spin cruise, it 'hammered' the pins of the high-gain antenna's deployment mechanism in a final attempt to release the antenna, but to no avail.

As it headed back towards Jupiter, Galileo checked out the scan platform and its remote-sensing instruments and began sampling low time-resolution particles and fields data, sending thrice-weekly reports on magnetic fields and dust. However, as the spacecraft had to keep its antenna aimed at the Earth in order to replay the atmospheric probe's data, the dust detector could not face Jupiter, was not really measuring the dust environment and so received few 'hits'. When the replay was completed in mid-April, only the high time-resolution particles and fields data that was recorded while the spacecraft had been within Io's orbit remained to be transmitted. However, because this data was so bulky, it had been decided to leave it until the compression software had been installed.

One week later, with all the probe's data recovered, the tape recorder was subjected to engineering tests. Over two days of trials, the tape was made to stick and then released in a number of circumstances to explore the limits of its 'safe' use. It had been decided that the first encounter sequence would be fairly conservative in its use of the tape recorder but, as Bill O'Neil put it, "the better we understand the tape recorder, and the way that it sticks, the more effectively we will be able to use it later on."

The first encounter was to be with Ganymede, the largest of the Jovian satellites. Because the closer-than-planned Io fly-by had advanced this first encounter by a week, the Ganymede fly-by would have to be contrived to re-establish the planned tour schedule. On 3 May, Galileo made a 1.3 metre per second retrograde trim to delay its arrival on 27 June by 35 minutes, so that when it passed 850 kilometres above the moon's surface its trajectory would be deflected by just the required amount – the orbital tour was to be a game of billiards. The spacecraft's energy would be either increased or decreased, depending upon whether it passed ahead of or behind a moon, and whether it did so before or after perijove. Galileo's tour would be controlled as much as possible by such slingshots in order to minimise the number of thruster burns and, thereby, preserve the small amount of propellant remaining to retain the option of extending the mission and returning to Io.

THE PLANNING PROCESS

On 13 May 1996, the Deep Space Network started to upload software to prepare Galileo for its first encounter sequence. It had taken three years to specify, write and test software to perform the primary mission without the high-gain antenna, and had involved more than 100 team members. O'Neil described this reprogramming of the spacecraft and the associated upgrading of the Deep Space Network as "one of the most remarkable in-flight failure work-arounds, ever."

The second half of 1996 was to see four fly-bys. The first two would be with Ganymede, then one with Callisto, and then the long-awaited close up view of Europa (no observations of which had been made during the run in to Jupiter).

During each encounter sequence, data from the various instruments would be stored on the tape recorder, which had a capacity equivalent to 100 images. With data compression, the low-gain antenna would be able to transmit two or three images per day. Without the upgrade, it would have taken a fortnight to transmit a single image together with its engineering and navigational information. If the high-gain antenna had been available, then the tape recorder would not have been used to store data for long periods. It would have been possible to transmit most of the data in real time, with the tape serving as a short-term buffer for the most intensive periods of observation, and it would soon have been 'flushed' to Earth. Now, Galileo was to transmit the tape during its long cruise out to apojove and back in again. However, unlike the capture orbit's 20 million kilometre apojove, and six-month period, the orbits of the primary mission were to last only two months so there would barely be time to replay the tape – even with data compression.

In contrast to the hoped-for flood of data in real time, the scientists would have to settle for a trickle. At any given time, the scientists were working at three levels: analysing data already received, receiving recent data, and planning the observations to be made on future encounters. This led to a form of science in which the discoveries were not forthcoming until long after the observations had been made. In many cases, the results were presented at conferences months in advance of formal publication in the journals.

For each pass through the inner Jovian system, Galileo was provided with a long sequence of commands detailing what it was to do. The encounter sequence would typically last a week. If it all went well, the spacecraft would execute the entire sequence autonomously. The science programme comprised three themes – atmosphere, satellites and magnetosphere – each of which was managed by a 'working group' in which the individual instrument teams argued for 'observing time' (or, more specifically, 'downlink time'). Once competing proposals had been resolved, the working groups submitted their lists of observations to the sequence integration team.

Leo Cheng was the chief sequence integration engineer – he headed a team of 25 people, each of whom laboured to produce a particular piece of a complex jigsaw puzzle. The first job was to plan all the activities that would have to be undertaken to keep the spacecraft healthy. Then individual segments were developed to make the observations requested by the science teams. Once Cheng had integrated all the segments into one timeline, all the various specialists pored over it to ensure that it called for no action that would cause a problem for his or her own particular subsystem. At a higher level, Cheng eliminated conflicts and identified opportunities to enhance the scientific yield. A typical conflict might involve scheduling observations by two instruments that shared a solid-state data buffer where one instrument would have overwritten the data from the other before it had been written to tape. If the sequence could not be changed to save the data in between observations, one would have to be cut. Each encounter sequence took about two months to prepare, and since this matched the period of Galileo's orbit, it was clear that Cheng's team was going to be kept busy!

On 3 June – now half-way back from its high initial apojove – Galileo's camera took a picture of Ganymede against the stars for a navigational check. The new software stripped the image down to the bare minimum number of bits prior to transmission. The tape recorder was 'conditioned' on 21 June in preparation for the first pass through the inner Jovian system.

This initial sequence was to focus on Ganymede, but every pass through the inner system was to include observations of Jupiter. Since the entire sequence had to be developed months in advance, planning Jupiter observations posed a problem. Whereas features on the satellites were predictable, and calculating where to aim the instruments was straightforward, features in Jupiter's atmosphere drifted and – even more frustrating – often disappeared! The ongoing monitoring by NASA's Infrared Telescope Facility in Hawaii was therefore crucial to planning the atmospheric observations. As Glenn Orton put it: "It was going to be pretty much up to me to judge when in each orbit we were going to be looking at the planet, and which location we were going to be looking at to get the atmospheric feature we'd planned on [observing] for the last two years." If the high-gain antenna had been available and imagery could have been sent in real time, Galileo would have been able to survey wide areas on each orbit, but the reduced data rate meant that the observing programme had to be very selective and the target would be missed if the timing did not work out.

Galileo started its encounter sequence on 23 June 1996. Its observations would run to the end of the month as it first penetrated and then withdrew from the realm of the large moons.

Io was too deep within the magnetosphere for Galileo to make another close approach, but distant imagery would be taken on each encounter sequence to monitor the degree of volcanic activity. The ongoing observations by the Infrared Telescope Facility and other observatories provided context for these spacecraft observations. This first encounter sequence also included opportunities for preliminary long-range studies of Europa and Callisto.

The particles and fields instruments were to perform near-continuous low time-resolution sampling throughout each encounter sequence in order to build up a comprehensive survey of the spatial and temporal variability of the inner magnetosphere. The magnetometer would measure the ambient magnetic field strength, and the plasma wave spectrometer would note variations in the electromagnetic waves. This low-rate data was to be transmitted in near-real time. In addition, when the spacecraft was close to whichever moon was the focus of a given sequence, the particles and fields instruments would switch to high time-resolution sampling and their data stored on the tape. Ganymede takes seven days to orbit Jupiter. The magnetosphere rotates with the planet every 10 hours. It was likely that the moon would create a 'wake' in the magnetosphere (note that in these circumstances the wake would be ahead of the moon) and the magnetometer team were eager to look for this. In addition, high time-resolution particles and fields data was to be taken at perijove to investigate the torus, and the nature of the interaction between Io and the magnetosphere. The ultraviolet spectrometer was also to make observations of Io's torus. Since the torus spanned Io's orbit around Jupiter, the extreme-ultraviolet spectrometer was to perform its observations from a distance to enable it to scan its full width and so, from one orbit to another, monitor its size and shape. By being in orbit, Galileo would augment the earlier 'snap shots' provided by its fly-by predecessors.

9

Tectonic Ganymede

VOYAGER'S GANYMEDE

At 5,250 kilometres in diameter, Ganymede is not only the largest of the Jovian moons, it is the largest satellite in the Solar System. In fact, being somewhat larger than the planet Mercury, it is really a small planet that happens to be orbiting Jupiter. The Voyager missions mapped 80 percent of Ganymede with a resolution of 5 kilometres or better per pixel.[1]*

A 'dark cratered terrain' comprising large circular and small polygonal dark features appears to be the oldest part of the crust. The largest of these areas – appropriately named Galileo Regio – is an oval feature 2,800 to 3,200 kilometres in diameter, and it spans about one-third of the anti-Jovian hemisphere. It is etched for hundreds of kilometres by parallel curvilinear 'furrows' which have been interpreted as grabens. It is heavily cratered, but its northern section is less dark than the rest, possibly indicating the presence of some kind of condensate. This dark oval had actually been seen by telescopic observers, but apart from this feature and the bright 'poles spot' the early maps bore little resemblance to the moon's surface.

The lighter 'grooved terrain' accounts for most of the polar regions and is characterised by parallel alternating grooves and ridges running for hundreds of kilometres. The troughs and crests are separated by 5 to 10 kilometres horizontally, and several hundred metres vertically. The ridge/trough pairs are 'bundled' to form lineations. In some places, these cut across one another at various angles, forming a 'reticulate terrain' which documents a complex record of tectonic activity. This terrain often includes small hills, and so presents a 'hummocky' texture. There are patches of remarkably smooth material within the grooved terrain, and the grooves warp around these 'smooth plains'. Since there is no albedo change over the boundaries, the smooth plains may be grooved terrain that has been melted from beneath, or flooded by fluid extrusions – cryovolcanism.

The grooved terrain cuts into the dark cratered terrain, and is therefore younger. Furthermore, edge-relationships between dark units separated by a strip of light

* For Notes and References, see pages 413–424.

This long-shot of Ganymede's anti-Jovian hemisphere is dominated by the large dark oval of Galileo Regio which is separated from Marius Regio to the southwest by the grey band of Uruk Sulcus. The fragmentation of Marius Regio by narrower sulci is highly suggestive of crustal spreading.

terrain imply that the dark terrain has been fractured and drawn apart by the upwelling lighter material. The tectonism is shown in stark clarity by the symmetry with which Marius Regio has been split from Galileo Regio by Uruk Sulcus, and by which Marius Regio has been fractured into polygonal blocks. Such 'spreading' is indicative of nascent plate tectonics.

The 2.0 g/cm^3 bulk density means that Ganymede is a roughly equal mix of rock and ice. Clearly, however, most of the rock is in the interior, so the tectonic forces would have acted upon a mainly icy crust, presumably as it was forming.

Tiamat Sulcus, which intrudes Marius Regio, is cut by a fault which extends from Kishar Sulcus. Intriguingly, the southern part of Tiamat Sulcus is somewhat wider than the northern part. In terms of grooving, Voyager counted 14 grooves in the north and 20 in the south. The grooves on different sides of the fault are also of different widths. This variation is believed to indicate that the two sides of the fault underwent spreading at different times.[2]

There are several relics of multiple-ringed basins on Ganymede. The largest is Gilgamesh, in the southern hemisphere. It has a 150-kilometre wide central depression which forms a smooth plain, surrounded by hummocky ejecta and disrupted terrain. The outermost ring (marked by a scarp) is 275 kilometres radius. Overall, the ring-structure is rather subdued. Its impact origin is evident from the secondary craters – some in characteristic short chains – and other forms of 'sculpture' ranging out to 1,000 kilometres. Superposition relationships show that

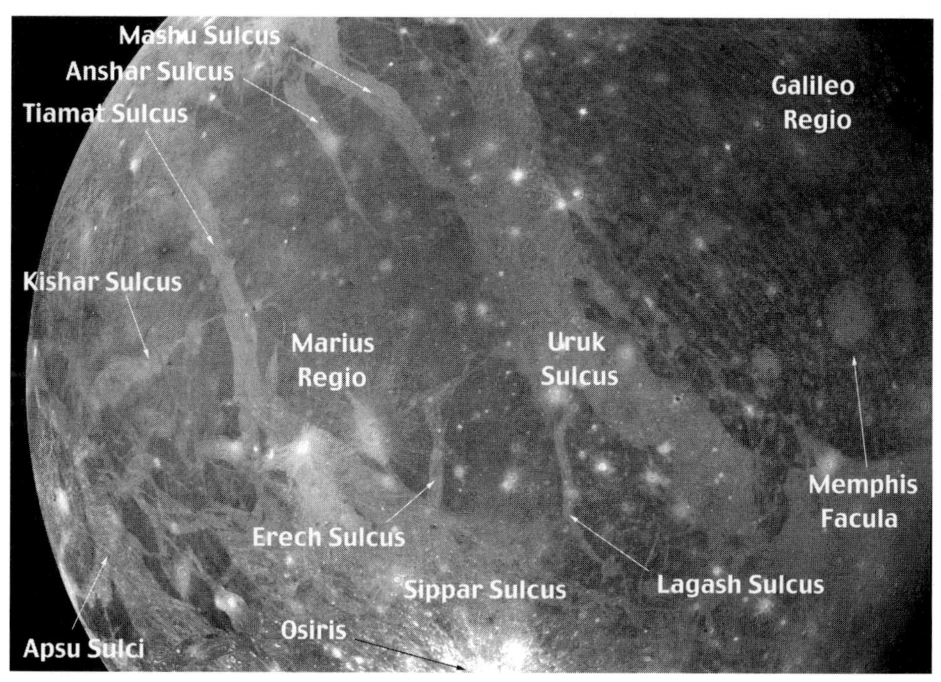

A Voyager close-up, showing Ganymede's fragmented Marius Regio.

Gilgamesh is the best preserved of Ganymede's multiple-ringed impact basins. In addition to creating concentric scarps, the massive impact laid down an annular blanket of ejecta. These Voyager mosaics show Gilgamesh's context (left) and a 'radial' running out from the central plain through the rings (right). This is Ganymede's version of the Orientale basin on the Earth's Moon.

Gilgamesh formed after the adjacent grooved terrain. This probably dates the tectonic activity that created that terrain to the tail-end of the Great Bombardment, some 3.8 billion years ago, and the sulci are therefore 'ancient'.

Unlike on the Moon and Mars, the craters on Ganymede become progressively flatter with increasing size. Actually, this had been predicted in 1973,[3] on the basis that an icy crust would be plastic and would flow to soften the vertical relief. The rim of a crater would gradually sink and the floor would rise to form a dome. This was isostasy at work, and the process would operate most rapidly on the largest craters with the most deeply excavated cavities and the most piled-up ejecta. It has transformed large impact craters into 'ghost craters' or palimpsests, the largest of which are 400 kilometres across. Ironically, these craters are still surrounded by small secondaries. Taken together, the palimpsests account for 25 percent of the mapped surface, so they represent a major part of Ganymede's early cratering history. They are preferentially (although not exclusively) found on the older more densely cratered dark terrain.

This Voyager image shows the light-toned palimpsest of Memphis Facula (on the left), superimposed on the furrowed terrain of eastern Galileo Regio. The smooth appearance of the 340-kilometre wide feature belies its origin as a massive impact. The crater walls have slumped, the floor has risen isostatically, and the remaining topography has been smoothed out by slush. Nidaba (right) is younger, slightly smaller, and better preserved so it shows a more rugged topography.

At 340 kilometres across, Memphis Facula is a fairly typical palimpsest in having a remarkably smooth surface. Except towards the centre, the bright material is probably only a few hundred metres thick, and may be slushy infill from the interior of the moon. These structures provide clues to the moon's early thermal history and composition. The better preserved impact basin, Nidaba, just east of the palimpsest is younger, slightly smaller, and shows a more rugged topography. As both are superimposed on the furrowed terrain of eastern Galileo Regio, they post-date the process which formed this negative topographic relief.

The morphology of the palimpsests exceeding 50 kilometres in diameter suggests that the icy crust was about 10 kilometres thick at the time of the impact. The smaller impacts failed to penetrate to the slush beneath, while the larger ones – which probably date to the tail-end of the Great Bombardment – did penetrate and allowed fluid to flood the cavity. One study[4] found that many of the smaller craters on Ganymede have ejecta blankets that end abruptly at a scarp – as if the crater is on a pedestal. Similar craters on Mars are thought to be impacts that hit permafrost and flooded the surrounding terrain with fluidised ejecta which rapidly froze solid.

The darker terrain was deemed to be ancient on the basis that it was more heavily cratered, and the light sulci were deemed to be 'clean' ice extruded as a result of tectonic activity. It was therefore concluded that Galileo Regio, Marius Regio, Nicholson Regio, Barnard Regio and Perrine Regio were the oldest terrains.[5] Although the surface is predominantly ice, it has been speculated that it grows 'dirty' with age. Certainly the brightest areas are the 'ray craters' such as Osiris, where fresh white ice has been excavated. However, if both the dark and light terrains are ancient (as indicated by the superposition of Gilgamesh on the sulci), the 'darkening' may have taken place only early on.

Although the furrows on Galileo Regio have a radius of curvature which is focused in the vicinity of the anti-Jovian point, suggesting that they may have been induced by forces extant while Ganymede's rotation was being captured, this may be simply a coincidence. One study[6] suggested that the furrows of the Valhalla multiple-ringed structure on Callisto resemble those that etch Ganymede's dark cratered terrain, and proposed that Galileo Regio may be marred by the rings of a basin near the anti-Jovian point which was obliterated during the tectonic processes that formed the light-toned terrain of Uruk Sulcus. The evidence is inconclusive, however.

Alternatively, all the tectonic faulting on Ganymede may be the result of crustal expansion. A study[7] proposed that both the furrows crossing the dark cratered terrain and the flat-floored grooves of the grooved terrain could be accounted for if the ice in the interior had undergone a phase change and expanded.[8] This model indicated that the crust was about 10 kilometres thick when the furrows formed, but only half as thick when the grooves formed. Since the grooved terrain formed after the dark cratered terrain, this crustal thinning is consistent with upwelling of warm ice.

GALILEO'S FIRST LOOK

Galileo 'tweeked' its trajectory on 24 June 1996 so as to make an 844-kilometre pass over Ganymede's surface – 70 times closer than Voyager. The observing plan called for medium-resolution regional views and six close views of preselected features with 10-metre resolution. Meanwhile, the infrared spectrometer would chart the chemistry of the surface, the ultraviolet spectrometer would search for an atmosphere, the integrated photopolarimeter and radiometer would map the moon's thermal characteristics, the magnetometer would search for an intrinsic magnetic field, and the Deep Space Network would monitor the doppler on the radio signal in order to refine the model of the moon's internal structure. The fly-by at 06:29 UT on 27 June was so accurate that the spacecraft's altitude was only 12 kilometres short of the objective.

As Galileo approached Ganymede's illuminated anti-Jovian hemisphere, it was able to see Galileo Regio and Marius Regio with Uruk Sulcus between them. Its closest point of approach was over the eastern fringe of Galileo Regio, a little before the evening terminator on the leading hemisphere at 30 degrees north and 112 degrees west. The best imaging conditions occurred 15 minutes prior to closest approach. The sub-Jovian hemisphere was studied during the exit, in darkness.

Because Ganymede had been well imaged by Voyager, Galileo's objectives were to secure high-resolution observations which would:

- characterise any cryovolcanism
- determine the nature and timing of any tectonic activity
- determine the history of formation and degradation of impact craters
- determine the nature of the surface materials.

For this first encounter, adjacent sites representing both types of terrain had been selected, centred on Uruk Sulcus. Galileo's higher resolution imagery "will show areas that have been tectonically ripped apart," Torrence Johnson predicted, and they "should tell us a lot about the forces that shaped Ganymede's crust".

The infrared spectrometer's overall objectives were to define the main compositional units, terrain types, crater types, polar effects and differences over the leading and trailing hemispheres, and to determine the composition of any atmospheric envelope. The principal objective was to chart the entire surface in 204 wavelengths at a spatial resolution of 100 kilometres in order to determine the chemical composition of the surface units.[9] The morphologies of representative features – including the dark- and light-toned areas and the craters, basins and palimpsests – were to be studied at higher spatial resolution. Because their ejecta is excavated from beneath the surface, craters were expected to characterise the subcrust. The comparison of the leading and trailing hemispheres was to test the hypothesis that the oldest terrain is dark due to charged particles circulating in the Jovian magnetosphere chemically modifying the surface materials.

As Galileo withdrew and started to replay its tape, the images came in at a rate of two or three per day. For the 'old hands' who had stood in awe as a new image from the Voyagers had flashed onto TV screens every few minutes during the most intensive periods, this was a frustrating time, but the results were certainly worth the wait. "It's been like Christmas every day," enthused Jim Head, a geologist from Brown University, Providence, Rhode Island, and one of the imaging team members. "As we come in, we open another packet from Ganymede."

"The results from the Ganymede encounter are remarkable," pointed out Bill O'Neil as he set the scene for the announcement of the first imagery, "absolutely stunning!".

URUK SULCUS

In the Voyager imagery, the light terrain of Uruk Sulcus had appeared to be fairly smooth. It had been interpreted as having been resurfaced by 'ice lava' from cryovolcanoes. However, even Galileo's regional view showed it to be strikingly

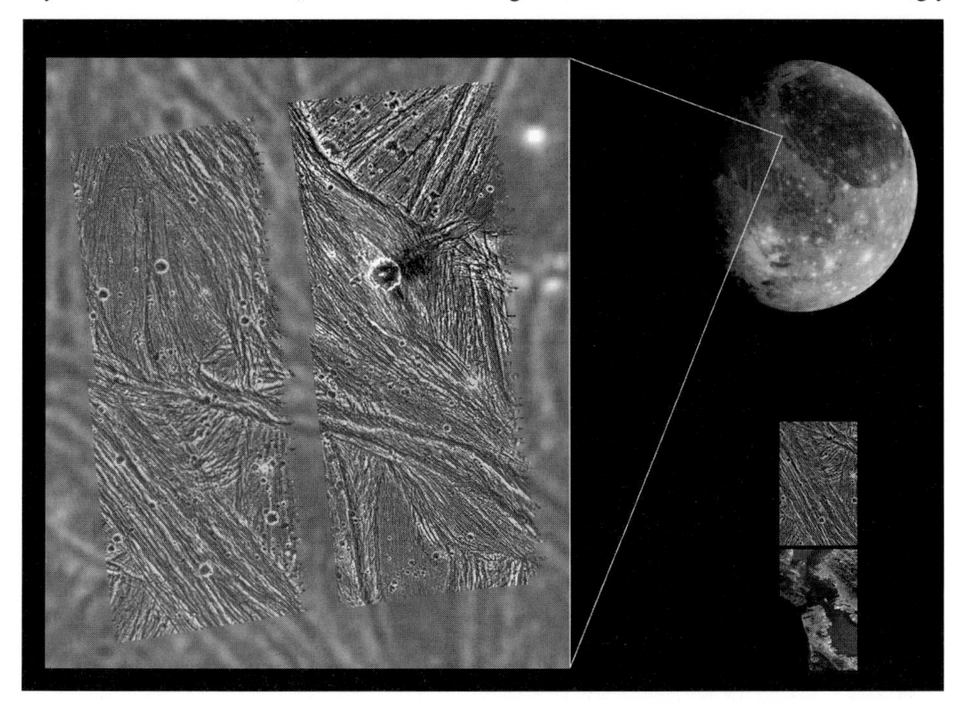

On its first orbit, Galileo took a look at a section of Uruk Sulcus. In Voyager 2 imagery with a resolution of 1.3 kilometres per pixel this light-toned terrain had displayed linear albedo variations suggestive of grooves. Galileo's 75-metre resolution view confirmed this and showed that the entire area is etched by small ridges. The patterns of ridges and grooves indicate that extension (pulling apart) and shear (horizontal sliding) forces have both shaped the icy landscape. The bright circular feature is an impact crater which has excavated some dark ejecta and superimposed it on the ridged sulcus. The terrain to the northeast of the crater is the ancient heavily cratered Galileo Regio. The San Francisco Bay Area is provided as an insert to provide a sense of scale.

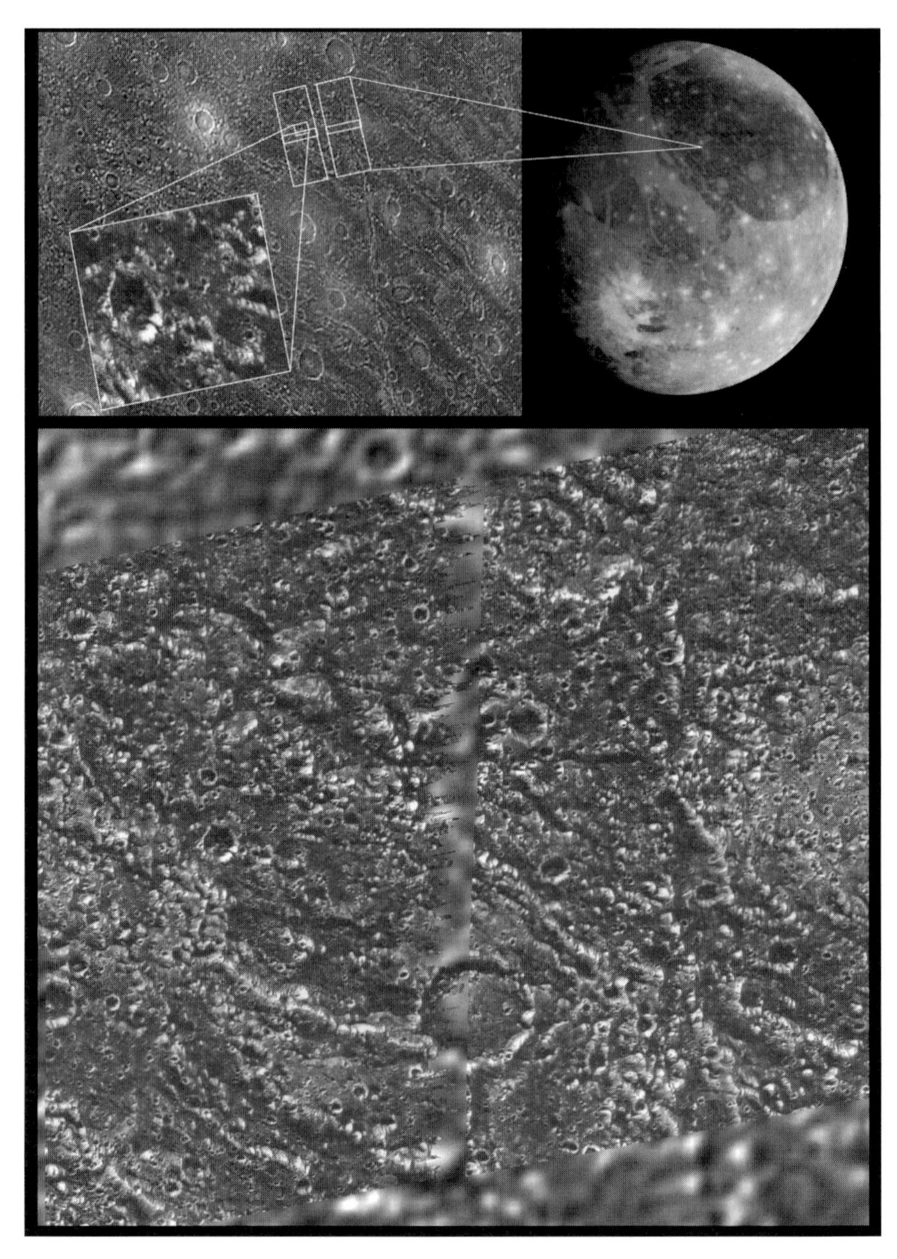

On its first orbit, Galileo took a look at a section of Galileo Regio with a resolution of 80 metres per pixel. The four-element mosaic is shown (top) in context, together with a close-up of part of one element, and in a 'rectified' view (bottom). This ancient heavily cratered terrain, with its broad curvilinear furrows, is striking different from the ridged Uruk Sulcus. There are distinctive variations in albedo ranging from the brighter rims, knobs, and furrow walls to the lower slopes and crater floors where a dark material may have accumulated.

ridged – as if it had been graded by a giant rake. It was grooved with dozens of long parallel ridges and valleys. This supported the hypothesis that the sulci resulted from tectonic action in which the water-ice crust expanded, wrinkled, and cracked. In effect, the crust had been torn apart by 'icequakes'. The 100 K surface temperature implied that the ice would be as hard as rock and probably brittle. The 75-metres per pixel high-resolution view highlighted the fine-scale topography. The pockmarked heavily cratered Galileo Regio to the north abutted the younger grooved terrain of Uruk Sulcus to the south, which was dotted by prominent ice mountains. A bright impact had superimposed dark ejecta on the ridges of the sulcus. The superposition relationships revealed a complex geological history.

GALILEO REGIO

Galileo Regio revealed several surprises, too, displaying at least three terrain types which were distinguished in terms of both albedo and morphology.[10] In addition to the anticipated fault-induced relief, it was littered with many small hills – it was a jumble of hummocky terrain with bright mountains and dark plains which hinted at cryovolcanism.

So, although the cratering indicated that Galileo Regio was very ancient, it had also clearly been reworked – or resurfaced – by repeated episodes of shearing, crustal rifting, and cryovolcanism. In fact, both the light and dark terrains were found to be rather more morphologically complex than expected. Some craters had been transected by faults, and torn apart.[11] Although such distortion complicated the task of measuring the diameters for compiling crater statistics, the fact that the craters were modified enabled the tectonism to be dated.

"Something dynamic is going on inside," noted Jim Head. The juxtaposition of fractured and faulted terrain was amazing, displaying "age relationships that turn our previous thinking upside down".

In terms of craters of around 1 kilometre in diameter, the cratering density on Ganymede was about an order of magnitude greater on the darker terrain than on the light-toned terrain.[12] Once the cratering statistics for small craters were merged with those for larger craters derived from the Voyager imagery, the cratering curve bore a striking similarity to that of the Earth's Moon, suggestive of an ancient surface. If Ganymede's youngest multiple-ringed structure, Gilgamesh, was assigned an age of 3.8 billion years, in order to anchor the curve, the ancient terrains would date back to the period 4.1 to 4.3 billion years.

THE BIGGER PICTURE

The Voyager hemispherical views provided context for the infrared spectrometer's multispectral imagery charting the distribution of different materials.[13] For example, the distribution of water-ice on the surface could be determined from measurements at given wavelengths and displayed in 'false' colours in which dark represented less

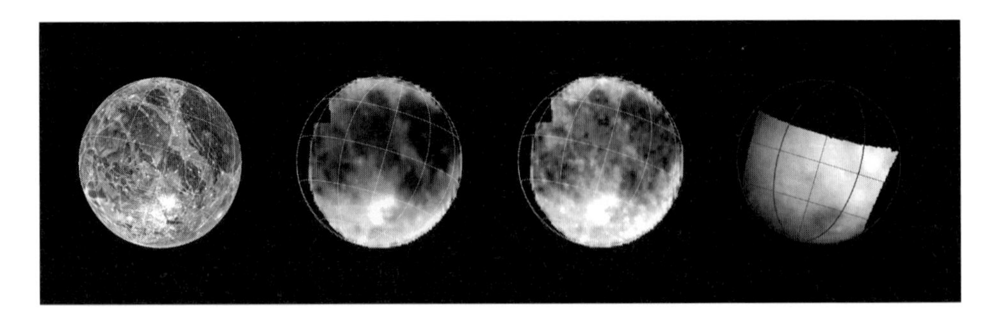

On its first orbit, Galileo's near-infrared mapping spectrometer documented Ganymede at wavelengths selected to identify the composition of the surface. The context is shown in a Voyager image (left). The distribution of ice (centre left) shows a higher proportion of water in the sulci. The distribution of non-ice material (centre right) indicates that the dark terrains have differing amounts of minerals. The integrated photopolarimeter and radiometer also mapped Ganymede (right), showing that day-time surface temperatures ranged from 90 to 160 K. The radiometer's map covers almost the same area, but rotated about 30 degrees.

water and bright represented more water, showing that the visually dark regions are less icy than the sulci. Similarly, the sizes of the ice grains could be inferred and charted. The spectrometer identified various minerals in the visually dark regions which were not present in the sulci.

The integrated photopolarimeter and radiometer scanned the temperatures of the surface on both the sunlit and the darkened hemispheres. The temperatures ranged from 90 to 160 K during the day, depending on time of day and latitude.

The real surprise was that Ganymede possessed an intrinsic magnetic field which created a magnetosphere. This was noted by the magnetometer and also by the plasma wave spectrometer. As the magnetometer was monitoring the southerly direction of the ambient Jovian magnetic field, it abruptly detected a five-fold increase in the field strength and the vector rotated to point at the moon. Simultaneously, the spectrometer indicated that it had penetrated a magnetosphere.

In contrast to the complex signal detected by the magnetometer as it passed Io on 'Arrival Day' in December 1995 (see p. 298), the Ganymede data was unambiguous – the moon has an intrinsic magnetic field one-thousandth of the strength of Earth's field.

The plasma wave spectrometer "indicates Ganymede is surrounded by a thin ionosphere, which suggests that Ganymede probably has a tenuous atmosphere", said Donald Gurnett of the University of Iowa, the instrument's team leader. In fact, it detected a 100-fold rise in the charged particle density at closest approach.

When the Hubble Space Telescope observed Ganymede on 21 June[14] it detected a tenuous oxygen atmosphere. The observations had been made spectroscopically in the ultraviolet (the ultraviolet region of the spectrum was selected because solar reflection from the moon's disk swamped the faint emission lines in the visual range). The oxygen emission lines were split, indicating the presence of a magnetic field. It was "very tentative evidence for polar aurorae", said Doyle Hall of Johns Hopkins University,

Baltimore, Maryland, who led the team. Such observations could not have been made from the ground because ultraviolet is absorbed by the Earth's atmosphere.

The prerequisites for an auroral display are a magnetic field, circulating charged particles, and a tenuous atmosphere. The Hubble Space Telescope's remote observation and those from the Galileo fly-by fit together like a jigsaw, providing convincing evidence that Ganymede has a magnetic field.

Astronomers had already[15] spotted a thin ozone 'frost' of recombined oxygen that had been liberated from the ice and Galileo's infrared spectrometer found surface features showing a variety of hydrated materials. "The surface of Ganymede is thought to contain about fifty percent water ice," pointed out Charles Barth of the University of Colorado at Boulder, and a member of the ultraviolet spectrometer team.

Hall's team had reported two years earlier that Europa has a thin atmosphere. Both Europa and Ganymede evidently release oxygen by dissociation of water-ice exposed on their surfaces. However, such atmospheres are no denser than the atomic oxygen that is present at the region of space in which the Space Shuttle orbits the Earth, so they are more appropriately referred to as 'exospheres'.

"Hydrogen atoms are being knocked off the icy surface by charged particles," said Barth. The process by which the energetic charged particles circulating in the Jovian magnetosphere bombard the surface of the moon and knock out atoms is known as sputtering. This maintains the moon's tenuous atmosphere. Because hydrogen is lighter than oxygen, it leaks to space,[16,17,18] leaving the oxygen behind, which forms the frost.

Some interaction between Ganymede and the Jovian magnetosphere had been anticipated, but a "magnetosphere within a magnetosphere", as Torrence Johnson put it, was unexpected. "We knew Ganymede was an interesting place. What we have just found makes it even more interesting."

The 'standard model' for an intrinsic magnetic field involves an iron core, but the inferred structure of Ganymede made this rather unlikely. Voyager implied a bulk composition which suggested that the moon was about half rock and half water-ice. The best model suggested an icy crust less than 100 kilometres thick capping a convecting mantle of 'soft ice' some 400 to 800 kilometres deep, englobing a large silicate body. If this contained an iron core, it was not expected to be very large. The verdict on the moon's internal structure would have to await the doppler data from a closer fly-by.

Once it was realised that Jupiter's magnetic field was inducing electrical currents in fluids within Europa and Callisto, and that these induced secondary magnetic fields (see pp. 178 and 196 for details), the magnetometer team set out to re-examine their Ganymede data to try to isolate any secondary magnetic effects superimposed on the intrinsic dipole field, but the data was inadequate to decide whether such an inductive field existed.

RETURN TO GANYMEDE

On 23 August 1996, with Galileo heading back towards Jupiter, the commands for the second encounter sequence were uploaded. Ganymede was to be the focus again, but this time with a 260-kilometre fly-by – in fact, this would be the closest approach to any moon in the primary mission. The next day, as the spacecraft was replaying the last of the taped data from its earlier encounter sequence, the active Command and Data Subsystem 'safed' itself. As the development of 'contingency packages' was running late, the prevailing wisdom at JPL was that it would take several weeks to recover from this fault. The return to Ganymede was only a week away so as Bill O'Neil put it, everyone "worked at flank speed".

A thruster burn had to be made on 27 August to tweek the trajectory for Ganymede, so for the first time this had to be done by the back-up C&D Subsystem. Two days later, the problem with the primary C&D was finally identified as a software error. The overflow of commands in a buffer had caused the processor to overrun the allotted time on a given task, and a timer had cut in to halt the processor in case it had a hardware fault.

"What a relief!" said O'Neil after new software had been sent to restore Galileo to full operation. In effect, the spacecraft was being incrementally adapted to match its working environment. If the condition recurred, the revised software would recover more gracefully.

NORTH POLE

Ganymede was in much the same orientation with respect to Jupiter and the Sun as before, but this time Galileo's trajectory took it over the moon's north polar region,

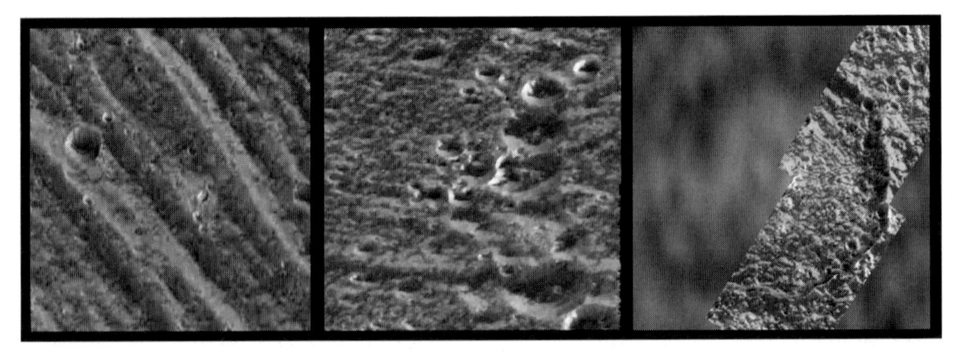

On its second orbit, Galileo inspected several areas near Ganymede's north pole at high resolution. The ridges (left) on the edge of the polar cap shows bright east-facing slopes and dark west-facing slopes, with troughs containing a darker material below the larger ridges. The large crater is 2.4 kilometres across. It has a bright north-facing rim slope. The Sun is shining from the south at such a high latitude, but the north-facing walls of the ridges and craters are brighter than the walls facing the Sun, suggesting that frost is forming on the cooler north-facing slopes. The floor of the 36-kilometre crater near the north pole (right) is partially brightened. The Voyager view prompted speculation that the crater had been flooded by cryovolcanism, but Galileo showed the brightening to be frost deposition. There is also evidence of frost deposition on protected slopes at a latitude of 60 degrees (centre). The resolution in these Galileo images is 45 metres per pixel.

with the point of closest approach being at 80 degrees north and 123 degrees west, somewhat north of Galileo Regio. The high albedo of Ganymede's polar regions vindicated the early telescopic mappers who had charted a 'polar spot'.

At high northern latitudes, the north-facing walls of the ridges and craters were somewhat brighter than those facing the Sun, even though the illumination was from the south. This has been interpreted as meaning that the north-facing slopes are covered with water-ice frosts. An 18-kilometre wide swath across a crater near the north pole charted a partially brightened floor. In Voyager image, the crater was thought to be flooded by cryovolcanism, but Galileo showed the brightening to be frost deposition. Fractures had cut across the floor of the crater, and two large blocks of ice had collapsed off the side of the steep northeastern wall.

Ridges on the fringe of the north polar cap have bright east-facing slopes and dark west-facing slopes, with troughs of darker material below the larger ridges. The bright slopes may indicate differences in grain size, differences in composition between the original surface and the underlying material, frost deposition, or illumination effects.

STEREOSCOPIC MODELS

Since the same hemisphere was illuminated, it was possible to re-image the same terrain to facilitate stereoscopic analysis of the high-resolution targets in Galileo Regio and Uruk Sulcus. As the human eye-brain has a tendency to over-interpret the alternating light and dark patterns of these heavily ridged terrains, stereoscopic

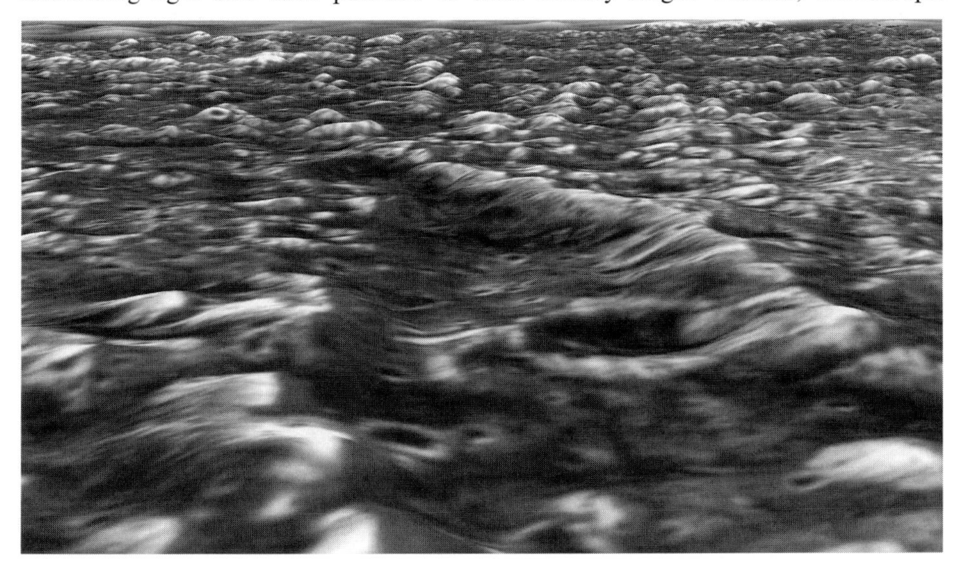

By combining imagery taken over Galileo's first two orbits, it was possible to construct a three-dimensional perspective showing the deep furrows and craters on Ganymede's ancient terrain.

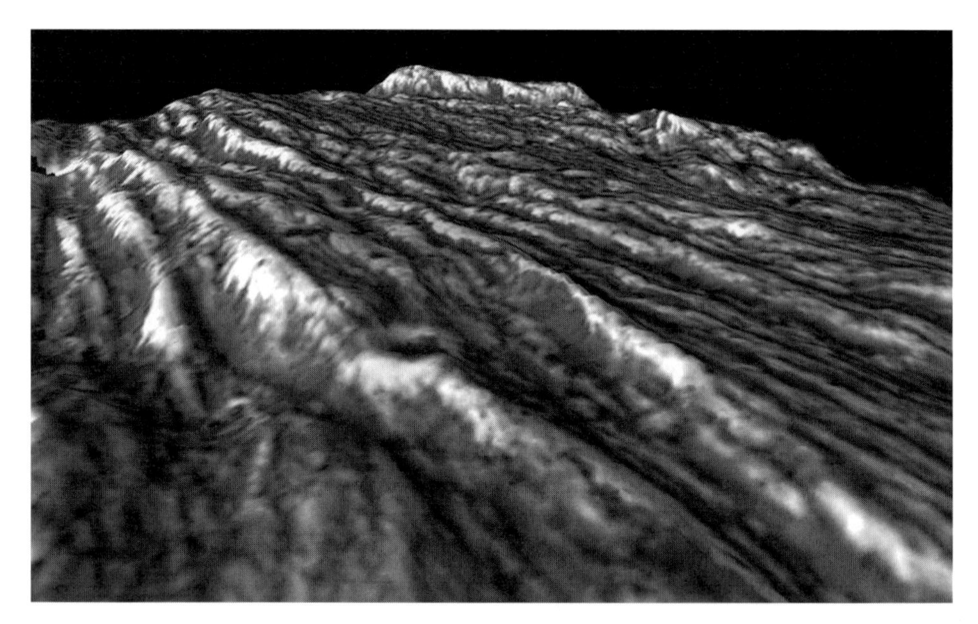

By combining imagery taken over Galileo's first two orbits, it was possible to construct a three-dimensional perspective of this intensely fractured surface. Bright icy material is exposed along the ridge crests, and dark material has collected in the adjacent troughs.

analysis would expose the topography, free of this bias. "The three-dimensional views will give us a better idea of what is 'paint' on the surface, so to speak, versus real topography," explained Torrence Johnson. Clarifying this distinction would facilitate models for the processes that produced each type of feature.

The topographic study of Uruk Sulcus dramatically showed that the icy surface is intensely fractured. The variation in elevation between the highest ridge and the lowest trough is a few hundred metres. The reconstruction demonstrated that the albedo variations are correlated with topography, in that bright icy material is exposed on the crests of the ridges, and dark material has collected in low-lying areas – suggesting that it may be the result of 'pooling'.

Impacts excavate material, so craters serve as 'drill holes' and their ejecta provides insight into the subcrust. Some craters on Ganymede – for example, Khensu – have extremely dark floors. About 13 kilometres in diameter, Khensu is located on Uruk Sulcus. Intriguingly, in addition to having a dark floor, it has a bright ejecta blanket, while, significantly, the larger nearby crater, El, does not have a dark floor. Khensu's dark floor may simply be the residue from the impactor, but it may be dark because the impact punched right through the sulci material and exposed a darker layer beneath – perhaps indicating that a sliver of the dark terrain had been caught up in the sulci, flooded and submerged.

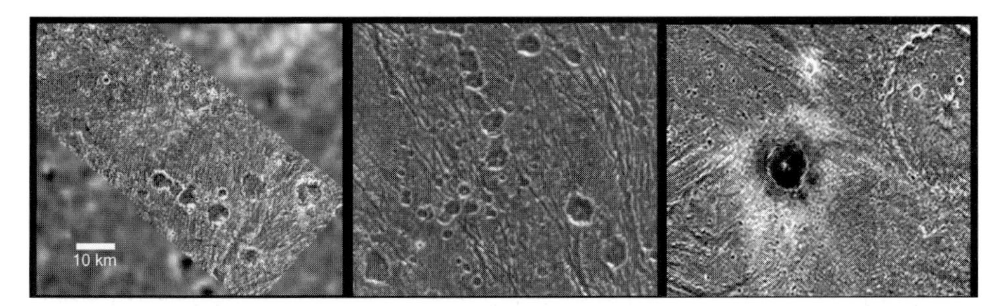

Crater morphology was one of the imaging objectives of Galileo's second orbit. Galileo imaged the fringe of a 350-kilometre wide palimpsest (left) at a resolution of 88 metres per pixel to document the transition to the surrounding terrain. The palimpsest's edge (which crosses the image from southwest to northeast) is visible only as a slight brightening. The fact that the chain of secondary craters radiating out to south-east can be traced back into the palimpsest (where they have been smoothed) demonstrates that the edge of the palimpsest is the fringe of the thick ejecta blanket, rather than the rim of the primary impact. In another case (centre) at a resolution of 190 metres per pixel, the margin of a palimpsest is so diffuse as to be barely identifiable, and most of the craters in the field of view are from a more recent impact to the south. In addition to studying palimpsests on the ancient dark terrain, Galileo imaged Khensu (right), a 13-kilometre wide crater on Uruk Sulcus, at a resolution of 111 metres per pixel. This combines an extremely dark floor with a bright ejecta blanket. Interestingly, the 'pit' in the centre of the larger crater El to the northeast also has a dark core.

PALIMPSESTS

The Voyager imagery recorded the location of palimpsests, but did not have the resolution to study their structure in fine detail. Galileo imaged a swath across the rim of a 350-kilometre wide unnamed palimpsest on grooved terrain at a resolution of 88 metres per pixel, showing the strings of the secondary craters formed by the ejecta from the ancient event. The rim of the palimpsest is visible, extending diagonally across the image but only as a slight albedo variation; it has no topographic features. The southwest–northeast trending grooves that are prominent outside the palimpsest continue for a short distance into the palimpsest, before being overwhelmed. This indicated that the edge of the palimpsest is the edge of the isostasy-eroded ejecta blanket rather than the rim of the crater.

Galileo also took a look at a palimpsest on an area of dark terrain. In Voyager imagery, the area is crossed by the diffuse 'fringe' of a large circular bright feature. In the higher-resolution view, the diffuse margin of the palimpsest is noticeable only as a progressive increase towards the west in the area covered by bright hummocks. A more recent palimpsest-forming impact to the south has peppered the entire area with chains and clusters of secondary craters.

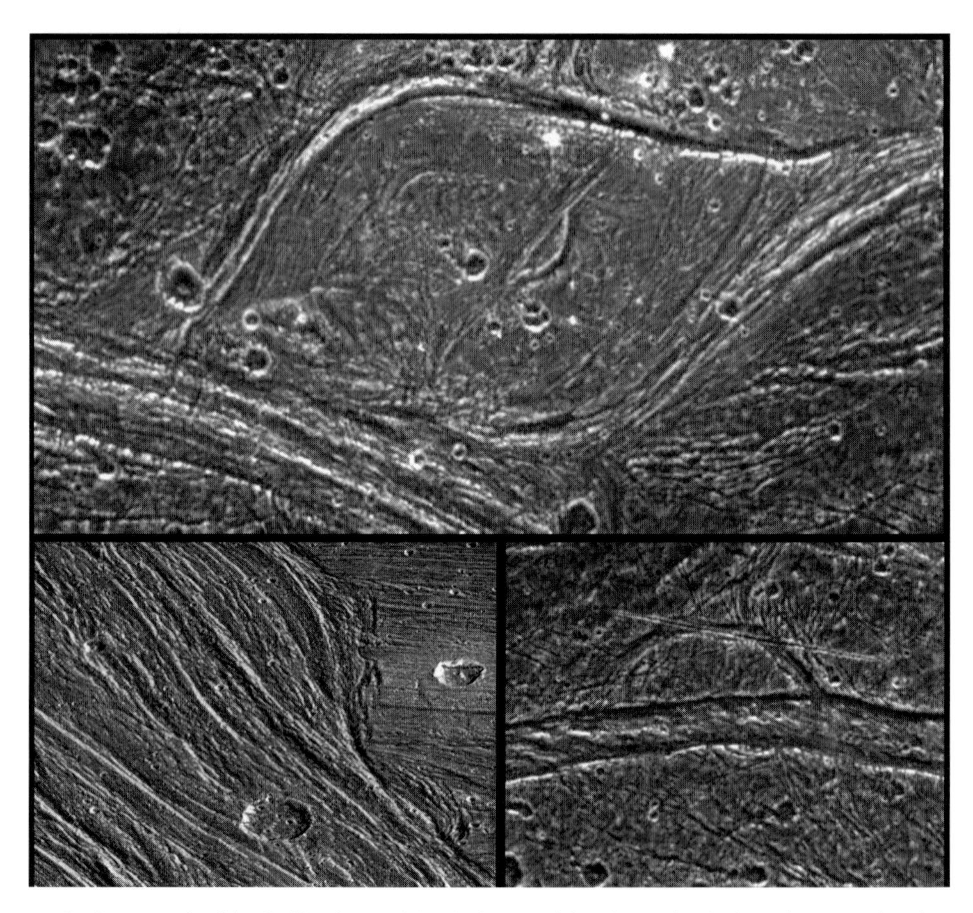

On its second orbit, Galileo imaged (top) the transition from the dark ancient terrain of Marius Regio to Nippur Sulcus at a resolution of 188 metres per pixel, demonstrating that the tectonism which created the sulcus to the south also tore apart the adjacent dark terrain forming an unusual 80-kilometre wide lens which shows signs of shearing and rotation. Interestingly, in several places around the border of the lens the bright ridges appear to turn into dark grooves. Another case (lower right) was recorded at the same resolution. The semicircular structure is about 33 kilometres across. The narrow linear feature which cuts across its northern extent appears to comprise a chain of very small craters. The ridges and grooves of Nippur Sulcus were documented at a resolution of 93 metres per pixel (lower left). The intersections of the grooves reveal complex age relationships. The sinuous northwest–southeast trending grooves truncate the older east–west trending ridges to the east. The crater towards the southern edge of the image is about 12 kilometres in diameter.

NIPPUR SULCUS

Galileo imaged a patch of Nippur Sulcus – which is northwest of Marius Regio – showing intersecting ridges and grooves revealing complex age relationships. Sinuous northwest–southeast trending grooves have cut through (and apparently destroyed) the older east–west trending features. The ridges and troughs are spaced 1 to 2 kilometres apart, and several broader ridges 4 to 5 kilometres across have smaller ridges running on top of them. The entire area is pocked by impact craters, the largest of which is 12 kilometres across (the dark ring at the base of its walls may indicate pooling of dark material which slipped from the steep slopes).

The northern fringe of Marius Regio, in the transition to Nippur Sulcus, shows numerous intriguing tectonic structures with small criss-crossing fractures. An arcuate structure about 33 kilometres across abuts a broad east–west fault system. Is this arcuate form part of a much larger feature? It is etched by a narrow linear feature that appears to comprise a series of small pits or craters. Can this be the result of internal processes, or is it a chain of craters resulting from the impact of a fragmented comet?

The transitional area also contains an 80-kilometre wide lens-shaped tectonic feature. The appearance of this feature is probably due to shearing of the surface where areas have slid past each other and been slightly rotated. In several places, especially around the border of the lens, bright ridges appear to turn into dark grooves. The presence of such pronounced tectonism on the fringe of the dark Marius Regio indicates that the forces which created the nearby sulcus acted on a fairly wide area. To this extent, Ganymede's sulci resemble the 'rift valley' systems which act to split continents apart and form proto-oceans on Earth.

DIFFERENTIATED INTERIOR

The results from the particles and fields instruments from the two Ganymede fly-bys were reported in December 1996. The plasma wave spectrometer's data was conventionally shown as a brightly coloured spectrogram. "The instant I saw the spectrogram, I could tell that we'd passed through a magnetosphere at Ganymede," reported Donald Gurnett of the University of Iowa, and the experiment's principal investigator.

The radio signal data was transformed into an audio signal by compressing 50 minutes of data into a 1-minute segment and dropping the pitch to suit a human ear. When this was played to the assembled Press, it sounded like "soaring whistles and hissing static". It was promptly dubbed the 'Ganymede Symphony'. The approach to Ganymede was quiescent, comprising a rather monotonous 'sizzle' in the Jovian magnetosphere. Suddenly, there was a crescendo of 'bangs' and 'pops' when entering Ganymede's magnetosphere. There was firstly a rising, then a falling, tone as the spacecraft approached and gradually withdrew from the moon, plus a background of low growling rumbles (called a 'chorus') typical of being in a magnetosphere. Finally, there was another crescendo when exiting Ganymede's magnetosphere to resume the Jovian 'sizzle'.

"The magnetic field is strong enough to carve out a magnetosphere with a clearly defined boundary within Jupiter's magnetosphere," confirmed Margaret Kivelson, the magnetometer leader. "It was thought possible that Ganymede's interior hadn't ever separated into different layers – and even if it had separated, its interior was thought to have frozen solid during the lifetime of the Solar System, so there was little reason to expect that it would have a magnetic field," she said later in a public lecture at the University of California at Los Angeles, "but we were wrong." The moon is far more 'evolved' than had been expected.

The very close fly-by also provided an excellent opportunity for the Deep Space Network to utilise doppler measurements to probe the internal structure. "These data show clearly that Ganymede has differentiated into a core and a mantle," reported John Anderson of JPL,[19] the celestial mechanics team leader. There is an 800-kilometres thick layer of warm ice beneath the warped and faulted icy crust, an equally thick mantle of rock, and a dense core.

"Combined with the discovery of an intrinsic magnetic field," Anderson said, "our gravity results indicate that Ganymede has a metallic core." Depending on whether it is pure iron or a mixture of iron and iron sulphide, it might comprise as little as 1.4 percent, or as much as 30 percent of Ganymede's mass. "Without its ice, Ganymede's interior looks a lot like Io's, with perhaps a little more rock and a little less core."

The chemical differentiation required to form such a metallic core indicated that Ganymede had undergone significant internal heating. In the case of an object as large as Ganymede – in effect a small planet – the accretional and likely radiogenic heating would certainly have been sufficient to invoke partial melting.[20] This may have been further stimulated by a pulse of heat from gravitational stresses as Ganymede's rotation was slowed to synchronise with its orbital motion. And the orbital resonance with Europa may serve to keep Ganymede's interior warm. There is no sign of current cryovolcanism, however.

Ganymede was found to bear a striking structural similarity to the Earth. This was a major discovery because it increased by one the number of 'terrestrial planets' available for students of comparative planetology. If volcanic Io was also included, this further improved the sample. Before the 'Space Age', the Jovian system – so far from the Sun – had been expected to be a frozen wasteland, but it was proving to be very different.

Galileo's orbital tour would not return to Ganymede's vicinity until April 1997. However, a 35,000-kilometre pass on the sixth orbit enabled the spacecraft to snap an image of the moon's trailing hemisphere, which had not been viewed very well by the Voyagers, and so fill a gap in the map with a resolution of about 3 kilometres per pixel, which was comparable with the best achieved by the Voyagers in other regions.

NICHOLSON REGIO

On the seventh cycle of its orbital tour, Galileo intercepted Ganymede after perijove rather than before perijove. Since the moon's axial rotation was tied to its motion

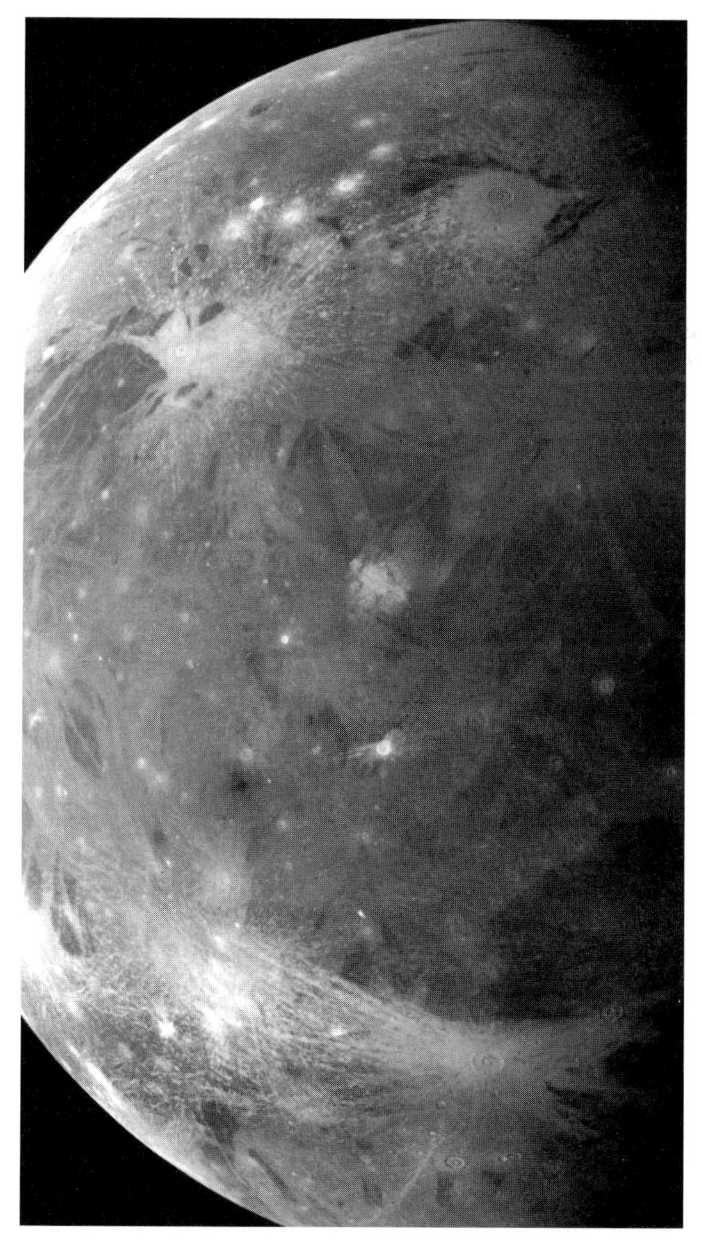

On its sixth orbit, Galileo took this hemispherical view of Ganymede from a range of 34,386 kilometres so as to document a longitude range which was poorly mapped by the Voyagers (the image is centred on 285 degrees, immediately west of Marius Regio). In addition to the fragments of dark terrain split by sulci, and several bright young ray craters, there is a linear chain in the equatorial area which might be the result of a strike in the style of Shoemaker Levy 9. The 3.6-kilometres per pixel resolution matches that of the Voyager map.

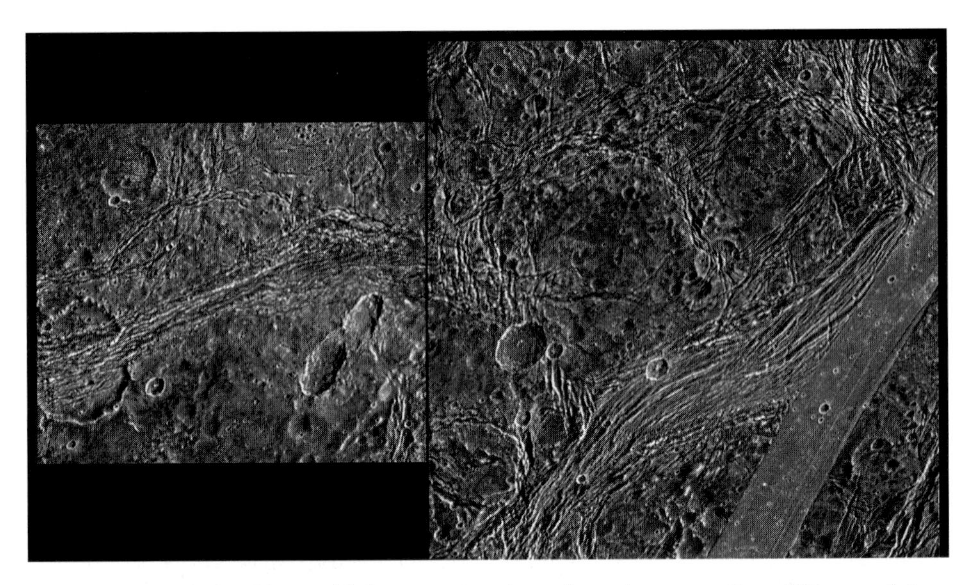

Galileo's seventh orbit provided an opportunity to investigate a section of Ganymede's Nicholson Regio at 180 metres per pixel. A strip of remarkably smooth-looking sulcus cuts across the southeastern part of the documented area. Just to the west, and curving to the south, is a heavily fractured lane which had appeared bright in Voyager's lower resolution view, but is evidently transitional in nature. The extensional stresses have disrupted much of the ancient dark terrain, in some cases apparently being focused by weaknesses created by older craters, transecting them. The large crater far to the west shows this particularly well. This relationship provides a means of dating the tectonic activity. The old crater towards the northwest has been partially buried by dark ejecta from another impact about 50 kilometres further north (beyond the edge of this image).

around Jupiter, the illumination had 'migrated' west by about 100 degrees, and so ran from the western fringe of Marius Regio across a broad swath of trailing hemisphere terrain which the Voyagers had not been able to map very well, and on over the sub-Jovian hemisphere to Nicholson Regio. This facilitated regional mapping of Nicholson Regio. The closest approach of the 3,100-kilometre fly-by was at 56 degrees north and 88 degrees west, northwest of Perrine Regio, just beyond the terminator, and so in darkness.

The dark terrain of Nicholson Regio is intensely fractured by tectonic forces, and at least one large crater has actually been torn apart. A lane of ridges and grooves, most probably extensional fault blocks, have distorted the crater's originally circular shape. A strip of particularly smooth sulci was nearby. Immediately to the west of this, a very heavily fractured lane of dark terrain curved towards the south. This transitional terrain had appeared relatively bright in lower resolution Voyager imagery. It was further evidence that the tectonic forces which created some of the sulci acted over wide areas and disrupted the dark terrain for a considerable distance.

A pair of oblong craters may designate the impact of a binary asteroid of the type that was proving to be fairly common, and the oblong shapes of these craters suggested that the impactors had struck the surface at a shallow angle. An older crater to the north had been partially buried by dark ejecta from an impact beyond the field of view. The age relationships of these features told a story of prolonged bombardment.

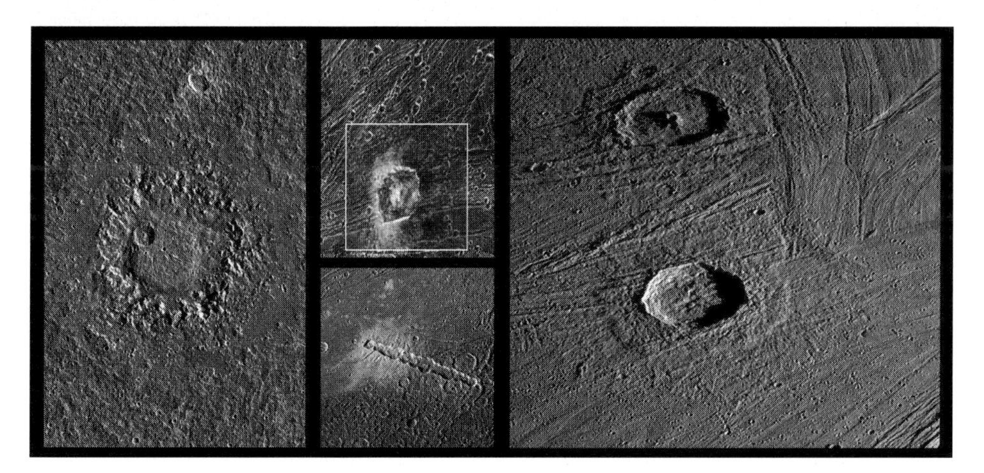

Galileo investigated a number of interesting craters on Ganymede on its seventh orbit. Neith (left) is 160 kilometres across. It was documented at a resolution of 150 metres per pixel. It has a prominent 45-kilometre wide 'dome' on its floor. The dome is surrounded by a wreath of rugged terrain. The wreath does not represent the original crater rim but the rim of a large central pit. The rim itself is barely visible and is located along the outer boundary of a relatively smooth circular area, assumed to be the crater floor, which in turn surrounds the wreath of rugged terrain. The rim is not circular but appears to be petal-shaped. In some parts along the rim, inward-facing scarps may be seen. Outside the rim, a continuous ejecta blanket may be discerned. The 13-crater chain of Enki Catena (lower centre) probably formed when a fragmented comet rained down, in the style of Shoemaker Levy 9. It crosses the margin from older dark terrain to a lighter-toned sulcus, which is marked by a narrow trough. The ejecta deposit surrounding the craters appears very bright on the sulcus. Even though all the craters formed nearly simultaneously, it is difficult to discern any ejecta deposit on the darker terrain, possibly because these impacts excavated and mixed dark material into the ejecta and the result does not stand out against the dark background. Kittu (top centre) was imaged at 280 metres per pixel. In addition to dark rays, it has a white central peak and rim. Its diffuse dark rays thinly mask the surrounding grooved terrain. The southern rim was straightened by collapse along the trend of a pre-existing fault. The oblique view (right) of a pair of fresh impact craters in the light-toned terrain near Ganymede's north pole at 175 metres per pixel shows 38-kilometre Gula (to the north) and 32-kilometre Achelous. The lobate ejecta deposits imply that these impacts melted substantial amount of ice, which froze as it splashed out. Craters with such prominent flows are sometimes referred to as 'pedestal' craters.

CRATERS

Galileo also took high-resolution imagery of a number of high-energy impacts for insight into the formation of craters on this icy surface. The Kittu dark-ray crater (which is located on the equator at 335 degrees) has a bright white central peak and rim and is surrounded by dark material. Diffuse rays of dark material cover the surrounding grooved terrain. Because the impactor struck the grooved terrain, a portion of the rim has collapsed along the trend of an older fault, giving the crater a straight edge.

At about 160 kilometres across, Neith (which is situated on dark terrain north of Barnard Regio) turned out to be an unusual impact structure. Impact features like this have been called 'pene-palimpsests' by some geologists and 'dome' craters by others and are considered to be transitional between craters and palimpsests. The large circular dome within Neith is some 45 kilometres across and is surrounded by a wreath of rugged terrain. However, this terrain does not represent the original crater rim; it is the rim of a large central pit. The rim itself is actually barely visible, located along the outer boundary of a relatively smooth circular area – the remains of the crater floor. In some parts along the rim, inward-facing scarps are in evidence. The rim is not circular, but appears to be petal-shaped, and outside the rim there is a substantial ejecta blanket. The morphology of an impact feature such as Neith results either from the immediate response of a relatively weak surface to a high-energy impact, or from its progressive viscous relaxation over a period of several hundred million years. It is difficult to say which.

Absolute ages derived from crater frequency measurements are model-dependent. Using a crater chronology model based on impacts dominated by asteroids, Neith might be old, perhaps even dating back to the period of the Great Bombardment that drew to a close about 3.8 billion years ago. However, using a model based on a more or less constant impact rate by comets, it may be 'only' about 1 billion years old. One of Galileo's tasks was to gather statistical evidence of cratering on the Jovian moons in order to make reliable estimates of the ages of their various surfaces.

Galileo took an oblique view of Gula and Achelous – a pair of fresh craters located in sulci terrain on the meridian to the north of Barnard Regio. At 38 kilometres in diameter, Gula, the most northerly, has a distinctive central peak. Achelous is 32 kilometres wide. Both have lobate ejecta deposits that extend out about a crater's radius from the rim, and terminate with scarps. Such craters, which appear to have excavated slushy ice which froze in place, have been called 'pedestal craters'. The star of the show, however, was nearby Enki Catena. This has a chain of 13 craters which were probably formed by a string of cometary fragments. It straddles the boundary between areas of bright and dark terrain northeast of Perrine Regio. Interestingly, a narrow trough marks the interface. Although the ejecta is very bright on the bright terrain, it is difficult to see on the dark terrain, perhaps because the impacts there excavated darker material, in which case the dark tone that characterises the regios is not simply a surficial 'weathered' mantle but extends for some depth.

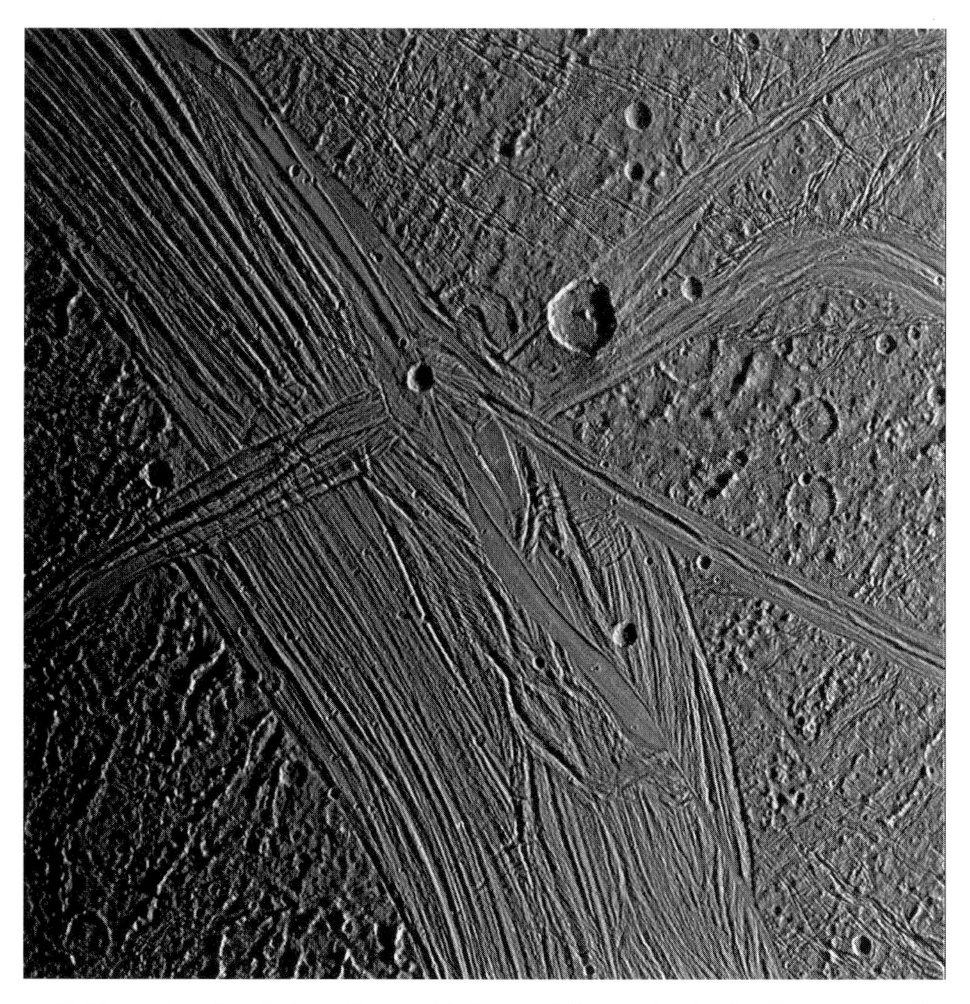

Galileo took a close look at Ganymede's Tiamat Sulcus on its eighth orbit. A study of the Voyager imagery had suggested that this sulcus was transected by a transform (or 'strike-slip') fault, and this proved to be the case. Galileo imaged it at 500 metres per pixel just after sunrise in order to highlight the grooved topography. Tiamat is divided by the east–west trending Kishar Sulcus. The southern part of Tiamat is wider than its northern counterpart, indicating that there has been a greater degree of extension south of Kishar Sulcus than north of it. The portion of Kishar Sulcus to the right of Tiamat Sulcus appears to have slipped horizontally along a northwest–southeast trending fault. The adjacent dark terrain is marked by furrows and fractures too.

MARIUS REGIO

Galileo's fourth Ganymede fly-by, which occurred in May 1997 on the eighth cycle of the orbital tour, marked the final opportunity to inspect this particular satellite up

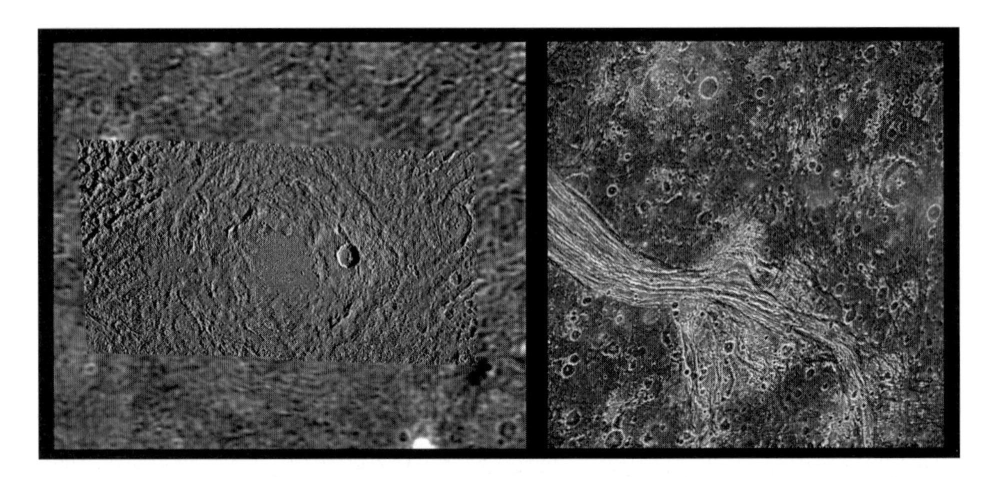

On its eighth orbit, Galileo imaged at a resolution of 288 metres per pixel the southern end of Lagash Sulcus (right), which transects Marius Regio. In this case, the tectonic stresses would appear to have been concentrated within the sulcus because (apart from the diamond-shaped bright area) there has been little deformation to the adjacent ancient dark terrain. A deep trough marks the actual interface between two types of terrain. On this orbit, Galileo also composed a mosaic spanning Buto Facula, a palimpsest located on Marius Regio, under low-angle illumination at 180 metres per pixel (left). A number of morphological zones can be distinguished working outwards from the centre of the structure. The centre of the palimpsest is about 45 kilometres across, and is dominated by a relatively smooth area which shows isolated small hills near the limit of resolution. The outline of the smooth area is roughly elliptical and in some parts it is petal-shaped. Inward-facing scarps occur along some parts of the outer boundary of this inner zone. The next outer zone is characterised by a much rougher surface and two to three almost circular ridges about 40 kilometres apart which do not form closed circles but concentric arcs instead. The outermost zone is somewhat less rough than the ridge-arc zone and shows vestiges of the furrows which characterise the dark terrain. There are also chains of secondaries out beyond the periphery. Overall, Buto Facula is about 300 kilometres in diameter.

close. However, the moon was on the far side of its orbit from the previous encounter and so faced its opposite hemisphere to the Sun. The exit trajectory required a fairly remote 1,600-kilometre fly-by over the darkened trailing hemisphere. By intercepting the moon slightly earlier prior to perijove than it had on its first two fly-bys, Galileo was able to record Marius Regio on the way in and, on the way out, to improve upon the Voyager coverage of a swath of the leading hemisphere ranging from Nicholson Regio westward to Galileo Regio. An occultation provided an opportunity to probe the physical properties of its tenuous atmosphere.

This time, a variety of features on Ganymede were to be imaged, including Osiris (a crater with very bright rays and a 'domed' floor), craters with dark floors, a multiple-ringed structure, a fault crossing Tiamat Sulcus, and a feature that appeared to be a site of cryovolcanic activity.

Galileo snapped a close-up of Tiamat Sulcus just after sunrise in order to enhance the relief of Kishar Sulcus which crosses perpendicular to the axis of the grooved terrain. The low solar illumination also highlighted the many furrows and fractures in the surrounding darker terrain. Tiamat Sulcus is divided in two by Kishar Sulcus and it is much wider in the south than it is in the north, indicating that there has been greater extensional stress south of Kishar Sulcus than north of it. Furthermore, the portion of Kishar Sulcus to the east of Tiamat Sulcus appears to have moved (with respect to its western counterpart) on a northwest–southeast strike-slip fault, marking this as the first clear evidence of a transform fault on Ganymede.

Lagash Sulcus is a highly fractured lane of grooved terrain cutting through eastern Marius Regio. Its boundary is marked by a deep trough, and the absence of fractures beyond the trough indicates that, in this case, the tectonic process was highly constrained to the sulcus itself.

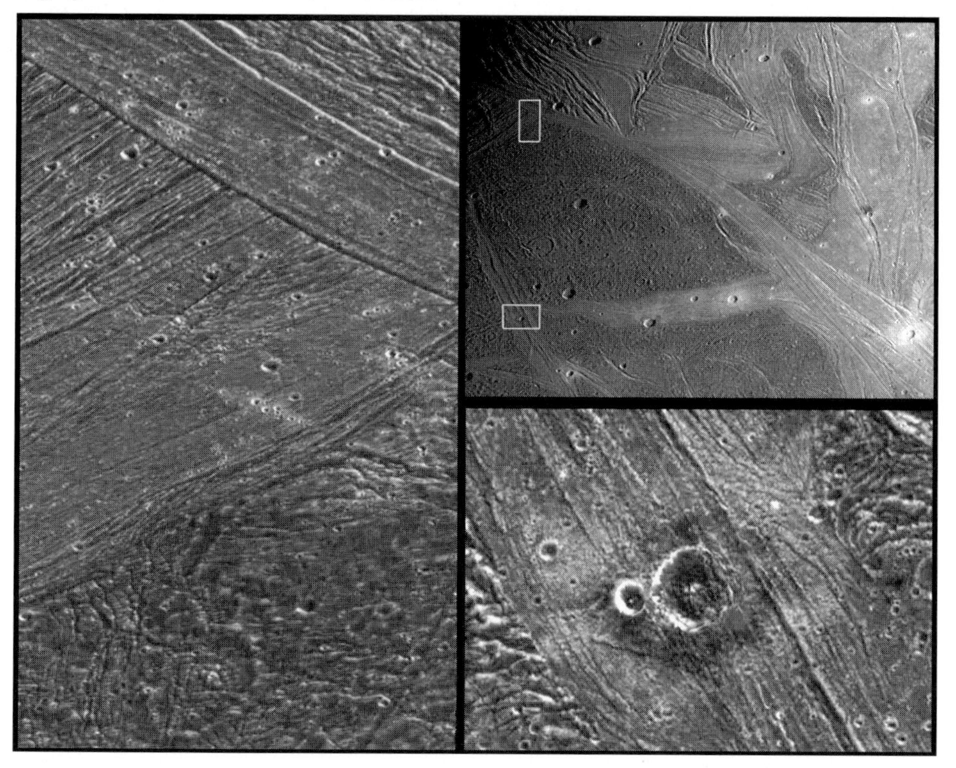

On its eighth orbit, Galileo took a medium-resolution view of the northwestern part of Marius Regio (top right) to provide context for a pair of high-resolution images it had taken on its second orbit. The 8-kilometre wide dark ray crater Nergal superimposed on Byblus Sulcus had been documented at 86 metres per pixel (lower right). This impact evidently melted the ice-rich sulcus and splashed it out, making a lobate ejecta blanket. The 54 × 90-kilometre mosaic (left) shows the abutment of the ancient dark terrain of Marius Regio (to the south in this image), the light-toned Philus Sulcus (to the west), and Nippur Sulcus (to the north). The cross-cutting relationship between the two sulci is stark.

Buto Facula is a palimpsest on Marius Regio. Galileo observed it with the Sun low on the eastern horizon, and with a resolution of 180 metres per pixel to refine the Voyager coverage. Several distinct morphologies are evident: the centre of the palimpsest – which is about 45 kilometres across – is dominated by a relatively smooth plain with isolated small hills visible near the limit of resolution; the outline of the plain is roughly elliptical, but in some parts it is petal-shaped; inward-facing scarps are present along some parts of the outer boundary of the plain; the surrounding zone is considerably rougher; there are several concentric arc-ridges about 40 kilometres apart; and the outermost zone, being somewhat less rough than the arcuate zone, displays vestiges of the underlying topography, including dark-terrain furrows and (off to the northeast) half of an older crater. The complete structure is 300 kilometres across, but this includes the eroded inner part of the ejecta blanket. The chains of secondary craters just outside the outer boundary confirm that Buto Facula really is an impact structure. The clusters of small craters inside Buto Facula are probably secondaries derived from the 20-kilometre fresh impact on one of the ridges.

Galileo took a regional view of Marius Regio and Nippur Sulcus near the terminator, in order to provide geologic context for smaller areas which it imaged at much higher resolution during the second encounter sequence. In addition to the furrowed and heavily cratered dark Marius Regio, this showed the abutting Byblus Sulcus, Philus Sulcus and Nippur Sulcus.

The prominent 8-kilometre crater in the close up of Byblus Sulcus is Nergal; the smaller crater to the west has yet to be named. Their distinctive ejecta blankets are darker close-in, and brighter further away. The inner sections of the ejecta possess lobate characteristics suggestive of frozen slush. Since such small impacts are unlikely to have broken through the icy crust to release slush from below, this fluid ejecta is more likely to be sulci material that was melted by the energy of the impacts.

Further north, the cross-cutting relationship between Philus Sulcus (the light-toned terrain immediately northwest of Marius Regio) and Nippur Sulcus (the light-toned terrain further to the north) is clearly illustrated – new terrain overlays older terrain, which overlays even older material. The ancient dark surface is tectonically deformed, as is an 18-kilometre impact crater. On the basis of the low-resolution Voyager imagery, the light-toned patch on Marius Regio adjacent to Philus Sulcus was interpreted as a cryovolcanic flow onto the dark surface, but Galileo found this brightening to be the result of intense fracturing of the dark terrain. Two bright patches to the north of this area were interpreted as fresh craters in the Voyager image, but were found to be chains of fresh secondaries from distant impacts.

Erech Sulcus cuts north–south across Marius Regio and is truncated at its southern end by the smoother bright terrain of Sippar Sulcus, which trends roughly east–west. Although both are light toned in comparison to the dark terrain, Erech displays prominent relief suggestive of tectonic stresses – the terrestrial East African Rift shows similar sets of faults. The relatively smooth appearance of Sippar Sulcus suggests that its raw relief is masked by extensive cryovolcanic flooding. Stereoscopic analysis[21] found that the sulci are typically depressed by up to 1,000 metres relative to the adjacent dark terrain, and that the least-grooved sulci are the most depressed.

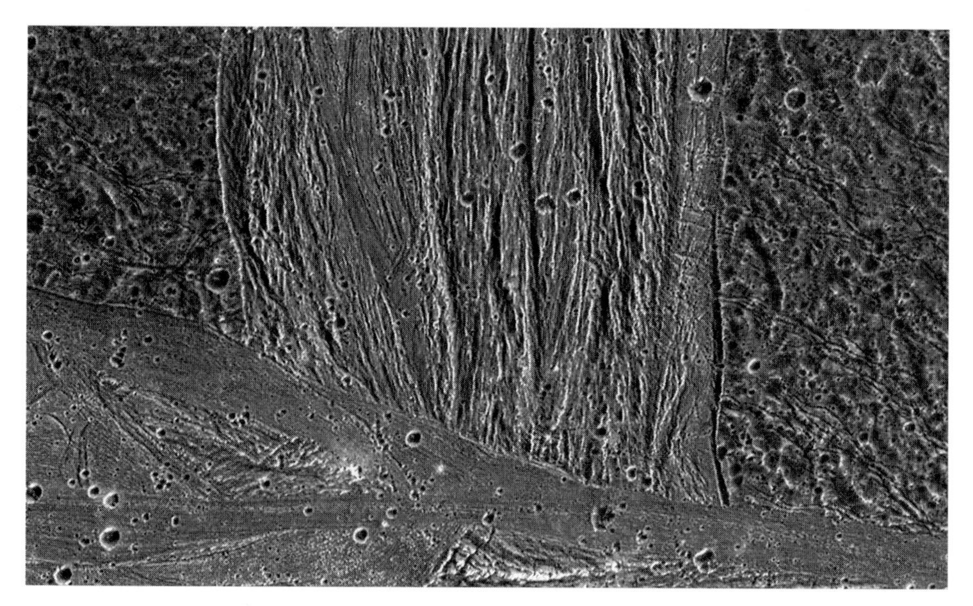

Galileo imaged the southern end of Erech Sulcus where it abuts Sippar Sulcus, on its eighth orbit, at a resolution of 150 metres per pixel. The grooving in Erech Sulcus, a 'rift valley' transecting the eastern tip of Marius Regio, is pronounced. A topographic analysis has indicated that the smoother sulci reside at a lower level than their heavily grooved counterparts (which are at or just below the level of the adjacent dark terrain). The smoother appearance of Sippar Sulcus suggests that a cryovolcanic flood has masked its grooves.

This supported the argument that low-viscosity cryogenic fluid had accumulated in the deeper areas. The 85-kilometre wide Erech Sulcus, which is intensely grooved, resides at almost the same elevation as the nearby dark terrain. The smoother terrain along the northern margin of Sippar Sulcus, however, is 700 metres lower. The embayment of the fringe of the dark terrain by the lighter material in Sippar Sulcus provides further support for the smoother sulci having been flooded.

ICE CALDERA

The Voyager imagery of Sippar Sulcus had recorded a horseshoe of curvilinear and arcuate scarps which appeared to contain a depression. Galileo's high-resolution view showed that it contains a lobate flow-like feature. Some 55 kilometres in length and 20 kilometres wide, the depression has scalloped walls and internal terraces, and has a lobate flow-like deposit some 10 kilometres wide within it with ridges which curve outwards – and downslope – towards a cross-cutting lane of grooved sulci terrain. The morphology indicates that a series of eruptions of fluid created a flow that then eroded into the icy surface and formed its own 'caldera'. This is considered to be the best evidence of isolated cryovolcanic activity on Ganymede. In fact,

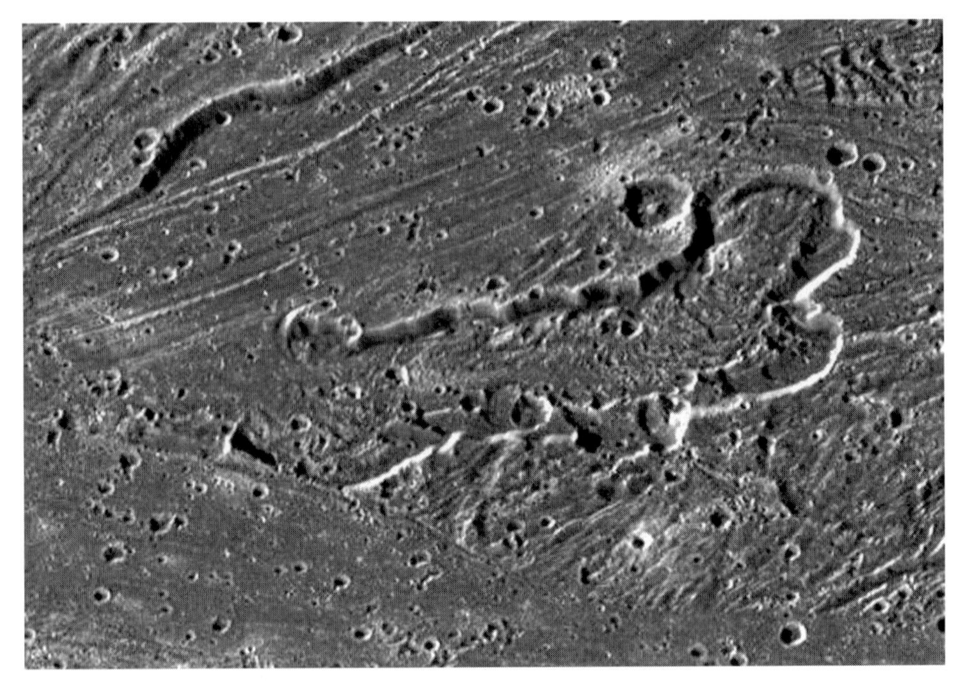

Analysis of the Voyager imagery of Sippar Sulcus identified a feature with curvilinear and arcuate scarps which suggested that it was the source of a cryovolcanic flow. On its eighth orbit, Galileo inspected it at a resolution of 175 metres per pixel and found a prominent depression 55 kilometres long. The scalloped and terraced walls enclose a depression with a lobate flow on its floor, supporting its interpretation as a cryogenic caldera.

Galileo had had time to inspect only one of eight such open-ended cuspate features on Sippar Sulcus which had been identified in the Voyager imagery.[22, 23, 24] They are all superimposed on the grooved terrain, and have given rise to swaths of smooth terrain that appear to be the youngest features in their locality.

CRYOGENIC GEOLOGY

The most intensely deformed reticulate bright terrains are the oldest parts of Sippar Sulcus, and are also the most elevated, standing 300 to 500 metres above the younger cross-cutting linearly grooved terrain, which is in turn a similar elevation above the smooth terrain (which is typically 1,000 metres below the level of the nearby dark terrain). Evidently, the tectonic forces which formed the fractures subsequently facilitated the upwelling of fluid which pooled in the lowest-lying territory and froze. The fact that some sulci are 800 kilometres long indicates that low-viscosity aqueous lava was readily available at shallow depth on at least a regional basis, and possibly even a global basis. The evidence from the palimpsests suggested that the ice was no more than, say, 15 kilometres thick at the time of their formation. Only one-third of the early surface persists in the form of the dark terrains. The brighter terrain is the result of resurfacing the rest of the moon.[25, 26] Such extensive crustal melting was undoubtedly a consequence of the interior undergoing early thermal differentiation.

Although no more close encounters with Ganymede were feasible for a while, Galileo was occasionally able to exploit remote passes to plug gaps in its overall coverage. On the twelfth orbit, for example, it imaged the Gilgamesh multiple-ringed structure from 17,000 kilometres for improved statistics of cratering, and a 920,000-kilometre pass on the fourteenth orbit provided a colour image to refine the moon's radius, shape, colour and albedo, as well as to survey the mobility of frost on its surface. With Galileo's attention directed elsewhere for a while, the Ganymede researcher settled down to a period of reflection.

The Voyagers had provided evidence of tectonic activity, and Galileo confirmed and refined this with striking examples of crustal spreading. There is a complex transition from the dark- to the light-toned terrains, with parallel ridges running for hundreds of kilometres. The icy crust evidently froze during the Great Bombardment, and cryovolcanic flows suggest that an ocean existed below the surface for some time thereafter, just as one seems still to exist on Europa.

If the well-preserved Gilgamesh multiple-ringed structure is presumed to date from the tail-end of the most intense period of bombardment, and it is assigned to 3.8 billion years in order to provide a calibration point, the oldest dark terrain dated to about 4.2 billion years, and the youngest of the light-toned sulci to 3.7 billion years, which implies that the tectonism occurred over a period of half a billion years. If there is any ongoing cryovolcanism, it does not seem to modify the surface to any significant extent. Any such activity is probably limited to small water geysers.

Ganymede's remarkable geological history derives from the moon's proximity to Jupiter and to its orbital resonances with Europa and Io. With a differentiated structure, Ganymede is remarkably similar to the Earth, and more of a 'terrestrial planet' than a frozen remnant from the formation of the Solar System.

10

Battered Callisto

COMMUNICATIONS

In October 1996, the Deep Space Network tried electronically linking an array of antennas. Integrating the received power of their signals would further enhance the effective transmission rate of Galileo's low-gain antenna. The 70-metre antenna at Goldstone in California was linked to the 70-metre and 34-metre antennas at Tidbinbilla near Canberra, Australia, and the 64-metre Parkes radio telescope antenna 160 kilometres away.

The antenna-arraying technique was declared operational on 1 November, to enable Galileo to raise its data rate, which was timely because the spacecraft had only just had time to transmit the data from its second encounter sequence using the revised software before it had had to set up for Callisto.

Reprogramming the spacecraft to edit and compress data prior to transmission had boosted the effective data rate by a factor of 10. Arraying the antennas, plus a variety of other detector improvements, boosted the effective rate by another factor of 10, so the low-gain antenna was now able to pump out 1 kilobits per second (it could not actually transmit at this rate of course, but the amount of data that it could send was 100 times greater than would have been possible in the absence of all of these modifications – the high-gain antenna, remember, was to have transmitted at *134 kilo*bits per second).

"As the Earth turns relative to Galileo's position in the sky, different arrays of antennas will 'hand off' the receipt of data over a 12-hour period," explained Leslie Deutsch, one of the JPL communications engineers who had helped develop the upgrade strategy. "For two hours a day, up to five antennas are pointing in unison to receive transmissions from Galileo."

"With our spacecraft software and various ground station improvements already in place," pointed out Neal Ausman, the Mission Director, "this new arraying capability is 'icing on the cake'."

However, the inherently low rate of Galileo's low-gain antenna meant that the Deep Space Network had to follow Galileo virtually continuously for extended periods, which restricted the use of the network for other spacecraft.

"Galileo gets the credit for giving arraying a large push," said Joseph Statman, a JPL Deep Space Network engineer. This was "the way of the future, because no more 70-metre antennas will be built for the DSN, only 34-metre antennas".

Whenever necessary, Goldstone's cluster of 34-metre antennas were to be arrayed so as to simulate a 70-metre antenna, otherwise they were to operate individually, and so serve a larger number of spacecraft.

"The methods used, and much of the equipment, will be especially useful for the new era of 'faster, better, cheaper' interplanetary spacecraft," said Paul Westmoreland, JPL's Director of Telecommunications and Mission Operations. "This opens the way for new mission developers to reduce the cost of future spacecraft and operations by using smaller spacecraft antennas and transmitters."

"We knew that the Deep Space Network would have to support an increasing number of upcoming missions, and we had several older antennas which were nearing the end of their useful lives," said Jeff Osman, the Deep Space Network Antenna Project Manager. "That is why we decided to build these new antennas now."

The ongoing upgrade called for adding three 34-metre antennas at Goldstone, all of which were now operational, one at Canberra (due to be commissioned in February 1997) and one at Madrid (October 1997). Similarly-sized antennas were already available with less sophisticated receiver technology. These had been built in the 1960s as 26-metre dishes, and then increased in the mid-1970s to 34-metres to serve Voyager. However, they were no longer cost-effective to operate and maintain. The Cassini spacecraft, which was then being prepared to explore the Saturnian system, is to exploit the fact that the new 34-metre antennas are capable of handling high-capacity Ka-Band (32 GHz) transmissions.

VOYAGER'S CALLISTO

Callisto, the outermost of the Galilean satellites, orbits at a distance of 27 Rj with a period of some 17 days. At 4,800 kilometres in diameter, it is the second largest of Jupiter's moons. Actually, being only slightly smaller than Ganymede it, too, is effectively a small 'terrestrial' planet that happens to be orbiting Jupiter. However, with a bulk density of just 1.83 g/cm^3, it has to be roughly half ice and half rock. Although the least reflective of the moons, its surface is nevertheless brighter than the average albedo of the Earth's Moon. The ice has probably a high proportion of impurities. It may be dark for the same reason that Ganymede's oldest terrain is dark and, if this is so, then the absence of bright sulci implies that Callisto has not been subjected to tectonic spreading.

The closest Voyager approach was 124,000 kilometres and some 80 percent of the moon was mapped at 5-kilometre resolution. Most of its surface is a monotonous cratered terrain, broken only by the large multiple-ringed impact structures. The fact that it is peppered with craters to the limit of the Voyager resolution indicates that it is a *very* old surface.

This Voyager image shows the vast scale of the Valhalla multiple-ringed structure. Although it is the Callistoan equivalent of Ganymede's Gilgamesh basin (in that it is the site of a major impact) its morphology is comparatively subdued.

The largest, and most striking topographic feature on Callisto is the Valhalla multiple-ringed impact basin. It comprises a smooth central plain 600 kilometres across, and a series of rings spaced 20 to 100 kilometres apart, extending to a radius of 2,000 kilometres. The spacing of the rings increases with increasing radius, a relationship that places a constraint on any model for the formation of the structure. The impact which excavated the central plain may well have come close to punching right through the icy crust.

The population density of smaller craters is low on the central plain, which implies both that there has been a low impact rate since the basin formed, and that the basin formed towards the end of the period of heavy bombardment. The density of craters increases with radial distance through the rings, to merge with the prevailing density beyond. The existence of older craters within the rings indicates that the ring-forms are 'frozen' relics of the shock which propagated radially through the crust.

Although Valhalla bears a striking resemblance to the Moon's Orientale Basin, the fact that Callisto's crust is icy means that the morphologies are different in detail. Orientale's rings are composed of fractured and upthrust crustal blocks, but Valhalla's rings are ridges, troughs and outward-facing scarps. The central plain (which is effectively a palimpsest, marking the extent of the impact's excavation) is not rimmed by an inward-facing scarp front, as is the case with the lunar basins.

Three distinct morphological zones can be identified within Valhalla. There is an inner zone comprising the central plain and the first few rings.[1]* The rings in this zone consist of bands of intermediate-albedo terrain which are sinuous and scalloped, and might really be ridges. In the transition zone, the rings are discontinuous. In the outer zone, the rings take several forms: to the south and east they are narrow sinuous troughs with light-toned floors; to the north and west they are outward-facing scarps with dark heavily cratered back slopes, and there is some light-toned material running along the base of the scarp and flooding into nearby small craters.

The Asgard basin is similar, but smaller. Its central plain is 230 kilometres in diameter and its rings extend out some 800 kilometres. There are indications of half a dozen other basins on Callisto, but they are even more subdued.

The Voyagers did not detect anything to suggest that Callisto's surface has been shaped by endogenic processes so, in contrast to its inner neighbours, apart from continuing cratering, it has evidently been inert for most of its existence. Callisto's state undoubtedly reflects both its greater distance from Jupiter and the fact that it is not tormented by any orbital resonances. In fact, its 'aged' appearance was its most distinctive feature! In a sense, Callisto was the least interesting of the Jovian moons.

GALILEO'S FIRST LOOK

Galileo's imaging objectives at Callisto were:

- to acquire high-resolution samples of typical cratered terrain and components of the Valhalla and Asgard basins
- to fill gaps in the Voyager coverage
- to make multispectral surveys of the surface at local, regional, and global scales
- to determine the photometric properties of the surface.

* For Notes and References, see pages 413–424.

Long-range infrared spectrometer observations on Galileo's second orbit established that there is carbon dioxide frost on Callisto's surface.[2] Follow-up observations[3] revealed localised correlations with ice-rich features, and a global distribution which concentrated on the trailing hemisphere – which implied that an exogenic agent (possibly the charged particles circulating in the Jovian magnetosphere) was active in forming the frost. Ironically, the polar zones were least frosty. The distribution of sulphur dioxide was rather more patchy, and the absence of a concentration on the trailing hemisphere suggested that the sulphur ions in the magnetosphere (derived from Io's volcanoes) were not interacting with the surface.[4]

Galileo's third encounter sequence was to focus on Callisto, with a 1,100-kilometre fly-by. The general objectives were to survey the composition of the surface features, to determine the relative age relationships of the surface features, and to search for signs of tectonic activity on this seemingly ancient surface.

The infrared spectrometer undertook general mapping in order to determine the composition and mineralogy of the surface on a regional basis. Of particular interest was the distribution of any 'non-ice' material. The spectrometer was ideally suited to detecting hydroxylated silicates, opaque minerals, and organics, but close fly-bys would be needed to combine the instrument's high spectral resolution with high spatial resolution to chart small-scale variability. Telescopic infrared spectroscopy had hinted at the presence of ammoniated clays on the surface.[5] During the primary mission, this instrument was to make charts at 204 wavelengths and 60-kilometre resolution covering approximately half of Callisto's surface (the characteristics of the orbital tour precluded doing so on a global basis). Although some sections of the Valhalla and Asgard structures were to be surveyed at 25-kilometre resolution, this would be at a reduced spectral resolution. The constraints imposed by having to use the low-gain antenna meant that detailed multispectral observations with a surface resolution of a few kilometres per pixel would be practicable only for small areas of special interest.

As Galileo approached Callisto, it saw the trailing anti-Jovian hemisphere in darkness. Its trajectory took it across the dawn terminator and then across the moon's orbit just ahead of the leading hemisphere. At 13 degrees north 78 degrees west, the point of closest approach was in daylight. After imaging Asgard on the way in, it closely inspected Valhalla's central plain, and then traced the far side of this ring structure off towards the evening terminator.

Galileo flew almost 100 times closer than had Voyager, so real revelations were expected. "With data from this encounter, we will find out more about why Callisto is so different from Jupiter's more lively moons," Torrence Johnson predicted.

When Kelly Bender of Arizona State University, a member of the imaging team, presented the 30-metre resolution shots of Callisto in the von Kármán Auditorium on 12 December 1996, even the 'old hands' of planetary exploration community were astonished.

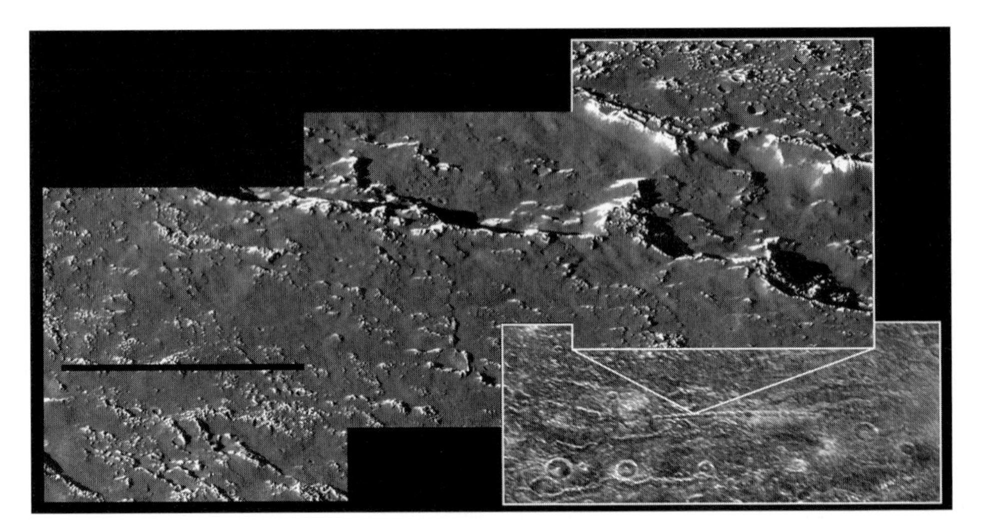

On its third orbit, Galileo investigated a chain of craters located in the northern section of Valhalla at a resolution of about 160 metres. It found a surprising lack of very small craters. Small-scale features appear to have been masked by a blanket of dark material. The bright slopes may indicate 'mass wastage' where the dark material has slipped and exposed fresh ice surface. The dark layer may be several tens of metres thick.

VALHALLA

The highlight was a portion of a chain of about 25 craters (apparently where the fragments of a disrupted comet had struck one after the other) forming an almost continuous trough. But the surprise was what was absent – there was a startling paucity of small craters. It had been expected that there would be an increasing profusion of ever-smaller craters. In fact, the local topography seemed to have been blanketed by dark material that formed an unusually smooth surface. What was this material? How had it been applied? Had the depressions been filled in by a 'splash' of fluidised ejecta from a large impact? If so, where was the impact? Surely, this ejecta would have to have come from one of the multiple-ringed basins? The chain of craters was situated on the northern fringe of the Valhalla basin, at 35 degrees north and 46 degrees west. Had the dark material been ejected by the impact that made *that* basin?

But Galileo revealed Valhalla's central plain to be smooth and dark as well.[6, 7] Since it was inside the impact's excavation radius, the central plain's smoothness suggested that a fluid had welled up through the badly fractured cavity floor. Intriguingly, the rims of some of the craters appeared to have been 'softened' by slumping – 'mass wastage' – that had exposed a bright ice-rich material which highlighted the steep slopes. This was what gave the plain its high albedo in Voyager's lower resolution imagery. Since the dark deposit covered the central part of the ring structure, it could not be basin ejecta, so neither could the material in the peripheral zones. Could it be non-ice material?

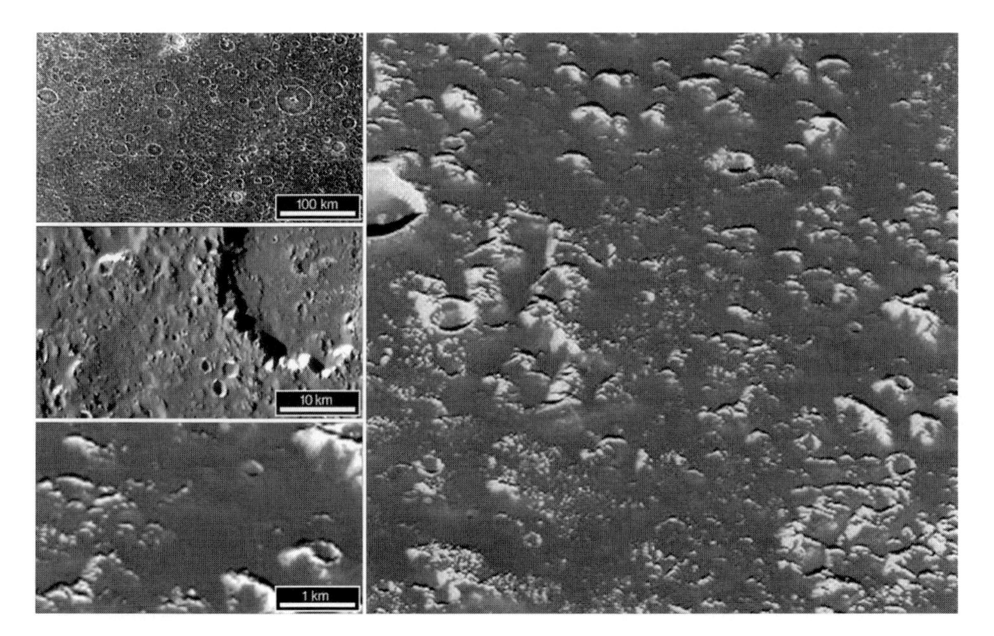

To Voyager, Callisto's battered surface had appeared to be saturated with craters right down to the limit of resolution, but Galileo discovered that there is a paucity of craters less than a few kilometres in diameter. Galileo documented the cratering density over a range of scales. At 1,000 metres per pixel (top left), which compares with the Voyager imagery, craters are the dominant landform, their rims appear bright, and interiors and the adjacent terrain are dark. At 100 metres per pixel (centre left), some of the rims are shown to be incomplete rings composed of bright isolated segments, and dark material has slumped from steep slopes to expose bright ice. At 30 metres per pixel (lower left), the dominant feature is the smooth dark blanket. Significantly, the 60-metres per pixel view of Valhalla's central plain (right) shows that it too is blanketed with dark material.

On its third orbit, Galileo recorded one of Valhalla's outer rings at a resolution of about 46 metres per pixel; it turned out to be a fault scarp.

Galileo also revealed a prominent fault scarp in the Valhalla multiple-ringed structure in striking clarity. Several smaller ridges were in evidence running parallel to the main feature.

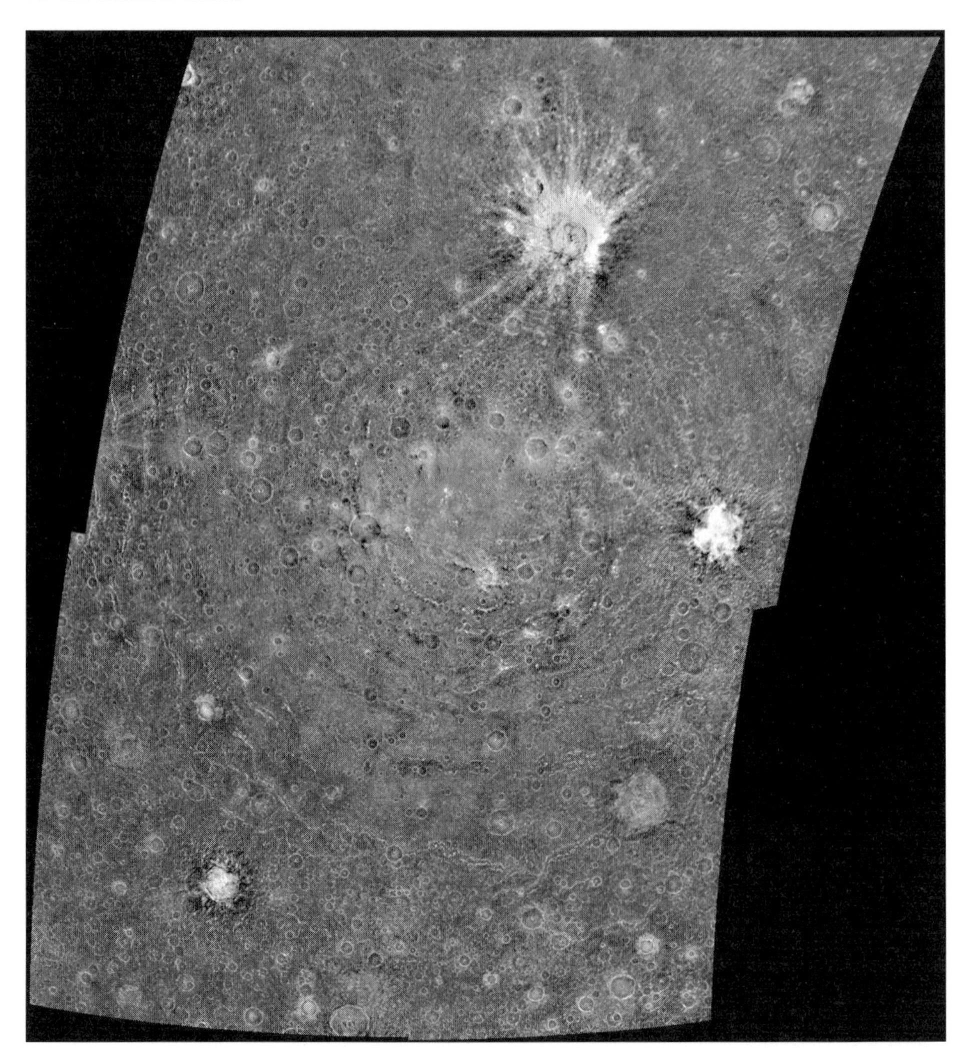

Galileo assembled a regional mosaic on its third orbit to chart the 1,700-kilometre wide Asgard multiple-ringed structure. The bright central plain is surrounded by a series discontinuous rings which are scarps near the central transition zone and troughs at the outer margin. The ray craters were formed long after the Asgard impact.

ASGARD

Asgard's bright central plain was revealed to be surrounded by discontinuous rings which included degraded ridges near the central zone and troughs out towards the outer margin. The subdued form of the rings suggested deformation of the soft icy crust as it settled isostatically after the shock of the impact. Although the infrared spectrometer showed that the inner plain was primarily 'clean' ice, there was a significant fraction of non-ice material in the peripheral ring-form. The structure has been disfigured by subsequent impacts. Doh is a 55-kilometre diameter crater on the bright central plain. It has a central 'dome' taking up half of its width. The morphology of such craters suggested that the impactor struck an already weakened shallow crust, and broke through to a 'slushy' zone. Burr is a 75-kilometre crater on the northern periphery of the central palimpsest, and its bright 'rays' of 'clean' ice imply that it is relatively recent.

CRATERING

There are two methods of estimating the impact rate for the Galilean satellites. One way is to assume that conditions in Jovian space have been similar to those of the inner Solar System and that the size-to-frequency curve derived from counting craters on the Earth's Moon can be applied. This presupposes that the large multiple-ringed structures were excavated by asteroidal bodies impacting during the bombardment which ended around 3.8 billion years ago, and that the population of impactors is now dominated by smaller bodies – with the smallest being the most numerous.

Applying a cratering rate calibrated in the inner Solar System to the Jovian environment is complicated by the fact that Jupiter's tremendously far-reaching gravity draws in asteroids and comets, thereby increasing the population of potential impactors. Furthermore, as this 'focusing' will produce more extensive cratering on satellites closer to the planet, the statistics have to be scaled to compensate for their positions in relation to the planet.

As another complication, the population of large asteroids diminished rapidly over time, but the population of comets did not. To derive an age estimate for a surface based on a population dominated by comets it is necessary (a) to determine the current population, (b) to assume that it is more or less constant, and (c) scaling for Jupiter's proximity, to run the rate backwards in time to calculate how long it would have taken to crater each moon to the observed extent. It sounds simple, but correlating the computer model with observation is not straightforward, so there are widely differing opinions.

Gene Shoemaker had concluded early on that the cratering rate calibrated in the inner Solar System could not be applied to Jovian space, where, he believed, the ongoing flux of comets predominates.[8] He conducted a search for comets in order to refine the statistics, and came to the conclusion that Europa's smooth surface is extremely young. If this rate is scaled to the outer moons, it implies implausibly young ages for the 'ancient' terrain on Ganymede and Callisto, and suggests that the multiple-ringed structures were formed a billion years after such massive impacts had ceased in the inner Solar System. Shoemaker's statistics were based on Voyager imagery, however, which could not resolve craters smaller than a few kilometres in diameter, so the curve was frustratingly truncated.

The imagery from Galileo's first few orbits provided a factor of a hundred increase in resolution over Voyager. Although Galileo saw only limited areas in fine detail, this facilitated counting craters down to 100 metres in diameter, and the statistics could be superimposed on the curves derived from Voyager to form a record covering craters up to 100 kilometres in diameter. For Ganymede and Callisto, these curves are consistent with an ancient bombardment by asteroids.[9] Galileo established that Callisto is typically more heavily cratered than Ganymede's dark terrain, and so is older. If Ganymede's well-preserved Gilgamesh multiple-ringed structure is assumed to date to the tail-end of the bombardment and the rate is scaled for Callisto, it implies that Callisto's most ancient terrain is 4.3 billion years old, which makes it the oldest planetary surface yet observed.

UNDIFFERENTIATED INTERIOR

Although the Voyager fly-bys of Callisto had been distant, the manner in which it deflected their trajectories prompted the suggestion that it possessed a lithosphere of ice and rock 200 to 300 kilometres deep, with a 1,000-kilometre thick mantle of convecting 'soft' ice englobing a silicate core. Galileo's first fly-by was expected to refine the model. The doppler data from the Deep Space Network enabled Callisto's moment of inertia to be estimated. This suggested that Callisto does *not* have a central mass concentration. The data suggested that the outermost 300 kilometres is predominantly ice, and that the interior is a homogeneous mix of 40 percent water-ice and 60 percent silicates, iron and iron sulphide in ratios reflecting solar abundances because the moon formed from the solar nebula.[10]

The magnetometer detected no evidence to suggest that Callisto possesses a magnetic field. On the other hand, from the viewpoint of the magnetometer it had been a fairly remote fly-by. Although the plasma wave spectrometer noted what Donald Gurnett, its principal investigator, called "a very minor response", this was "no evidence of a magnetic field, or a magnetosphere". Being only slightly smaller than Ganymede, Callisto can be expected to have undergone only slightly less accretional and radiogenic heating. Even so, Ganymede is evolved and Callisto is not. The fact that the magnetometer did not find a magnetic field is consistent with the absence of an iron core. Evidently, the outermost Galilean satellite never experienced sufficient heating to induce significant chemical differentiation. Callisto would have acquired a thick lithosphere early on as it rapidly cooled, so its surface preserves the tail-end of the accretional process and the multiple-ringed basins have been modified only by localised isostatic adjustment and by later, lesser impacts.

"Callisto", reflected John Anderson of the celestial mechanics team, "has had a much more sedate, predictable, and peaceful history than the other Galilean satellites, and therefore it is a more typical Solar System object." When the Cassini spacecraft starts exploring the Saturnian system, the large moons there may turn out to be smaller cousins of Jupiter's Callisto.

RETURN TO CALLISTO

After providing a tantalising early glimpse of Callisto, Galileo's trajectory did not facilitate another close encounter until its ninth orbit. However, on the seventh it made a colour global mosaic and the infrared spectrometer made a low-resolution mineralogical survey, but it approached no closer than 636,000 kilometres. The second fly-by was on 25 June 1997, when Galileo intercepted the moon earlier in its orbit than it had previously, and as it withdrew it was able to look back to draw its regional coverage east of Valhalla to the sub-Jovian point, which had previously been in shadow. At 2 degrees north and 259 degrees west, the point of closest approach was on the trailing hemisphere, in darkness. And at 416 kilometres, this closer pass would hopefully determine whether the "very minor response" that had been registered by the plasma wave spectrometer was of any significance.

SUB-JOVIAN POINT

The constraints imposed by having to employ the low-gain antenna meant that Galileo had to be selective in its high-resolution coverage. A mosaic of the heavily cratered terrain near the sub-Jovian point revealed a few sinuous valleys emerging from the southern rims of irregular craters, and deposits resembling landslides in the southern and southwestern floors of many of the larger craters. The smooth intercrater plain seemed to be the evidently ubiquitous dark material.

Nearby was the 50-kilometre diameter double-ringed structure, Har, which has a large dome on its floor. The image included the western rim of Tindr, a larger irregular crater to the northeast. Tindr's floor shows numerous irregular pits of a type observed in some other craters, and on the dark plains, and such modification is thought to be due to sublimation of subsurface volatiles. Isolated chains of hills make up its eastern and southeastern rim, while a continuous ejecta blanket covers a few of the older craters to the northeast. This ejecta merges into the surrounding cratered plain without a distinct morphological or albedo boundary, which in turn implies that the dark material was emplaced after Tindr was excavated. If the overall cratering has been dominated by early asteroidal bombardment, Tindr may be as much as 3.9 billion years old, but if a more or less constant rate of cometary impacts has been the dominant factor, it might be younger – possibly 'only' 1 billion years old. Tindr's ejecta has etched Har's eastern rim, so Har is even older.

On its ninth orbit, Galileo assembled a mosaic to record the distribution of craters in an area 215 × 315 kilometres near Callisto's sub-Jovian point (left). The local smoothing by dark material is evident in the plains between the craters in the southeastern area. A number of sinuous valleys emanate from the southern rims of 10- to 15-kilometre wide irregular craters in the west-central area. There is evidence of landslides in the southern and southwestern floors of many craters. The nearby 50-kilometres wide 'double ring' crater Har (lower right) has a massive 'dome' on its floor. It was subsequently pocked with secondaries from Tindr to the northeast (see insert). The 20-kilometre crater that disfigured its western rim formed later. Har's southeastern rim has been 'sculpted' by Tindr's ejecta. Tindr's eastern and southeastern rim appears to be degraded to isolated hills and chains of hill. To the northeast, Tindr's ejecta has filled in older craters. The fact that Tindr ejecta merges with surrounding plain without a distinct morphologic or albedo boundary indicates that the seemingly ubiquitous dark material which blankets Callisto's surface was emplaced after Tindr had formed. The floor of the crater shows numerous irregular pits which are believed to be caused by sublimation of subsurface volatiles.

ASGARD RADIAL

Galileo intercepted Callisto on 17 September 1997 at virtually the same point in its orbit as it had a few months earlier, but this time it made its 538-kilometre fly-by over the illuminated leading hemisphere, with its closest approach at 5 degrees north and 79 degrees west.

In addition to continued probing for a magnetic field by the particles and fields instruments, the ultraviolet spectrometer sought evidence of a surrounding cloud of neutral particles and the infrared spectrometer continued mapping the composition of the surface. The targets for high-resolution imaging included a long chain of craters called Gipul, and a swath running through Asgard's rings.

Galileo's high-resolution radial out through the Asgard multiple-ringed structure turned up a few surprises.[11] An earlier study had identified three distinctive morphologies – the central bright plain, a zone of inward-facing cliffs or scarps, and a zone of discontinuous concentric troughs. The inner rings turned out to be degraded ridges, rather than inward-facing cliffs. Also, in the outermost rings it was evident that dark non-ice material had slipped down the walls of the troughs and made their floors darker than the surrounding cratered plains.

The 'inner' end of the 700-kilometre long radial was centred on the 50-kilometre diameter crater Doh, which is located on Asgard's bright central plain. Since it is within Asgard's excavation radius, Doh post-dates the basin's formation. Its 25-kilometre wide dome probably indicates that the impactor penetrated the weakened icy crust which readily rose isostatically.

The highlight of the radial mosaic was taken towards the inner edge of the transition zone, and it revealed large landslide deposits about 3 kilometres long within two large impact craters. The slides may have occurred when the shock of a nearby impact caused the crater wall to fail. The fact that the material flowed so far in the absence of an atmosphere – or other fluid which might have lubricated the flow – could indicate that the surficial material is very fine-grained. If so, and it was 'fluffed up' by electrostatic forces, it may have acted as a fluid and so flowed in vacuum. The remote-sensing instruments confirmed that Callisto is indeed coated with fine dust.

Small closely spaced bright 'humps' were in evidence throughout the innermost part of the transition zone, creating a more finely textured appearance than that seen on many of the other intercrater plains on Callisto. At low resolution, the icy humps make Asgard's centre brighter than the surrounding terrain. The origin of these humps is not known. This close in, the 'ring' is actually a ridge. Out beyond this ring, the surface changes significantly. Although it is still peppered with craters, the number of icy humps decreases, while their average size increases and the texture is smoother – possibly because material from positive relief features (such as ridges) has slumped and smothered small-scale features.

The image taken about half way out through the radial presented a mystery. A closer look showed what seemed to be very small craters immediately alongside, and on the floors of the larger craters. Many of these small depressions are not circular. They may be partially eroded secondaries, but the similarity to unclassified 'pits'

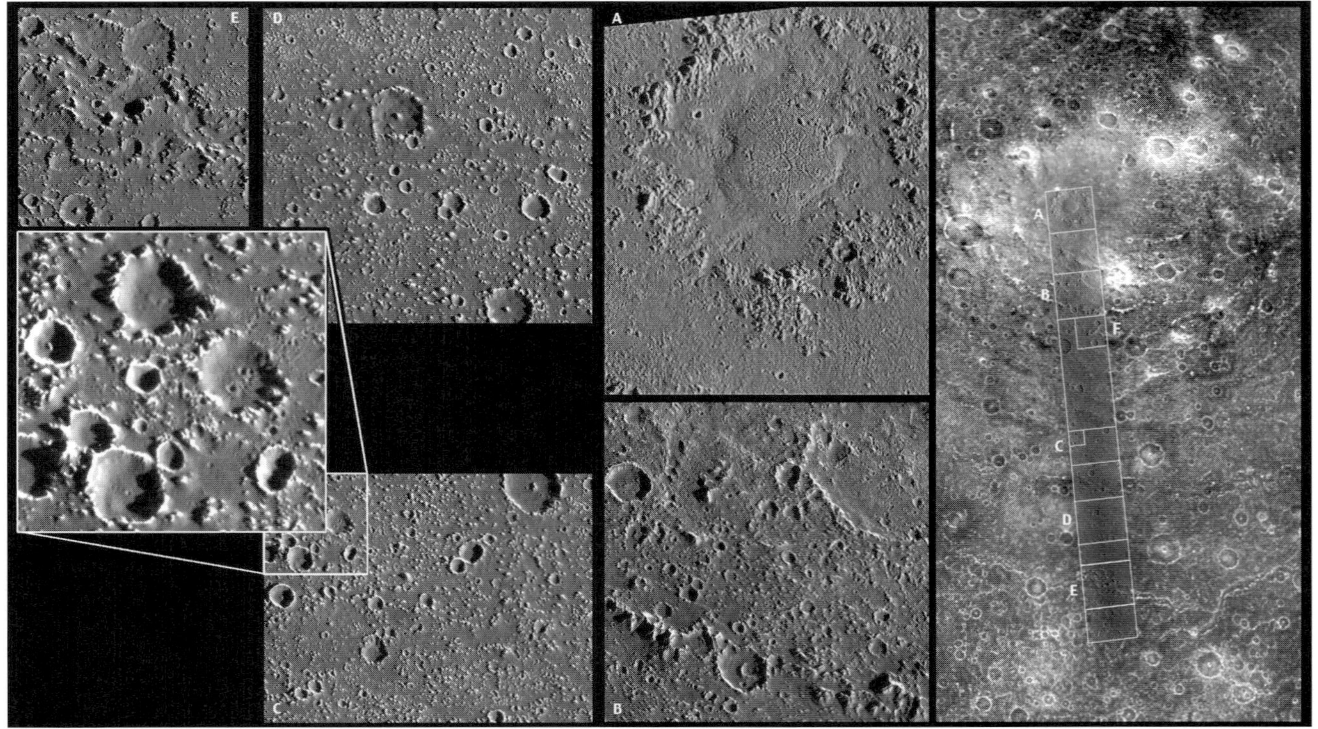

On its tenth orbit, Galileo made a radial survey from Asgard's central plain out through its ring structure (right). The 'dome' crater on the south-central section of the central plain (A) is Doh. The rings in the transition zone (B) are ridges. At 100 metres per pixel, the inner part of the basin is finely textured because of small, bright, closely spaced bumps. At low resolution, these icy bumps make Asgard's centre brighter than the surrounding terrain. Exterior to the ring, the surface changes significantly. Although still peppered with craters, the number of icy bumps decreases and their size increases. Midway out through the ring structure (C) very small 'pits' are evident (see blow-up) on the floors of some larger craters, as well as in the immediately adjacent terrain. Further out in the radial (D) the surface is dominated by the dark blanket. In the structure's outer margin (E), the rings are irregular troughs.

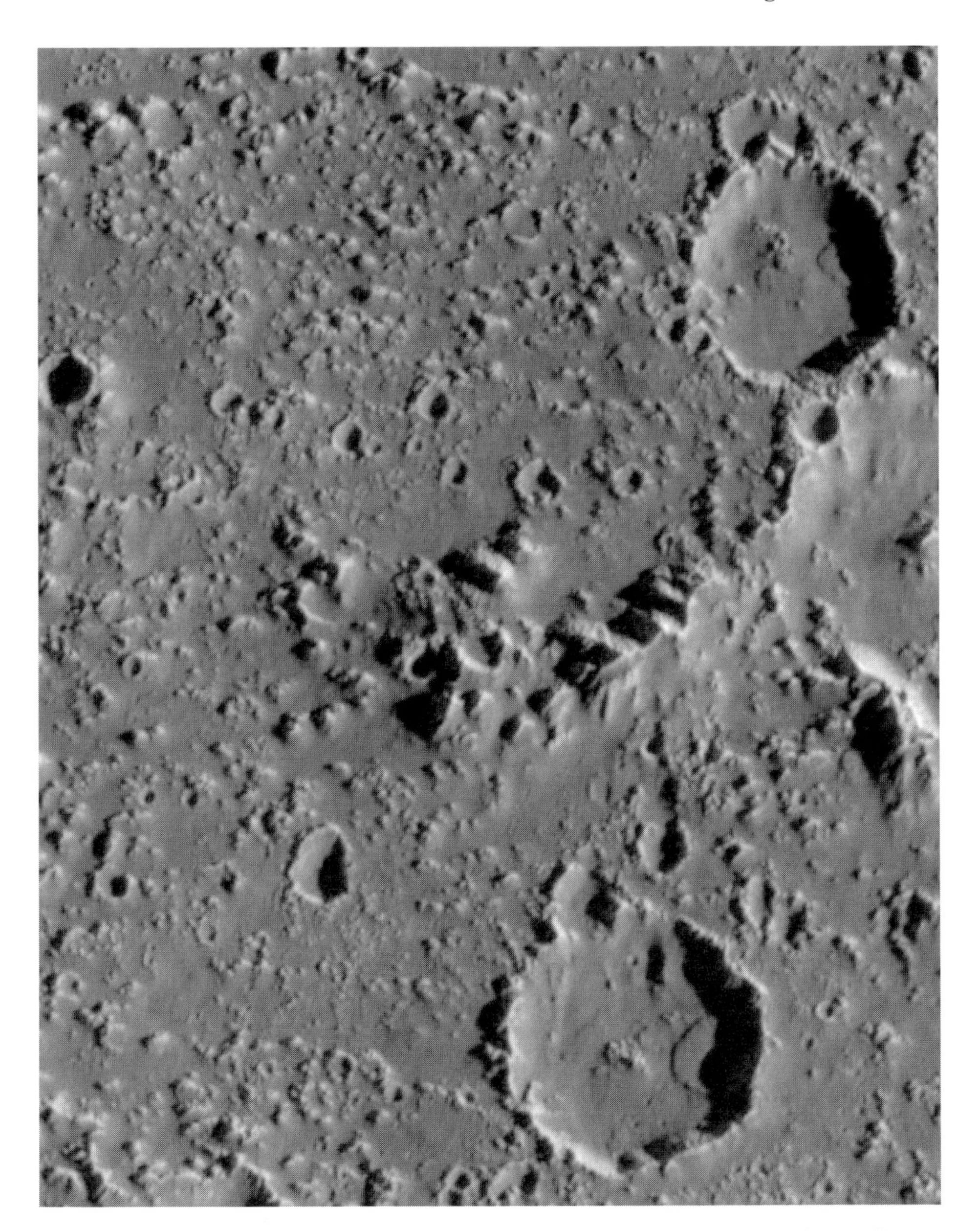

This 44 × 55-kilometre section in Asgard's transition zone (labelled 'F' in the previous figure's contextual view) revealed landslides in two craters. Some 3.5 kilometres long, they evidently formed when rim material failed under the influence of gravity and slumped into the craters.

noted elsewhere raises the possibility that they may be of endogenic origin. An analysis of the orientation of the pits and the clusters of smaller craters in relation to larger impacts should show whether the pits are related to larger impacts and thus are ejecta. The 'outer' end of the radial was typical of the rest of the moon's battered surface.

Galileo's radial survey through Asgard's rings served as a tantalising illustration of what could have been achieved if the high-gain antenna had been available – the whole of Callisto, and indeed all of the Galilean satellites, would have been surveyed at this resolution, or better, and imagery would have been supplemented by multispectral data to chart the mineralogy on a global basis, with radiometry and polarimetry providing the thermal and physical properties of the surficial materials.

EXOSPHERE

The ultraviolet spectrometer detected hydrogen atoms escaping from Callisto, which meant that there was oxygen from dissociated water molecules bound up in the crust, and by implication that there may be an oxygen 'frost' on its surface. In 1996, the instrument had found evidence of oxygen on the surface of Ganymede and in that case the active agent was probably the bombardment of charged particles in the Jovian magnetosphere. However, Callisto is much further out from Jupiter, and the active agent is more likely to be solar ultraviolet.

"The surface of Ganymede is thought to contain about fifty percent ice," explained Charles Barth, a former director of the Laboratory for Atmospheric and Space Physics at the University of Colorado at Boulder and a member of the instrument team, "while the ice on the surface of Callisto is thought to comprise less than twenty percent of the planet's surface."

When the infrared spectrometer had made remote observations on the second orbit, it had found evidence of a carbon dioxide frost. On a subsequent encounter, Galileo spotted gaseous carbon dioxide. "Callisto's atmosphere is so tenuous that these carbon dioxide molecules are literally drifting around without bumping into one another," said Robert Carlson, the infrared spectrometer's principal investigator. "An atmosphere this thin is easily lost due to ultraviolet radiation from the Sun – which breaks the molecules into ions and electrons which are swept up by Jupiter's magnetic field." For an exosphere to exist, there has to be a steady release of carbon dioxide from the surface. While some of this may be sublimated frost, the geological evidence of erosion raises the possibility that carbon dioxide may be periodically vented from the beneath the surface.

REVELATIONS

The model for Callisto's internal structure was revised in light of doppler tracking from the two close fly-bys in late 1997.

The early Galileo data had suggested that the moon was essentially undifferentiated. "This new information suggests it has a strange interior," explained John Anderson. "It isn't completely uniform, nor does it vary dramatically. There are indications that the interior materials – most likely compressed ice and rock – have settled partially, with the percentage of rock increasing towards the centre."

"The fact that Callisto is the only one of the four large Jovian moons that is not completely differentiated raises an intriguing possibility," pointed out Gerald Schubert, a multidisciplinary scientist at the Department of Geophysics and Planetary Physics at the University of California at Los Angeles. Over time, Io, Ganymede and Europa have separated into layers because they have been heated by tidal stresses. "Because Callisto is farther from Jupiter, it is 'half-baked' compared to the other moons – with its ingredients somewhat separated but still largely mixed together." This raised the possibility that cryovolcanism may still be active on Callisto, despite the ancient appearance of its surface.

After the 1,100-kilometre Callisto fly-by on the spacecraft's third orbit, the magnetometer team had reported that the moon did not possess a magnetic field, but once it was realised that Europa not only had a weak magnetic field, but that this was due to Jupiter's magnetosphere inducing weak electrical currents to flow within a layer of salty fluid at shallow depth beneath the icy surface, the Callisto data was reanalysed and a very weak 'signature' was discovered. Something was conducting an electrical current which, in turn, generated the magnetic field and this interacted with the Jovian field. An icy crust would not be a very efficient conductor, even if contaminated by electrically conducting 'non-ice' minerals. As was the case for Europa, the electrical current within Callisto flowed in opposite directions at different times. "This is a key signature," explained Krishan Khurana, a member of the magnetometer team at the University of California at Los Angeles, "consistent with the idea of a salty ocean, because it shows that Callisto's response – like Europa's – is synchronised with the effects of Jupiter's rotation" (see p. 196).

The prospect of Callisto possessing a subsurface ocean came as a surprise. "No one talked about it as a likely situation, based on what they had seen of the surface," Margaret Kivelson, the leader of the magnetometer team, admitted. "However, it does 'fit' the data so well!" The correlation of the magnetic field's fluctuations with the planet's rotation was the final proof.

The ocean – possibly just a transition zone of slushy ice – must be fairly deep below the surface of Callisto because there is no geological evidence of the crustal fractures that occurred on Europa. Indeed, the fact that the multiple-ringed basin structures are still in evidence indicates that it is an ancient surface. Nevertheless, the basins are subdued, indicating that the crustal material was sufficiently plastic to slump. This lack of rigidity is borne out by the fact that Callisto does not possess any high massifs. Although the surface is rifted, there are no cryovolcanic flows in evidence. When the fractures formed, the fluid must have been too deep to make its way to the surface. However, the morphology of the largest impact scars strongly suggests that there was an icy slush at shallow depth at the time of these ancient impacts.

A detailed analysis of the magnetometer evidence found that Callisto's ocean is global in extent, some 100 to 200 kilometres down, and of the order of 10 kilometres deep. The heat to maintain it is probably radiogenic. Because Callisto is so far from Jupiter, and is free of orbital resonances, gravitational tidal stress is weak. However, it would have accumulated its share of radioactive elements during accretion, and in its undifferentiated state these will be uniformly distributed. Over time, as it loses heat to space, the moon will have progressively frozen from the outside in, so it will only be warm far below the surface. Furthermore, any salt in the slush will act as a natural anti-freeze.

"We thought Callisto was a dead and boring moon, just a hunk of rock and ice," reflected Kivelson. "The new data certainly suggests that something is hidden below Callisto's surface and that 'something' may well be a salty ocean." The electrical current required to produce the observed magnetic field would not need a fluid any more salty than the Earth's oceans.

As Ron Greeley pointed out, this discovery was "making all of us go back to our models, and think about our understanding of these icy bodies".

PERIJOVE REDUCTION

In May 1999, a month after it emerged from solar conjunction, Galileo made the first of a series of Callisto encounters designed to reduce its perijove in order to facilitate a return to Io.

The Callisto fly-by was on the way out, with the spacecraft passing in front of the moon's night-time hemisphere, with its closest point of approach 1,315 kilometres over the equator at 102 degrees longitude. It was therefore able to make observations of the daylit part of the sub-Jovian hemisphere on the way in, and of the anti-Jovian hemisphere on the way out, covering much of the hemisphere opposite the Valhalla and Asgard multiple-ringed structures.

The major geological theme of these Callisto encounters was to be the deficit of very small craters. "Craters less than about 1 kilometre across seem to have been partially obliterated, or 'disaggregated', by an unknown process," noted Torrence Johnson. "On planets where there has been weathering or erosion, obliteration of small craters is expected. But Callisto doesn't have an obvious source of erosion. So what's happened to the craters? It's a real puzzle." The objective was to chart variations in surficial composition that might shed light on this mystery.

Three sites were documented at different spatial resolutions in order to gather statistics on the sizes and distribution of craters. A statistical survey would help to derive the cratering rate in the Jovian environment. While near Callisto, the dust detector sought evidence for material which had been blasted off the moon by meteoric impacts.

A single image was taken of the 100-kilometre crater Bran in order to serve as a reference for coordinated investigation of the composition of the surface by the infrared and ultraviolet spectrometers. This bright ray crater was being used as a 'drill hole', and its prominent ejecta was a 'window' down into the crust. The spectrometers also observed Callisto's bright limb to investigate the composition – in addition to carbon dioxide – of the recently confirmed exosphere.

Prior to the perijove reduction campaign, Galileo's perijove was 9.4 Rj. The gravitational slingshot of this Callisto fly-by reduced the spacecraft's orbit, but because the encounter took place on the way out the effect would not be felt until the return to Jupiter. However, upon its return in late June, the Callisto fly-by on the way in slowed it further, lowering the perijove to 7.3 Rj.

With Callisto on the far side of its orbit, this pass over the illuminated leading hemisphere provided an opportunity to view Valhalla with the Sun high overhead. High-resolution images were taken a few minutes before closest approach to study variations in the appearance of the dark material near this ring structure to seek additional insight into the processes that modified the surface. High-resolution scans of this area by the integrated photopolarimeter and radiometer measured the 'brightness temperature' to infer the physical properties of the surface materials, and the infrared spectrometer took imagery combining high spectral and spatial resolutions in order to identify the composition of the surface materials. The 15 metres per pixel imagery on this encounter was the most detailed view of Callisto so far.

The perijove reduction campaign's third encounter with Callisto was a virtual repeat of the first, with Galileo crossing the darkened leading hemisphere as it drew away from Jupiter, but this time around the moon was ignored. "We're not doing much observing at Callisto on this encounter," said Duane Bindschadler, the head of the science planning and operations team, "just radio science to probe Callisto's atmosphere and its internal mass distribution." This slingshot reduced the perijove to 6.5 Rj, and the fourth and final fly-by in September set the spacecraft up for the Io encounter.

ODD ONE OUT

Although Callisto is almost as massive as Ganymede, their internal structures and surface geologies are strikingly different. In Ganymede, the rock and metals have separated from the ice to form a dense core. Apart from a tendency for the outer 200 kilometres to be mainly icy, Callisto is undifferentiated. Although they both have large tracts of dark cratered terrains, Ganymede's ancient terrain has been modified by furrows and scarps, whereas such features on Callisto are associated only with the multiple-ringed structures. Also, of course, fully two-thirds of Ganymede's surface has been completely resurfaced by the tectonism which produced the sulci. Even so, as noted earlier, Callisto is probably more 'normal' for the outer icy bodies of the Solar System, and as such it serves as a valuable point of reference.

11

Europan enigma

VOYAGER'S EUROPA

Although at a radius of almost 1,500 kilometres Europa is the smallest of the Galilean satellites, it is nevertheless comparable in size to the Earth's Moon. With an albedo of 0.64, it is one of the most reflective bodies in the Solar System, but the reason for this was not confirmed until the Voyager fly-bys of 1979. The best imagery was taken by Voyager 2 from a range of 204,000 kilometres. While this provided a resolution of 2 kilometres per pixel, only a small part of the moon's surface was covered, so the mapping effort was severely constrained. Nevertheless, it was clear that the moon is enshrouded in ice.

Europa is 'billiards ball' smooth. In fact, there is so little topographical relief that the entire surface is confined within a vertical range of a few hundred metres. Even though the ice is ubiquitous, it presents albedo, colour, and textural variations.[1]* One physiographic analysis[2] identified two principal surface units, referred to as 'plains' and 'mottled terrain' (the mottling was divided into 'brown' and 'grey') and isolated irregular spots were classified as maculae.

Four types of plain were distinguished. The 'undifferentiated plains' are smooth, tend to be gradational with adjacent terrains, and are cut by numerous linear features. 'Bright plains' are located preferentially towards high latitudes, and are criss-crossed by a variety of types of lineation. 'Dark plains' resemble the light plains, but are darker. 'Fractured plains' appear to have been shattered and they bear curved grey streaks and numerous brown spots.

As to the mottling – the brown mottling is moderately textured and forms sharp contacts with adjacent terrains, implying that it is younger. In contrast, the grey mottling is smooth and has diffuse boundaries. Intriguingly, the brighter mottled terrain is on the leading hemisphere, and the darkest is on the trailing hemisphere. So what is the texturing on the brown terrain? It seems to be hummocky. Is this related to the hue? What is the brown material? Is it endogenic in origin? Or is it exogenic?

* For Notes and References, see pages 413–424.

The Jovian magnetosphere rotates with the planet, so the trailing hemisphere of the slowly-orbiting moon is bombarded with charged particles and the leading hemisphere is shielded to some extent by being within the magnetospheric 'wake'. Is the difference in albedo between the hemispheres due to this irradiation? If the ice is darkened by exposure to charged particles, then this would imply that any light plains close to dark plains must be younger.

This Voyager 2 mosaic of a crescent Europa shows the long dark linea, mottled terrain, irregular dark maculae, and the intensely fractured zone near the anti-Jovian point. The sharply defined terminator indicates that there is very little surface relief.

The most striking features on Europa, however, are the various types of linearity. The 'triple bands' that predominate in global views comprise a bright stripe – possibly a ridge – running down the centre of a dark band. Although they run for thousands of kilometres, they are typically less than 15 kilometres wide. They often start or end near dark circular spots, or brown mottled terrain. The fact that some bands extend almost half-way around the moon and trend northwest in the northern hemisphere and southwest in the southern hemisphere suggests that they are the result of stresses induced by orbital eccentricities. In contrast, 'dark wedges' are typically 300 kilometres long and span 25 kilometres at the open end of the wedge. The way that they cut across older features suggests that the icy crust has fractured and separated, with the gaps being sealed by upwelling. Could such spreading imply incipient plate tectonics? They cut across many other features and are therefore relatively recent. The 'grey bands' are cut by all other types of lineation, so they formed first. Imagery of the south polar region near the terminator showed many arcuate features – variously called 'flexi' or 'cycloids' – which cut across all other features, and so are the youngest features of all.

The generally accepted geological history of Europa, based on Lucchitta and Soderblom's analysis of the Voyager imagery, is that a bright plain formed when an ocean froze over. It has been disfigured by the dark and fractured plains and the grey mottled terrains. Some time later, fluid oozed out as the brown mottled terrain formed. The various lineations were added still later. However, deriving the relative ordering is only half of the solution; what is needed is the timescale. How old is the youngest feature?

Impact craters are spectacularly lacking, but there are some. The craters fall into two types. There are a few palimpsests 100 kilometres across with concentric fractures and low surface relief which may have been caused by large impacts, plus a dozen or so examples of craters with rims up to 25 kilometres across. The subdued craters imply that the ice was thin at the time of the impact. The overall paucity of craters suggests that the surface is *extremely* young. The prospect that the surface froze over relatively recently raised the amazing possibility that there might be an ocean beneath the ice.

GALILEO'S FIRST LOOK

"A major goal of Galileo's study of Europa", said Ron Greeley, a member of the imaging team at Arizona State University in Tempe, "is to search for signs of past or current activity to help to resolve the question: is there a liquid ocean? We want to go back to some of the areas which suggest soft ice or liquid water under the ice, and test some of the questions that we're asking now."

The infrared spectrometer's programme[3] for Europa was to have started with a study of the south polar region during the approach to Jupiter in December 1995, but the fault with the tape recorder had led to these observations being cancelled. The primary mission's orbital tour was to enable the instrument to chart of the anti-Jovian and trailing hemispheres – which included the region best observed by the

Voyagers – with a resolution of about 50 kilometres per pixel. Although the longitude range 0 to 90 degrees west would not be documented at anything better than 300 kilometres resolution, this would be able to be ameliorated if the orbital tour could be extended beyond the baseline two-year mission.

The spectrometer's Europa objectives were grouped into three 'campaigns'. Polar mapping would provide an opportunity to search for frozen volatiles. Global mapping would determine the composition and distribution of the various surface materials, paying particular attention to the 'non-ice' material and the extent to which the trailing hemisphere has been modified by the charged particles circulating in the Jovian magnetosphere. If plate motions could be discerned, indicating that global tectonism had been active, the spectrometer would conduct a study of the compositional difference across the margins between the individual plates. The composition of the fluids that emerged from crustal fractures would provide an insight into the materials in the interior – the brown surface deposits in particular were believed to be of endogenic origin. A study of fracture patterns in palimpsests would provide an indication of how thick the icy crust had been at the time of the various impacts. Tyre, a large circular feature near the northern fringe of the region imaged by Voyager, was to be studied in detail, combining 408 wavelength spectral resolution and 10-kilometre spatial resolution. But these observations would not be practicable until the spacecraft's trajectory facilitated a low pass over this part of Europa, and Tyre was lit with the Sun high in the sky in order to enhance the reflection spectrum. The observational programme was therefore largely defined by orbital dynamics.

On 27 June 1996, after its first perijove, Galileo imaged Europa from 155,000 kilometres. Although this was only 20 percent nearer than the best Voyager fly-by, Galileo's camera was much better. The highest resolution was only 1.6 kilometres, but this yielded a major surprise. The regional mosaic included intensely fractured terrain southwest of the anti-Jovian point, a number of triple-bands in the northern hemisphere, and a large patch of mottled terrain on the trailing hemisphere.

In the Voyager imagery, the edges of the triple-band linea had appeared to be distinct, but Galileo found the edges of both the bright median and the dark lateral bands to be diffuse and patchy. Also, the icy plains comprised at least two physically distinct regions, and infrared data revealed that the 'blue' ice crossed by Minos Linea is almost pure water-ice.

One theory explained the triple bands by proposing that gravitational tidal stresses made a long and deep-seated fault that penetrated right through the icy crust. A 'cryogenic lava' then oozed out onto the surface and promptly froze in the vacuum, sealing the fault. This explained the dark band as silicate-enriched water (or perhaps 'warm ice'). Then, if subsequent activity allowed 'clean' water to ooze out, this would have made the distinctive bright median. Such cryovolcanism corresponds more to a terrestrial fissure eruption than to the localised vent of a volcano or caldera. "The scale of these fractures – extending distances equivalent to the width of the western United States – dwarf the San Andreas fault in both length and width," said Greeley.

The irregular 'beaded' margins discovered by Galileo prompted Greeley to

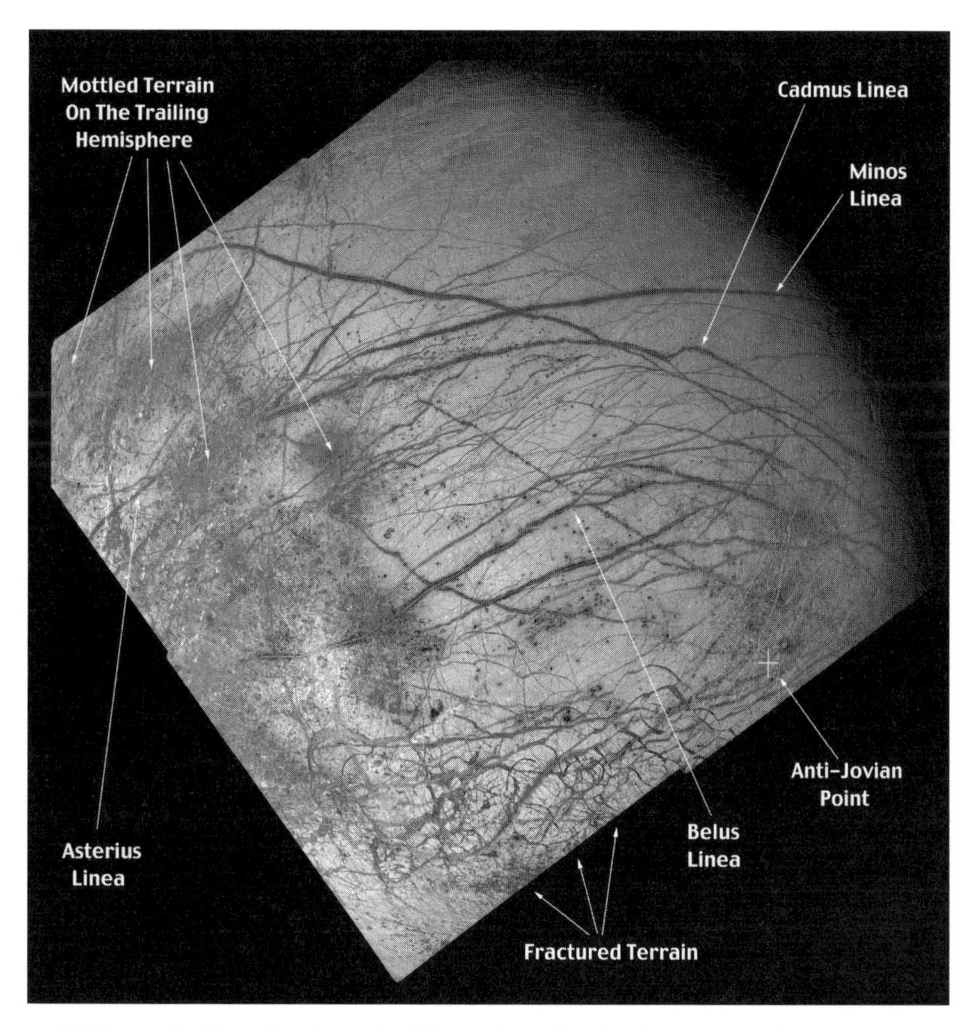

Galileo made this regional mosaic of Europa's trailing hemisphere on its first orbit from a range of 156,000 kilometres. In addition to the dark linea which extend for thousands of kilometres across the icy plain, the intensely fractured plain near the anti-Jovian point is evident.

propose that a string of geysers had explosively vented along the nascent fault, and painted the surface with 'dirty' ice fall-out. The median would have been produced later, but with 'clean' water. This explained the patchy and diffuse form of the bands in terms of the release of volatiles under pressure, rather than the extrusion of fluids, and it raised the prospect that geysers might still be active, but it was early days and more observations – at much higher resolution – would be needed before it would be practicable to make a definitive statement.

There was a newly discovered 30-kilometre wide impact crater to the south of Belus Linea. It excavated the ice and threw white ejecta across the surrounding plain

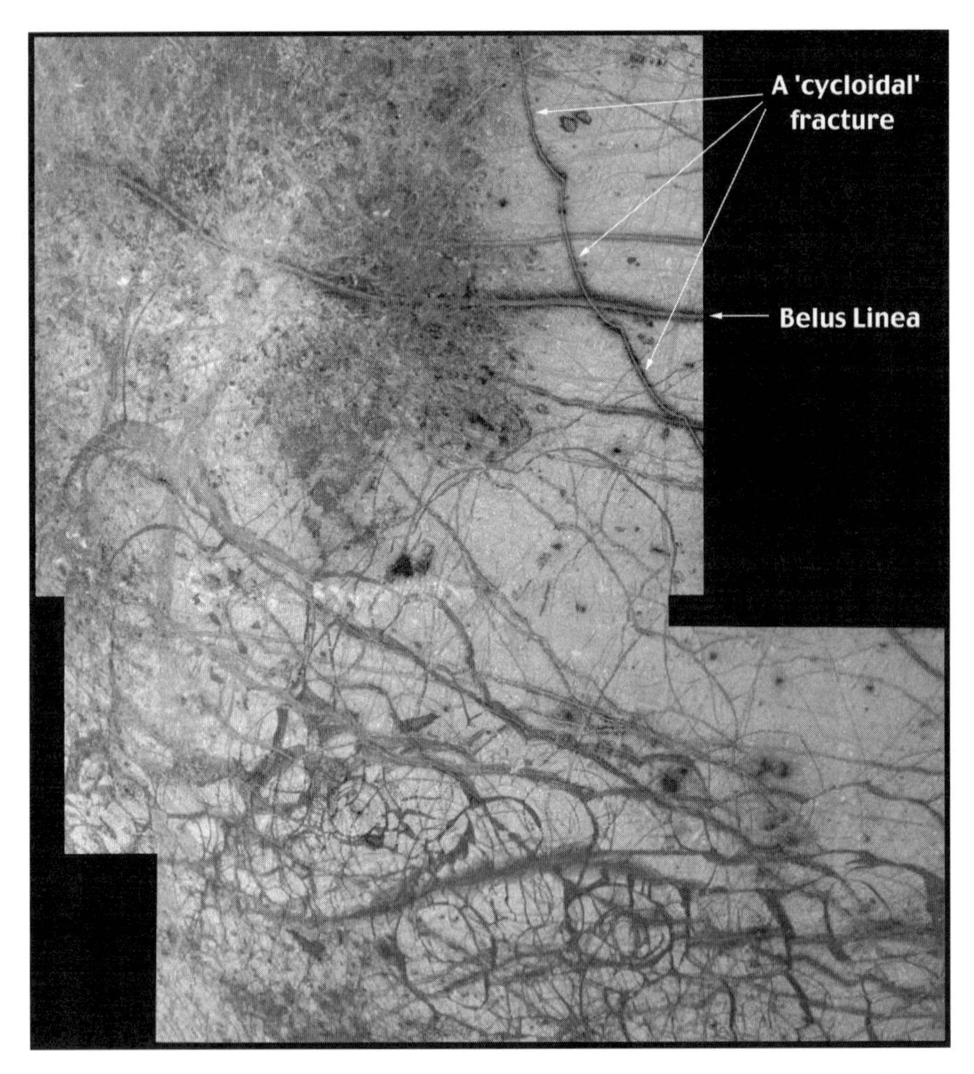

This mosaic shows the intensely fractured plain near Europa's anti-Jovian point, where the ice has been broken into slabs as much as 30 kilometres across. Note that the Belus Linea 'triple band' is patchy and has diffuse margins. The impact crater to the south has excavated the ice and deposited it on the surrounding plain. The curved X-shaped 'grey band' to the southwest marks where the crust fractured, spread, and infilled with slush which froze in place. The fractured terrain displays features resembling ice flows in the Earth's polar seas. Galileo recorded this mosaic, which has a surface resolution of 1.6 kilometres per pixel, on its first orbit.

and over the linea. The 'X'-shaped grey band south of the crater is a system of spreading ridges, where the crust has been pulled apart and the crack progressively sealed by slush contaminated with rocky debris which froze in place as soon as it encountered the vacuum of space.

The intensely fractured terrain revealed jostled blocks of ice which are strikingly similar to 'pack ice' at the Earth's poles, but on a gargantuan scale. "The ice is broken into large pieces that have shifted away from one another," Greeley reported, "but which obviously fit together like a jigsaw puzzle." Dark linear, curved and wedged-shaped bands have fragmented the icy crust into blocks up to 30 kilometres across which have separated and rotated.

"These fantastic new images of an icy moon of Jupiter are reminiscent of the ice-covered Arctic Ocean," said Dan Goldin. "The lack of craters, the cracks, and the signs of movement, all indicate that this might be young ice on a dynamic surface."

The presence of the ice flows indicated that the icy crust is very thin. "Perhaps only a few kilometres," suggested Greeley. This strongly supported the case for a substantial – maybe even global – ocean beneath the ice. If the triple-band linea were produced by geysers, there would necessarily have been a source of volatiles at shallow depth all along their length. It was beginning to appear that water in Europa's interior exploited every opportunity to escape from its confinement. "The pictures are exciting and compelling, but not conclusive," warned Goldin.

A Europan ocean would freeze from top to bottom. Unless there was a source of internal heat to maintain its base in a liquid state, the ocean would freeze solid in several million years. Although some heat could be derived from the decay of radioactive elements deep within the silicate interior, Europa is more likely to be kept warm by gravitational tidal stress. Its orbital period is twice that of Io and half that of Ganymede, so it is in a double resonance. As Europa is farther than Io from Jupiter, it will endure less stress than Io, and so it is unlikely to be as differentiated. Europa has not boiled off its volatiles, and so it is enshrouded with ice. It may be volcanically active, but this will occur in the silicate lithosphere which forms the floor of the ocean. Hydrothermal activity on the seabed could provide an energy source for any primitive form of life which may have developed. This led to Europa being added to the short-list[4] of places where life may have developed.

A thermal chart of the illuminated hemisphere produced by the integrated photopolarimeter and radiometer on Galileo's second orbit established that the surface is either a porous 'fluffy' ice, or is composed of finely powdered grains of ice.

Shortly after perijove on Galileo's third orbit, the spacecraft flew by Europa at a range of 34,000 kilometres (nearly as close as the 'lost' encounter on the run in to Jupiter in December 1995) and secured some imagery with a resolution of 400 metres. One image of the plain near Argiope Linea showed the dark wedges in unprecedented detail. The symmetric ridges within the wedges confirmed that these features are sites of spreading. The central feature marks the spreading axis itself. The ridges were formed in succession, as the crust was pulled apart and fluid oozed out and froze, so their age increases with distance from the axis. The fact that the wedges are pointed at one end and open at the other suggests a similarity to lithospheric plate tectonics on Earth which is characterised by both separation and by rotation around an 'Euler pole'. A variety of knobs and irregular dark patches within the wedge suggested that ices and gases had been erupted later.

To the extent that the dark wedges on Europa involved cryovolcanic tectonism,

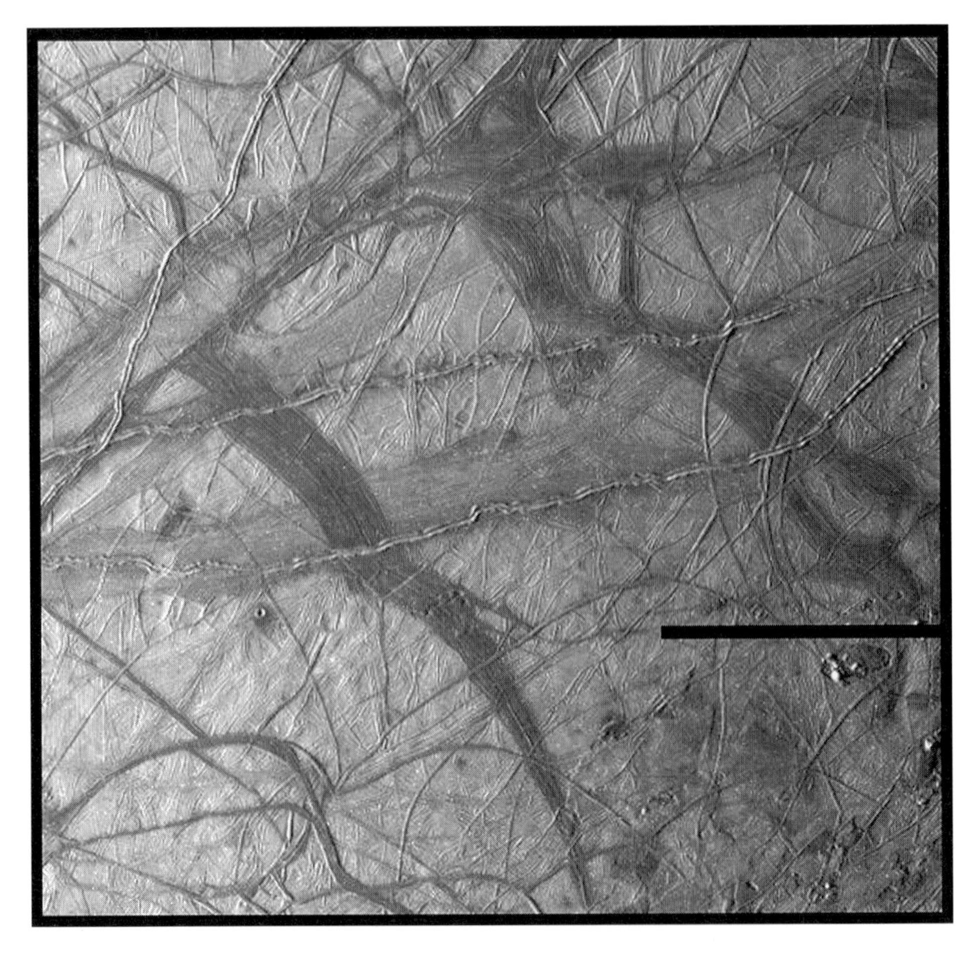

During its third orbit, Galileo documented an area of 225 × 238 kilometres centred at 16 degrees south at 196 degrees. The plain has been modified by fractures and ridges. The prominent symmetric dark wedge-shaped feature marks where the ice has fractured, spread and been infilled with slush which froze.

they were similar to Ganymede's light-toned grooved sulci, but the spreading activity on Ganymede was on a grander scale.[5]

EUROPA UP CLOSE

The fourth encounter sequence focused on Europa in December 1996. Even with the recently upgraded communications, the spacecraft had just enough time to replay its Callisto data over the low-gain antenna (because the high-resolution imagery of the richly-textured moon could barely be compressed), but Galileo started taking the particles and fields data to survey the inner magnetosphere on schedule, on 14

December. The next day, however, the tape recorder developed a fault and the spacecraft 'safed' itself. This problem was "terribly subtle" pointed out Bill O'Neil. A slight misalignment of the start-of-track marker had fooled the machine into thinking that it was at the end of the track rather than the beginning and the fault protection software had suspended operations. The engineers were able to restore the tape recorder with only an hour to spare before its first scheduled use.

The 692-kilometre Europa encounter – fully 300 times closer than Voyager – occurred just after Galileo made its perijove passage: it viewed the trailing hemisphere in daylight on the way in, saw the low surface relief casting early-morning shadows, crossed the terminator near the sub-Jovian point, and made its closest point of approach at 37 degrees west longitude in darkness. The spacecraft flew behind the moon as viewed from Earth, and the radio science team had an opportunity to test for signs of an ionosphere during the occultation. The leading hemisphere was viewed in darkness on the way out. The features that could be imaged at high resolution, and under the most favourable lighting for resolving surface relief, were therefore defined by orbital dynamics – in this case, the equatorial region east of the sub-Jovian point. Since this hemisphere had not been well presented to the Voyagers, this was effectively *terra incognita*. The capacity of the low-gain antenna was too low to enable the camera to assemble large high-resolution mosaics, therefore medium-resolution regional mapping was done on the way in. The highest-resolution imagery was taken within 15 degrees of the dawn terminator – which was at 340 degrees west. With a 20-fold improvement in resolution and a very low sun-angle, it was possible to resolve 'house-sized' features.

An image of an area near the terminator at 6 degrees north taken about 3 hours

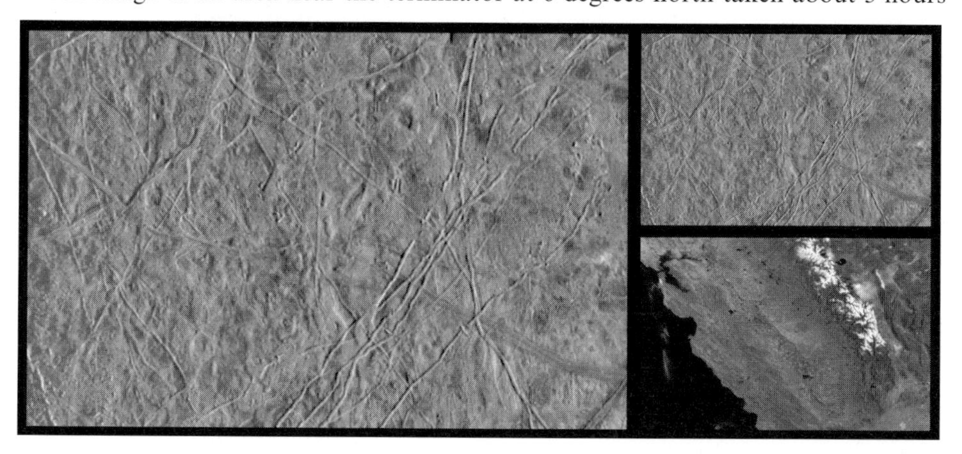

This image of Europa's icy surface, spanning 252 × 393 kilometres, was taken from a range of 62,000 kilometres on Galileo's fourth orbit. It shows several ridges, patches of smooth low-lying darker materials, and plateaux about 10 kilometres across. Some ridges are broken, and textural differences indicate that the missing sections may have been swept away by cryovolcanic flows. The paucity of impact craters suggests that this surface froze 'recently' (in geological terms). The scale of the Europan surface is evident from the Californian coastline (right).

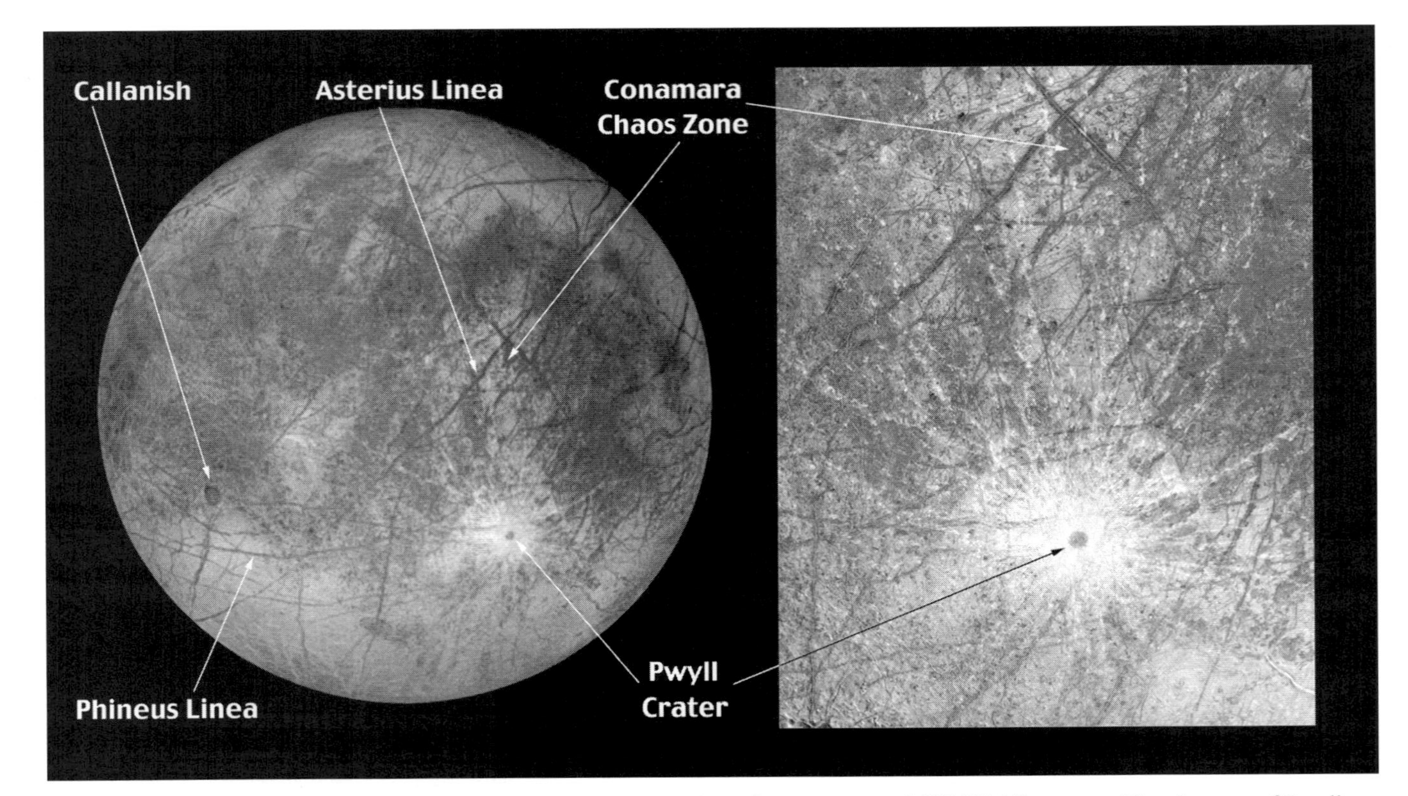

On its second orbit, Galileo documented Europa's trailing hemisphere from a range of 677,000 kilometres. The close-up of Pwyll area (right) was documented on the fourth orbit, during its first close fly-by. The bright rays from Pwyll are fresh fine water-ice particles which have 'painted' streaks for a thousand kilometres. The fact that the rays are superimposed on so many different types of terrain indicates that they are younger, and that the impact was fairly 'recent'. The patch of mottled terrain near the intersection of two prominent linea, due north of Pwyll, was selected for close inspection on a later orbit.

out (from a range of about 60,000 kilometres) provided a resolution of 'only' 1.6 kilometres, but it revealed the ridges, plateaux, and patches of smooth low-lying darker material in stark clarity. Many of the ridges are interrupted by gaps, and in some cases appear to have been either partially buried or subdued by viscous glacier-like ice flows.

Europa has only a few impact craters, and they possess a variety of morphologies. It had been speculated that this variety reflected differing thickness of the icy crust when a crater was formed. A comparative study of craters might indicate either regional trends in the thickness of the ice, or possibly, if the craters can be reliably dated, global trends in crustal thickness over time.

Brilliant rays radiating from a circular spot in the southern hemisphere left no doubt that it was an impact which excavated white ice and scattered it around (this crater was subsequently called Pwyll by the International Astronomical Union).

Upon closing to 12,000 kilometres, Galileo snapped an area about 15 degrees south, just north of Phineus Linea, and recorded a 100-kilometre macula with a resolution of 240 metres. Did this fractured zone (later called Callanish) mark the site of an impact? If so, then either the impactor punched right through the icy crust or, as the ice isostatically adjusted to the stresses induced by the range of vertical relief, the crater was transformed into a central patch of rugged terrain surrounded by a zone of circular fractures. However, another proposal posited that the macula marks where a diapir of 'warm ice' rose within the icy crust, formed a dome which first made the cracks and then, once the volatiles had been vented from the cracks and released the pressure, collapsed to leave the badly fractured spot. The volatiles could have been water, methane or ammonia which, after being vented, would have sublimated to gas and been lost to space. More examples would have to be investigated before the nature of such macula would be resolved.

Whatever its origin, Callanish sits on a plain that is criss-crossed by double ridges several kilometres wide. The younger ridges overlap the older ones in a mind-

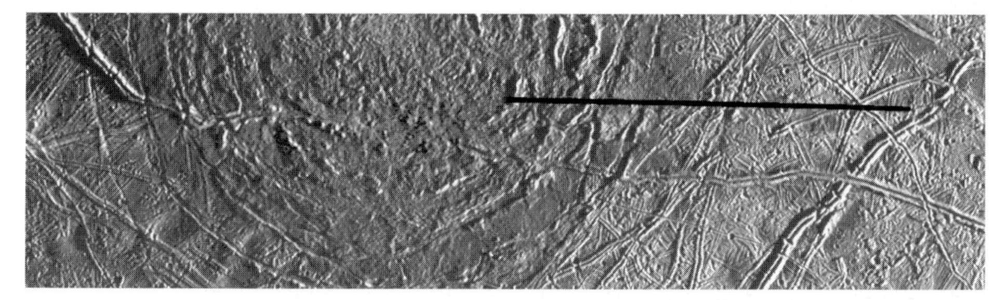

On its fourth orbit, Galileo took this close-up mosaic of a 100-kilometre wide dark spot (see the hemispherical context on the previous figure) which, upon being identified as a multiple-ringed structure, was named Callanish. The original impact crater in the ice has been modified by isostatic forces to create a central patch of rugged terrain, surrounded by a series of circular fractures. Several linea cross the ring structure. One prominent example can just about be followed running east-to-west through the southern fringe of the central plain, at which point its form has been degraded. The image resolution is 240 metres per pixel.

numbing 'ball of string' complexity recording their formation over an indeterminate period of time. Irregular patches of richly textured terrain have erased sections of the older features, so there is clearly a complex geological history etched on the Europan plains. Indeed, one prominent east–west double ridge appears to have been modified by Callanish's presence.

The infrared spectrometer scanned the trailing hemisphere half an hour out, to make a high-resolution compositional map (in this instrument's terms, this meant 10 kilometres per pixel), and it detected hydrated minerals in the water-ice crust.

As Europa loomed large, Galileo imaged an area at 6 degrees north with a resolution of 100 metres. This remarkable image further highlighted the structural complexity of the overlapping ridges and fractures. Offsets in the ridges indicate lateral faulting. Missing segments mark obliteration by the emplacement of new material. Patches of ridged 'chaotic' terrain co-exist close alongside patches of smooth craterless terrain which is relatively free of ridges. Sinuous rille-like features and the knobby terrain are evidence of processes as yet ill-understood. Small craters ranging in size from less than 100 metres to about 400 metres were in evidence. Robert Sullivan of Arizona State University at Tempe, a member of the imaging team, noted that the "slot-like trough" resembled a volcanic vent and he likened the flow that emanates from it to the lava flows of the Snake River Volcanic Plain in Idaho.

This was an enormous advance over the Voyager view. If Galileo's high-gain

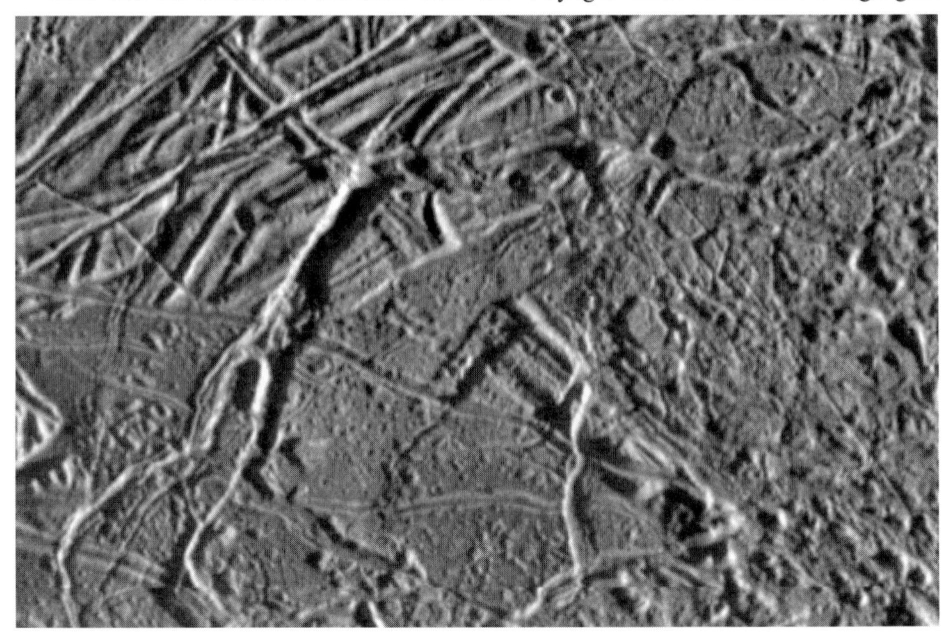

During the low fly-by on its fourth orbit, Galileo documented a 10 × 16-kilometre area centred 6 degrees north at 325 degrees. At 100 metres per pixel, the plain is revealed to have been etched by a complex series of cross-cutting ridges, grooves, sinuous rille-like features, and knobby features. Note the prominent wedge-shaped ice flow.

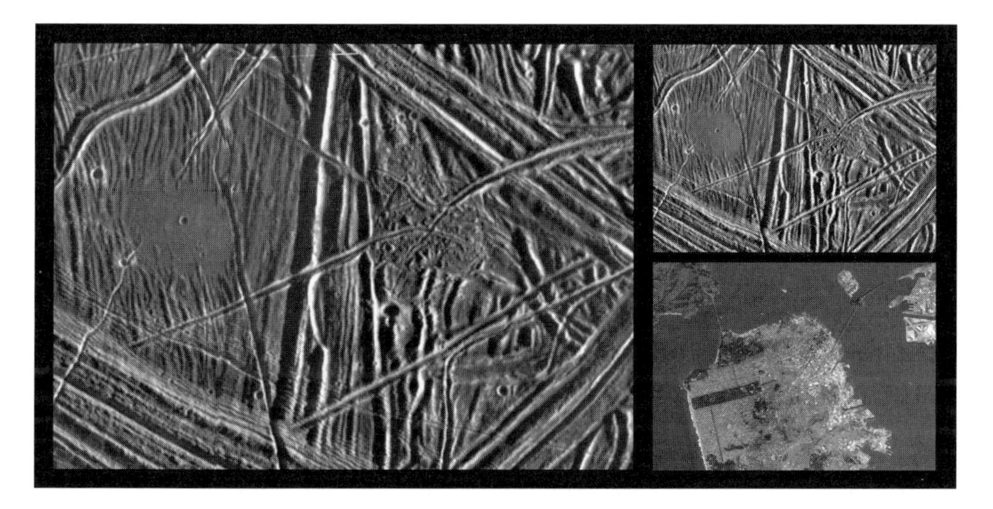

In documenting this 11 × 16-kilometre area of intensely ridged terrain at 26 metres per pixel during the low pass on its fourth orbit, Galileo found an 'ice rink' 3.2 kilometres across. The fact that the peripheral ridges can be traced into the feature implies that this is a depression which has been flooded by an extruded fluid. The smooth area contrasts with the similarly sized rugged patch east of the prominent ridge, where the surface has been fragmented. Although small impact craters are most obvious in the 'ice rink', they actually occur throughout the ridged terrain. The San Francisco peninsula is included to provide a sense of scale.

antenna had deployed properly, then it would eventually have been possible to map the entire moon at this resolution.

The closest image was of an area of 11 × 16 kilometres at a resolution of 26 metres. This shows a remarkably flat smooth area some 3 kilometres across which seems to mark flooding by a fluid that has erupted onto the surface and swamped the pre-existing ridges and grooves. "Although these images do not show currently active ice volcanoes or geysers, they do reveal flows of material on the surface that probably originated from them," Ron Greeley said. "This is the first time that we've seen actual ice flows on any of the moons of Jupiter." The smooth patch, dubbed an 'ice rink', implied that there had once been sufficient heat in the interior to drive fluid to the surface. The dark 'paint' on some of the nearby faults and ridges might be fall-out from geysers. The smooth patch contrasts with the distinctly rugged terrain to the east, on the opposite side of a prominent ridge system. The rugged chaotic site is a localised disruption of the pre-existing ridges and this may be further evidence of heat leaking through the icy crust. The most noticeable impact craters are those on the smooth patch, but they are also present on the ridged terrain. The prominent crater in the middle of the 'ice rink' is about 250 metres across. A photoclinometric analysis[6] of the manner in which the peripheral ridges project into the smooth patch concluded that it is a more or less circular depression in which fluid has pooled to a depth (in the centre) of no more than about 50 metres.

Several patches of 'peculiar terrain' appear to have been shaped by an erosive agent. One theory posited 'sublimation erosion' in which water and other volatiles sublimated to gas and sculpted the surface. This was disputed, but as Greeley pointed out, "*Something* is destroying the topography."

"Ridges are visible at all resolutions," observed Sullivan. "Closely paired ridges are most common. However, when viewed at higher resolution, ridges previously seen as singular are revealed to be double." Some ridges appear to have formed along extensional faults and warm material has oozed out and frozen to form the ridge as the crust has pulled apart. Other ridges appear to have formed along compressional faults where two large crustal blocks have jostled one another and buckled their edges into a 'compression ridge', or 'crumple zone'.

Considered together with the wedge-shaped spreading margins, it was clear that Europa's icy crust has been extremely active. It was notable, however, that the Europan tectonism is markedly different to that on Ganymede, which implies that the two moons are different internally.

It was now beyond dispute that Europa *once* had liquid water beneath its icy crust. This question was: is it still in the liquid phase? The circumstantial evidence from the surface features was undermined by ignorance of the geological timescale. What was needed was insight into how the interior is *today*.

The doppler data from this first close fly-by established that Europa has a layered structure with a distinct core. "Before Galileo, we could only make educated guesses about the structure of the Jovian moons," reflected John Anderson, the leader of the celestial mechanics team. Galileo's fly-by showed how the material is distributed within Europa and suggested an outer shell of water 100 to 200 kilometres thick. "We know that Europa had a very deep layer of water in some form, but we don't yet know whether this is liquid or frozen."

When in close to Europa, the magnetometer detected what Margaret Kivelson, the leader of the magnetometer team, referred to as "a substantial magnetic signature" with a strength almost one-quarter that of the field possessed by Ganymede.[7] This discovery was a welcome surprise, but further data would be required to characterise the nature of this field. An intrinsic magnetic field could indicate the dynamo effect in a differentiated iron core, but the doppler data had not been sufficiently pronounced to support this, so attention switched to exploring the intriguing possibility that the magnetic field was induced by electric fields flowing through 'salty' water.[8] If so, then this would indicate that most of Europa's outer shell of water is still in the liquid phase; but more data would be needed to *prove* that this magnetic field was being induced in a fluid by the rapidly rotating Jovian magnetosphere.

The refraction of the radio signal when Galileo was occulted by Europa established that the moon has an ionosphere. The presence of an ionosphere strongly suggested that Europa has an exosphere, in which case active dissociative surface processes would be required to sustain it. "Most likely," noted Arvydas Kliore of JPL, "charged particles in the Jovian magnetosphere strike Europa's surface with sufficient energy to knock atoms and molecules off the moon's surface." The Hubble Space Telescope had detected oxygen emissions from Europa in 1996, and this had hinted at a tenuous atmosphere. In fact, exospheres had been identified on almost all of the Solar System bodies studied to date.

ICEBERGS!

When Galileo returned to Europa in February 1997, it encountered the moon a little before rather than a little after perijove. Nevertheless, the interception was similar and the anti-Jovian hemisphere was observed in daylight inbound, and the departure was over the sub-Jovian and leading hemispheres in darkness. The closest approach was at an altitude of 586 kilometres, at 324 degrees west, in darkness. "I think that this fly-by may provide additional clues regarding the prospect of liquid water oceans on Europa," predicted Robert Mitchell, the Galileo Mission Director.

On the run in, Galileo imaged most of the daylit hemisphere at medium-resolution both for global mapping purposes and to continue the search for the palimpsests which would provide clues to the nature of the icy crust at the time of the impact. The high-resolution objective was Argiope Linea, near 17 degrees south. It was hoped that a close look at this triple-band would provide conclusive evidence of a process that was being referred to as 'stress-controlled cryovolcanic eruption'.

David Seidel, JPL's Outreach Supervisor, moderated a panel of experts at a Press Conference on 9 April which presented the astonishing results.

The southern hemisphere crater Pwyll was measured to be 26 kilometres in diameter. The dark zone immediately surrounding it was interpreted as being material excavated from a few kilometres below the surface. Earlier lower resolution imagery had shown that the bright rays of excavated near-surficial ice extend for a thousand kilometres over the surface. Galileo later took stereoscopic imagery to inspect Pwyll's topography. Unlike most young, deep impacts, the crater's floor is at the same level as the exterior plain, and the central peak complex is 600 metres tall, and thus rises far above the rim.

However, the 'star' of the show was a high-resolution view of a 100-kilometre wide patch of mottled terrain some 1,000 kilometres due north of Pwyll. It was almost a case of 'X marks the spot', because the 'chaos' (as this type of terrain was dubbed) is immediately south of the intersection of two prominent linea.

This particular patch of chaotic terrain was soon to be named Conamara. Galileo revealed that the surface has been fractured into polygonal 'rafts' of ice individually 3 to 6 kilometres across. The western fringe has been 'painted' by one of Pwyll's bright rays. As with the linea, the chaos is stained brown, suggesting the presence of minerals of endogenic origin. Notably, the northwestern edge of the chaos is aligned parallel to, and offset a few kilometres from, Asterius Linea. Also, the process which broke the icy crust and jostled the rafts was localised, because the surrounding area is seemingly unaffected.

"We're intrigued by these blocks of ice," noted Ron Greeley, because they are "similar to those seen on Earth's polar seas during springtime thaws." Paul Geissler of the University of Arizona's Lunar and Planetary Laboratory was insistent that the motions of the rafts could not be explained by convection in ice; only a fluid medium could produce such rotation and tilting. Michael Carr of the US Geological Survey pointed out that the motion could not have resulted from sliding down an incline, the rafts had clearly been caught up in a strong current in a fluid medium, which was almost certainly water. In fact, because they had been floating, they were true

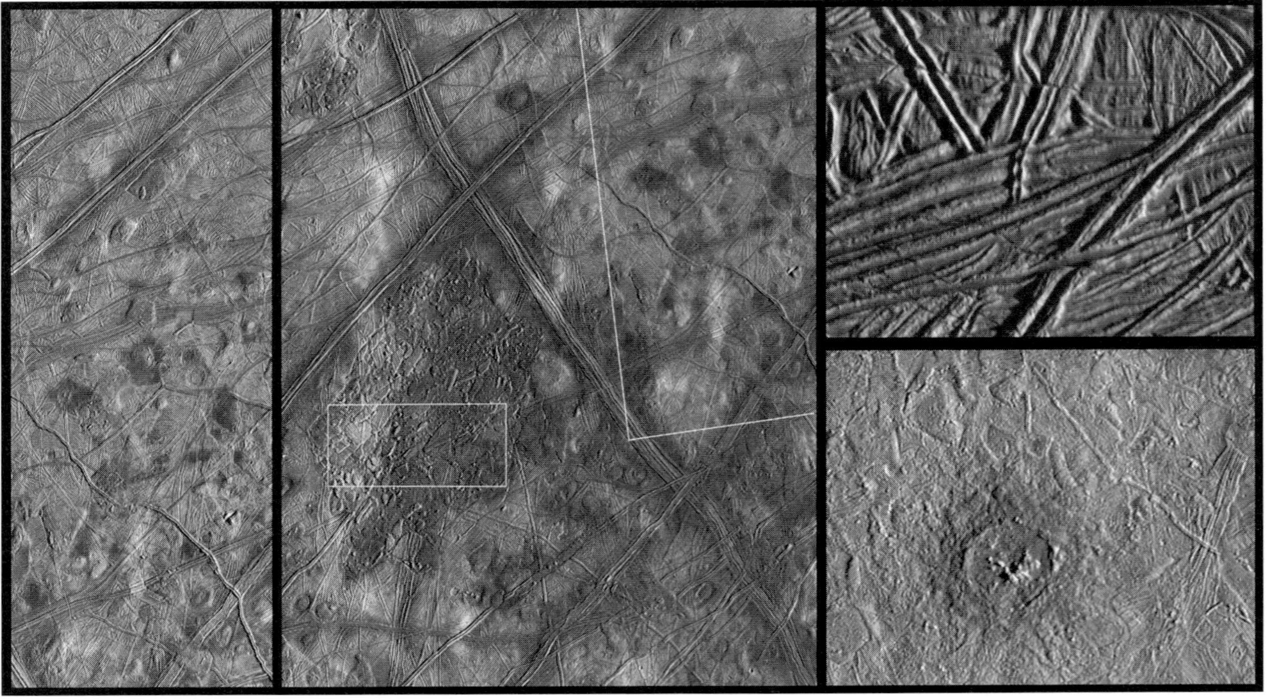

On its sixth orbit, Galileo investigated the patch of mottled terrain on Europa's trailing hemisphere near intersecting linea (centre), and discovered that the rich texture derives from the fact that the icy crust has been disrupted and the blocks have 'rafted' into new positions. The bright white areas are part of a ray from Pwyll, 1,000 kilometres to the south. The northeastern part of this site is shown (right) in close-up. It shows numerous clusters of hills and low domes up to 10 kilometres wide, many associated with dark non-ice material which may be of endogenic origin. The crater Pwyll (lower right) was also further investigated. The 40-kilometre wide splotch of dark material in and around the crater is believed to be non-ice which was excavated from several kilometres under the surface. The 12 × 15-kilometre section of ridged terrain (top right) off to the north of Conamara shows cross-cutting ridges, many of which are cut by transform (or 'strike-slip') faults, indicating that the crust has undergone lateral as well as compressional and extensional stresses.

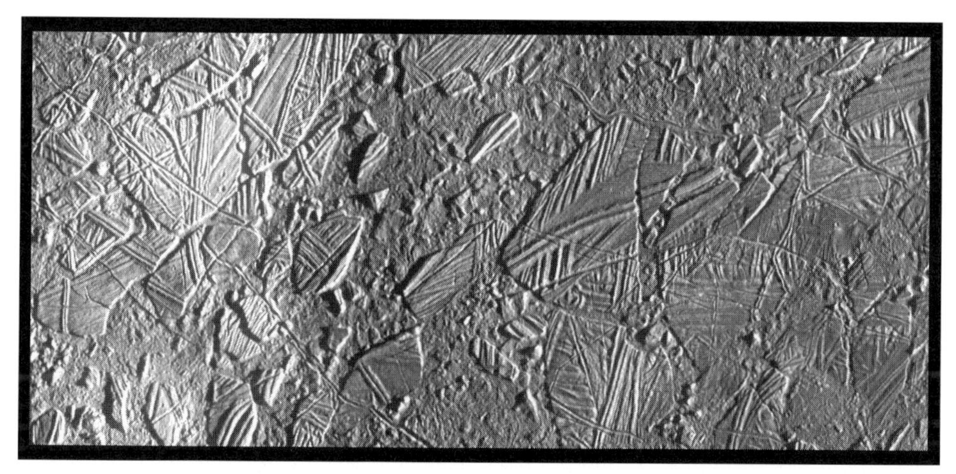

During the extremely close 200-kilometre fly-by of its sixth orbit, Galileo recorded the displaced and rotated 'rafts' in a 30 × 70-kilometre strip of the chaotic mottled terrain (later named Conamara) at 54 metres per pixel. The western section has been 'painted' white by a ray from Pwyll and there are several craters up to 500 metres across which are almost certainly secondaries formed when large blocks of ice from Pwyll's impact rained down. Analysis established that the fractured crustal blocks are true icebergs, in that they were adrift in a fluid matrix which rapidly froze and locked them in position.

icebergs and most of their bulk would have been below the water level.[9] Max Coon of the Northwest Research Association presented the panel with a number of pictures of polar pack ice on Earth which were similar. He posited[10] that the processes of formation were analogous, with the wedge-shaped and grey bands corresponding to 'leads' where large sheets of ice had parted, and the linear ridges marking where compression between two sheets had buckled-up the margins. In contrast to the fracture-and-geyser theory for the formation of the triple bands, he proposed that they resulted from compression and isostatic adjustment. As a compressional ridge formed, isostasy would cause it to settle. In addition to the ice's tendency to flow down slope, the sheer mass of the ridge would force the entire structure to sink to its natural level of buoyancy, and a thin crust would not be able to support a ridge taller than a few hundred metres. He therefore proposed that the dark outer strips of a triple band marked where compression led to downwarping and flooding of the depressed zone alongside the axial pressure ridge – the median stripe – which is fresh ice. One way to test this theory for the triple bands would be to examine the edges of the dark outer strips at high resolution, to determine whether there is a distinct 'water-line', or whether, as the early remote imagery of Minos Linea suggested, the transition is diffuse.

The icebergs were the "smoking gun", Carr summed up, proving that Europa had once had liquid water exposed at the surface – at this site, at least. One study[11] of this and other chaos zones found that the thickness of the ice increases from the equator to mid-northern latitudes, varying from 2 to 6 kilometres. In fact, the Conamara ice may still be just a few kilometres thick. The doppler data confirmed that Europa's rocky interior is covered by an outer shell of water 100 to 200 kilometres thick, so a few kilometres of ice represents only a thin and brittle frozen shell on an ocean so deep that the outer shell is decoupled from the underlying silicate lithosphere.

Intriguingly, some rafts carry segments of ridges. Attempts to 'reconstruct' the disrupted ridges revealed that more than half of the original surface is 'missing', and it could only have been melted. It was as if some internal heat source – maybe a volcano far below on the ocean floor – had created a plume of warm water which had melted through the icy crust and shuffled the rafts before freezing over again. Max Coon opined that because open water would be exposed to vacuum, it would simultaneously boil and freeze. The gap in the ice could remain open only if the water was turbulent. Significant sublimation to vapour would occur while it was turbulent, and this vapour would subsequently condense, fall as snow, and 'paint' vast tracts brilliant white. However, once the water settled down, freezing would dominate, the sublimation would cease as soon as a 'skin' formed on the calm water, and, as this thickened, it would lock in the floating rafts.

In retrospect, it seems likely that the sharp-edged patch of rugged terrain which is located 5 kilometres from the 'ice rink' imaged on the fourth orbit is actually a small rafted chaotic zone. Perhaps that chaotic feature is the result of a 'sudden' collapse of the crust that refroze leaving a mass of icebergs, while the 'ice rink' – not necessarily formed at the same time – is the result of a more gradual and more complete melting, so that there were no icebergs left when it refroze. Both types of feature strongly supported the contention that the icy crust is relatively thin. In fact, there are also several rounded mounds a few kilometres across, and a few isolated patches of

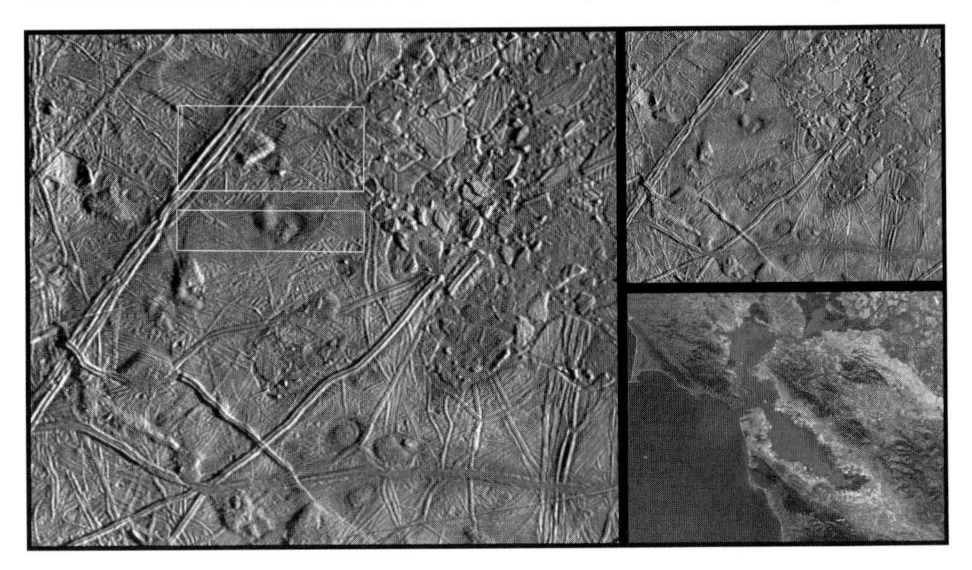

This image of the area between the western tip of Conamara and Asterius Linea shows that although the boundary between the chaotic area and the plain is fairly well defined, the nearby plain is peppered with isolated features of likely endogenic origin. The San Francisco Bay area is included to provide a sense of scale.

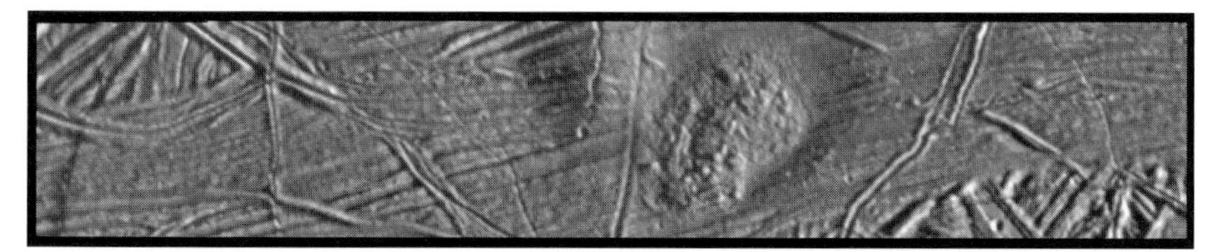

This is a close-up of a peculiar mound about 6 kilometres across on the plain west of the Conamara chaos. It has a hummocky texture, and appears to reside in a depressed area. Such structures probably represent uplift of the icy crust by endogenic processes.

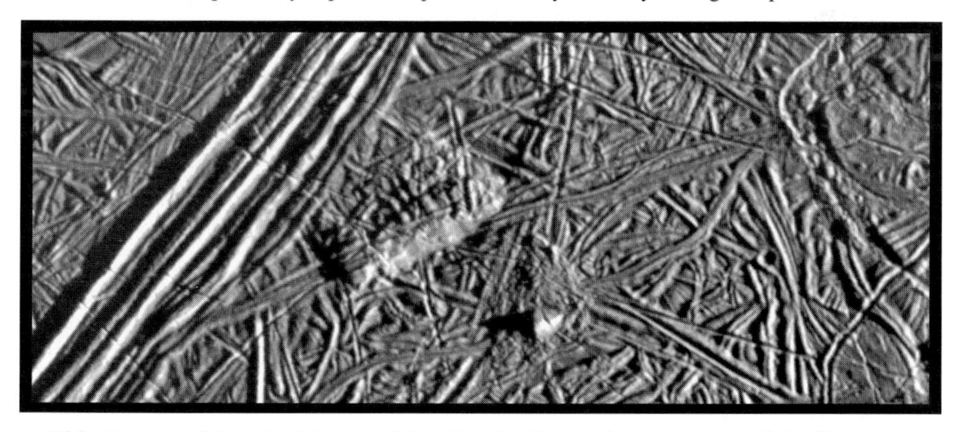

This close-up of Asterius Linea and the plain leading to the western tip of the Conamara chaos shows that although originally referred to as a 'triple band', the linea comprises a system of ridges spanning 6 kilometres. Some of the ridges reach heights of about 180 metres. Other features on the plain include a rectilinear block bearing ridges on its back which seems to have been uplifted, and an isolated icy hill which, judging from its long pointed shadow, is 480 metres tall.

rafting on the plain near the main Conamara chaos, and these may be surface manifestations of dynamic processes within the icy crust.

Galileo's high-resolution coverage included the segment of Asterius Linea immediately to the west of Conamara. In the earlier lower resolution imagery this linea had appeared to be a typical triple band, but was now revealed to be a series of parallel ridges about 6 kilometres in width, rising 180 metres above the plain (the vertical relief could be measured from the length of the shadows). In contrast, an isolated hill on the icy plain between Asterius and the western edge of the Conamara chaos rose 480 metres which, for Europa, made it a veritable mountain. What force had caused such a large block of ice to rise from the plain?

On the other hand, a smaller linea to the north is an extremely well-defined double ridge about 2.5 kilometres wide, rising about 300 metres. It is noteworthy that neither Asterius nor this linea shows any indications of painted terrain produced by geysers. The complex terrain to the north has been cut by a number of right-lateral transverse faults which have offset both some of the smaller ridges and sections of the smoother banding immediately to the north of the double ridge, but have not impinged upon the

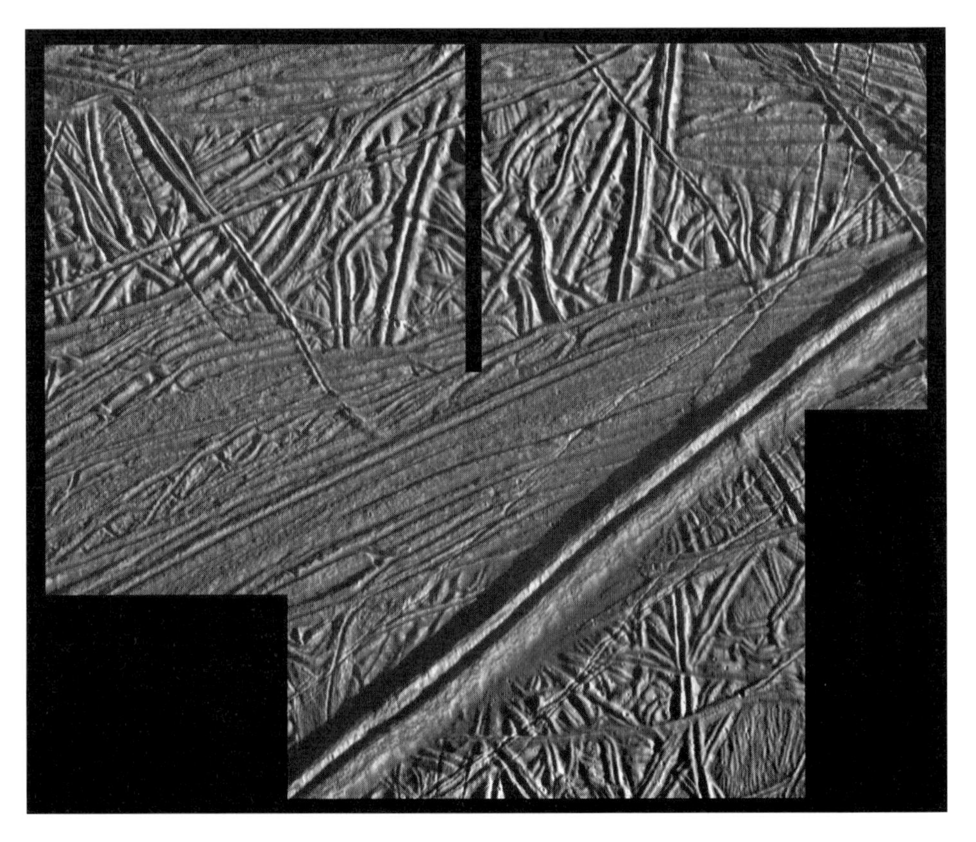

A section of ridged terrain just north of the Conamara chaos was imaged at a resolution of 20 metres per pixel. It shows a double ridge about 2.6 kilometres wide, rising some 300 metres above the adjacent terrain. The complexity of the ridges shows that parts of the crust have been repeatedly modified by intense faulting. The sequence of formation of the ridges can be identified by using the principle of cross-cutting relationships, with the more recent features cutting older ones. Note the displacement of some of the older ridges by transform ('strike-slip') faults. The 'double ridge' is virtually pristine, so it is fairly recent.

double ridge. The relative ages of the various features could readily be inferred from a detailed analysis of the cross-cutting relationships.

THE AGE OF THE ICE?

Of course, the age of the crust is contentious. Clark Chapman of the Southwest Research Institute in Boulder, Colorado, argued that the smooth regions devoid of craters indicate that the surface is much younger than had been speculated earlier. "We're probably seeing areas a few million years old, or less!" Although there is a paucity of the medium-sized craters which could be expected on any ancient surface, the recent close-up imagery revealed a sprinkling of very small impacts. The distribution of these 'craterlets' could be correlated with relative ages derived from the cross-cutting relationships of nearby features. "Although we cannot pinpoint exactly how many impacts occurred in a given period of time," Chapman pointed out, the 'ice rinks' have so few craters that "we have to think of the surface as young." A million years ago is 'yesterday' in terms of geological processes, and these processes are almost certainly still active. Chapman has observed that many small dark spots seen in medium-resolution imagery of Europa which had been counted as craters had turned out, upon closer inspection, not to be impact craters. "This confusion has resulted in crater counts about 100 times greater than now observed!" By any assumed cratering rate, overestimating the crater population will lead to overestimating the age. A recent study by Gene Shoemaker of an impactor population dominated by comets concluded that Europa's ice froze 10 million years ago.[12]

Michael Carr, on the other hand, argued that Europa's surface could be a billion years old. Although this could be considered to be fairly 'recent' in terms of the age of the Solar System, any endogenic processes active then would undoubtedly have expired by now. Ironically, like Chapman, Carr based his conclusion on cratering. He noted that the largest impacts on nearby Ganymede, which can reasonably be attributed to the tail-end of the Great Bombardment some 3.8 billion years ago, have not been disfigured by smaller impacts to the same extent as basins on the Earth's Moon, which implies that the rate for smaller impactors was lower in the Jovian environment than in the inner Solar System. The paucity of craters on Europa could therefore not be assumed to imply *extreme* youth. "There are just too many unknowns," Carr reflected.

As Galileo accumulated cratering statistics for Ganymede, a comprehensive study[13] that assumed an early asteroidal impactor population calibrated by the Gilgamesh basin concluded that Europa's surface froze 3 *billion* years ago. Fitting the size of Europa's impacts to the curve scaled from Ganymede enabled the age of the ice at different times to be estimated. At the time of the impacts which produced the Tyre and Callanish palimpsests, the ice was thin enough to be punctured, but it had evidently thickened by the time Pwyll struck.

It is notable that the debate over the age of Europa's surface spans a range of *three orders of magnitude*.

If Europa's surface really is only a few million years old, then this fact will have to be established independently. "We want to establish whether Europa's surface has changed since the Voyagers," said Ron Greeley, "or even during the time of the Galileo mission." If Europa is still geologically active, there may be evidence of it. "We want to look for current activity – possibly erupting geysers." Galileo was making limb observations in search of geyser plumes, but to date had not seen anything.

Geologists are also poring over imagery from one fly-by to the next, searching for signs of changes on the surface due to cryovolcanism, but the search has been severely constrained by the fact that without the high-gain antenna it is not possible for Galileo to return thousands of high-resolution images per orbit – as had been intended when the mission was devised. The search is therefore limited to taking medium-resolution regional views and a few high-resolution shots of specific features. If it had been possible to map the entire surface several times over at high-resolution, a methodical search for evidence of surface modification would have been feasible. There may well be active sites but, without comprehensive and repetitive mapping, detection will be a hit-and-miss process.

In an attempt to increase the scope, Cynthia Phillips and Alfred McEwen of the University of Arizona at Tucson are comparing Galileo imagery with that from the Voyagers. This search is hindered by the fact that Europa received the least-complete coverage of the Galilean moons, and most of the imagery is of low resolution.

To confirm the existence of an ocean beneath the Europan ice would be a truly momentous discovery. "How often is an ocean discovered?" asked Richard Terrile, a planetary scientist at JPL. "The last one was the Pacific, which was discovered five hundred years ago by Balboa." John Delaney, a submarine volcanologist at the University of Washington in Seattle, observed that recent research had led to the belief that life would spontaneously develop whenever there was volcanic activity and liquid water, even within solid rock. "As best we can tell," reflected Terrile, "life on Earth appeared within the first 700 million years." This implies that terrestrial life originated during the Great Bombardment. From its beginning, the Europan ocean, rich in dissolved minerals and organics, would have been suitable for the origin of life, and it "simply begs further exploration".

TYRE

As the data from the sixth orbit trickled in, the communications engineers had a surprise for the scientists. "Bad news today," reflected Tal Brady in his diary for 24 February 1997. "The magnetometer instrument software seems to have stopped running prior to the encounter, and the real-time data is probably no good. The recorded data may be okay? We'll just have to wait and see." It was later discovered that the instrument had been disabled by a radiation-induced glitch. This was particularly frustrating because this orbit had been the primary mission's closest approach to Europa. Although it was not possible to confirm the previous announcement that this moon had its own magnetic field, it had nevertheless been a remarkable encounter.

Since the point of closest approach was on the night-side, the infrared spectrometer and the integrated photopolarimeter and radiometer collected multi-spectral and thermal data. One target for the spectrometer was 'Tyre Macula', a circular structure some 140 kilometres across which Voyager had imaged from afar. Once Galileo revealed it to mark where an asteroid or a comet had struck, Tyre was reclassified as a multiple-'ringed structure' like Callanish.

The next orbit provided a respectable 24,600-kilometre fly-by of Europa, so the integrated photopolarimeter and radiometer made a thermal map of the day-side hemisphere as part of a long-term campaign to record the thermal characteristics of the surface both in daylight and in darkness in an effort to identify any 'warm spots' that could indicate 'thin ice'. As the Tyre ring structure was illuminated, Galileo was finally able to study it in detail by imaging in both visible and near-infrared.

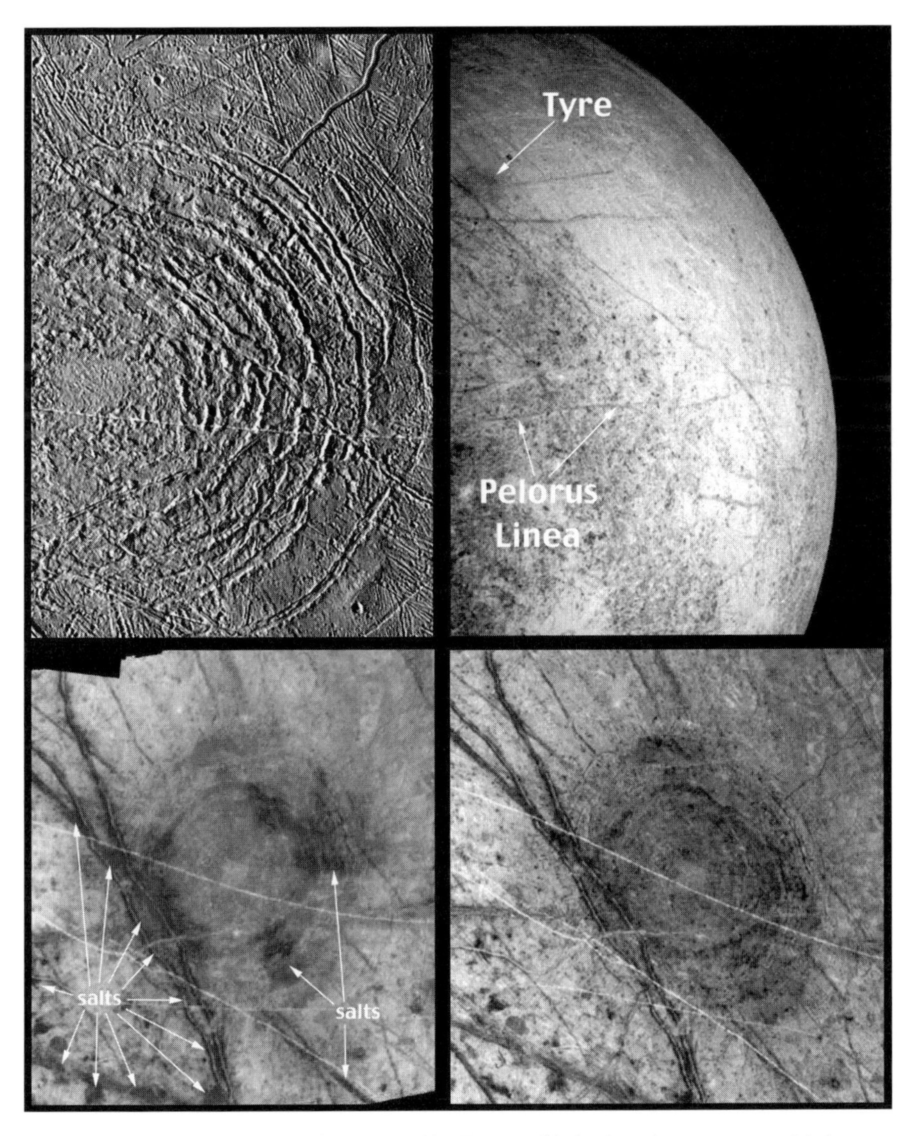

These images show the 140-kilometre wide Tyre multiple-ringed structure at 34 degrees north and 146.5 degrees. As the general context (top right) shows, several prominent linea run close to the structure. The closer view (below) shows that these run through the outer part of the circular fracture zone, and a narrow bright fault bisects the entire structure. The infrared spectrometer's view (bottom left) highlights the composition of the surface deposits. The dark sections are believed to be salty fluids which leaked out from the deep faults. Notice that the deposits along the linea to the west are patchy and have diffuse margins, suggesting that the material was spewed out by geyser activity. These two images were taken on Galileo's seventh orbit, and have a resolution of 600 metres per pixel. The shadowed late-afternoon view taken on the fourteenth orbit (top left) highlights topographic relief of the eastern part of the structure at 170 metres per pixel.

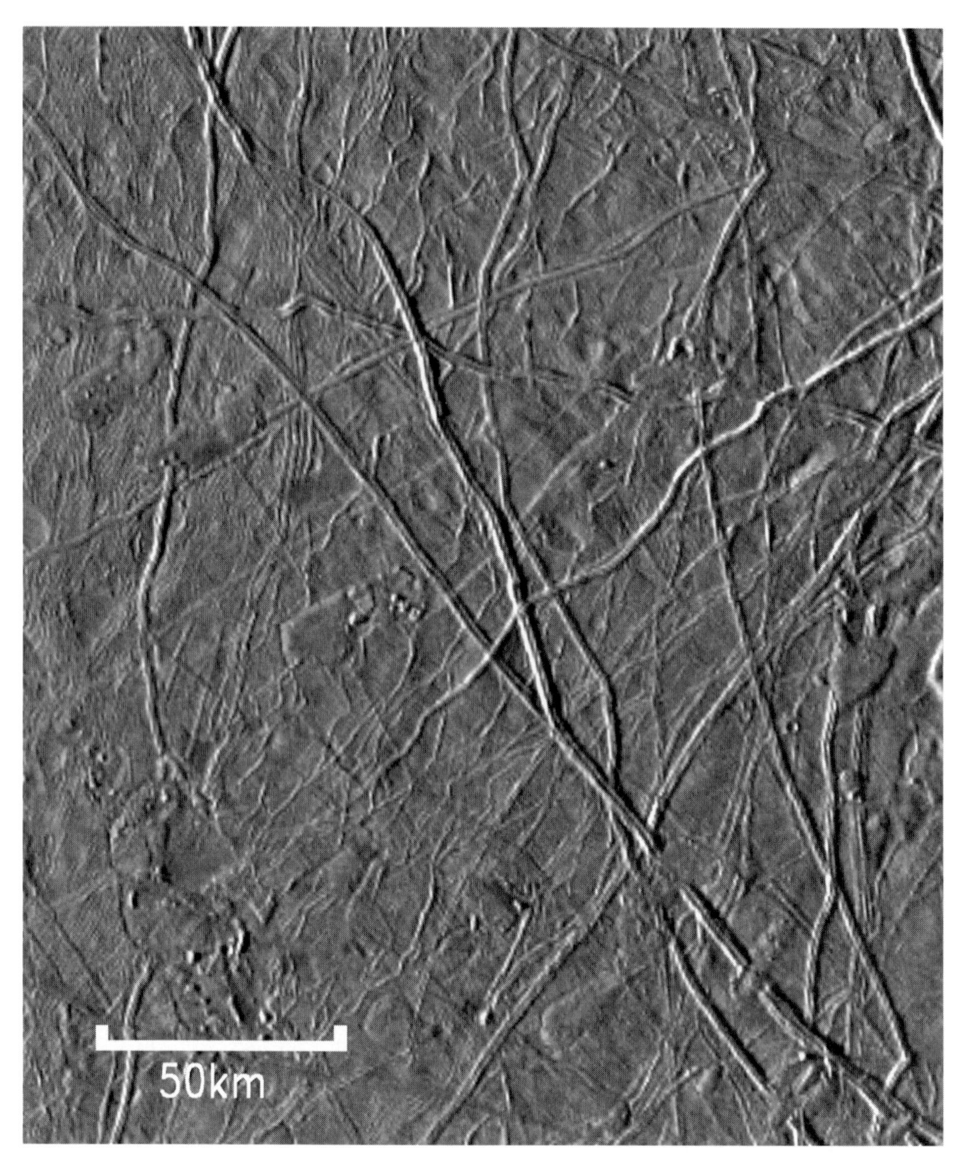

On its seventh orbit, Galileo recorded this 180 × 215-kilometre patch of the icy plain centred near 23 degrees north and 107 degrees. The 1-kilometre per pixel resolution is sufficient to show what would appear to be ice flows from cryogenic volcanoes. They are most evident where they have encroached upon the linea. One particularly striking example is near the southern end of the north–south trending linea towards the western side of the image, but a careful inspection reveals that they are pervasive. Their shapes resemble terrestrial silicate lava flows. There are intriguing concentrations of 'pits' on, and adjacent to, many of the flow features.

The infrared spectrometer identified areas with a high surface abundance of water-ice, and others with concentrations of what may be mineral salts similar to the evaporite deposits in Death Valley, California. The surface at the centre of this impact seemed to be composed of a coarse-grained ice. The impact masked the Minos Linea triple band, which ran southeast across the western rim of the palimpsest. There is a cluster of features 800 kilometres to the southeast whose shapes strongly resemble terrestrial lava flows. In some cases, these have submerged sections of ridges, so evidently the cryovolcanic activity post-dated the formation of the ridges.

Galileo took this 460-metre resolution image of a 335 × 365-kilometre area of mottled terrain near 3 degrees south and 234 degrees on its eleventh orbit. In addition to a patch of chaos (in the northeast corner) it shows a prominent 'grey band' linea which marks where the icy crust split, separated, and was sealed by slush which oozed up and froze. The parallel lines running along the length of this spreading zone document successive phases of such activity. Note that subsidiary spreading activity on axes perpendicular to the main axis has further fragmented the crust.

MOTTLING AND CHAOS

Although Galileo's eleventh encounter sequence was the final orbit of its primary mission, the fly-by on 6 November 1997 was constrained by the need to establish a return to Europa to enable the mission to be extended to further investigate the enigmatic ice moon.

Galileo intercepted Europa just prior to perijove, so it made a mid-morning approach to the trailing hemisphere. After tracking east over the anti-Jovian hemisphere, it crossed the terminator near the centre of the leading hemisphere. As it withdrew, the spacecraft was able to look back at the sub-Jovian hemisphere in darkness. The main theme this time was the nature and age of the mottled terrain.

One area on the equator on the trailing hemisphere showed that the splotchy appearance is the result of the bright plain having been disfigured by a profusion of isolated chaotic patches that have released a dark endogenic material. It is also peppered with numerous isolated peaks and depressions. The extent of the mottling implies that whatever internal processes disrupted the icy crust, it operated on a grand scale.

This area is also crossed by a smooth grey band. Like the dark wedges, this is

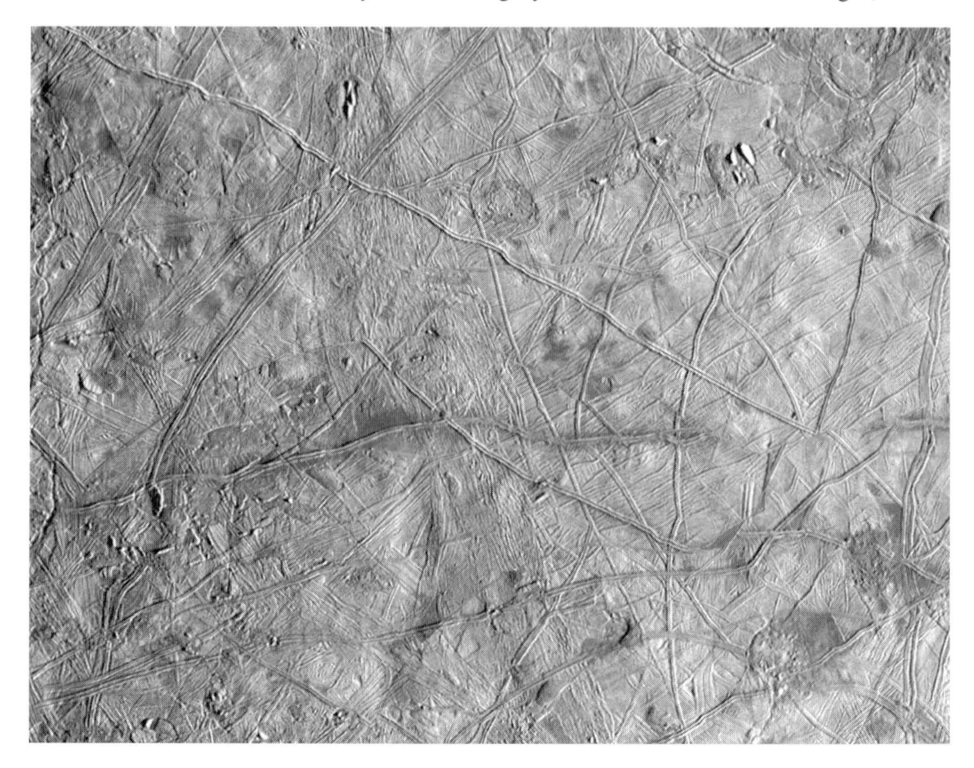

By 'editing out' the prominent spreading zone in the previous figure, the configuration of the icy plain prior to the opening of the fault can be reconstructed.

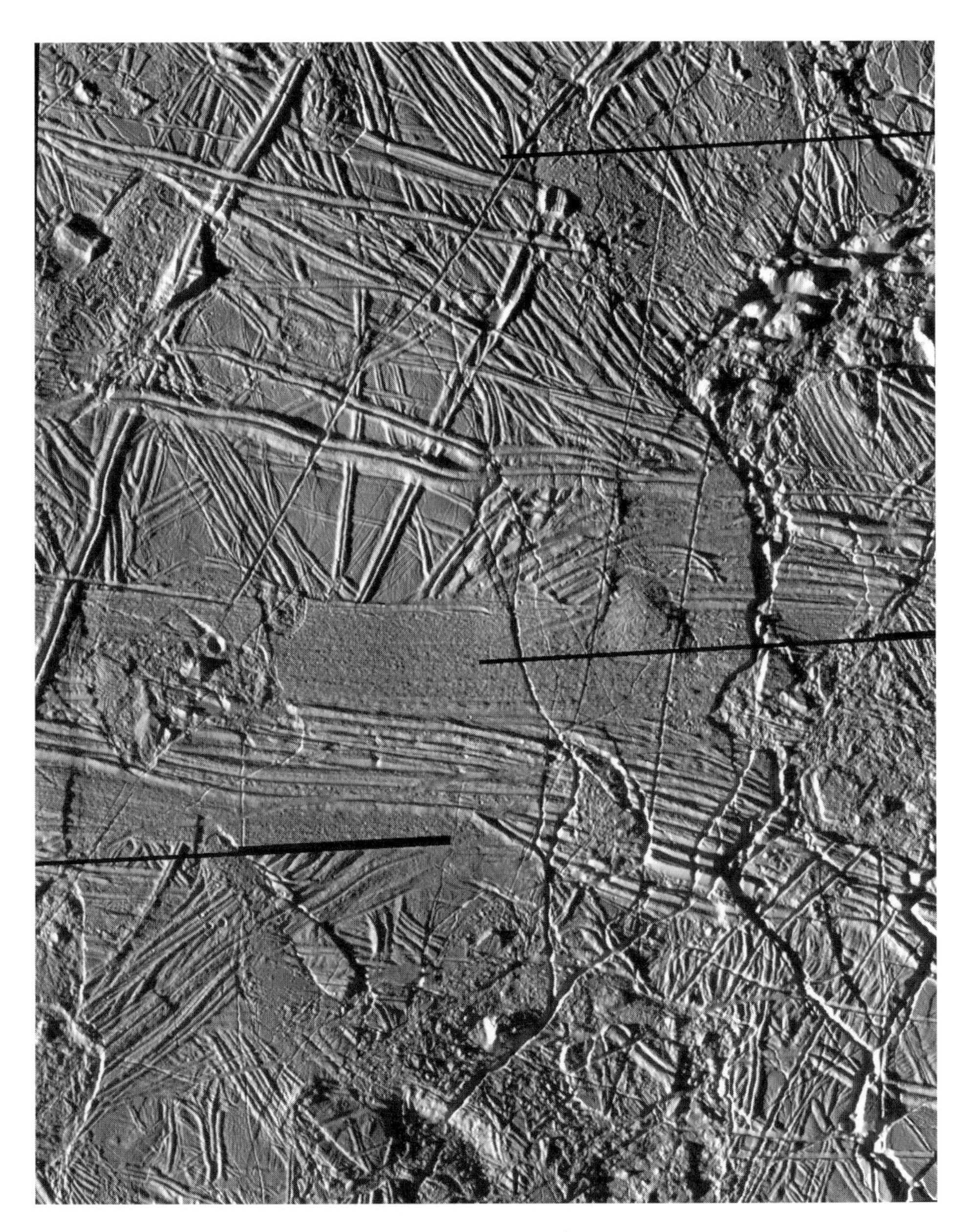

On its eleventh orbit, Galileo assembled this 68-metres per pixel mosaic of a 90 × 108-kilometre area at 35 degrees north and 87 degrees. The southeastern part of the ridge-etched plain has been disrupted, possibly indicating the onset of the melting believed to produce the 'rafting' of a chaotic zone. The plain is dotted with soft mounds and an ice flow has encroached upon the prominent linea to the west. A prominent icy peak in the south is catching the Sun. It is about 500 metres tall. There is a cluster of peaks on the fringe of the disrupted area in the northeast part of the image.

evidently a spreading margin which has been sealed by material which oozed up through the fault as the crustal blocks were pulled apart. The 'tributary-like' banding indicates that the crustal blocks suffered rotational as well as extensional forces, and hence split and split again. Although the grey bands had been thought to be relatively old, 'editing out' this one indicates that the plain had already been etched by a complex network of linearities.

A high-resolution mosaic of an area on the leading hemisphere is littered with hundreds of ridges whose complex cross-cutting relationships record many episodes of ridge creation. In lower-lying areas flooded by flows of fluid, the ridges have been submerged. There are also irregular patches which contain an intensely fractured chaos-like terrain suggestive of melting. As in the case of the area around Conamara, there are some isolated ice-mountains ranging up to 500 metres tall.

HYDROGEN PEROXIDE

Unfortunately, by sheer bad fortune, six minutes prior to the moment of closest approach, the Madrid Deep Space Communications Complex suffered a power failure. Goldstone in California was able to pick up the spacecraft five minutes beyond closest approach but doppler data from this fly-by was not secured. It was a relatively distant 2,000-kilometre fly-by, however, and much closer passes were planned for the extended mission, so the loss of this data was not serious.

As Galileo cut across Ganymede's orbit on its way out, the infrared spectrometer scanned Europa's leading hemisphere – its 'cleaner' hemisphere. "These distant observations reduced the radiation-induced noise," explained Robert Carlson, the instrument's team leader, "so we can get good spectra in the 3.5 micron region."

In fact, the observations[14] turned up a surprise – the Europan ice is laced with hydrogen peroxide. Because hydrogen peroxide is colourless when dissolved in ice, it does not stain the surface. Nevertheless, the spectrometer could detect it by its absorption band. Subsequent observations demonstrated that the icier areas have more hydrogen peroxide. Other chemicals had already been detected, including carbon dioxide, sulphur dioxide (probably pollution from Io) and – possibly – hydrated salts of endogenic origin.

"Hydrogen peroxide is a really weird chemical", noted Carlson, "that reacts strongly with almost everything." As a water molecule with an extra oxygen atom attached, it can readily be dissociated into hydroxyl ions. These are highly reactive, so the only way that an appreciable amount of hydrogen peroxide can be maintained is if the processes of creation and destruction are in equilibrium. It must be created continuously by a process of radiolysis, as the energetic charged particles in the Jovian magnetosphere interact with the ice on the surface. Fragments of disrupted molecules can recombine in alternative species. Almost as soon as the hydrogen peroxide forms, it is either altered by reacting with other chemicals or dissociated by solar ultraviolet. If the free-radical by-products are dissociated before they can react, the result will be oxygen and hydrogen. The hydrogen will escape more readily than the oxygen – some of which will react with the ice to make hydrogen peroxide and so complete the cycle.

With free-radicals present, there may well be more bizarre species on the surface. "We're interested in watching changes in chemical composition over short periods of time," Carlson said. "By studying chemical processes on Europa – and the other moons of Jupiter – we'll be able to learn more about how these moons interact with Jupiter, and how similar processes occur elsewhere in the Solar System."

EXTENDED MISSION

Although Galileo's baseline mission had called for two years in Jovian orbit, it had always been hoped to extend this and, in fact, this had proved practicable. In terms of administration, 7 December 1997 marked the end of the primary mission.

"I've been involved with the Galileo mission since its beginning in 1977, and have been at the helm since 1990 for the flight to Jupiter, the first-ever outer planet atmospheric entry, and orbit insertion, and the primary mission's tour of the Jovian system," Bill O'Neil reflected. "I feel extraordinarily fortunate to have had this priceless, truly unique experience. But it is time for new challenges. I'm delighted to turn the reins over to Robert Mitchell. Having worked closely with Bob for more than 25 years, I know that he'll do a superb job leading the team."

Since joining JPL in 1965, Mitchell had worked on trajectory design, mission design, and navigation for planetary exploration projects such as the Mariner and Viking mission to Mars. As the Galileo Mission Design Manager from 1979 to 1988, he led the numerous redesigns of the mission as the launch date slipped from January 1982 to October 1989, and he headed the team which developed the innovative VEEGA trajectory. Prior to being appointed as Galileo's Mission Director in 1996, Mitchell had managed the science and sequence office.

Funding for the extended mission would have an annual budget only 20 percent of that for the primary mission. With the 'reduction in force' that this would impose, many of the Galileo engineers had to move on.

Propulsion engineer Todd Barber was offered the task of designing the propulsion system for a Mars sample-return vehicle, but he opted to join the team which was to operate the soon-to-be-launched Cassini mission. "Cassini just looked too exciting to pass up!" Development of a new spacecraft was one thing, but operations was something else. "I absolutely love mission operations – flying a spacecraft."

Considering the degree to which its communications had been degraded, Galileo's primary mission had been extremely productive. "We look forward to providing even more fascinating science results over the next two years," Mitchell said. "Maintaining the pace with the reduced resources is going to be a real challenge," he acknowledged, "but we have an excellent team in place, and I'm looking forward to it."

An extended mission was possible only because the spacecraft had healthy propellant and power margins. By their nature, RTGs slowly cool down. They operate by transforming the heat from radioactive decay into electricity, and as they consume radioactive material at an inverse exponential rate, the power output diminishes. Galileo's problem was exacerbated by the fact that its RTGs had been on the shelf for so long while the mission remained grounded, and had then to sustain it during the six-year cruise to Jupiter. A propellant margin would not have been practicable without very accurate navigation within the Jovian system. The prospects for extending the orbital tour for several more years were therefore excellent.

"The Galileo orbiter is performing flawlessly," pointed out O'Neil, "and all eleven of its sophisticated science instruments (and two radio science investigations) are still providing excellent data."

Although the stuck filter wheel on the integrated photopolarimeter and radiometer had been successfully released, the plasma wave spectrometer's functionality was slightly degraded and the infrared spectrometer had lost two of its 17 channels. Nevertheless, despite having been in space for so long, and having been exposed to so much radiation in the Jovian magnetosphere, the spacecraft's ability to undertake integrated science studies was still high.

"A great bounty of Jupiter system science has been obtained and the continuing study of these data will surely add many important discoveries," O'Neil reflected. "While not all of the original objectives could be met due to the high-gain antenna failure, I believe that the overall science return from Galileo will easily exceed what was envisioned at the project's inception, twenty years ago, because our team of scientists and engineers has done such a superb job of capturing the most important observations." In fact, upon reflection, the original concept had been remarkably limited. "The original, official project plan for Galileo had no encounters of Europa or Io, and those of Ganymede and Callisto were to be at (or above) 1,000 kilometres altitude."

EUROPAN CAMPAIGN

The extended mission was divided into three 'campaigns'. The first was a series of eight Europa fly-bys, then four encounters with Callisto in the summer of 1999 to halve Galileo's perijove in order to facilitate a return to Io and thereby make up for the opportunity forsaken on the initial run in to Jupiter. The twin campaigns focusing on Io's volcanism and Europa's icy tectonism prompted the popular description of the extended mission as 'fire and ice'.

The objectives of the Europan campaign were to understand the moon's geological history in greater detail by addressing specific questions:

- How thick is the icy crust?
- When did the surface freeze?
- If the ice is ancient, why is it so free of craters?
- Is there currently liquid water beneath this crust?
- Is there ongoing cryovolcanism?

Due to the difficulty in dating Europa's surface using the crater-counting technique, it is at least possible that the chaotic iceberg zones represent an ancient ocean that froze, rather than a still-liquid ocean beneath a global crust of ice which broached the surface and then refroze.

The search for ongoing cryovolcanism in the form of flows (and possibly even geysers) would continue, and the contaminants that coat the surface would be mapped to seek evidence of endogenic processes, but what was really needed was very-high-resolution magnetometer evidence that Europa's magnetic field is due to currents induced in a salty ocean, as this would indicate the state of the moon's interior *today*. The fly-by on 16 December 1997 was exceptionally close – only 200 kilometres – in order to probe its magnetic field, and to generate high-resolution gravitational data to refine the model of the moon's internal structure.

The geometry of the encounter was similar to the previous one, but Galileo intercepted the moon a little after, rather than before, perijove and this time it flew over the illuminated rather than the dark hemisphere, with the point of closest approach being 8 degrees south and 225 degrees west.

Because the orbital tour would not provide a closer encounter, the high-resolution imagery from this pass would be the best of the entire mission. In fact, the 'context shots' would yield 50-metre resolution, and so rival the best previously achieved.

SPREADING IN CLOSE-UP

On its third orbit, when Galileo got its first good look at Europa, it took an image showing a dark wedge-shaped tectonic spreading zone on the anti-Jovian hemisphere. It now recorded several sites within this area at high resolution to further investigate the wedge and the plain it etched.

An image where a more recent double ridge cuts across the wedge includes three types of terrain – the dark wedge, the bright ridge, and a segment of the plain. The wedge displays a linear texture trending along the spreading axis. The orientation of these lines changes slightly to follow a bend in the wedge. The fact that the east–west trending double ridge transects the wedge indicates that is more recent – as is evident from its fresh appearance and exposure of bright ice. Stereoscopic images indicate the ridge system's topography. The ridge crests rise 300 metres above the adjacent furrowed plain. In addition to the 1.5-kilometre wide median valley marking the fracture, there are numerous small terraces and deep troughs. The brighter material – probably pure water-ice – is evidently exposed near the ridge crests and on steep slopes. The dark material – 'dirty' ice containing either silicates or hydrated salts believed to have welled up through the fracture – is concentrated in the troughs.

A pair of images document the transition between the plain and another wedge, and here a second bright double ridge is in evidence. In this case, the wedge is rather more complex, and there is a relatively smooth patch approximately 30 square kilometres where warm ice appears to have welled up from below to make an 'ice rink' which had an isolated block 'adrift' within it. The double ridge is similar to its neighbour, but at 5 kilometres across it is wider. Its walls – both inner and outer – show bright and dark material streaming downslope and some of it forms broad fans. An older ridge running north–south is narrower and relatively flat. This has fine-scale ridges and troughs running along its length and is transected by several deep fractures which then run on across the plain. The dark ridge continues south onto the adjoining image.

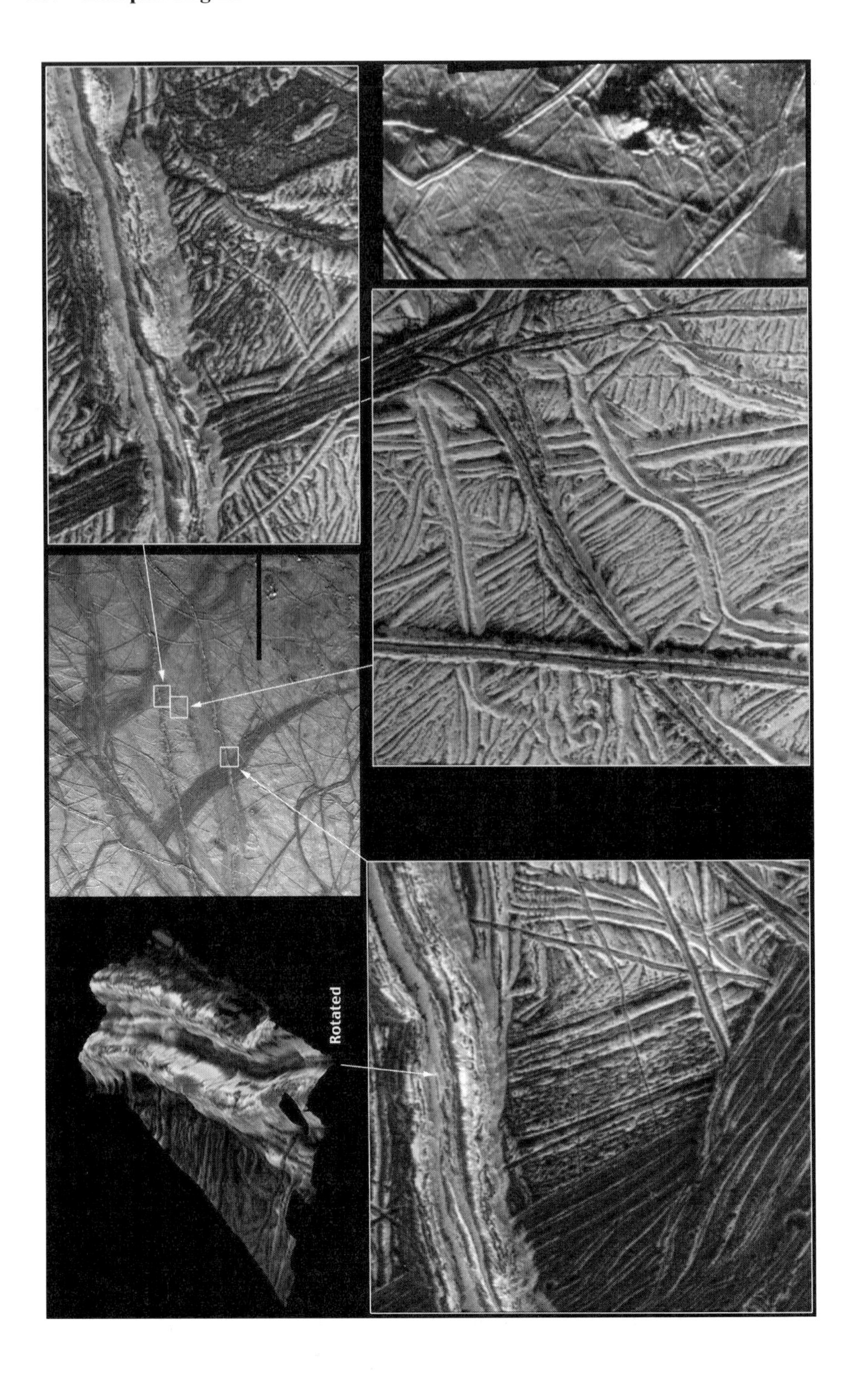
Rotated

By now, a trend had become evident – with each increase in imaging resolution (the new ones had 26 metres per pixel) the icy plains were revealed to be etched by ever-finer fractures.

The task of trying to identify the processes that formed each type of feature is not simply a matter of identifying how the ice became fractured. There is the question of how the crustal blocks draw apart. The moon has a fixed surface area, so the blocks cannot move freely. They are tightly locked. As a spreading zone opens, what happens at the far side of the block? How can this process create two spreading margins close together, and almost parallel to one another? What happens to the strip in between? What is the significance of the arc-like shapes? In essence, the task facing the geologist is to discover how tectonism operates on an ice world – more particularly, one on which the ice is probably just a thin crust on a very deep ocean.

CONAMARA'S MATRIX

Galileo's discovery of rafting in the Conamara chaos had led to speculation concerning the nature of the 'matrix' within which the icebergs had drifted. This very close fly-by offered an opportunity to take a series of 6-metre per pixel images along an east–west line running through the chaos.

The matrix is topographically lower than the rafts, and its rich texture suggests that a jumble of variously-sized chunks of ice were swirling around in a turbulent fluid. There could no longer be any doubt that the rafts had been floating buoyantly in – and even to some extent submerged by – this fluid, which may have been liquid water, warm mobile ice, or a slush of ice and water. Since it froze over, the matrix has been transected by fractures which, if they were to be spanned, would require something similar to New York's Brooklyn Bridge.

◀ On its twelfth orbit, Galileo closely investigated a pair of arcuate dark-wedge spreading features which it had first imaged (top centre) on its third orbit. In each case, it focused on where the wedge is transected by a bright 'double ridge'. The southern case (lower left) actually includes sections of the dark spreading zone, the light plain and the bright double ridge. The change in orientation of the linear texture within the wedge matches the apex between the two arcuate sections. The double ridge includes terraces, and troughs containing dark material. A computer-generated three-dimensional perspective (top left) revealed that bright material (probably pure water-ice) is exposed at the ridge crests and steep slopes and most dark material (perhaps ice mixed with silicates or hydrated salts) is confined to lower areas such as the deep central trough. The crests of the ridge system reach elevations of more than 300 metres above the plains, and are parted by a trough about 1.5 kilometres wide. At 5 kilometres across, the double ridge crossing the other dark wedge (top right) is rather wider. Its inner and outer walls show bright and dark debris which has slumped, in some places forming broad fans. The relatively smooth rectilinear dark area (in the eastern part of the top right image) may mark where 'warm' ice broached the surface. Notice the isolated bright block of ice within this feature. The resolution of these images is 26 metres per pixel. The cluster of ice mountains located nearby (lower right) was imaged at 440 metres per pixel on the third orbit. Notice that the more recent imagery looks different from those taken earlier because the Sun was higher in the sky on the twelfth orbit.

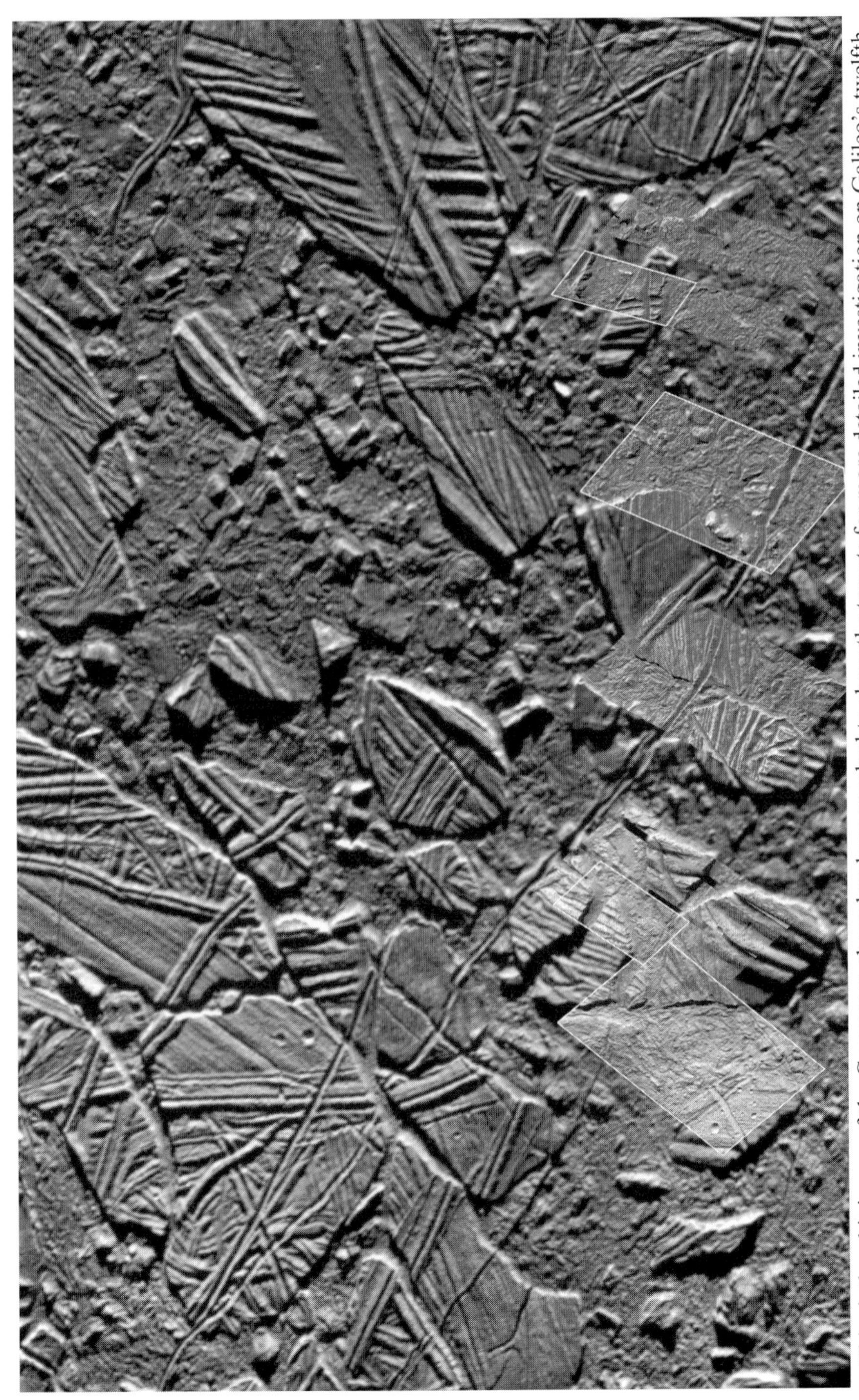

This sixth-orbit image of the Conamara chaos has been marked to show the targets for more detailed investigation on Galileo's twelfth orbit.

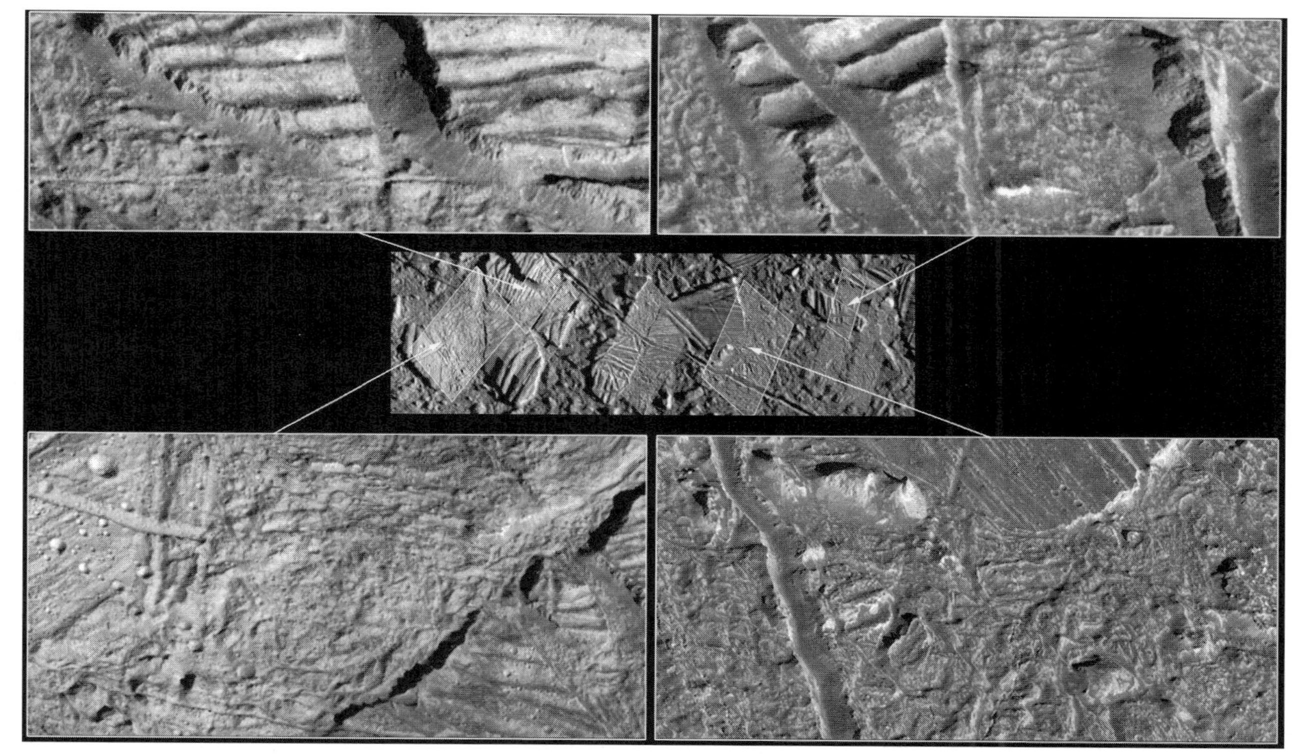

On its twelfth orbit Galileo took this series of close-ups running east-to-west across the Conamara chaos in order to investigate the nature of the matrix. Its rich texture suggests that a jumble of variously sized chunks of ice were swirling around in a turbulent fluid. Cliffs form the edges of the 'rafts', which are riding high in the matrix because they are icebergs. Some of the rafts are fairly flat, but others are heavily ridged – reflecting the features extant on the disrupted plain. Some of the most heavily 'corrugated' rafts have been fractured with steep broadly scalloped slopes marking the transitions, and in some cases there are house-sized blocks of debris piled up at the base the cliffs. The cluster of craters in the westernmost image are secondaries from Pwyll. The resolution in all these images is a mere 6 metres per pixel.

The buoyant rafts ride remarkably high, and their boundaries are marked by cliffs. Some of the rafts are fairly flat, but others are heavily ridged – reflecting the features extant on the disrupted plain. Some of the most heavily 'corrugated' rafts have been fractured, with steep broadly scalloped slopes of the order of a hundred metres tall marking the transitions between levels. House-sized blocks of debris are piled up at the base of some of the cliffs, and an inspection of the shadows revealed a twin-peaked block of ice projecting 250 metres up from the matrix just south of one fairly flat iceberg etched by fine lineaments; it is very rough terrain.

The improved resolution also demonstrated that, in addition to painting the western part of Conamara with a ray of light-toned material, Pwyll had peppered the area with a large number of secondaries ranging in size from 30 to 450 metres in diameter. Prior to employing cratering statistics to infer the age of a surface, it is necessary to distinguish craters produced by primary impacts from those excavated by ejecta from larger impacts. The view of one of the rays from Pwyll therefore provided insight into how secondaries are distributed. The prominence of the ray system indicates that Pwyll is a 'recent' impact – estimated at between 10 and 100 million years. Since the ejecta is superimposed on the chaos, the rafting cannot be any younger.

"Together," pointed out James Head of Brown University, Providence, Rhode Island, the spreading ridges, the icebergs in the chaotic zones, the flows of ice, and the large craters with subdued topography "support the hypothesis that in Europa's most recent history liquid (or at least partially liquid) water existed at shallow depth in several different places." The fluid need not be global – like oil in the Earth's crust, it may exist in isolated pockets. In fact, by making thermal scans in daylight and in darkness to measure how rapidly the ice cooled, the integrated photopolarimeter and radiometer had found indications of 'warm' spots suggesting that the ice is thin at these sites.

"I think there's no question that the ice shell you see is very thin," opined Michael Belton, the leader of the imaging team.

If the existence of the subsurface ocean is accepted, then the next question is the source of heat which maintains it. Both gravitational tidal action from the orbital resonances with Io and Ganymede, and the decay of radioactive elements could contribute. The doppler data from the exceptionally close fly-by refined the model of Europa's structure; in fact, the metallic core could well account for half the moon's radius. Such a large core implies that the moon has suffered considerable tidal heating. As this has been sufficient to induce differentiation in the interior, there might be ongoing igneous activity in the silicate lithosphere which forms the bed of the ocean, and perhaps this energy prompted the tremendous melting which created the extensive mottling on the trailing hemisphere.

CONJUNCTION

As Galileo completed its encounter sequence, an attitude-control problem left it misaligned by about 10 degrees. The engineers suspected that one of the two gyroscopes had drifted, and by the end of January 1998 they had confirmed that there was a hardware fault in the system.

On its thirteenth orbit, Galileo was passive because Jupiter was close to solar conjunction. Although the spacecraft's signal had to pass through the solar corona, the updated Deep Space Network was able to continue to receive data and on 10 February, at the time of the fly-by of Europa, the spacecraft was transmitting an image of the crater Pwyll from its previous fly-by. Doppler data was collected, but the 3,562-kilometre pass was too distant to offer insight into the moon's structure. This passive orbit provided sufficient time to transmit the richly textured high-resolution imagery that was difficult to compress. In fact, as a luxury, some old imagery was reprocessed onboard to highlighting different aspects of the data.

On 10 March, the attitude-control system was submitted to comprehensive diagnostic tests, to determine whether the faulty gyroscope had further degraded during the recent pass through the inner magnetosphere – which it had. The engineers decided to deactivate the gyroscopic system for the upcoming encounter and to use the star scanner instead, even though this would not be able to hold the scan platform stable against any dynamic oscillations that would interfere with the infrared spectrometer's 'push broom' mode of operation.

RHOMBOIDS

The 1,645-kilometre Europa fly-by on 29 March 1998 was just after perijove with Galileo crossing from within to outside Europa's orbit. With its closest point at 12 degrees north and 228 degrees west, it flew over the daylit hemisphere and caught the longitude range 120 to 300 degrees illuminated. This provided the long-awaited opportunity to take a close look at the Tyre ring structure, which had been imaged from afar on the seventh orbit. In the Voyager imagery, Tyre had been just a fuzzy spot. There was now no doubt that it was a large impact. The crater itself is only 40 kilometres wide, but the ring structure is considerably larger. The outer rings comprise troughs and ridges.

However, the highlight of this encounter was an intensely fractured terrain south of the anti-Jovian point. A regional Voyager mosaic shows several chains of 20-kilometre wide dark rhomboids extending across the fractured terrain. Galileo imaged a section of one of these chains with a resolution of half a kilometre per pixel.

Although the rhomboidal bands mark where the icy crust has parted and been filled in by dark slush which oozed up through the fracture, they differ from the wedge-shaped spreading margins by being linked together by transform faults.[15] The displacement on such lateral faults would have opened up the bands and slowly drawn them out, incrementally growing the dark band as the fault moved. Greg Hoppa of the University of Arizona's Lunar and Planetary Laboratory suggested that these faults had been activated by tidal stresses upon the thin icy crust. An analysis of the faults revealed how extensional forces had ripped the crust apart, forming both the rhomboids and the thin wedges projecting from them.

Galileo also inspected a mysterious 'dark spot' on the equator at 225 degrees west

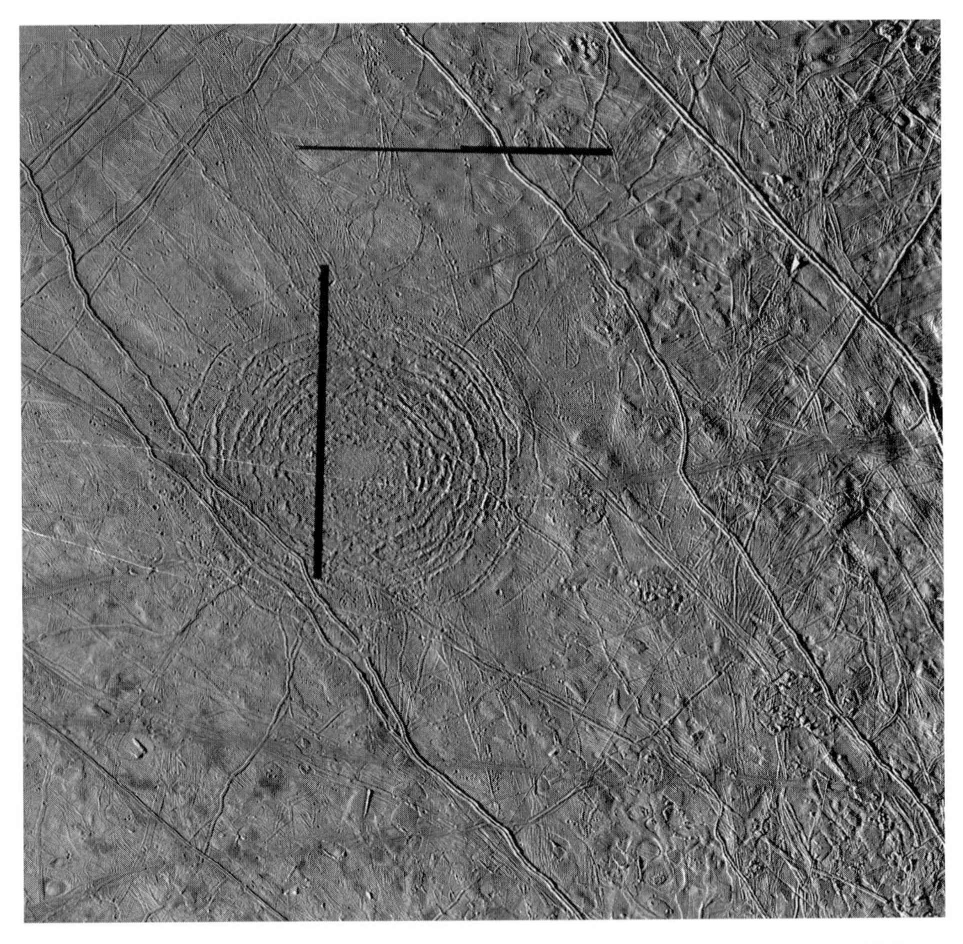

This 170-metres per pixel mosaic on Galileo's fouteenth orbit puts the Tyre multiple-ringed structure into context. Although the overall structure is about 140 kilometres wide, the crater was confined to the 'bull's-eye', but the weakened crust has since been modified by isostatic forces. The concentric rings consist of troughs and ridges. The deep faults would appear to have allowed fluid to leak out from the interior.

that had been noted in the regional coverage taken on the final orbit of the primary mission.

Stereoscopic analysis established that the dark spot is a smooth patch below the level of its surroundings. It is bounded to the west by a terraced cliff, which might have been formed by 'normal' faulting – that is, by the block on one side of a deeply dipping fault dropping below the level of the adjacent plain. If so, the dropped block is tilted, because there is a gentle slope descending to the east, into the dark spot. The dark material shows every sign of having been emplaced as a fluid, or as an icy slush which flowed into the depression. In fact, it has almost submerged the bright ridges which protrude intermittently from the 'ice rink'.

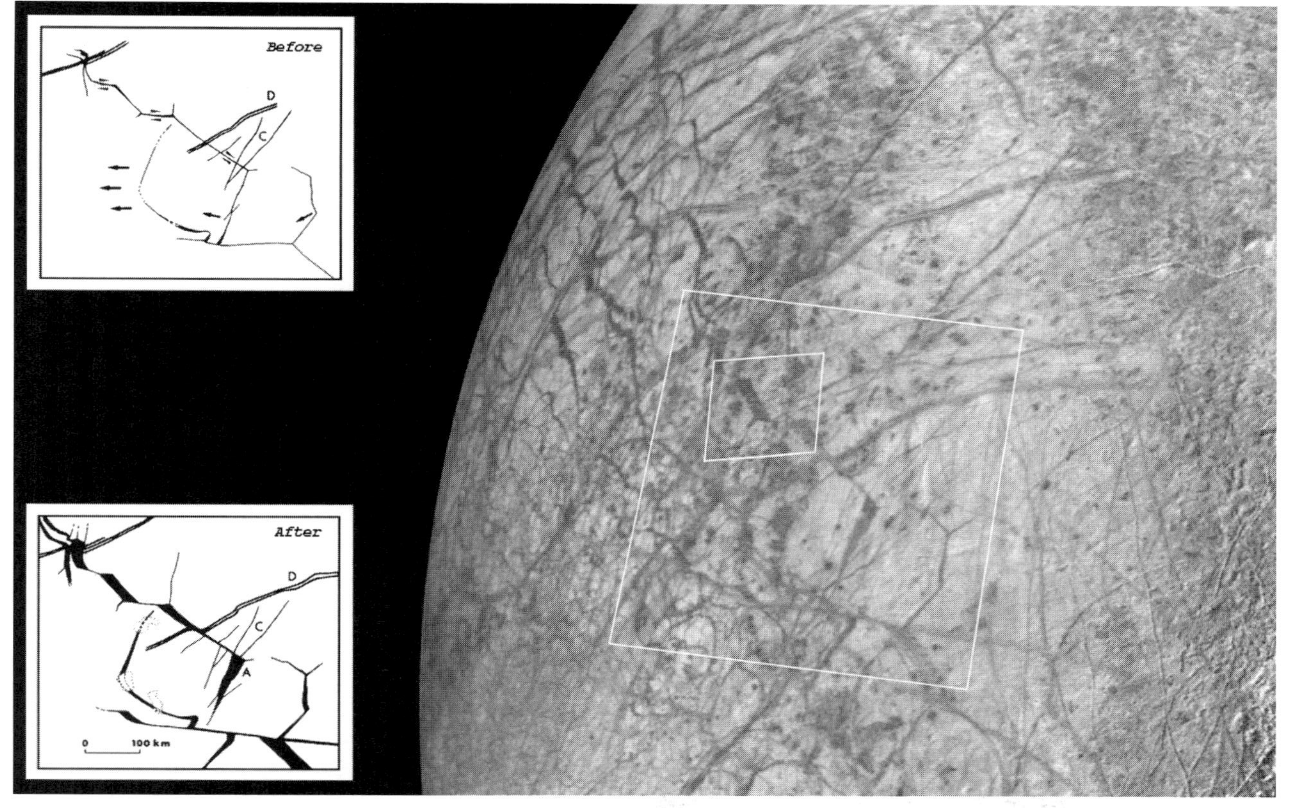

This Voyager mosaic shows the intensely fractured terrain around the anti-Jovian point. The outer box corresponds to the area represented by the inserts, which depict how one particular set of fractures formed. Notice how the wedge-shaped spreading zones work with transform ('strike-slip') faults to open rhomboidal 'gaps' which are in turn sealed by slush that oozes up and freezes. Several such chains can be seen in the image. The area in the inner box is expanded in the following figure. (Inserts courtesy of Paul Schenk of the Lunar and Planetary Institute, Houston)

On its fourteenth orbit, Galileo inspected a section of a chain of rhomboidal features in the intensely fractured terrain around the anti-Jovian point. The transform ('strike-slip') faults that link together the dark rhomboids are readily identified. See the previous figure for the context. The image resolution is about 250 metres per pixel.

A few impact craters are in evidence – some in groups, implying that they are secondaries from a larger impact event, possibly nearby Mannann'an. Interestingly, some of these craters have also been flooded by the dark material, indicating that the flow was more recent than the cratering. As the evidence accumulated, it was becoming possible to demonstrate that melting and extrusion of fluids had occurred at many different times in Europa's history.

The integrated photopolarimeter and radiometer scanned the temperature across the moon, and this data was correlated with chemical composition provided by the infrared spectrometer and with surface morphology to search for recently active cryovolcanism. Although it could now be done at higher resolution than during the primary mission, the search was a trade-off between detail and coverage. The ultraviolet spectrometer sought an increase in the exosphere which might indicate gaseous emission from a geyser – the holy grail for the prospective Europan oceanographers – or perhaps ice vaporised by a significant impact.

Immediately after the encounter sequence, the engineers ran further tests on the gyroscopic system. A specific circuit was identified as the probable cause of the anomalous behaviour. This particular chip had seemingly received more radiation than had similar chips. It was decided to revise the spacecraft's software so that it would monitor the faulty unit's performance, and if it detected an anomaly, would

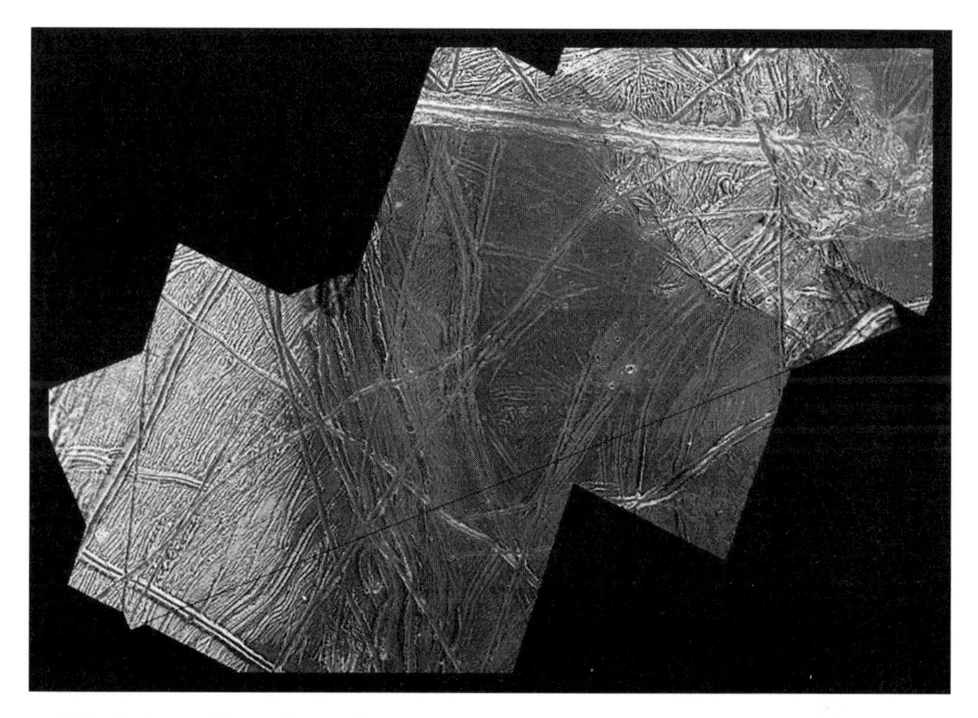

This 'dark spot' located near the equator at 225 degrees was documented by Galileo on its fourteenth orbit at a resolution of 50 metres per pixel. A topographic analysis found that the western margin of the spot is bounded by a cliff and terraces, which represent 'normal' faulting in which the eastern block dropped, creating a depression which was then flooded by a dark fluid of endogenic origin, and submerging some of the bright terrain and ridges near the centre of the spot. The clusters of small craters may be secondaries from nearby Mannann'an. Significantly, the flooding has encroached on some of these craters, implying that the extrusion of fluid is fairly recent.

apply a 'correction coefficient'. The gyroscope would have to be calibrated from time to time to refine the coefficient. The reprogramming effort on 6 May ran into difficulties – the code was correctly installed, but it failed to 'correct' the signal, and a series of tests were conducted to determine the reason. A week later, a second set of software was loaded, and this time the system seemed to function, so it was retained for the forthcoming encounter sequence.

CONSOLIDATION

The 2,516-kilometre fly-by on the fifteenth orbit occurred after perijove, with the closest point of approach in daylight at 134 degrees west and 15 degrees north. A pair of images were taken for stereoscopic analysis of the region around the anti-Jovian point, particularly a feature called Cilix. Although it had been interpreted as a crater on the Voyager imagery, a long-range view on Galileo's first orbit had prompted speculation that it might actually be a mound. This close look confirmed it to be a 23-kilometre wide impact structure with a rim projecting half a kilometre above the plain. Its floor appears to have rebounded and domed, so the central peak is almost as high as the rim, the western part of which has suffered pronounced slumping.

Galileo took stereoscopic imagery of an unusually rugged area several hundred kilometres southeast of the Tyre multiple-ringed structure. In addition to irregular chaotic patches, the plain is criss-crossed by prominent ridges and is peppered by

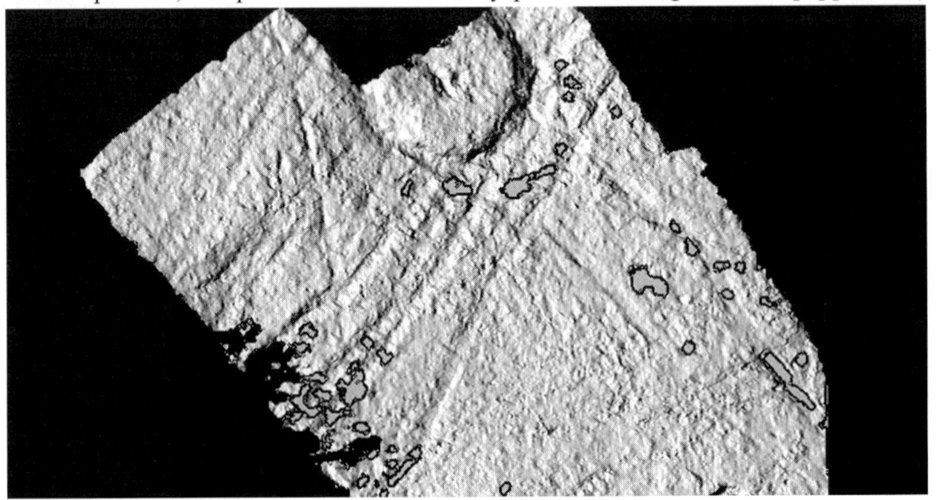

This shaded relief depicts relative elevations around the Cilix impact crater. The graphic was created by shading a model of the surface. A number of images at various viewing angles were combined to derive a three-dimensional model, and this has been provided with a mid-morning illumination. The rim of the 23-kilometre diameter crater rises 500 metres above the plain. The floor appears to have rebounded and domed, so the central peak is almost as high as the rim. The western rim has suffered pronounced slumping. The uniformly grey patches outlined in black show where the data was insufficient for modelling. The linearities outside the crater might be double ridges (although they were insufficiently resolved to show their structure).

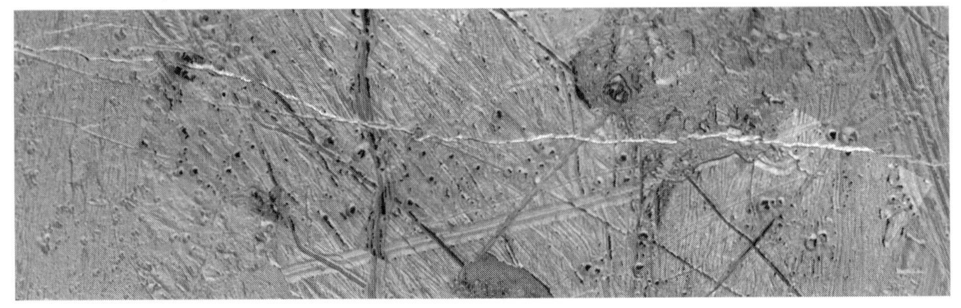

This terrain, southeast of the Tyre multiple-ringed structure, was documented on Galileo's fifteenth orbit at a resolution of 30 metres per pixel. In addition to having patches of chaos, the plain is intriguingly 'pitted'.

On its fifteenth orbit, Galileo assembled a 350 × 800-kilometre mosaic of an area of ridged plain centred at 40 degrees north and 225 degrees (north of Minos Linea), at a resolution of 230 metres per pixel. In fact, the image was secured using a variety of filters so as to be able to create 'false colour' views to identify the composition of the icy plain and the surface materials and so it is best viewed in colour. (It is 'PIA01641' in JPL's Planetary Image Archive, so download a copy.)

circular-to-oval pits which are probably secondaries from the Tyre impact. This provided another view of how the ejecta from a major impact distorts the statistics of small craters. The plain itself represents several types of terrain – lineated and ridged. The ridged plain comprises many small ridges which trend from northwest to southeast. The lineated plains appear to be more finely textured, with rather more closely-spaced ridges running north to south. In addition to the blocky chaos that contain rafts several kilometres across, there is a hummocky type of chaos made almost entirely of a finely textured matrix with only a few small rafts, as if it had been more thoroughly melted but had been too turbulent to form a smooth 'ice rink'.

In light of the low-gain antenna's transmission limitations, imaging a single area repeatedly using different filters limited the total number of sites that could be observed, but viewing a few selected targets in colour provided a valuable context for interpreting similar terrain which had been documented only in monochrome. For this reason, the highlight of this pass was the false-colour image of an area spanning some 800 kilometres east-to-west, located just north of the western end of Minos Linea. The background smooth plain is virtually pure water-ice. The double ridges are distinctly brownish, indicating that they extruded mineral-rich water that froze upon being exposed to vacuum.

On a world seemingly encrusted by ice, it was important to identify anything that was not water-ice. Previously, the infrared spectrometer had made multispectral studies of the trailing hemisphere (which is continually blasted by charged particles in the Jovian magnetosphere) to identify the heavy ions which 'pollute' the ice, and to try to distinguish between these and the salts erupted from the linea and the chaos zones which appear to create the pervasive mottling. Now the instrument made high-resolution scans of the leading hemisphere to identify non-ice components on the pristine surface in the moon's magnetospheric 'wake'. However, a glitch prompted the spacecraft to switch off its gyroscopic system, so the instrument's observations on the way out were marred.

In early June, as Galileo withdrew from the inner magnetosphere, the engineers re-ran the test to determine whether the balky gyroscope had suffered during its perijove passage. It was important to characterise how its performance degraded, because the radiation exposure would significantly increase during the coming perijove reduction campaign.

A CHANGE OF HANDS

On 4 June 1998, Robert Mitchell became Cassini Program Manager and Jim Erickson took over Galileo. Erickson had managed the science and sequence office and had served as deputy manager of the Galileo engineering office prior to being appointed deputy project manager for Galileo's extended mission, so he was well positioned to maintain the pace.

UNLUCKY SIXTEEN

The sixteenth encounter sequence was pre-empted by a glitch when Galileo was deep inside the magnetosphere. At 18:05 UT on 20 July, on the run in to perijove, the primary Command & Data (C&D) Subsystem suffered a fault, so the spacecraft switched to the back-up. It also detected a fault, so it terminated science operations and 'safed' the spacecraft. Communications were restored at 06:35 UT on 21 July, but would not be possible to restart science operations until the problem had been diagnosed and overcome. The C&D – which was effectively Galileo's 'computer' and comprised about 1,000 solid-state chips on 28 circuit boards – was a complex system. Analysis suggested that a signal line had been shorted out by a particle of debris, and this had induced a series of 'resets'. Spurious signals occurred from time to time, and the hand-over to the back-up system had facilitated recovery, but this was the first time that the spurious resets had persisted.[16] The first signal had prompted the primary C&D to hand control to its back-up, and the next had prompted the secondary to 'safe' itself – it had no back-up. Restoring the spacecraft to operations took two days; unfortunately, the Europa fly-by was a few hours after perijove, and no observations were secured because the spacecraft was inert.

The illumination had been similar to that on the fourteenth orbit, but this time the point of closest approach was in the southern hemisphere, at 226 degrees west. Its imaging targets had included Agenor Linea (which was unusually bright for such a feature), Thynia Linea (a grey band spreading feature that was believed to be very old, yet had been remarkably unaltered by subsequent activity), the irregular dark macula Thrace (which was also to have been observed by the infrared and ultraviolet spectrometers in order to characterise the surface materials), and the impact crater Taliesin. The loss of data was frustrating, but most of these targets would be favourably presented on the next encounter.

The gyroscopes acted abnormally on 24 September, prompting the fault protection system to deactivate them. With less than 24 hours remaining before the next encounter sequence, the engineers had little option but to leave the spacecraft to rely upon its star scanner.

The 3,600-kilometre fly-by on the seventeenth orbit was similar to that of the fifteenth orbit, except that the closest point of approach was at 43 degrees south and 139 degrees west. The imaging objectives included a global observation to help to define the shape of Europa's tidal bulge, regional mosaics on both the sunset and sunrise terminators covering a variety of terrains which included features supporting the case for an ocean beneath the ice, and terrain in the south polar region – this in coordination with surface studies by the infrared and ultraviolet spectrometers and the integrated photopolarimeter and radiometer. The targets for high-resolution imaging included Agenor Linea, the Thrace and Thera maculae and nearby Libya Linea, Thynia Linea, Astypalaea Linea, and the impact crater Rhiannon.

The infrared spectrometer plugged gaps in its regional map to characterise the composition of the moon's surface, this time paying particular attention to areas of non-ice and diffuse dark material. Mapping global spectral characteristics produced information on the composition of the surface and regional variations. The ultraviolet spectrometer undertook coordinated studies to establish the extent to which the surface was influenced by micrometeoroids and by charged particles in the Jovian magnetosphere. It also monitored the exosphere for signs of outgassing from ongoing cryovolcanism. The integrated photopolarimeter and radiometer continued to map thermal anomalies – it had established that although the day-time temperatures were as expected, and they varied considerably from place to place over the night-time hemisphere, the anomalies did not seem to be correlated with either albedo or surface morphology.[17]

ASTYPALAEA

In 1996, when Randy Tufts of the University of Arizona in Tucson was looking back over the Voyager images of Europa's southern polar region, he suspected Astypalaea Linea of being a transform fault – also sometimes called a strike-slip fault – where enormous blocks of the icy crust have slipped past one another horizontally, without vertical displacement. "Astypalaea Linea is simply a beautiful structure," reflected Tufts. "I think this thing is gorgeous."

On its sixth orbit, Galileo had made a regional mosaic in the southern hemisphere running along Agenor Linea's 1,500-kilometre length. Now the encounter geometry and illumination permitted Galileo to take a close look at Astypalaea Linea. "Comparison between this Europan fault, and faults on Earth, may generate ideas we can use in studying crustal movements here on our own planet," explained Tufts. On Earth, transform faults are an integral feature of plate tectonics. The most widely known terrestrial example is the San Andreas. This projects north from the Gulf of Mexico, along the coast of California, and then heads offshore just north of San Francisco. The coast of Southern California has been transported almost 600 kilometres north by the fault's action.

At some 800 kilometres long, Astypalaea Linea is comparable in size to the

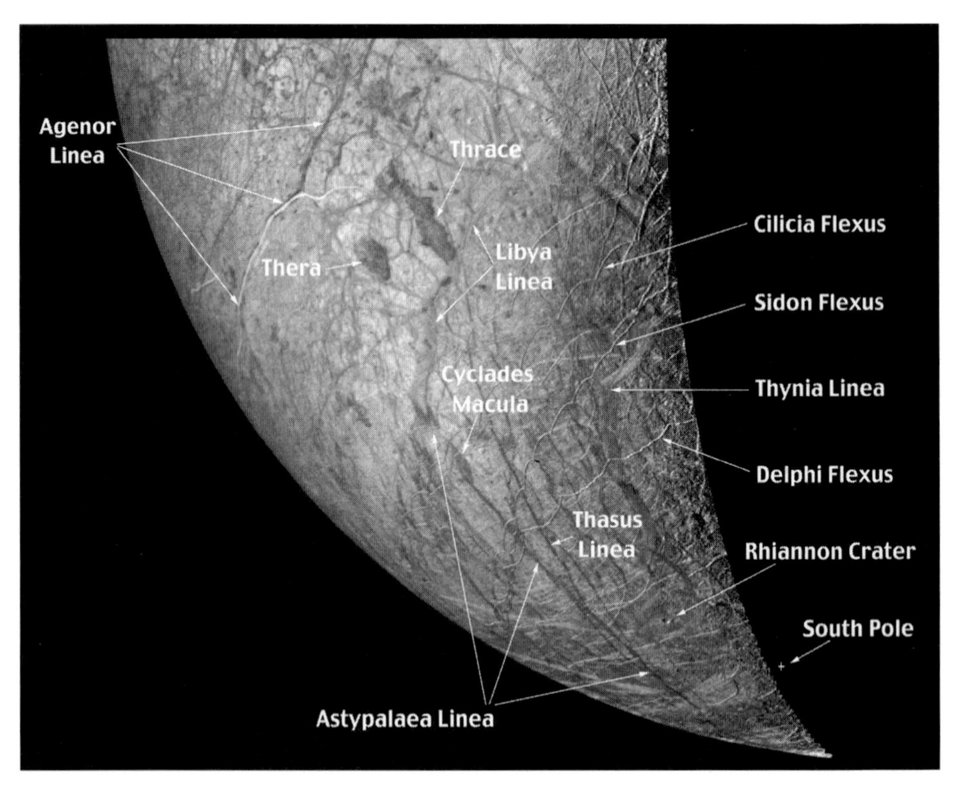

This definitive Voyager mosaic shows Europa's southern anti-Jovian hemisphere.

Californian part of the San Andreas fault. Galileo made a mosaic running along the northern 300 kilometres or so of it, and confirmed its identification. Along much of its length the actual fracture is marked by a ridge. In places, the structure forms a 'staircase', the opposite sides of which can be matched to recreate the terrain which has been displaced by the fault. A correlation of the features on each side showed that it has undergone about 50 kilometres of motion. As the fault moved, its zigzagging line created openings and warm ice welled up to create areas displaying a distinctly rhomboidal shape, similar to those in the fractured terrain near the anti-Jovian point. A similar upwelling has been identified within deep troughs such as Death Valley in California and in the Dead Sea in the Middle East, except that there the intruded material is buried by sediments. On Europa, the process is exposed for inspection. The growth is recorded as a succession of strips – rather akin to tree rings – when material periodically oozed out and froze. The symmetry with respect to the fault (which gently curved up the centre of the upwelling) is evident in the close-ups.

Although both Earth and Europa possess long transform faults, the processes which create them are different. In the terrestrial case, the lithospheric plates are driven by convection in the silicate mantle, but this cannot be the case on Europa. Its crust will be particularly susceptible to cracking under gravitational stress because it is only a thin brittle shell on a very deep ocean.

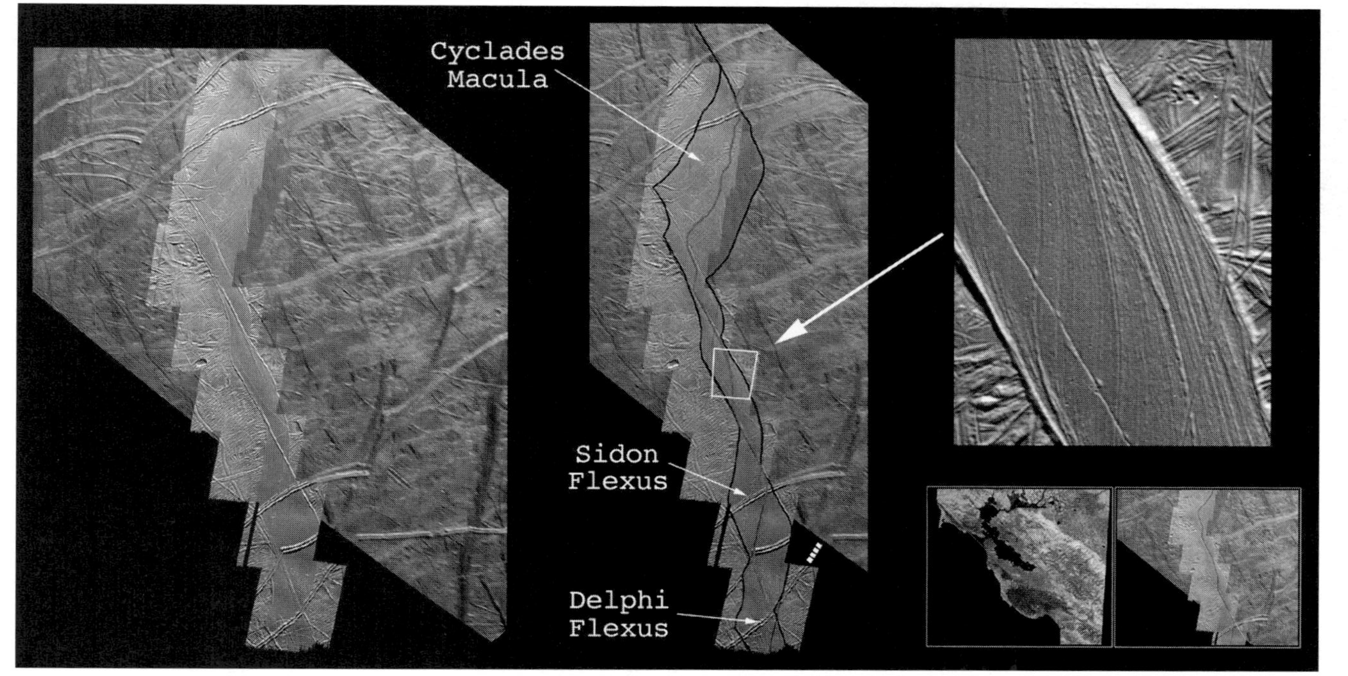

Astypalaea Linea is a transform ('strike-slip') fault extending for 800 kilometres across the south polar region (see previous figure for the general context). On its seventeenth orbit Galileo assembled a 300-kilometre long mosaic of its northern section (left). It is similar in form and in scale to the San Andreas fault in California. Overall, the fault has produced 50 kilometres of displacement. The motion has been on a continuous narrow crack, but its curved form has opened up a band through which slush oozed and froze. In the annotated graphic (centre), the fault is marked by the central line and the lines to either side delimit the zone of upwelling. The fault itself can be seen gently curving up the band in the close-up (top right). This 40-metres per pixel image also documents the fine structure as it spread by successive upwellings. Note the similarity to the intensely fractured zone at the anti-Jovian point, where the lateral motion of transform faults has produced rhomboidal upwelling zones. Astypalaea Linea is transected by a number of prominent linea, so it has not moved recently. The Californian coast provides a sense of scale (lower right).

One idea is that convection currents in the putative subsurface ocean could have dragged the sheets of icy crust, driving the fault in much the same way as mantle currents drag tectonic plates of the Earth's lithosphere. However, the boundaries of the icy plates should show signs of twisting and shearing where they rub past each other. "We don't see them," Tufts reported. This supported an alternate hypothesis, in which the fault's motion was induced by the diurnal[18] gravitational tides.

Although Europa's rotation is synchronised so that it turns on its axis once per orbit of its primary, its orbit is a little elliptical because it is caught in orbital resonances with both Io and Ganymede. When Europa is closer to Jupiter, the planet's pull is greater, and the water in the ocean rises towards the sub-Jovian point. As the moon moves out, and the pull relaxes, this build-up of water ebbs. These tides are awesome. The Moon raises tides in the Earth's oceans with an amplitude of two metres. Europa is comparable in size to the Moon. Jupiter, however, is three hundred times more massive than the Earth. When the crust was forming, Jupiter would have raised tides 30 metres high. In the 'morning', tidal tension would have opened a pre-existing fault, and the build-up of stress would have made it move in one direction. In the 'afternoon', when the extensional force relaxed, the fault would have closed up and locked, preventing it from returning to its original position. In this way, the tidal cycles produced a progression of offsets – referred to as 'walking'. "The analogy describes perfectly what we think happened," Tufts added, "a steady accumulation of lengthwise offset motions."

The fact that Astypalaea Linea is itself cut by many fractures indicates that it has not been active recently. A detailed analysis of the entire length of Astypalaea may show whether it has been 'locked' by a specific feature, because a distinctive stress pattern should be present in the ice at that point. On the other hand, it may well be an ancient structure which ceased moving because the icy crust thickened sufficiently to inhibit the tidal forces from making it 'walk'.

If Astypalaea Linea slipped like the San Andreas fault, did it also trigger the equivalent of earthquakes? "We've no idea," admitted Tufts. Ice and rock behave very differently.

Even though Europa's rotation was long-since captured, which would have been a stressful process, the small stresses arising from the orbital eccentricity could still induce similar activity in particularly brittle parts of the icy crust. As the 'walking' mechanism was explored it detail, it became evident that it could explain other mysterious features.

CYCLOIDS

About this time, a mystery was finally resolved.[19] The Voyager imagery of Europa's south polar area showed chains of scalloped lines linked arc-to-arc at their cusps, extending over the icy plains for hundreds of kilometres. When Galileo arrived, the origin of these 'cycloids' (or 'flexi' as the International Astronomical Union referred to them) was still a mystery. The early regional views revealed that they are more generally distributed and take the form of ridges as well as cracks. When the process that gave rise to them was finally identified, it was further evidence for there once having been a deep ocean beneath the icy crust. Evidently, the cuspate form is the result of pressure on the underside of the thin icy shell induced by diurnal tides in the enclosed ocean.

"This caused Europa's icy shell to flex," explained Greg Hoppa of the Lunar and Planetary Laboratory at the University of Arizona. From the viewpoint of the thin crust undergoing these forces, of course, Jupiter's immense gravity caused the ice to bulge over the period of an orbit, and to do so cyclically. Once the tidal force exceeded the tensile strength of the ice, it began to crack. The crack propagated relatively slowly across the ever-changing stress field, following a curving path. As soon as the stress fell below the tensile strength of the ice, the propagation of the fracture halted. Later, when the stress built up again, the crack restarted, but it did so along a new curve. The distinctive scalloped appearance resulted from the fact that successive curved fractures shared cusps.

Significantly, while this model accurately accounted for the shape of the cycloids, it could do so only if the tidal bulge slid freely over the interior, which implied that there was a deep ocean separating the thin icy shell from the underlying silicate

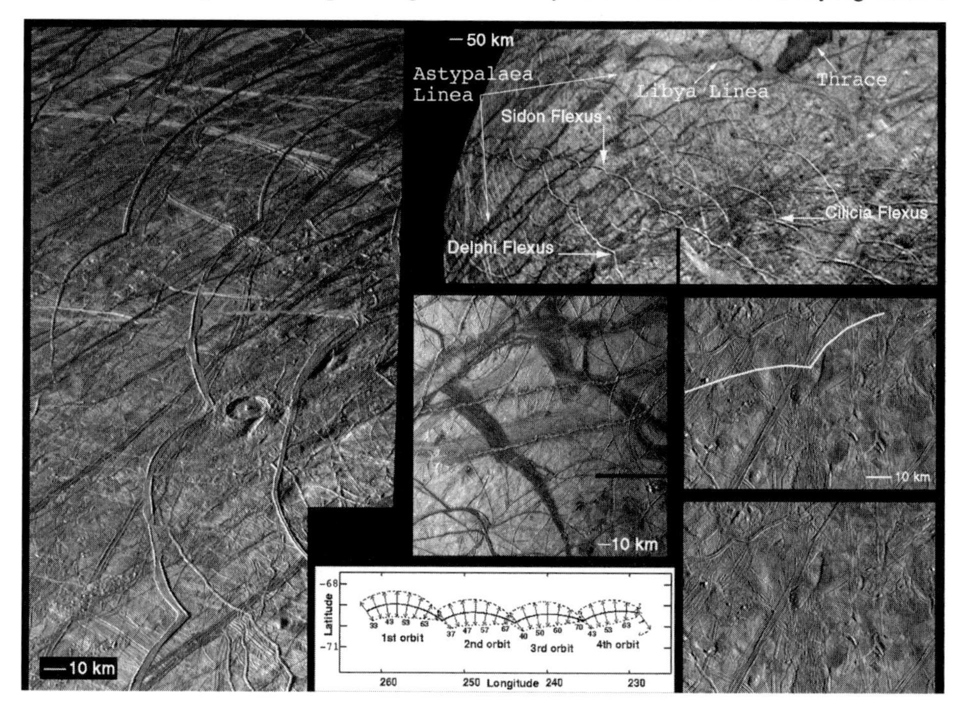

One of the outstanding mysteries of the Voyager coverage of Europa was the nature of the systems of linked arcuate ridges that were pervasive in the south polar region. Even as Galileo accumulated imagery showing them to be widespread, their origin remained unknown. It was eventually realised, however, that they derived from diurnal stresses induced in the thin brittle icy crust by the ebb and flow of tides in the subsurface ocean. (Courtesy Greg Hoppa, University of Arizona)

lithosphere. "The only way this model works," Hoppa emphasised, "is if there is a significant water layer below the ice." And since the cracks probably extended only a few kilometres into the ice, the shell may have been as little as 15 kilometres thick. Considering that Galileo's doppler data had established that the silicate structure was enveloped by a layer of water of the order of 100 kilometres thick, it was evident that the icy shell would have been free to flex independently of the solid body inside.

"What amazes me about this is just how long these features have been a mystery," Hoppa noted. "The reason this took 20 years to figure out, was that nobody had calculated the stress due to tides that Europa feels every 85 hours. Once we made this calculation, we noticed that the stresses rotated by 180 degrees every orbit. Randy Tufts then came up with the idea that these features may form in response to the diurnal flexing of the tidal bulge. We made a few assumptions as far as what type of stress it would take to break the ice and the speed that the crack would have to propagate. We plugged the parameters into our model, and it was pretty amazing how a cycloidal shape just popped right out of it."

One striking conclusion of the analysis was that each arc of a cycloid formed over a single diurnal cycle. "We can look at a crack that has four or five cusps (one formed every 85 hours) and know that the entire chain formed in about two and a half weeks," Hoppa reflected in awe. "It seems what they have been telling us all along is that an ocean was there when these things formed." Such a crack would have propagated very slowly, at about 3 kilometres per hour. A human explorer could probably have walked along with the advancing tip, as it was opening, and although there would not be enough air to carry the sound, he would definitely have felt the vibration in his boots as the crack propagated.

Arcuate segments ranging from 75 to 200 kilometres in length built up features extending for thousands of kilometres. The cracks are edged by ridges, which are thought to result from water that froze in place after oozing out. Cracks which oozed sufficient material evolved into prominent ridges. The direction that any given crack propagated can be inferred from the hemisphere it is in and the shape of the arcuate segments. Remarkably, there is some evidence that the entire icy shell (being decoupled from the silicate lithosphere) has rotated with respect to the moon within.

The paucity of impact craters on the plains which the cycloids etch suggests that these are 'recent'. Since the cycloids imply that there was an ocean then, and there is no obvious reason why it should 'suddenly' have frozen, it is likely still present. "We think there is a fairly good chance that there is liquid water under the surface," predicted Hoppa.

If a new cycloid was observed in the process of formation, this would be evidence that the ocean tides are still active. However, the absence of activity would not imply that the ocean no longer exists, only that the icy crust is now thick enough to resist the continuing tidal stresses. Nevertheless, as Jeffrey Kargel of the US Geological Survey pointed out, "There's a nagging feeling that we really should be seeing changes."

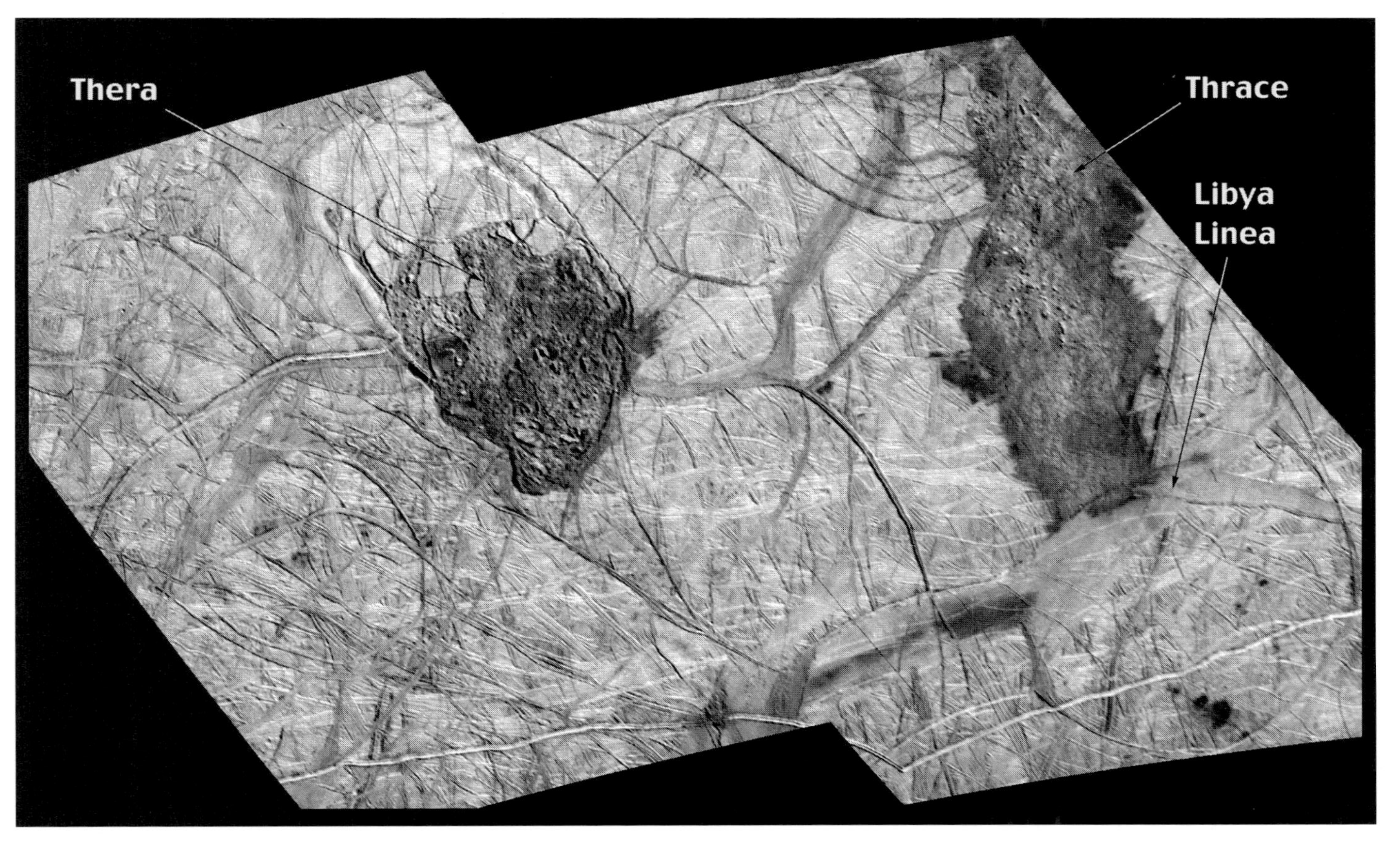

Galileo assembled a mosaic showing the irregular dark maculae Thrace and Thera in the southern hemisphere at 220 metres per pixel on its seventeenth orbit. It would seem that endogenic processes disrupted the icy plain, creating zones of chaos and staining it with dark brown fluid. The mottling is best viewed in colour. (It is 'PIA 02099' in JPL's Planetary Image Archive).

SALTS

The irregular dark features Thrace and Thera Maculae are north of Astypalaea Linea. They were suspected of being similar to the chaotic terrain in which icebergs were trapped in a fluid matrix – only in this case with the matrix being very dark. Galileo took a close look in colour to find out. It found that Thera – at about 80 kilometres across, the smaller of the two – is slightly lower than the level of the surrounding plain. The curved fractures along its periphery suggest that its creation may have involved collapse. Thrace, in contrast, appears to stand at or slightly above the surrounding plain.

Thrace's southern margin abuts Libya Linea. In the Voyager imagery, the macula appears to terminate abruptly at the linea, but Galileo showed that the grey band is stained with the dark macula material. A study[20] of Thrace's southern margin indicated that it has been flooded by an upwelling of salty fluid. The central area consists of broken and disrupted blocks, and hence it is a chaos zone, but the blocks have not been shuffled around. This is in contrast to Conamara, which has been extensively melted and is 60 percent matrix material. Around Thrace's periphery, the topography of the surrounding plain can be traced through a transition zone poking through the dark flooding material.

Several theories have been suggested to explain the formation of these chaotic areas. These ranged from a total melt-through of the icy crust by the ocean beneath, to a partial melting and disruption of the surface by an upwelling of warm ice. The lobate character of the maculae had prompted the proposal[21] that they might be the result of extensive flooding following the ascent and eruption of large quantities of water magma, but these high-resolution views contradicted this. The process appeared to have been more gentle. By providing evidence that the surface is salty in places,[22, 23] the infrared spectrometer supported the proposal that the crustal melting was the result of an infusion of brine into the ice from below – the salt would act as an antifreeze and make 'warm' ice, which would melt.

"The best spectral match is magnesium sulphate, better known as Epsom salts," said Bob Pappalardo of Brown University. "When warm convecting ice rises up and hits a salty area, it melts at a lower temperature than the surrounding plain. This might explain regions where we see what appears to be melting and liquid-like flows across the surface." It also established a correlation between these salts and the rusty-brown discoloration of Thera and Thrace and at the cryovolcanic ridges. "The problem is that Epsom salt is not brown – the best candidates for the brown material are iron compounds or some sort of sulphur compound. Both iron and sulphur are relatively abundant in the Solar System and have a red appearance; plus, we know that there is sulphur dioxide on Europa." On the other hand, of course, there was every chance that the sulphur dioxide represented contamination from Io. There was general agreement that this signature was a hydrate, but for a while its specific form remained debatable.[24]

AGENOR

Agenor Linea is an unusually bright triple-band. Galileo had recorded its context during its sixth orbit, and now it took a close look at a short section near its eastern end, where it divides just north of the Thera and Thrace complex. Most of Europa's ridges and bands are relatively dark, so a major objective was to determine why sections of Agenor are so much brighter than the 'bright plain' across which this 1,500-kilometre long feature runs.

The high-resolution mosaic revealed that Agenor Linea is not a ridge, it is

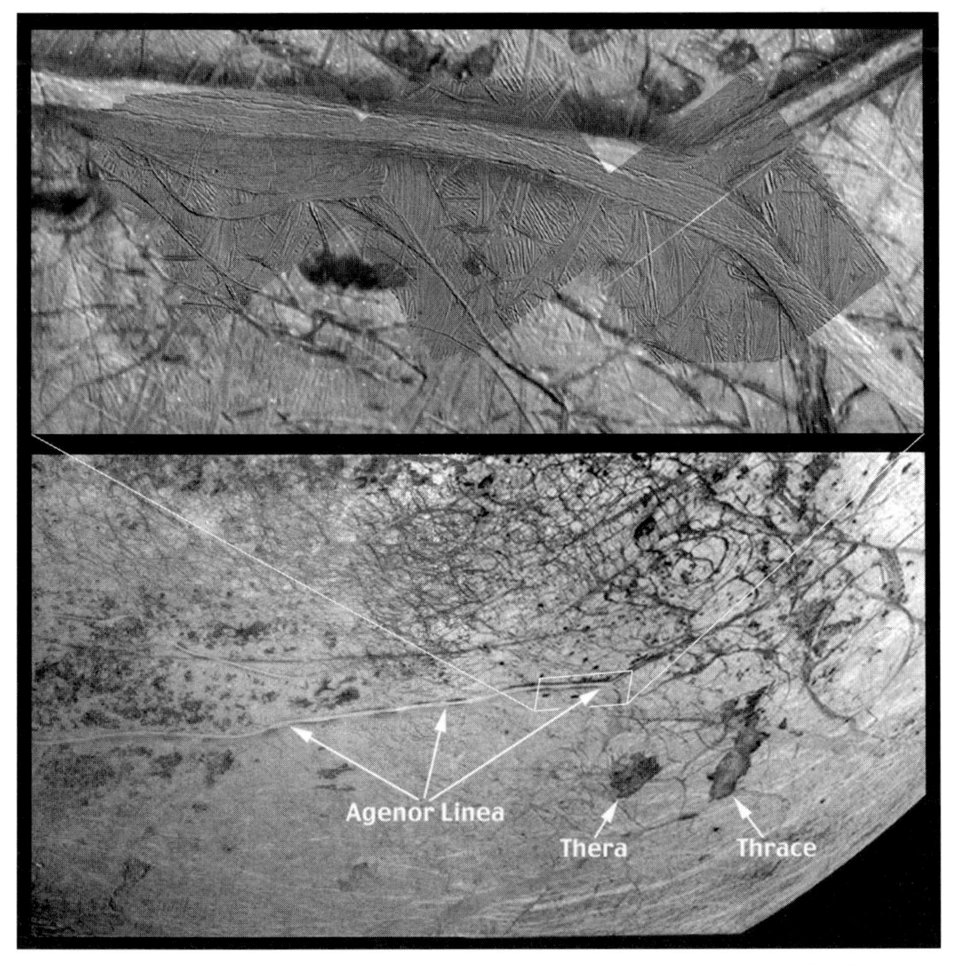

The oddly bright Agenor Linea was mapped by a southern hemisphere mosaic taken on Galileo's sixth orbit (bottom). The box indicates the eastern section (where it forks) which was charted in detail (top) at 50 metres per pixel on the seventeenth orbit. This showed that it is not a ridge, it is fairly flat and comprises several long bands, just one of which is very bright. The fine striations along each band suggest that this feature is the result of crustal spreading. It is flanked by brown material, possibly of endogenic origin.

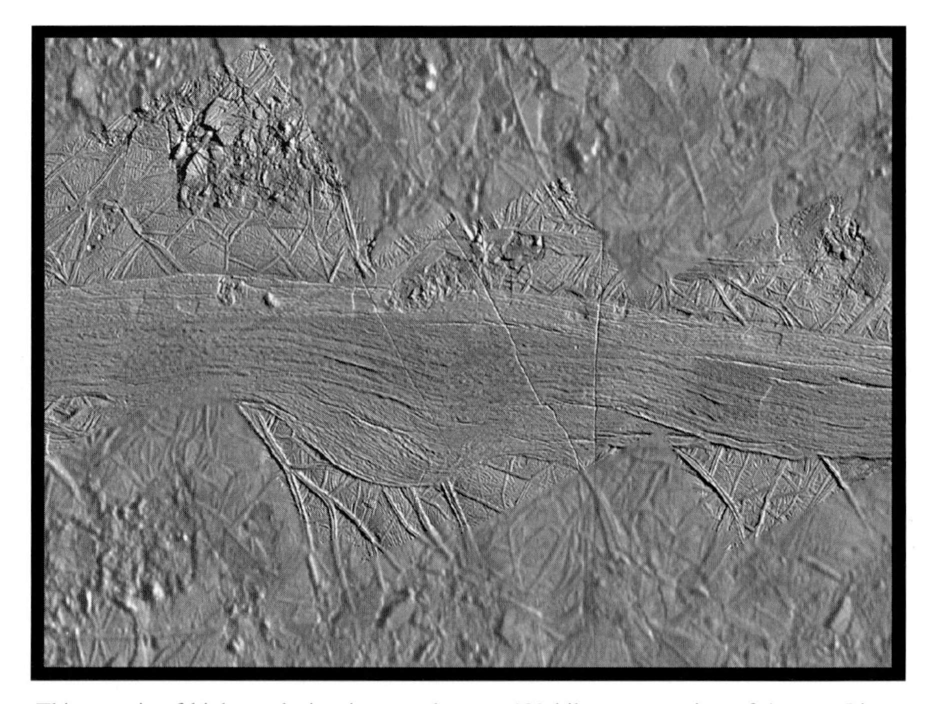

This mosaic of high resolution images shows a 130-kilometre section of Agenor Linea somewhat to the west of the section in the previous figure, at a resolution of 50 metres per pixel. It is cut by narrow fractures, and by some small subcircular 'lenticulae'. Chaos zones seem to be eroding its periphery. Agenor's brightness had prompted speculation that it was very young, but its evident relationships with these other features indicate that it cannot be.

relatively flat and consists of several long bands, only one of which is very bright. Extremely fine striations run lengthwise along each of the subsidiary bands. On the basis of its brightness – interpreted as fresh ice – Agenor Linea used to be believed to be one of the youngest features, but Galileo's close-up inspection prompted a rethink. Rough chaotic terrain immediately to the north and to the south seems to have eroded the edges of the feature, and it is not only cut by a number of narrow lineations but is also marred by small 'lenticulae' and even a few very small craters – therefore it is not recent. Intriguingly, in places Agenor is flanked by a dark reddish material. A comprehensive analysis of this complex structure concluded that it is a transform fault which formed in three stages by a combination of fracture, separation and shearing.[25]

SPURIOUS SIGNALS

The eighteenth encounter was set for a few hours after perijove, but both of Galileo's C&D Subsystems suffered spurious reset signals while in the inner magnetosphere, so the spacecraft was inert when it flew by Europa.

The ultraviolet spectrometer had already made an observation in search of outgassing, the infrared spectrometer had mapped the composition of the surface from afar, and the integrated photopolarimeter and radiometer had made a polarimetry scan to determine the texture of the icy surface, but the rest of the planned observations were lost.

The fly-by would have provided illumination similar to the sixteenth orbit (which had also been lost) but the point of closest approach was north of the equator at 220 degrees west. The imaging objectives had included Rhadamanthys Linea, the transition from a bright plain into a dark plain, the bright north polar plain, some mottling which – rather uncharacteristically – appeared to *predate* the ubiquitous ridges and bands, and the domed crater Tegid.

In effect, the last few encounters had been set up so that the closest approach of those after perijove would be west of the anti-Jovian meridian (in the longitude range 220 to 230 degrees) in some cases passing north and sometimes south of the equator. The fly-bys prior to perijove had had their lowest point near 135 degrees with one pass north and one south of the equator, but the observations on the post-perijove encounters had been frustrated by spurious reset signals which had induced the spacecraft to 'safe' itself.

All of these passes had been in daylight, and as none had offered a decent opportunity to make a thermal scan of the sub-Jovian region, on which the tidal stress was most intense, the final fly-by of the Europan campaign had its point of closest approach at 31 degrees north and 330 degrees west, near the meridian on the darkened hemisphere.

The imaging targets included the north polar plains, Rhadamanthys Linea, observations to determine the extent to which Europa's synchronous rotation suffered libration effects which would induce tidal stresses and so sustain tectonism, and a limb observation to try to look for outgassing. Furthermore, the infrared spectrometer was to make a novel observation to try to prove that cryovolcanic activity was ongoing.

FORMS OF ICE

Just as Eskimos have dozens of names for different types of snow, ice can also exist in various forms. The shape of an ice crystal is determined by the temperature at which it forms. If it had frozen long ago at the extremely low temperatures at Jupiter's distance from the Sun, the ice on Europa's surface should either be a non-crystalline form (an amorphous glass) or have cube-shaped crystals. Ice formed from water vapour at 100 K is glassy. Ice takes a cubic crystalline form if it freezes at 140 K, and adopts a hexagonal crystal if it freezes above 170 K.

"If we can find evidence for hexagonal ice on Europa, it would mean that the ice formed fairly recently from liquid water, or warm vapour," pointed out Robert Carlson, the leader of the infrared spectrometer team. Hexagonal ice would be convincing evidence of recent geyser activity.

But how was the crystalline form of the ice on Europa's surface to be inferred by remote sensing?

On Earth, when the Sun – or indeed, the Moon – is viewed through humid high cirrus cloud, the hexagonal ice crystals produce a 22 degree refraction halo. A similar effect can be seen when sunlight glints off a freshly formed snowy surface. If Europa's ice is hexagonal, sunlight striking it at a near-grazing incidence should be refracted and create the tell-tale ring-like effect. Even if hexagonal crystals make up only 10 percent of the ice, the resulting glint ought to be detectable. The effect would be most evident in the infrared, hence the use of the infrared spectrometer rather than the solid-state imager.

"NIMS is a spectrometer," Carlson pointed out, "but we're not using its spectral capability for this work." The observation involved searching for light intensity variations as a function of solar phase angle. "The geometry was crucial for the experiment. There was only one orbit when we could do it with the Sun only 22 degrees above the grazing incidence reflection point. The spot we observed doesn't have a name – it was a small patch of ice about 100 kilometres across, surrounded by darker material at about 48 degrees longitude." Two observations were made as Galileo withdrew. "If we don't see the halo, this doesn't necessarily mean anything," Carlson warned. "That's because crystalline ice exposed to particle radiation over a period of time becomes amorphous."

The trailing hemisphere is bombarded by the charged particles that circulate in the Jovian magnetosphere. Fortunately, the encounter geometry permitted the site for this test to be on the 'shaded' leading hemisphere.

"No one knows the time scale for this transformation," Carlson explained. "There's been some lab work at temperatures less than 80 K, but Europa's surface is warmer than that, and the time scales are sure to be longer at higher temperatures."

In fact, if hexagonal ice is discovered, and the laboratory work can be extended to measure the transformation rate for the Europan environment, the transformation could provide a useful dating technique for cryovolcanic activity. If no hexagonal ice is detected, but it is concluded that a specific feature was formed by cryovolcanism, the transformation rate would indicate a minimum age for it.[26]

ACID

For some time, the infrared spectrometer – operating in spectrometer mode – had noted a chemical signature on the surface whose identity had been a mystery, but extensive laboratory testing eventually established[27] that the Europan surface is laced with sulphuric acid, which is a highly corrosive chemical.

"This finding solves a puzzle that has been nagging me for a long time," Carlson admitted. "I kept wondering, 'What the heck is this stuff?' Sulphuric acid occurs in nature, but it isn't plentiful. You're not likely to find sulphuric acid on Earth's beaches, but on Europa it covers large portions of the surface." It is concentrated on the trailing side of Europa, the face of the moon which is struck by sulphur ions from Io carried by the rapidly rotating magnetosphere. Europa has evidently been

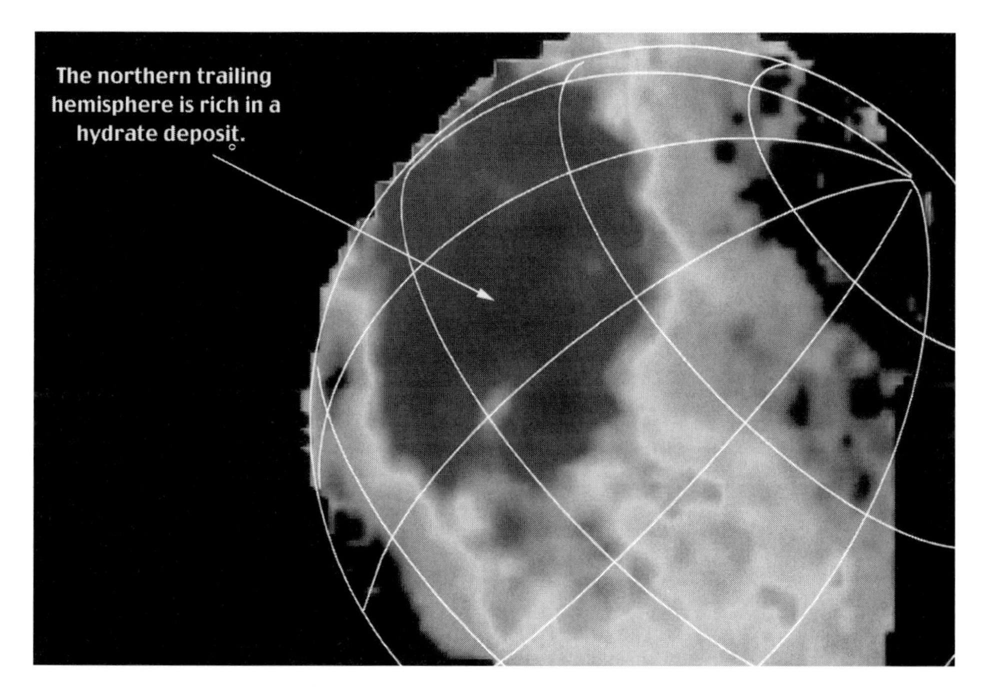

On Galileo's first orbit, its infrared spectrometer scanned the trailing hemisphere and found a concentration of non-ice on the surface in an area centred north of the equator between 240 and 270 degrees. This was initially suspected of being a hydrated salt of endogenic origin, but later analysis identified it as sulphuric acid, which may be either of endogenic or exogenic origin.

'painted' by sulphur, whereupon it forms sulphuric acid on the ice. However, long narrow features which criss-cross the surface also display sulphuric acid and this may be derived from sulphurous material either vented from geysers or oozed from very deep fractures. "This demonstrates once again that Europa is a really bizarre place," Carlson reflected.

Robert Johnson, a member of the infrared spectrometer team at the University of Virginia in Charlottesville, built upon the earlier indication of sulphate salts and suggested that sodium and magnesium sulphates might have reached the surface and evolved into sulphuric acid and other sulphur compounds upon being exposed to radiation. The fact that the sulphuric acid is concentrated on the trailing hemisphere is consistent with both the delivery of sulphur by the magnetosphere, and with irradiation-induced chemical evolution of endogenic materials. The Io-origin theory might be able to be tested by searching for sulphuric acid on the next moon out, Ganymede, to determine whether it too has been contaminated.

Furthermore, as Margaret Kivelson, the magnetometer team leader pointed out, "sulphuric acid would serve as a fine electrolyte in fluid form, or it could be dissolved in water." Thus, if this surface coating is of internal origin, an ocean of dilute sulphuric acid could well carry the electrical currents which are induced by the

Jovian magnetosphere. The various strands of the Europan mystery were clearly starting to pull together.

PROBLEMS, PROBLEMS

Luckily, this final encounter of the Europan campaign occurred prior to perijove, because the spacecraft 'safed' itself about an hour after perijove. This happened while Galileo was deep within the inner magnetosphere, and in this case the 'significant event' that had prompted the protection software to intervene was not a spurious reset signal but a spacecraft turn that took longer than expected. The spacecraft had to change its orientation frequently. Sometimes these were small adjustments to ensure that the low-gain antenna aimed at the Earth. Larger turns – known as 'science turns' – were programmed during an encounter sequence to present the various instruments with an optimal view of their targets in sequence. Galileo had been in the process of turning to reacquire its lock on the Sun after making a Ganymede observation but – as later analysis established – two of its sensors had degraded by prolonged exposure to radiation in the Jovian environment, and it was unable to complete the manoeuvre before the monitoring software decided that something was wrong. The post-perijove part of the encounter sequence was lost. The recovery process – over the next several days – was complicated by the fact that it was first necessary to determine the spacecraft's orientation, and then order a correction to aim the low-gain antenna at the Earth and restore full communications. For a fortnight in late March and early April 1999, Galileo's playback was paused as Jupiter passed behind the Sun.

On 20 April, the Deep Space Network uploaded revised software to enable Galileo to deal with the spurious reset signals that had recently knocked out its C&D subsystems and pre-empted science operations. The engineers had reasoned that the cause was progressive accumulation of fine particulate debris in the electrical connectors in the spin-bearing assembly. Once sufficient debris had built up to generate an electrical short, this prompted the spurious signal. They suspected that the debris was due to wear-and-tear on the 'slip rings' that carried signals between the parts of the spacecraft. After all, Galileo had been in space for considerably longer than its designers had anticipated.

On 3 May, as Galileo approached perijove on its twentieth orbit, it was struck by another reset, but the revised software recognised it for what it was and ignored it, so the infrared and ultraviolet spectrometers were able to make remote scans of Europa while it was eclipsed, to search for gases from ongoing geological activity.

"We're so thrilled that our efforts paid off," pointed out Nagin Cox, the systems engineer who led the team which provided the 'bus reset patch'. Galileo was proving to be remarkably adaptable to its operating environment, and was in good condition to start its perijove reduction campaign.

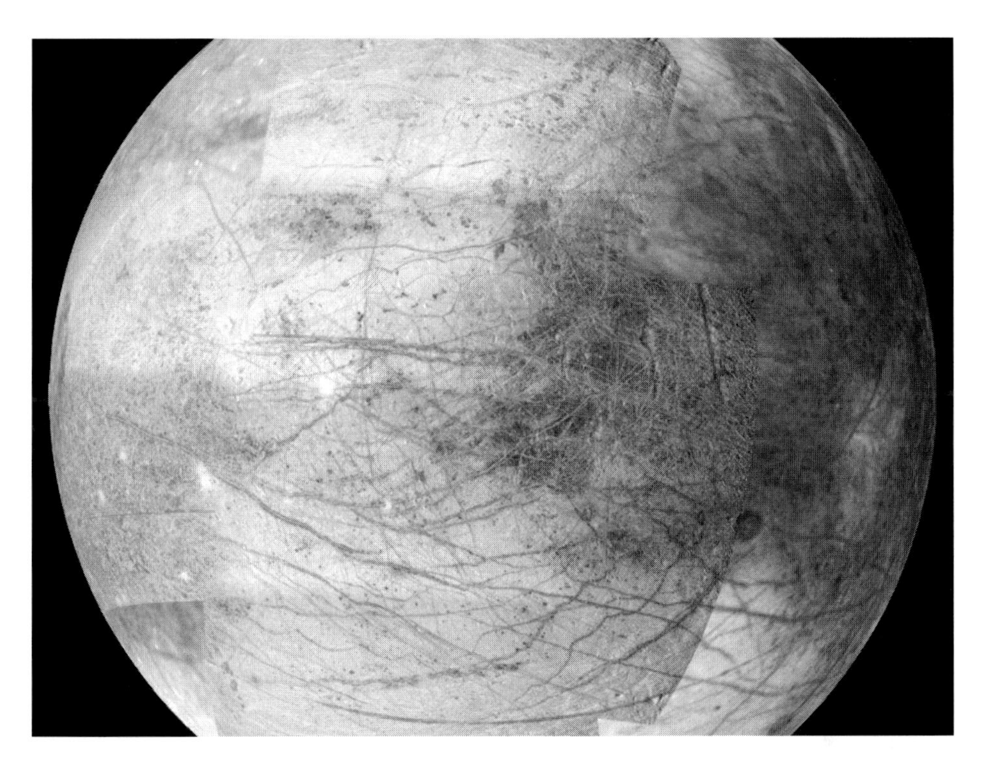

Galileo's twenty-fifth orbit provided the first opportunity to document the sub-Jovian hemisphere. This 1-kilometre per pixel mosaic was assembled from a range of almost 100,000 kilometres, as the spacecraft headed in towards Io. The mottled terrain which dominates the trailing hemisphere is evident on the right. The prominent dark circular feature is the Callanish multiple-ringed structure. Tidal stresses may have produced the profusion of fractures near the sub-Jovian point.

SUB-JOVIAN PERSPECTIVE

After successfully reducing its perijove and making the long-awaited Io fly-by, Galileo's twenty-fifth orbit provided an 8,642-kilometre Europa encounter with the first opportunity to document tectonic features on the poorly observed sub-Jovian hemisphere, and so virtually complete mapping at 1 kilometre per pixel.

AN OCEAN – OFFICIAL

Galileo began the new millennium with a 343-kilometre fly-by of Europa on 3 January. It suffered a spate of spurious resets but ignored them and completed the planned observation programme smoothly.

Although this was not the closest fly-by,[28] it offered the magnetometer team an opportunity to test their hypothesis that a salty ocean generated the magnetic field; a radio occultation investigated the moon's tenuous gaseous envelope; and the Deep Space Network's doppler data refined the model for the moon's internal structure.

The axis of Jupiter's magnetic field is inclined with respect to the planet's spin axis, so the ambient field in Europa's location periodically reverses its polarity. If Europa's weak magnetic field was produced by electrical currents induced by this changing magnetic field, the moon's magnetic poles ought to be near the equator, and they ought to migrate. This fly-by was set up specifically to determine whether the alignment of the magnetic pole varied.

Galileo found that the north pole was changing alignment. "In fact, it is actually reversing direction entirely, every five and a half hours," noted Margaret Kivelson. Other than seeing a cryovolcanic flow underway, this was about as definitive as the circumstantial evidence was likely to get. "I'm cautious by nature, but this new evidence certainly makes the argument for the presence of an ocean far more persuasive." There is no other likely candidate. "Currents could flow in partially-melted ice beneath Europa's surface, but that makes little sense," said Kivelson dismissively. "Europa is hotter toward its interior, so it is more likely the ice would melt completely. In addition, as you get deeper toward the interior, the strength of the current-generated magnetic field at the surface would decrease." The fluid had to be near the surface.

So that was it: the magnetometer provided a real-time probe into the state of the interior of the moon, and the data was unambiguous. There is indeed an ocean beneath Europa's frozen surface – an ocean 100 kilometres deep.[29] What of conditions on the seabed which, being the silicate lithosphere, is this ice-enshrouded moon's *real* surface?

HYDROTHERMAL VENTS

When Voyager showed Europa to be 'billiards ball' smooth, planetologists speculated that there might be an ocean beneath the shell of ice. The interest focused on geological processes, however; the prospect of life having developed on this frozen moon seemed to be a topic for science fiction[30] rather than serious consideration by scientists.

The Voyager fly-bys coincided with the discovery of life-forms near hydrothermal vents on the Earth's ocean floor, far below the depth to which sunlight could penetrate. The existence of organisms relying upon geochemical rather than photochemical processes for energy came as a surprise.

The species colonising the hydrothermal vents are giant forms of clams and tube worms. These sophisticated life-forms eat the bacteria which live off the heat and nutrients (that is, the chemical energy) in the superheated water emerging from the vents.

Hardy bacteria have been found living in a wide variety of conditions previously believed to be lethal. These 'extremophiles' have been found in 10-million-year-old ocean sediment, in algae mats in perennially ice-covered Antarctic lakes, in basalt lava flows kilometres inside the Columbia River Plateau in Washington State, in the hot geysers of the Yellowstone caldera in Wyoming, in hot water enriched with alkaline salts at Mono Lake in California, in deep caves devoid of light, in 400,000-year-old ice, in 5-million-year-old permafrost in Siberia, and even on the surface of nuclear fuel rods. A new species of worm was discovered in frozen natural gas (which is mostly methane) half a kilometre under the Gulf of Mexico in 1997, living in a symbiotic relationship with methane-consuming bacteria – the worms digested a by-product of the bacteria.

Recent research has indicated that these extremophiles link with all other terrestrial life by way of a thermophilic sulphur-consuming bacteria referred to as 'archaea' (meaning literally 'old one'). Archaea is similar to a prokaryote (common bacteria) in that it has its genetic material floating freely inside a single-cell structure, but its DNA more closely resembles the eukaryotes (which include all plant and animal life on Earth) which have their genetic material confined to a nucleus within the cell. It therefore represents an early branch in the 'tree of life' on Earth.

All terrestrial life is based on so-called right-handed amino acids, and right-handed amino acids have been identified in a carbonaceous chondrite meteorite. The Solar System may have been 'seeded' by organics delivered by comets. If life has developed independently elsewhere, it may more closely resemble archaea than eukaryote. Finding extraterrestrial life – even if it is now extinct – would be a discovery of the highest order because, as Philip Morrison of the Massachusetts Institute of Technology once observed, it would "transform life from the status of a miracle to that of a statistic".

As we know it, life uses carbon as a 'building block' to manufacture everything from cells to DNA. Many organisms obtain their energy from carbon-based molecules such as sugar, and an energy source is required to free the carbon atoms from their chemical bonds. Algae and flora use photosynthesis to make organic molecules from carbon dioxide, which they take from the atmosphere or the ocean. Although liquid water, oxygen and sunlight provide the basis for the current biosphere, life formed without oxygen and photosynthesis. The primordial Earth's atmosphere and oceans were anoxic. In fact, an oxygen-rich atmosphere is an *indication* of life, not a condition for its development. Life developed when the only source of energy and nutrients available was the superheated and chemically enriched water from surface geology. Intriguingly, such an environment might currently prevail on the floor of Europa's ocean. If this ocean is salty (as indicated by the magnetometer data) it may support a halophilic (literally, salt-loving) bacterial life-form.

A conference co-sponsored by NASA's Solar System Exploration Division's Exobiology Programme and the National Science Foundation's Ocean Sciences Programme took place in November 1996 in San Juan Capistrano in California. The interdisciplinary conference drew together planetary scientists, volcanologists and oceanographers for the first time, to discuss the significance of a Europan ocean, possible hydrothermal activity on its floor, the prospects for biological activity, and ongoing and future robotic exploration of Europa.

MISSIONS TO EUROPA

Several establishments are proposing spacecraft to explore Europa. Two candidates are an orbiter and a lander.

The recently formed Centre for Space Science and Exploration at the Los Alamos National Laboratory, New Mexico, is currently developing an Ice Penetrating Radar that could be used by the Europan orbiter. As ice is transparent to radar, a triple-frequency radar would be able to compare reflections from the surface and from the ice-water interface to infer the thickness of the ice along its groundtrack in order to conduct a global survey over a period of about 30 days. Of course, interpreting the data will be complicated by the need to isolate radar reflections off fractures in the ice crust, salty and non-salty ice, and the seabed. A laser altimeter would chart the surface topography, so that the radar data could be processed to derive a three-dimensional model of the icy shell. "Unless it is very, very deep, or there is just a little of it, we will be able to find water," assured Robert Staehle, NASA's manager for planning advanced missions.

As it mapped the topography, the laser altimeter would also accurately measure the size of the tidal bulge and then characterise its variation over a period of several of the moon's orbits around Jupiter. If water is lying under a thin shell of ice, the tidal distortion will be more significant than for a thick shell. "If the tidal bulge is small, then we'll know that there is little or no water beneath the ice," explained Brad Edwards of the Los Alamos Space and Remote Sensing Sciences Group. In fact, if the tidal bulge is less than about 30 metres, it would mean that the ice extends all the way down to the silicate lithosphere.

Galileo's observations suggest that in some places the ice may be as thin as a few hundred metres. Once the orbiter had identified the thinnest ice, the lander would set down and melt its way through in order to release a 'hydrobot' to explore the ocean below.

Of course, an orbiter is likely to carry a suite of instruments. An imager to map the surface at a resolution of 10 metres would almost certainly be included, and it would employ a variety of filters to determine composition of non-ice components. The orbiter's solid-state electronics would have to be especially thoroughly 'hardened' against the radiation, because in a month of mapping it would accumulate 25 megarads!

A TERRESTRIAL ANALOGUE

In 1974, when scientists were performing an airborne radio-sounding and altimetry survey of the Soviet Union's Vostok Station, located 1,000 kilometres from the South Pole in Eastern Antarctica, the radio-sounding instruments detected an expanse of water beneath the ice. Later studies found that this covered an area of 10,000 square kilometres. Since this was comparable to Lake Ontario in North America, the water within the ice was promptly dubbed 'Lake Vostok'. Although it is in one of the coldest locations on Earth, it is believed that geothermal activity on the 'bed' of this lake keeps it liquid. Unless it is laced with salts from hydrothermal activity, it should be one of the largest bodies of fresh water on Earth.

A workshop held in April 1997 discussed 'field testing' the various ice-penetrating robots proposed for Europa by exploring Lake Vostok. As the ice capping the terrestrial analogue is 3,700 metres thick, the exercise would be realistic. "We've had a mixture of scientific disciplines at this workshop," said Frank Carsey, an oceanographer at JPL. "Oceanographers, biologists, glaciologists, planetologists, and experts in building the instrumentation." An interdisciplinary approach was important. "If we're going to explore Europa, there will have to be development and testing on Earth. And that programme itself will be a great opportunity for Earth science."

"We perceive this as a really integrated programme of science exploration and technology development," explained the late Jurgen Rahe, the former Director of Solar System Exploration in the Office of Space Science at NASA Headquarters. "This concept fits in well with NASA's programme of comparative planetology." Intriguingly, one of the issues discussed at the workshop was how to recognise life!

"There's a natural convergence between technology and science," said Nancy Maynard of NASA's Earth Observation System (initially called the Mission to Planet Earth) "and between space science and Earth science. We're looking at many of the same intellectual questions – just on different planets."

It would have to be a step-by-step effort, however. "Getting a probe into Lake Vostok will probably take about three or four years," explained Carsey.

Having established a Europa/Lake Vostok Initiative, JPL teamed up with the Woods Hole Oceanographic Institute in Massachusetts and the University of Nebraska to investigate technologies that would be required to explore Lake Vostok. "We're at the stage of developing our science requirements," noted Joan Horvath, formerly of JPL's advanced concepts office, who was appointed to lead the new initiative.

On 12 May 1998, JPL's Outreach Programme hosted 'A Day On Europa', in the form of a series of events to explain to the public the significance of an ocean beneath Europa's icy crust. Arthur C. Clarke, who has referred to Europa as "the human race's lifeboat", implying that we could relocate there in the event that the Earth becomes inhospitable, participated in a panel by 'webcam' from his home in Colombo, Sri Lanka.

CALLISTO TOO?

On Galileo's early Callisto fly-bys, the magnetometer suggested that this moon had a salty fluid below the icy surface, but in contrast to the Europan ocean, Callisto's fluid would appear to be a few hundred kilometres down, and only 10 kilometres thick. While Europa is tidally heated by virtue of being closer to Jupiter and being caught in orbital resonances with both Io and Ganymede, the undifferentiated state of Callisto's interior indicates that it has been heated only radiogenically, so the prospects for life are considerably diminished. Nevertheless, there are salty brines on

Earth that teem with life. The brine pools of Owens Lake in California, for example, bloom with halophilic bacteria.

EBB AND FLOW

As Galileo found ever more convincing evidence for an ocean on Europa, and institutional inertia developed to devise a mission to explore it, a number of interdisciplinary studies were undertaken to assess the prospects for life.

"Can an ocean persist for 4.5 billion years, and not have life in it?" wondered Christopher Chyba, a planetary scientist at the Search for Extraterrestrial Intelligence Institute, in Mountain View, California. It would have been enriched with pre-biotic organics by impacting comets.

The Europan ice would seem to be of the order of 10 kilometres thick, so very little of the Sun's life-sustaining energy would penetrate it. An analysis[31] by geobiologists at the California Institute of Technology concluded that life is unlikely because almost all forms of energy used by life on Earth are unavailable. "One must be careful when doing comparative planetology. It is not a safe assumption to use Earth as an analogy," observed Eric Gaidos. "A liquid-water ocean on Europa does not necessarily mean that there is life there."

Advanced terrestrial life derives chemical energy either directly from sunlight by photosynthesis, or from the oxygen that is a by-product. The system is not closed, even for the organisms living under the ice sheets, because energy from above is indirectly available to organisms under the ice. Oxygen reaches even the super-hot hydrothermal vents on the ocean floor. But a Europan ocean would have to be a closed system, dependent on internal energy sources. "Even in pure ice," Chyba pointed out, "the wavelengths you need to drive photosynthesis will not penetrate 15 metres."

However, as Gaidos acknowledged, their study assumed that Europan life was "based on the same energy sources used by life on Earth". Simple organisms might derive biochemical energy from oxidised iron (rust) that may exist beneath the ice. "But we're talking about very simple organisms that can live on these energy sources; these are not multi-cellular creatures." As Joseph Kirschvink, another member of the team, observed, "You're not going to find fish swimming in the Europan ocean."

Bruce Jakosky, a planetary scientist at the Laboratory for Atmospheric and Space Physics, at the University of Colorado at Boulder, and Everett Shock, a biologist researching chemical energy processes at the Department of Earth and Planetary Sciences at Washington University in St. Louis, analysed geochemical reactions from rock weathering and concluded that life on Europa is unlikely.[32] They compared the early Earth, Mars and Europa in terms of their hydrothermal processes, and estimated the amount of chemical energy that would have been available. On Mars, they concluded that the amount of life that could have been supported was minuscule in comparison to the biomass of the early Earth, primarily because terrestrial life soon learned how to utilise far more efficient photochemical processes. They estimated that the energy available on Europa would be much less than on

Mars. In fact, they opined that the most likely site for life on Europa would not actually be in the ocean, but within the rocks of the lithosphere – where the heat sources are.

In September 1998, following closely upon the creation of NASA's Astrobiology Institute, the National Science Foundation, in collaboration with the University of Washington in Seattle, started a new programme to train graduate astrobiologists. The contributing disciplines ranged from astronomy and atmospheric sciences to oceanography and microbiology. "We want to study life in extreme environments on Earth," said James Staley, the Institute's microbiologist director, "such as in high-temperature hydrothermal vents, freezing sea ice, and subterranean basalt formations. These systems serve as excellent models to study microbial life, and may be similar to environments on other planetary bodies."

The American Geophysical Union's San Francisco meeting in December 1998 included a report on the joint Russian–American–French project to drill down to sample Lake Vostok. By this point, a 3,623-metre long 'ice core' had been extracted from above the lake. Although this was virtually all the way through (there was less than 100 metres of ice remaining) the drilling had been halted, pending approval of the procedures to be employed to extract a sample of the water without risking compromising the lake's pristine state. The ice core was itself a valuable environmental tracer that could be correlated with the trends in temperature, greenhouse gases, and climate change gleaned from other sources. The lower part of of the ice core might well be half a million years old.

What everyone really wanted to know, however, was whether there was life in the lake. In early 1998, Richard Hoover of NASA's Marshall Space Sciences Laboratory and S.S. Abyzov of the Russian Academy of Sciences used an electron microscope to inspect the lowest section of the ice core, and received a pleasant surprise "We have found some really *bizarre* things," Hoover said. "There are all sorts of microorganisms in the ice. Some are readily recognisable as cyano-bacteria, bacteria, fungi, spores, pollen grains, and diatoms, but some are things we have never seen before."

In November 1999, the Committee on Planetary and Lunar Exploration of the Space Studies Board hosted by the National Academy of Sciences proposed that Europa be assigned a priority for exploration matching that of Mars. This high-level recommendation surprised some. "Mars is a good bet for the possibility of past life, for many reasons," admitted Ed Weiler, NASA's Associate Administrator for Space Sciences. "Europa's a long shot, and much, much more difficult to get to, to survive at, and to study."

"It's not as bad as I first thought," reflected Robert Carlson, the head of Galileo's infrared spectrometer team, upon finding that areas of Europa's surface are coated with sulphuric acid. His initial reaction was that this corrosive chemical would diminish the prospects for life. "In fact, it might be good!"

Terrestrial bacteria thrive in an acidic environment. "To make energy, you need fuel and something in which to burn it," explained Kenneth Nealson, the chief of JPL's astrobiology unit. "Sulphur and sulphuric acid are oxidants, or energy sources, for living things on Earth."

What about the fuels that would be burned? It was possible that they would be derived from hydrothermal vents on the sea floor, as an endogenic source, but there was also an exogenic source. A study[33] by Christopher Chyba proposed that the charged particles circulating in the Jovian magnetosphere are likely to make both fuels and oxidants on the surface.

When the fast-moving charged particles strike the icy surface, chemical reactions are likely to occur, transforming frozen molecules of water and carbon dioxide into organic compounds. Hyphomicrobium (one of the most common terrestrial bacteria) uses formaldehyde as its sole source of carbon, and Chyba proposed that Europa's ocean may well be teeming with similar formaldehyde-feeding microbes. The rain of charged particles would readily produce oxygen and hydrogen peroxide, which are oxidants in which carbon-based fuels could be burned. So, by an ironic twist, the intense radiation, in itself lethal, might indirectly support "a substantial Europan biosphere". However, as Chyba pointed out, the molecules produced on the surface would be "biologically relevant only if they reach the ocean".

If the Europan bacteria are to feast on formaldehyde, there has to be a way of transporting that compound through the dense layer of ice and into the liquid beneath. Some features of the surface appear to have undergone rapid melting. Such melting would have infused the ocean with organics formed on the surface and thus promoted a microbial 'bloom'. The population would fluctuate wildly, but episodic melting may be able to sustain a limited biosphere over geological time.

However, as Chyba admitted, no matter how promising the prospects, we won't know for sure until we send a probe to find out.

12

Jupiter from orbit

REMOTE SENSING

Jupiter's electromagnetic spectrum is characterised by solar reflection (which peaks in the visible wavelength range), infrared from re-radiation by gases of the energy absorbed from sunlight, and heat leaking from the interior.

The depth to which a remote-sensing instrument can view depends upon the opacity of the gas, which is dependent upon the chemical composition. Different gases condense at different temperatures and pressures, both of which increase with depth. Water will condense as it rises within Jupiter's atmosphere and passes through the 4-bar pressure zone. Ammonia condenses at much colder temperatures, higher in the atmosphere, at about 0.7 bar. As Jupiter's atmosphere is not cold enough to condense methane, it remains in the gaseous phase. Because the methane is well mixed, and because it has convenient spectral absorption features, it can readily be used as a 'tracer' for determining the vertical distribution of clouds of other species.

At visible and near-infrared wavelengths shorter than about 4 microns, Jupiter's spectrum is basically solar reflection, so we view the top of the ammonia cloud layer. There is a strong absorption feature in the 3- to 4-micron range due to the methane above the level of the ammonia clouds, which gives the disk a dark appearance. Thermal emission from the interior predominates longward of 4 microns.

The 7-micron view corresponds to the 0.01- to 1.0-bar range, which spans the stratosphere (in which radiative transfer is the dominant process), through the tropopause at about 0.1 bar (where convection becomes significant), into the upper part of the troposphere (where convection is the primary means of energy transport). At 5 microns, the outer atmosphere is effectively transparent, making this wavelength a 'window' into the deep troposphere.

The pressure levels of clouds can be inferred from images taken in the near-infrared. An image taken at a wavelength which is strongly absorbed by methane can penetrate no deeper than 1 bar; a wavelength which is less strongly absorbed can penetrate a little deeper; and a wavelength which methane does not absorb will penetrate down to about 8 bars. Given assumptions regarding opacity, such imagery can be processed to infer the pressure, and hence the depth at which a prominent feature resides. By 'stacking' the three images it is possible to infer the vertical extent of specific clouds.

In Galileo's case, such observations utilised filters at 0.889 micron (at which methane is strongly absorbed), 0.727 micron (at which its absorption is weaker), and 0.757 micron (at which it is negligible). The infrared spectrometer's range included these wavelengths, but the solid-state imaging system's greater spatial resolution meant that individual clouds could be studied in detail.

FIRST LOOK AT THE GREAT RED SPOT

When seventeenth-century astronomers began to observe Jupiter on a regular basis, they noted a mysterious reddish spot.[1]* Once it was realised that this was not a transient feature, it was named the 'Great Red Spot'. Spanning fully 20 degrees of longitude, it is an oval 13,000 kilometres on the north–south axis and 26,000 kilometres in the east–west axis, so it is much larger than the Earth! Although its hue has varied over the years, and at times has dramatically faded, it has been prominent throughout the 'Space Age'. Telescopic and Voyager observations of the anticlockwise air flow within it established that the 'spot' is really an anticyclonic vortex with a central upwelling column which yields to subsidence around its periphery. The central part of the feature rotates in about seven days.

The Atmosphere Working Group made the Great Red Spot the primary objective for the first encounter sequence of Galileo's orbital tour.

A typical 'observation' would involve imaging a target near the western limb, soon after it had crossed the dawn terminator; then again in the centre of the disk, at local noon; and finally as it approached the evening terminator. Jupiter rotates in 10 hours, so such a series of images would enable individual features to be tracked over a period of several hours in order to measure the wind speeds. Adding another image from either the previous or the subsequent rotation of the planet would provide a longer time-base. A pair of images taken on successive revolutions with a spatial resolution of 30 kilometres per pixel would enable winds speed to be determined to within 1 metre per second. Since this represented an accuracy of 1 percent for the typical jetstream, hopes were high for achieving major insight into the dynamics of the atmosphere.

Galileo took two six-element mosaics of the Great Red Spot on successive rotations of the planet, doing so at a wavelength at which methane absorption is negligible so as to be able to see deeply into the atmosphere. It revealed that the central part of this feature is relatively quiescent, but the anticlockwise winds reach 150 metres per second on the northern periphery where it interacts with a westward jetstream, which happens to be the strongest jetstream on the planet; the eastward flow to the south is rather milder.

* For Notes and References, see pages 413–424.

The oval of Jupiter's Great Red Spot, being some 26,000 kilometres east–west, spans almost 20 degrees of longitude. The core of the anticyclonic system rotates in 7 days. On its first orbit, Galileo took a series of images to identify thunderheads caught in the turbulent periphery, in order that they could be tracked to determine local wind speeds. Jupiter turns on its axis in just under 10 hours, so imagery taken on successive 'days' provided sufficient time-base for the resolution of 30 kilometres per pixel to determine wind speeds to within 1 metre per second. The results showed that the core of the spot is quiescent, but the northern peripheral winds peaked at 150 metres per second. Each of these images is a six-element mosaic taken at 0.757 micron, a wavelength at which methane absorption is negligible.

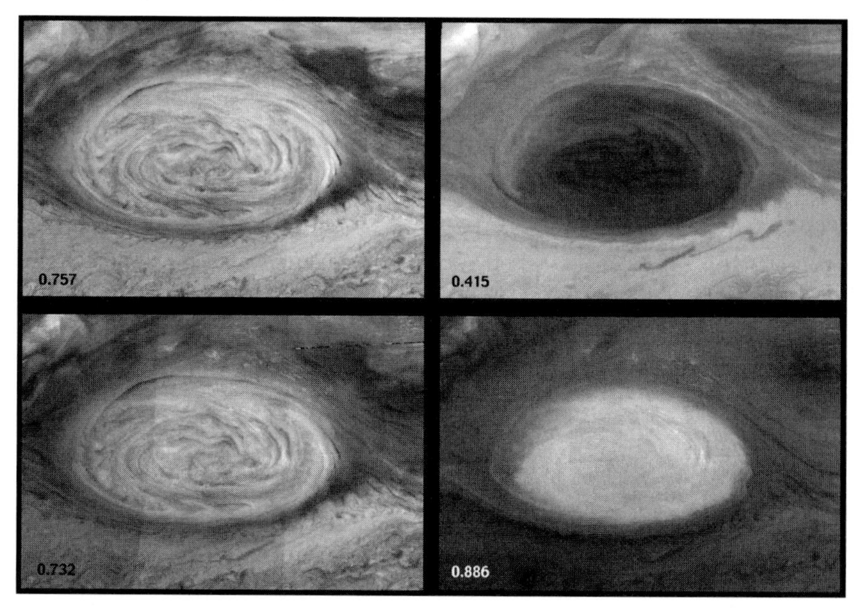

On its first orbit, Galileo imaged the Great Red Spot at several wavelengths in order to infer the altitudes of the various features: 0.757 micron (top left), 0.415 micron (top right), 0.732 micron (lower left) and 0.886 micron (lower right). The depth to which sunlight penetrates and is scattered or absorbed by different atmospheric constituents is wavelength dependent. Methane absorption is negligible at 0.757 micron, fairly weak at 0.732 micron, and strongly at 0.886 micron. The diffuse haze on top of the Great Red Spot is most apparent on the 0.886-micron view. The relative abundances of the chemicals which stain Jupiter's atmosphere are evident in the violet (0.415-micron) view. The imaging resolution is 30 kilometres per pixel.

Several thunderheads caught up in the periphery of the Great Red Spot were identified in 30 kilometres per pixel imagery taken by Galileo on its first orbit (lower right). The two enlargements (top right) were were taken 70 minutes apart so as to measure the winds. The close up (left) shows a much larger storm to the northwest of the Great Red Spot. The white cloud is the top of an 'anvil' projecting 25 kilometres above the surrounding ammonia cloud. Its base is at a pressure level at which water condenses. It is similar to a terrestrial thunderhead except for the fact that it is 75 kilometres tall and the top of the anvil is 1,000 kilometres wide – some five times the size of the largest counterpart on Earth.

In addition to documenting the Great Red Spot's rotation, the 30-kilometre resolution of the new imagery was sufficient to identify individual storms in its periphery.

The top of one enormous thunderhead just northwest of the Great Red Spot was found to rise almost 25 kilometres above most of the surrounding wispy ammonia clouds at the 0.7-bar level. The base of the storm was at least 50 kilometres deeper, at about the 4-bar level where the only available condensate is water.[2]

The energy driving Jupiter's winds had been a mystery until Galileo's atmospheric probe established that they persist to great depth, indicating that the weather system is powered by heat leaking from the planet's interior. Such massive thunderheads were clear evidence of the rising energy.

The Voyagers had observed convective clouds of this type near the Great Red Spot. They erupted once every 10 days or so and persisted for a few days. However, the spectral range of the Voyager camera system had not been capable of near-infrared imaging, and it had not then been possible to determine the vertical structure of these storms.

Such storms are analogous to terrestrial thunderstorms, with the tall bright portion being comparable to the familiar 'anvil' cloud on Earth, but the tallest anvils in terrestrial storms rise 18 kilometres and are 200 kilometres across, whereas the

Jovian storm was 75 kilometres tall and was 1,000 kilometres across its base. If this storm could be transplanted into the Earth's atmosphere, it would project out into space!

Terrestrial hurricanes develop over the tropical ocean when the water is heated by the Sun to 27 °C, prompting evaporation. The heated air rises. Once the low-pressure zone forms, the surrounding air at sea level is drawn in to fill the void. It is the Coriolis effect which makes the air flow rotate. Lighting occurs when moist air rises in a rapid updraft, and latent energy is released when this vapour condenses to rain and falls out. Although the Earth's weather system is driven by solar energy and Jupiter's is driven by heat from the interior, the active processes – convection and condensation – are the same. The longevity of the Great Red Spot derives from the fact that Jupiter's atmosphere is so deep. In the absence of a solid surface to dissipate the storm's energy (as happens when a storm makes landfall and is deprived of warm moisture off the ocean) the Spot has settled into a semi-stable state.

At perijove, the infrared spectrometer's resolution was several hundred kilometres, so this instrument could not investigate individual clouds. It was sensitive out to 5 microns, however, and was thus able to study 'infrared hot spots'.

The early observations by the infrared spectrometer were made at a range of wavelengths designed to probe the internal structure of the Great Red Spot.[3] It found that its base is at the 0.7-bar level in the ammonia layer, and that it rises 20 kilometres to the 0.24-bar level. It also discovered a ring of cloud surrounding the main column, set at a level a few kilometres below its top and slightly higher in the east than in the west.[4]

However, when Galileo penetrated the intense radiation in the inner magnetosphere on the way in for its 11 Rj perijove on 28 June 1996, a malfunction disabled its infrared spectrometer, so this instrument was not able to take data during the final phase of the sequence, and the team was particularly frustrated to have missed the

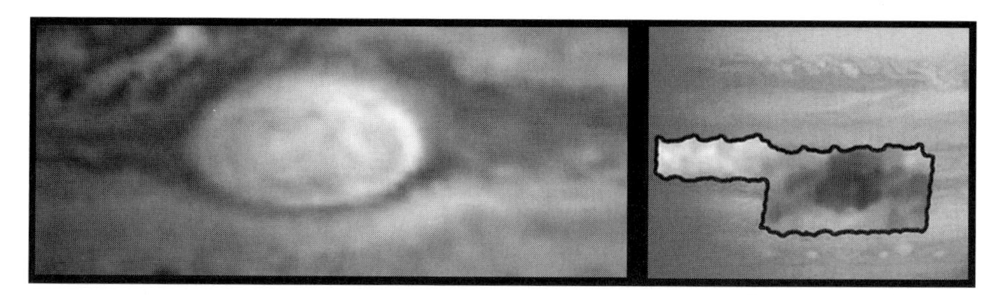

On Galileo's first orbit, the Great Red Spot was observed by the infrared spectrometer (in the near-infrared) and the integrated photopolarimeter and radiometer (in the far-infrared). The Spot's brightness in the infrared spectrometer's view (left) indicated that its core is higher than its surroundings. It also shows a spiral structure within the Spot itself. The radiometer's temperature map (right) of the atmosphere at the 0.25-bar pressure level (a level dominated by ammonia clouds) indicated that the centre of the Spot is much colder than the surrounding clouds. The turbulent zone to the northwest is much warmer.

high-resolution views of the centre of the Great Red Spot. "Our instrument software stopped!" noted an amazed Marcia Segura, the spectrometer's science coordinator. "It was not an 'Oops, we should have known better' anomaly but an 'Oh my God!!' anomaly."

Even at perijove, the best resolution of the integrated photopolarimeter and radiometer was 2,000 kilometres. However, it was sensitive out to 45 microns. It could perform a number of studies across wide areas of the atmosphere but, being a three-in-one package, it could operate in only one mode at a time. When it switched from radiometry to photopolarimetry, it had to change filters. Unfortunately, as the spacecraft approached perijove, the filter wheel jammed.

"This was fairly devastating news," reflected Glenn Orton. "We'd wanted to make lots of measurements of temperature to see what powered the winds. We also wanted to see how the cloud-tracked winds corresponded with these predictions. This would tell us what other sorts of forces were acting on the winds and cloud fields." Until a procedure could be developed to free the filter wheel, the instrument would be restricted to charting the thermal structure of the atmosphere in the 0.25- to 0.5-bar pressure range.

The radiometry established[5] that the inner part of the Great Red Spot "probably represents gas moving upward rapidly", and the central column at the 0.25-bar level is "quite cold" – in fact, it is so far above the ammonia cloud tops that it is most likely the coldest part in the entire atmosphere. It is slightly warmer to the immediate east and west, where the updraft is weaker, warm to the south, and much warmer in the turbulence to the northwest where the descending air is heated in response to the increasing pressure.

Simultaneous failures in two independent systems when the spacecraft was deep within the magnetosphere suggested that the solid-state electronics had been affected by a radiation surge. This was confirmed in the case of the infrared spectrometer. Three weeks of intensive analysis found that the solid-state memory in which the control programme resided had been corrupted by a stray particle of radiation, and so the processor had 'hung', as computers are prone to do. Once its programme had been reloaded, the spectrometer restarted and was fully functional. To make the instrument more robust, it was decided to have the spacecraft's computer reload the instrument's programme at key points during each perijove passage in order to restore it to life if it had stopped.

As half-expected, upon Galileo's return, the infrared spectrometer experienced anomalous behaviour leading up to perijove on 7 September, confirming that the previous glitch had been induced by the radiation. This time, however, as the spacecraft moved clear of the worst of the radiation the main computer reloaded the instrument's programme from its own solid-state memory, clearing 'flipped bits' in the instrument and restoring it so that it could take data on the outbound leg of the encounter sequence.

With the integrated photopolarimeter and radiometer's filter wheel jammed, the scheduled observations were cancelled and the instrument was switched off for the second orbit. "This was to ensure that the instrument's attempts to move the stuck filter wheel would not do itself further harm," explained Orton. There was therefore

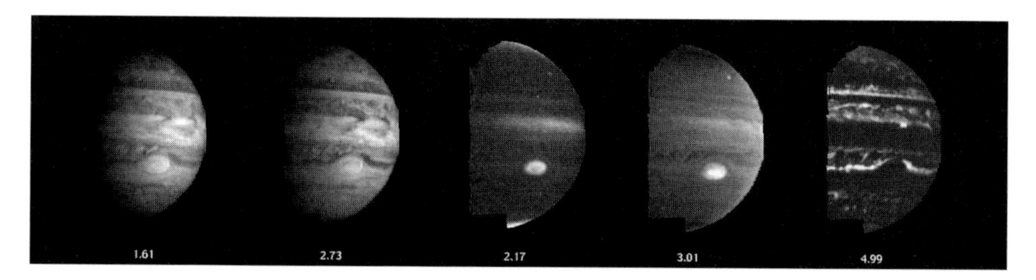

As it approached Jupiter on its second orbit, Galileo's infrared spectrometer imaged the hemisphere containing the Great Red Spot from a range of 2 million kilometres in order to investigate the vertical structure of the east–west 'zonal' weather system. The images at 1.61 microns (far left) and 2.73 microns (second left) are fairly clear views of the deep atmosphere, with clouds down to a level about 3 bar. The 2.17-micron view (middle) is severely affected by the absorption of light by hydrogen gas, the main constituent of the atmosphere, and it shows only the highest clouds, notably as the haze above the Great Red Spot. The 3.01-micron image (second right) records deeper clouds dimly against gaseous ammonia and methane absorption. At 4.99 microns (right) the bright areas indicate heat leaking from the deep atmosphere.

no data on temperature fields to complement the other instruments studying the atmosphere. Once again, the trends had to be derived from the monitoring by the Infrared Telescope Facility in Hawaii. In fact, these observations were "breath-taking", said Orton. "In some cases, they were as good as some of the 'distant' NIMS global observations of the planet!"

As the spacecraft withdrew from the planet, an attempt was made to release the stuck filter wheel by heating its mechanism, but this failed, so it was decided to activate the instrument on the next orbit and collect thermal data using the available radiometry filter.

IO'S POLLUTION

The 'plasma sheet' is a thin disk of low-energy plasma in the plane of Jupiter's magnetic equator, ranging out from Io's orbit. Material derived from Io's plumes is concentrated in the sheet. It carries very strong electrical currents, and there are dynamic interactions between the plasma and the planet's magnetic field. The disk swirls around with the inner magnetosphere, in synchrony with Jupiter's rotation. However, it is inclined about 10 degrees with respect to the plane in which most of the satellites orbit because the magnetic field axis is offset from the planet's rotational axis, so the disk washes over the satellites as it precesses. Where the solar wind slams into the magnetosphere, it compresses it. When Galileo made its original approach to Jupiter it crossed this bow shock some 140 Rj out. Although the solar wind sweeps Jupiter's outer magnetosphere 'downwind', drawing it out to form the magnetotail, the plasma sheet is confined to the symmetrical inner magnetosphere. A major event for the magnetometer team was the spacecraft's passage through the plasma sheet at a distance of 25 Rj as it drew away from Jupiter after its second encounter sequence. These observations were particularly welcome because they would shed light on how the volcanically generated plasma is transported out by the planet's magnetic field so that it can escape from the magnetosphere – as it must, for if it did not the magnetosphere would long ago have become saturated with Io's 'pollution'.

HAZES

It had been planned to focus on Jupiter's 'white ovals' during the third encounter sequence, but this was impracticable. "We could not find any features which we could predict would be stable over the eight-week period between planning for our next observations of Jupiter and actually making them," explained Glenn Orton.

It was decided instead to scan a swath ranging from 3 degrees north to 13 degrees

On its third orbit, Galileo assembled a 0.757-micron mosaic spanning from 3 degrees north to 13 degrees south to document the interface between the alternating east–west zonal jetstreams of the Equatorial Zone and the South Equatorial Belt.

south – spanning the Equatorial Zone to the northern fringe of the South Equatorial Belt – in order to make a synoptic study of zonal winds. The maximum speed of the Equatorial Zone's eastward jetstream was measured to be 128 metres per second.

Galileo imaged the limb at one wavelength strongly absorbed by methane and at one with negligible absorption and found that the stratospheric hazes at different latitudes are strikingly different. The haze is approximately 50 kilometres higher in the polar zones. The polar haze is produced by charged particles in the magnetosphere flowing down the magnetic field lines to interact with the upper atmosphere. The lower latitude haze is produced by photochemical reactions; it is a photochemical smog of hydrocarbon droplets.[6] The clouds that form the visible 'surface' resemble terrestrial cirrus, but are made of ammonia rather than water-ice, and the ammonia crystals are a hundred times smaller than water-ice.

AURORAL ACTIVITY

The existence of Jovian auroral activity was first inferred from the particles and fields data collected by the Pioneer probes. It was first directly observed by the International Ultraviolet Explorer, launched in 1978; the Voyagers had imaged it close-up; the Hubble Space Telescope had imaged it from afar in ultraviolet light; and terrestrial telescopes had seen it in the infrared.

While Galileo was behind Jupiter, the ultraviolet spectrometer searched for auroral activity on the dark side of the planet. The activity is concentrated in oval ribbons, one around each magnetic pole.

"Jupiter's auroras are a lot like the Earth's," said Scott Bolton of the plasma spectrometer team, "but we now know that the auroral arc on Jupiter is thin and patchy and we can estimate its altitude between 300 and 600 kilometres" above the 1-bar reference level. Nevertheless, the source of the plasma is different. Terrestrial aurorae occur when the solar wind penetrates the 'polar cusps' of the magnetosphere. Jupiter's aurorae are induced by Io's presence within the inner magnetosphere, and are the result of charged particles – mostly electrons – in Io's flux tubes interacting with the planet's upper atmosphere.

LIGHTNING

While in Jupiter's shadow, Galileo's sensitive solid-state imaging system sought lightning. The camera was progressively scanned across the dark hemisphere from just beyond the dusk terminator to the dawn terminator, and it recorded a chain of flashing thunderheads just south of the westward-moving jetstream at 46 degrees north. Almost all the lightning detected by the Voyagers had also been near the latitude of a westward-moving jetstream.

The camera's scanning action enabled the individual lightning flashes in the same storm to be isolated. In fact, the lightning was flashing far below the visible ammonia cloud, which was acting like a translucent screen and diffusing the light upwards.

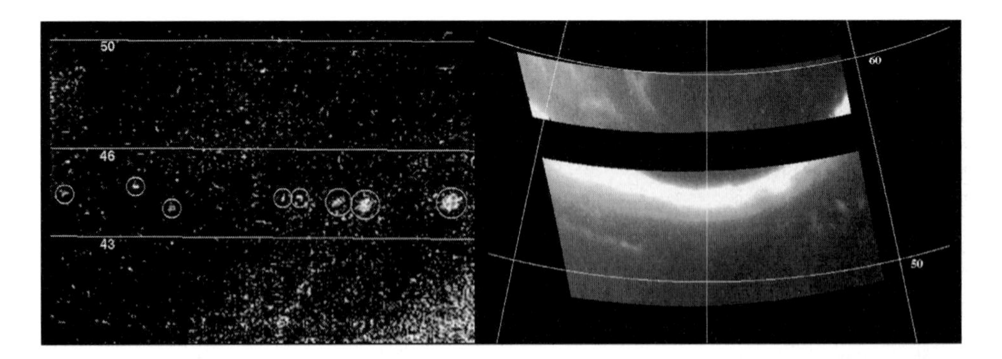

As it passed through Jupiter's shadow on its third orbit, Galileo operated its solid-state imaging system in 'line scan' mode over most of the dark hemisphere at mid-northern latitudes to search for flashes of lightning. The planetocentric latitude lines indicate that the lightning (circled) was concentrated just south of the westward-moving jetstream at 46 degrees (almost all of the lightning seen by the Voyagers was near a westward jet). The atmospherically active region between 36 and 46 degrees north is one of the zones where lightning is most likely. The diffuse background glow is probably reflection of light from the moons illuminating the cloud tops. While it was behind Jupiter, Galileo also looked for (and recorded) the oval auroral display around the north magnetic pole, which is tilted 10 degrees from the axial pole.

The apparent width of the flash could be used to infer its depth, which, at approximately 75 kilometres, is consistent with a water cloud. The most intense of the recorded storms was about 500 kilometres across, which matches the thunderhead near the Great Red Spot that Galileo observed on its first orbit.

HOT SPOTS

A high priority was to investigate an 'infrared hot spot' similar to the one that the probe had penetrated. "The more observations we had of these hot spots, the better we could understand the probe data," explained Glenn Orton. This was the benefit of combining an orbiter with an *in situ* probe.

Galileo's encounter sequences, however, needed to be defined months in advance, and hot spots are ephemeral features. When planning for the fourth orbit swung into high gear, there was an ideal candidate in the Northern Equatorial Belt. However, scheduling an observation was not straightforward. "Jose Luis Ortiz, a Spanish postdoc, had determined that a hot spot's location can be predicted in a limited sense. If you move a longitude system at a particular 'drift rate' with respect to Jupiter's interior – whose rotation rate is known via the variations of the magnetic field – there are some longitudes which are much more likely to have hot spots than others. This was something of a gamble, but the best we could do."

In early November, the candidate hot spot was still present, and on track, and the Infrared Telescope Facility in Hawaii was assembling a 'history' of it to provide a

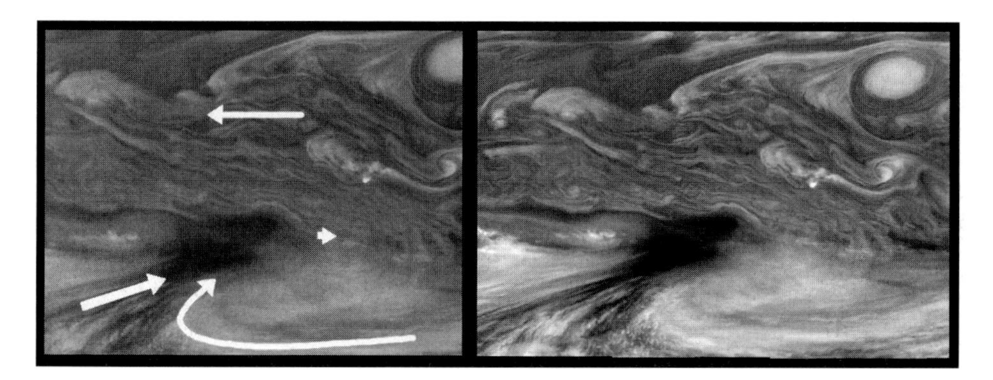

On its fourth orbit, Galileo took time-lapse images of an equatorial 'infrared hot spot' to measure the wind patterns. An infrared hot spot is actually a hole in the ammonia clouds through which energy at a wavelength of 5 microns can readily escape. It appears dark at visual wavelengths because there is no ammonia to reflect sunlight. The atmospheric probe had penetrated a similar feature. Jupiter's circulation is dominated by alternating east–west jetstreams. The mosaic spans 20 degrees of latitude. The upper half part lies within the North Equatorial Belt, which moves westward. The lower part of the image is the Equatorial Zone's fast eastward jetstream. The results (arrowed, scaled for wind speed) indicate that dry air is converging and sinking, thereby sustaining the gap in the clouds. Interestingly, there is little motion away from the hotspot in any direction. The fastest winds near the hotspot are about 100 metres per second.

context for Galileo's observations, but by the end of the month major changes had taken place and the hot spot had disappeared. Or perhaps it had been eased aside, for there was now a smaller hot spot nearby.

Rather than leave Galileo to image whatever happened to be at the location that had been programmed into the encounter sequence, a last-minute revision was sanctioned. To minimise this change, the timeline would be retained but the instruments would be aimed differently, to where the new object was likely to be. "This meant lots of work, which no one really needed, but everyone understood the importance of this particular feature." These observations would go a long way towards making up for those forsaken on the run in to Jupiter, a year earlier. In the event, Ortiz's prediction was perfect.

Galileo's mosaic spanned from the equator to 20 degrees north, documenting the eastward-moving Equatorial Zone and the southern edge of the westward-moving North Equatorial Belt.

A 'hot spot' appears bright in the infrared because heat from the interior is leaking out through a hole in the upper layers of cloud. However, Galileo observed it at a wavelength at which the ammonia cloud tops reflected sunlight, so the hole appeared dark.

Although the hot spot resided in the easterly jetstream, there was little cloud motion away from the hot spot in any direction. To the northeast, the clouds barely reflected the prevailing wind and the flow had actually been reversed to the southeast. The fastest winds were those associated with the hot spot itself – almost

100 meters per second. Considered together with the probe's composition data, this circulation pattern confirmed that these features mark where cold dry air is converging and being forced to descend. In effect, the prevailing eastward flow was pouring straight down the hole, maintaining its form.

Because the infrared spectrometer can sample at many wavelengths simultaneously, it can make a spectrum for each pixel in an image, and hence chart the distribution of chemicals. As the data accumulated, it revealed that the humidity in the vicinity of a hot spot ranges from 0.02 to 10 percent, with the lowest value in the centre. This banished all remaining doubt that the probe's entry site had been a meteorological anomaly. As the hot gas rises from deep in the atmosphere, the various volatiles precipitate out as rain. As the dry air 'turns over' at the top of the atmosphere, the winds converge and descend. However, by this point there are no volatiles left to condense to form clouds, so a dry clearing is created. As the cold air is forced down, the pressure rises and it is heated. "The dry spots may grow and diminish," explained Glenn Orton reflecting upon Ortiz's model, "but they recur in the same places, possibly because of the circulation patterns."

The infrared spectrometer also established that in a 'typical' area, the atmosphere is rather wetter, as had been supposed before the probe's data prompted second thoughts, and there are thunderstorms, lightning, and rain at a depth of about 80 kilometres below the visible surface. Local water abundance varies widely. In terms of percentage, the variation of concentration is comparable to that in the Earth's atmosphere. Nevertheless, although there is commonality of active processes, the two circulation patterns are very different.

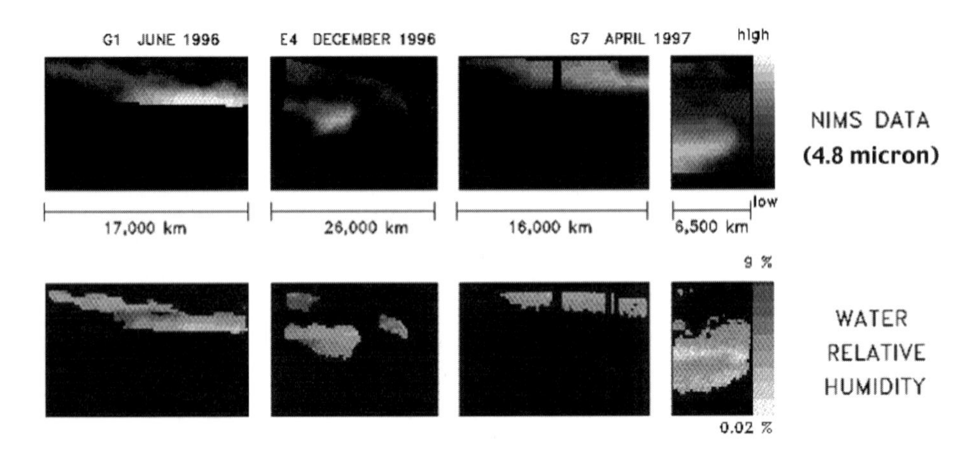

By observing at different wavelengths, Galileo's infrared spectrometer could measure the water concentration within the atmosphere. The bright features at 4.8 microns (top) are 'infrared hot spots' where thermal radiation is escaping from about 100 kilometres below the visible cloud tops. The instrument can measure a spectrum for every pixel in one of its images, and hence use absorption features to identify chemicals, and thereby infer the humidity (bottom). The results confirmed that with relative humidities of 0.02 to 10 percent such hot spots are arid.

The imagery of the equatorial hot spot was processed to generate a three-dimensional visualisation of a 11,000 × 34,000-kilometre area looking northeast towards where the air flow is diving, forcing a hole in the undulating ammonia cloud deck.

FAULTS AND FIXES

The infrared spectrometer continued to suffer radiation-induced glitches when deep in the Jovian magnetosphere, but by being dynamically reloaded it was securing most of its allotted observations. It suffered permanent damage on the third orbit when one of its 17 photovoltaic diode detectors was disabled. The cause of the failure is not known but it is possible that one of the wire bondings may have become separated. The loss of detector number 8 denied the instrument the 2.4- to 2.68-micron range. Fortunately, this particular range was not a crucial one, because the weak water absorption band it contained was not the only indicator of water and detector number 9, which was immediately to the longward, was able to make up for the loss.

On the other hand, the filter wheel on the integrated photopolarimeter and radiometer that had jammed on the first perijove passage had finally been released.

As Larry Travis, the instrument's co-principal investigator from the Goddard Institute for Space Studies in Greenbelt, Maryland, explained, "From examination of the PPR science and housekeeping data at the time that the wheel became stuck, we could deduce that it was not a mechanical problem, but rather a logical fault. The encoder that monitored the position of the wheel was indicating an incorrect position, namely position 20 rather than position 21 on the 32-position wheel." The analysis concluded that the pin for the least significant 'bit' was not making contact with the encoder assembly's disk, prompting the microprocessor to choose a mode for the stepper motor that delivered no torque, so the wheel did not advance to the next position. "The position at which the filter wheel was stuck is a radiometry position," reflected Travis, "so we were faced with the possibility that we'd be able to perform only radiometry at that single spectral band unless we could get the wheel unstuck. Although we didn't (and still don't) know why the pin was not making contact with the encoder disk, we tried the obvious, heating and cooling the instrument." The cooling regime in December was successful. "Under the circumstances, we decided that it was not prudent to use our planned modes of operation, which would have involved frequently stepping through position 20, so we adopted a strategy that would let us cycle back and forth over just a few wheel positions. In that way, we would be able to do portions of the radiometry, photometry, and polarimetry, without going back to position 20 on the wheel." As with the balky tape recorder, a conservative strategy was called for in using the photopolarimeter and radiometer while its status was fully evaluated.

SAD LOSS

As 1996 drew to a close, Carl Sagan succumbed to cancer. Planetary exploration had "lost one of its most gifted minds and eloquent voices", announced JPL's Director, Ed Stone. Sagan was renowned for successfully engaging the public in the excitement of space exploration. He was a member of the original team that had promoted the Galileo mission within NASA, and he had helped to rescue it from the Congressional axe. "He was one of the greatest intellects behind the genesis of space exploration generally, and specifically the Galileo mission," Torrence Johnson said in summary.

CONJUNCTION

Because Jupiter was nearing conjunction – an alignment which occurred every 13 months – the replay of the data from the fourth encounter sequence was delayed for several weeks. Furthermore, because the tape would still have a lot of data when the spacecraft returned to the inner system, no observations would be able to be made during the fifth orbit.

Transmission so close to solar conjunction would not have been feasible if it were not for the recent upgrade to the Deep Space Network to extract the low-gain antenna's weak signal from the background noise of the solar corona. The task was complicated when a 'coronal mass ejection' blasted out a vast blob of plasma at 1,000 kilometres per second across the line of sight to the spacecraft. This was frustrating for the engineers attempting to recover the data but it was a fascinating bonus for Richard Woo, a member of JPL's solar radio science team, who used the signal's degradation to investigate the region where the solar wind is accelerated. In any case, the replay would not be completed until the end of February 1997. Delaying the replay would have obliged Galileo to remain passive for the next *two* perijove passes.

Although Galileo started its second year in Jovian space with its engineering redundancy intact, solar conjunction was a worrisome time for its operators. "I always worry when we're out of contact with the spacecraft," confessed the Control and Data Subsystem Manager, Tal Brady. "You never know what happens when you are not watching."

While the spacecraft took a rest, the Galileo scientists attended a conference to discuss the first year's results. 'The Three Galileos' conference was held in Padua, Italy, where Galileo Galilei had lived and it celebrated the man himself, the Galileo spacecraft, and Italy's recently commissioned 3.5-metre Galileo National Telescope, located in the Canary Islands. This was "the most comprehensive collection of results from Project Galileo in one place", Bill O'Neil reflected. It was also a memorable occasion for the team members invited to an audience with Pope John-Paul II, who "encouraged the continuing exploration of the Universe".

As Galileo dived back into the inner Jovian system for the sixth time, it was approaching the mid-point in its primary mission.

WHITE OVALS

In 1939, a trio of dark brown streaks developed in the South Temperate Belt, which was abnormally bright at that time.[7] Over the ensuing years, observers methodically

This Voyager mosaic shows the turbulent eddies extending off to the northwest of the Great Red Spot, and one of the trio of large 'white ovals' which drift around the South Temperate Belt. Like the Great Red Spot, the ovals are anticyclonic systems.

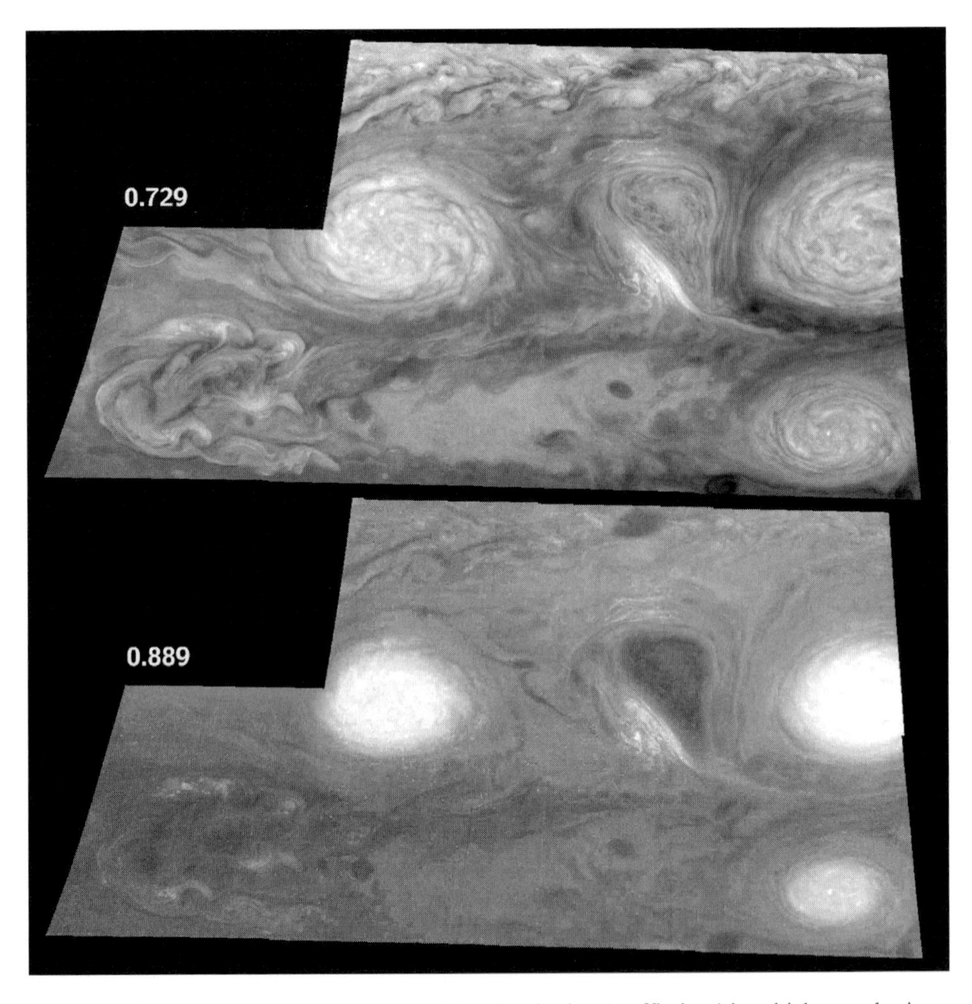

On its sixth orbit, Galileo documented the developing 'traffic jam' in which a cyclonic system was being squashed between two white ovals. Since methane moderately absorbs at 0.727 micron we see the ammonia cloud and upper-tropospheric haze, with higher features appearing brighter (top). Methane is strongly absorbed at 0.889 micron so we see only the hazy cloud tens of kilometres above the ammonia cloud deck, such as over the white ovals (bottom).

recorded the longitudes of the leading and trailing edges of the brown features, labelling them arcanely as 'AB', 'CD' and 'EF'. By 1948, it was evident that the intervening bright regions of this belt were evolving into distinct systems, becoming progressively more oval as their longitudinal extent diminished, so they are now referred to as 'white ovals'. However, because the ovals were anti-correlated with the dark features, the scheme resulted in the ovals being designated 'FA', 'BC' and 'DE'. At about 10,000 kilometres across, they are considerably smaller than the Great Red Spot but as anticyclonic systems (like the Great Red Spot) they rise far above the ammonia cloud tops. The ovals have remained within the latitude range 31 to 35 degrees, drifting eastward. Also because they migrate independently of one another, from time to time they jostle one another.

Over recent months, 'BC' and 'DE' had closed up and trapped a smaller storm in between. Having had its plan to observe the white ovals on the third orbit frustrated, the Atmosphere Working Group decided to try again to record the developing 'traffic jam'. The earlier plan to observe the ovals had been cancelled because their drift rates were too difficult to predict, but the Infrared Telescope Facility had monitored them during the run up to solar conjunction and everyone was confident.

Both the infrared spectrometer and the integrated photopolarimeter and radiometer were to complement the camera's imaging during this campaign. However, about midway through the sequence, the infrared spectrometer lost a second channel. The loss of detector number 3 was more serious than the loss of number 8 because, in addition to some water bands, the 1- to 1.3-micron band covered a broad absorption feature being used to map Io's surface composition (number 4, immediately to the longward, included the wing of this feature, so some information would still be able to be gained). However, the fact that the symptoms of the two failures were similar was worrying, because it raised the possibility that the instrument would be progressively disabled by a generic fault.

Galileo observed the 'traffic jam' through several filters. The interaction between the ovals had produced thick cumulus-like clouds in the southern section of the trapped system. However, whereas the haze above the white ovals showed that they penetrated the stratosphere, the absence of haze over the cyclone implied that this system was located deeper within the atmosphere.

DAWN-SIDE PLASMA SHEET

The attempt to collect data while crossing the planet's magnetic equator was frustrated by a magnetometer failure. However, on 30 March 1997, as it made its way back in, the spacecraft crossed the planet's magnetic equator at 46 Rj.

The impingement by the solar wind creates very different environments on the dawn-side and dusk-side of the inner magnetosphere. This was Galileo's first opportunity to study the plasma sheet as it was compressed rotating from the night-side to the day-side of the magnetosphere.

THE GREAT RED SPOT'S SPIRAL STRUCTURE

During Galileo's seventh perijove passage, the integrated photopolarimeter and radiometer scanned the temperature field across the Great Red Spot, with the highest resolution image yet recorded. Gradients in the temperature field prompt air movements, so this chart indicates strong upwelling and downwelling winds in the upper part of the atmosphere where the winds are at their most intense. The greatest temperature gradient in the survey is between the cold central column of the Great

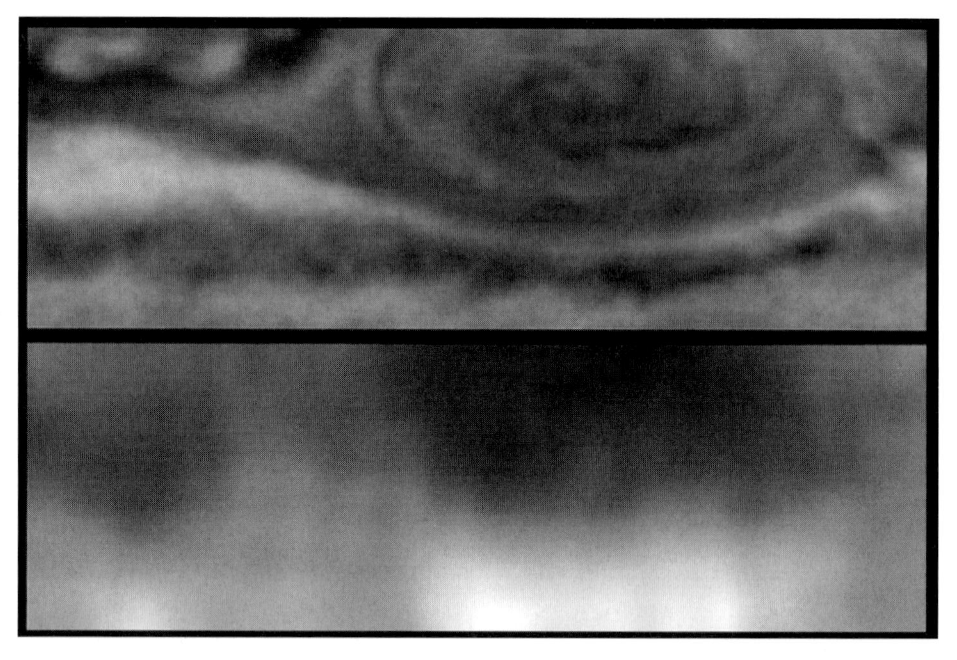

On Galileo's seventh orbit, the Hubble Space Telescope's Planetary Camera imaged the Great Red Spot (top) as the spacecraft's integrated photopolarimeter and radiometer (in radiometer mode) determined the temperature field across the southern part of the feature (bottom). Although the resolution of this chart was the highest yet achieved, even at perijove the instrument's spatial resolution was only 2,000 kilometres. Even so, it clearly indicates the forces which power the Jovian winds, and differentiates between areas of upwelling and downwelling in the upper atmosphere. The Spot's column rises far above the surrounding ammonia cloud deck, and is extremely cold. The winds in the Spot are more complicated than a simple counterclockwise rotation around the core. The strong eastward jetstream to the south is more than 10 degrees warmer (Hubble image, courtesy STScI)

Red Spot (which projects to the highest altitude) and the edge of the eastward-moving jetstream about 5,000 kilometres to the south, where it is 10 degrees warmer.

The fact that the southern-central section of the Great Red Spot is not as cold as the rest of the feature indicates that the winds at the fringe of the anticyclonic system are complex. Most astronomers had believed the Great Red Spot to be a dense cloud mass, but it was found to possess a 'spiral arm' structure of clouds with gaps between. Furthermore, it is 10 kilometres taller in the centre and (as the outer ring of cloud discovered earlier demonstrated) the entire structure is tilted to the east, like a crooked spiral staircase.

In some ways, this atmospheric pattern in the Great Red Spot is similar to what happens in a terrestrial hurricane – but on a far larger scale – with moist gas from deep in the atmosphere rising rapidly in a relatively narrow column and then spilling out far above the ammonia cloud tops. The gaps in the spiral structure enable the instrument to 'see' through to the levels below.

SYNOPTIC COVERAGE

The zonal circulation system with its alternating eastward and westward jetstreams prevails in the Jovian atmosphere all the way to the polar areas. On its seventh and

On its seventh orbit, Galileo assembled a mosaic ranging from 10 to 50 degrees north, showing the alternating jetstreams of the zonal circulation system.

On its eighth orbit, Galileo assembled a mosaic ranging from 25 degrees south all the way down to the south polar area.

eighth orbits, Galileo documented wide areas on alternating rotations of the planet in order to measure the wind speeds within each latitudinal band.

In addition to the overall zonal structure of the atmosphere, these synoptic mosaics record localised features such as white ovals, bright spots, dark spots, vast interacting vortices and intriguing turbulence.

TURBULENCE

In planning the ninth orbit's observations of the Great Red Spot, the Atmosphere Working Group had to trade-off a number of science objectives.

The infrared spectrometer team wanted to retain the focus on the centre of the feature in order to recover the observations that had been lost on the first orbit (when their instrument had been disabled by radiation). The imaging team wanted to switch attention to the turbulent region off to the northwest, because this was deemed to be more likely to provide significant insight into the dynamics of the atmosphere. However, some dynamicists wanted to shift the focus over to the 'leading edge' of the Great Red Spot rather than examine the turbulence in its 'wake'. The integrated photopolarimeter and radiometer team, however, having concluded their Great Red Spot programme, argued for tackling a completely new target. All the instruments were on the scan platform and this could be aimed only at one point at a time so, from the point of view of the polarimetry team, time spent aimed towards the Great Red Spot was time that would not be available for looking at something else.

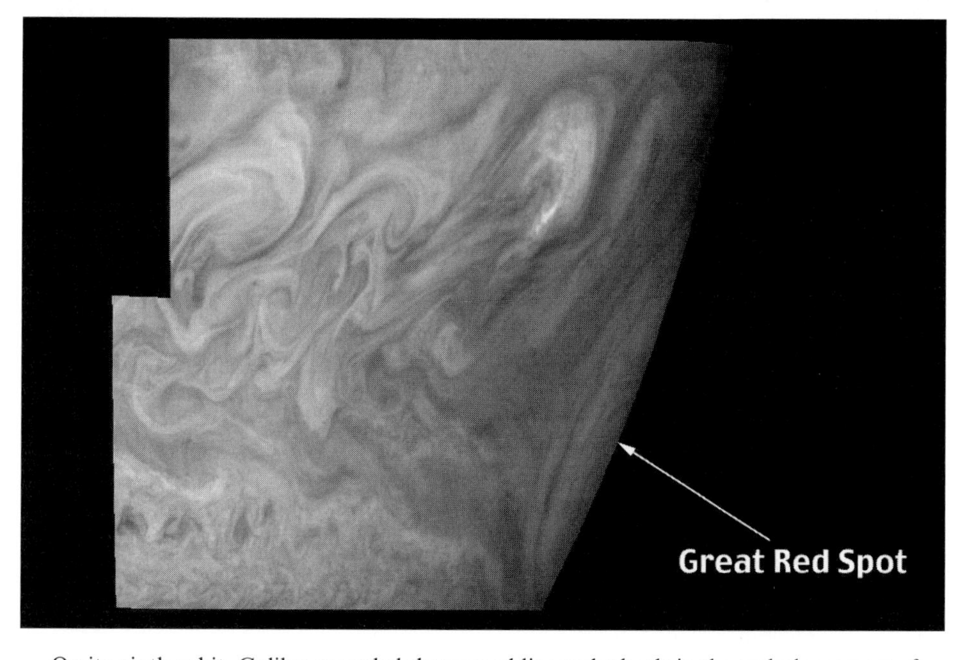

On its ninth orbit, Galileo recorded the vast eddies and whorls in the turbulence west of the Great Red Spot (the Spot itself is just beyond the limb). The turbulence results from the collision of a westward jetstream which the Spot deflects north into an eastward jetstream. The mosaic combines imagery at 0.756, 0.727 and 0.889 micron to highlight variations in cloud height and thickness. The fact that the eddies northwest of the Spot are bright is an indication that this area is intensely convective and is creating precipitate-laden clouds.

It was decided to concentrate on turbulence west of the Great Red Spot where the jetstream immediately to the south of the spot is deflected north and interacts with the one in which the spot resides, in the process creating a series of huge eddies.

MINI-TOUR

After moving away from Jupiter, Galileo was to perform an extended orbit, both in terms of apojove and orbital period so as to undertake a 'mini tour' of the Jovian magnetotail. The second and third orbits had produced high apojoves, but now the alignment of the orbit with respect to the Sun meant that even when the spacecraft was at its 143-Rj apojove it was still well within the tail.

The ninth encounter sequence was "a development nightmare", observed Laura Barnard, a science and sequencing technical engineering aide who regarded herself as a 'Jill-of-all-trades' who assisted engineers, programmers and mission support staff

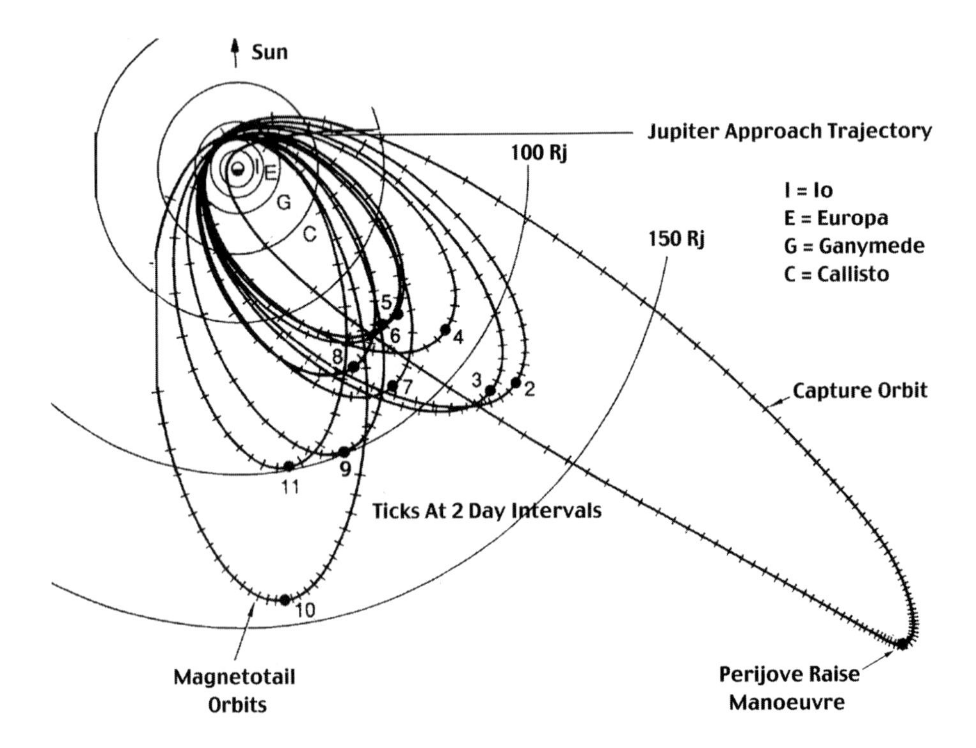

A diagram of Galileo's two-year primary mission orbital tour. The apojove of the initial 'capture orbit' was about 20 million kilometres from the planet. It led to the first fly-by of Ganymede (on the first orbit of the tour). Each close encounter set up the next orbit. The 'mini tour' of the magnetosphere was conducted during the tenth apojove passage, which was on the opposite side of the planet from the Sun. (Courtesy, NASA)

on a daily basis. "It's such a long orbit, just about twice as long as the others," she pointed out. "That means conflicts with other projects that need to use the Deep Space Network, and problems in allocating spacecraft resources."

It was not simply a matter of doing the same as before, only more so; a new procedure for the tape recorder was required to store the high time-resolution particles and fields data at key times during the long orbit. Usually, the spacecraft barely had time to replay the data from an encounter sequence before having to start its next sequence, but such a long orbit provided the advantage of time, so once a section of the tape had been downloaded the high time-resolution particles and fields measurements were to be written onto this part of the tape. Low-rate data was to be reported in near-real time throughout in order to provide a context for the detailed observations.

This was an important opportunity to investigate how the magnetosphere interacts with the solar wind. Trends in the magnetic field, electric currents, and the flow patterns of the plasma in the inner and outer areas of the magnetosphere were monitored. The results showed that the disk-like plasma sheet extends at least 100 Rj from Io's torus (from which it derives) and that the thickness of the undulating disk varies significantly from one 10-hour rotation to the next. "It is possible that all of these dynamical effects are due, in large part, to the variable volcanic activity of Io," suggested Bill O'Neil. "The magnetic signatures which suggest the merging of magnetic field lines in the sheet – releasing large amounts of energy – have been tentatively identified." It is likely that such processes are responsible for magnetic substorms. "There are very puzzling decreases in the hot plasma densities," he noted, comparing the results with those from 1979, "and changes in the compositions of the ions during Galileo investigation."

This "extraordinary magnetotail orbit", as O'Neil put it, was one of the key objectives for the Magnetosphere Working Group.

In fact, the solar wind may actually draw out Jupiter's magnetotail as far as Saturn's orbit, over 650 million kilometres further from the Sun, so the spacecraft's 10 million kilometre apojove could be considered to be fairly 'local' in terms of the enormous size of the structure.

LIMB SOUNDING

A one-off alignment opportunity presented itself on 8 September 1997. Now on its way back to Jupiter from down the magnetotail, Galileo approached the planet's night-time hemisphere, and as it passed into and out of the planet's shadow it was able to employ 'limb sounding' to measure the sizes and populations of the particulates in the atmosphere, in the same manner as some environmental satellites monitor the state of the Earth's atmosphere.

AURORAL FOCUS

During the tenth perijove passage, the integrated photopolarimeter and radiometer and the ultraviolet and infrared spectrometers observed the polar regions in order to investigate 'hazy' clouds thought to be produced directly beneath auroral activity. At this time, the plasma wave spectrometer was damaged. The fault introduced some 'noise' in the sensor that distinguished electrostatic waves from electromagnetic

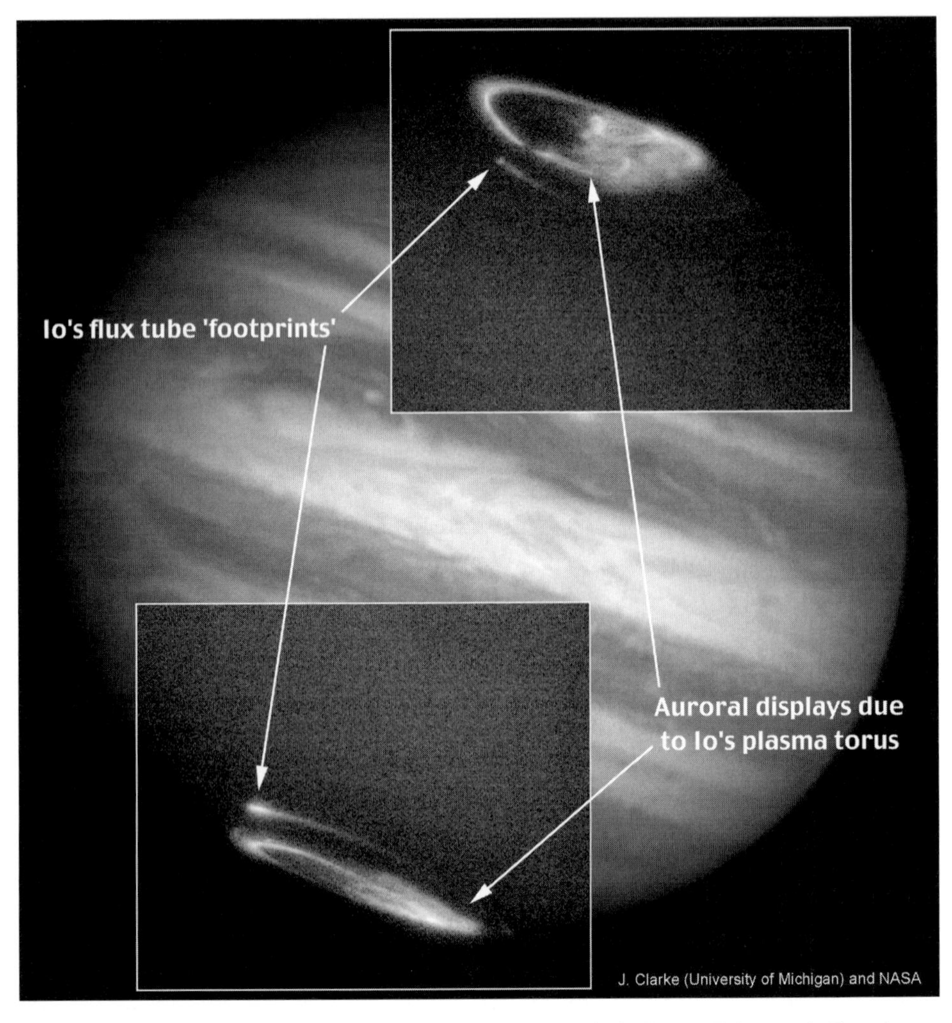

On 20 September 1997, the newly commissioned Hubble Space Telescope's Imaging Spectrograph documented Jupiter's aurorae in ultraviolet light. It was able to record the oval auroral curtains, the 'footprints' created by the energetic particles flowing in Io's flux tubes, and the 'comma-like' afterglow tails. The background image was taken by the Wide Field and Planetary Camera 2. (Courtesy STScI and John Clarke, University of Michigan)

waves. Henceforth its ability to discriminate between these two types of wave would be reduced.

On 20 September (during Galileo's encounter) the Hubble Space Telescope's Imaging Spectrograph made a series of ultraviolet observations of aurorae on Jupiter and Saturn. This telescope had observed Jovian aurorae before, but the imaging spectrograph (which had been installed in February 1997) represented an order of magnitude improvement in sensitivity. It was able to use shorter exposures, and so avoid motion-blur resulting from Jupiter's rapid rotation. Overall, it produced 2 to 5 times better resolution, which was sufficient to discern a 'curtain' of auroral light projecting several hundred kilometres from the limb, towards the auroral activity.

Each of Io's auroral footprints has a comma-like 'tail' because the charged particles continue to excite the atmosphere for some time after Io has 'passed overhead' – although, of course, it is actually Jupiter which turns beneath the slowly-orbiting moon. The Hubble Space Telescope had first observed the footprints in 1994, but this was the first time it had seen the residual glow.

MESOSCALE

As Galileo drew away from Jupiter on 5 October 1997, it was eclipsed for 19 hours, so it was able to undertake an extended observation of the level of night-side weather activity and scan for lightning. The lightning survey picked up again in early November as the spacecraft made its eleventh perijove passage.

In the Voyager 'movies' the Great Red Spot spawned a series of rapidly expanding white clouds resembling large thunderheads, and these were swept away by the prevailing jetstream. Galileo's detection of lightning confirmed that these are intensely convective features. This is consistent with the multiple-filter observations[8] of a cloud in one of these amorphous cyclonic regions. Being at a pressure level of 4 bars, this cloud had to have been rich in water. It was attached to a towering 'anvil' which rose to about the 0.4-bar level, which indicated that it was 60 kilometres tall.

"We even caught one of these bright clouds on the day-side, and saw it flashing away on the night-side less than two hours later," reported Andrew Ingersoll. Multiple lightning strikes confirmed that this was a site of moist convection in the saturated environment at a depth of about 75 kilometres. "This is a big step toward understanding how Jupiter's weather system gets its energy." The bright cloud resembled a site of convective upwelling in the Earth's atmosphere. The dark cloud-free region immediately to the west of the storm was a region of downwelling.

Although widely distributed, lightning is particularly common between 46 and 49 degrees north. It is controlled by the large-scale atmospheric circulation associated with low-pressure systems. Galileo did not detect lightning in high-pressure systems such as the Great Red Spot, which probably means that they are not drawing energy from below by convection; they are sustained differently.

Some of the results from the early orbits of the extended mission were reported at the 30th Annual meeting of the American Astronomical Society's Division of Planetary Sciences which was held in Madison, Wisconsin.

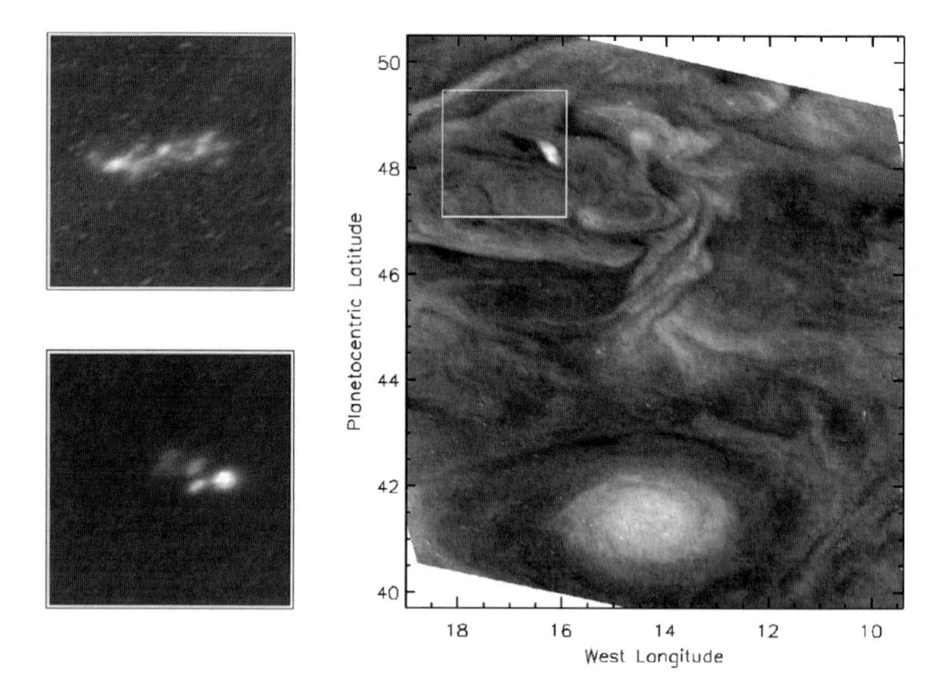

Galileo followed up its study of Jovian lightning on its eleventh orbit, by imaging sites of intense convective activity as they approached the evening terminator and then again two hours later in darkness, so as to be able to definitively locate the sites of lightning. In this case it recorded multiple lightning strikes in images taken about 4 minutes apart. In the contextual view, the bright cloudy area in the box is an upwelling and the darker (cloud-free) area immediately to the west is a downwelling. The lightning is below the ammonia cloud, which acts as a translucent screen and diffuses the light, and the width of the spot indicated that the lightning was taking place in the water zone 75 kilometres below the ammonia cloud tops.

The moist convection in the deep atmosphere confirmed that the model that was developed prior to Galileo's arrival was basically correct.[9]

"There is a lot of activity we see on Jupiter that we see on Earth," said Peter Gierasch of Cornell University. "We see jetstreams, large cyclonic elements, large anticyclonic elements, and many elements of unpredictability and turbulence." A terrestrial thunderstorm is simply a 'cell' comprising a distinct cumulo-nimbus cloud. The clusters of thunderstorms west of the Great Red Spot would be called 'mesoscale convective complexes' on Earth. These are large clusters of cells.

The main difference between the formation of a mesoscale complex and that of a cyclonic system – a hurricane – is its source of energy. Cyclones are driven by evaporation of moist air from the warm ocean; mesoscale complexes result from atmospheric instability, inversions where the air near the surface is warm and the air aloft is cool and instabilities spawn clusters of intense thunderclouds which can last for hours, even days, and dump unusually large amounts of rain. Such activity on

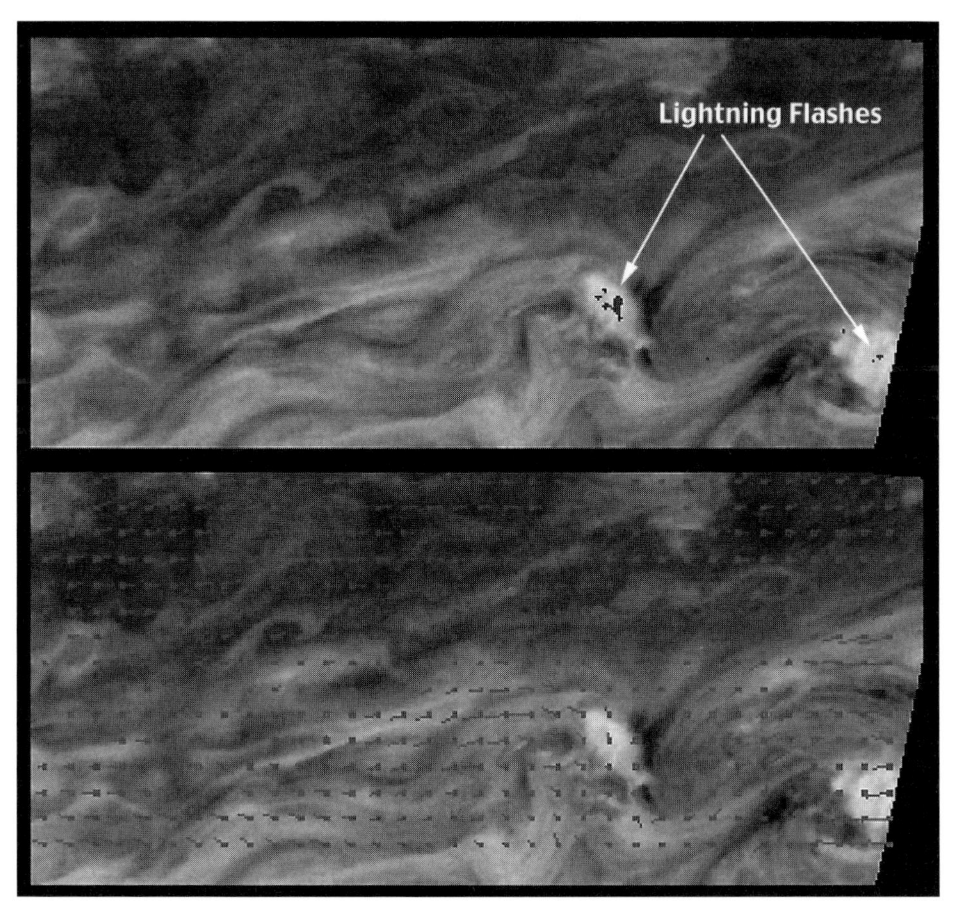

On its twentieth orbit, Galileo's continuing study of lightning focused on an area about 15 degrees south and caught two lightning clusters. The daylight context is shown with the lightning events projected as dark specks (top) and with the wind patterns (bottom). (Courtesy Don Banfield, Cornell University)

Earth is driven by heat from the Sun. Insolation at Jupiter's distance from the Sun is only 4 percent of that available to the Earth, but Jupiter is emitting 70 percent more heat than it receives from the Sun, and as it makes its way to the surface this energy drives the Jovian weather system.

Night-time observing conditions were significantly improved if there was 'moonshine' from at least one of the large satellites. On dark nights Galileo's images showed small blobs of glow from the lightning, but nothing else. If a moon was illuminating the ammonia clouds, lightning could be correlated with the atmospheric structures.

The individual lightning bolts on Jupiter are an order of magnitude more powerful than the terrestrial equivalent.[10] The lightning storms were clearly associated with the vast eddies that supply energy to the large-scale weather system.[11] The anticyclonic pattern of the eddies was consistent with their being the outflow from the top of convective systems. This lent credence to this interpretation of the larger-scale weather system.

The Voyagers had shown that Jupiter's zonal jetstreams and long-lived storm systems are sustained by soaking up the energy of smaller eddies. Galileo demonstrated that the eddies are themselves fed by thunderstorms in the water-saturated zone a hundred kilometres beneath the visible surface of ammonia cloud. It is these eddies that sustain features such as the Great Red Spot.

EXTENDED MISSION

As it withdrew from the planet in November 1997, Galileo concluded its primary mission. However, achieving this milestone was merely an administrative formality because the tour of the Jovian system had been extended for another two years. Over successive orbits, it settled into a routine of observations designed to monitor long-term trends and the extended mission was a period of consolidation. However, as a bonus, there were a number of highlights along the way.

MERGER

As the three white ovals drifted around the South Temperate Belt, they occasionally jostled one another. Recently, as one oval had encroached on another, it had squashed a low-pressure system in between. Galileo had recorded this traffic jam on its sixth orbit. The two ovals had merged in February 1998, while Jupiter was in conjunction. The ovals had been labelled 'BC' and 'DE', so the result of the merger became 'BE'. The pear-shaped cyclonic system appeared to have been consumed during the merger.

"The merged white oval is the strongest storm in our Solar System, with the exception of the Great Red Spot," pointed out Glenn Orton. "This may be the first time humans have ever observed such a large interaction between two storm systems."

The fact that the programme for each encounter sequence had to be developed months in advance meant that the first opportunity Galileo had to inspect the aftermath of the merger was in July, on its sixteenth orbit. Although this was cut short by a glitch deep within the magnetosphere, the integrated photopolarimeter and radiometer had by then made its early observations, so there was some data to ponder.

The integrated photopolarimeter and radiometer data indicated that the cyclonic system that had held the two white ovals apart had 'lost power', opening the way for the blocked white oval to continue its drift towards its companion. The merged white oval displayed an atypical temperature profile, perhaps indicating that it was a transitionary stage. "We can see it at visible light and some infrared wavelengths," reported Orton, "but we cannot see this new white oval at certain infrared wavelengths that peer underneath the storm's upper cloud layer."

Whereas the low-pressure systems operate in the same way on both the Earth and Jupiter, the high-pressure systems are different. A terrestrial high-pressure system forms when dry air which has emerged from the top of a cyclone moves clear, cools, and descends, increasing the surface pressure. Because it is dry, the descending air cannot make cloud. As this air radiates out near the surface, it gives rise to an anticyclonic circulation.

On Jupiter, however, a high-pressure region marks a convergence centre which consumes cyclonic systems, and it is this inflow that produces the increase in pressure. Since pressure is inversely proportional to altitude, the centre of the convergence is obliged to rise to relieve the pressure. This creates an upwelling column. As the air emerges from the top of the column, it spreads out and descends. So, instead of being a descending airflow, a high-pressure system on Jupiter towers far above the cloud tops. Nevertheless, the bright clouds of the white ovals are believed to be composed of ammonia. The overall area of the merged oval was rather less than the sum of the areas of the contributing features, which suggests that the merged system had been drawn out in the vertical direction, and that the conservation of angular momentum had induced it to rotate faster.

Galileo had established that most of Jupiter's anticyclonic spots reach to more-or-less the same altitude (and therefore the same pressure level) and thus are at the same temperature, but the Great Red Spot and the white ovals rise to the 0.2-bar level (recall that the ammonia cloud tops are around the 0.7-level). It had already identified an aerosol haze in the chilly thin air in the centre of the Great Red Spot, and the merged oval had a similar a haze. "With 'mature' white ovals, we can see the

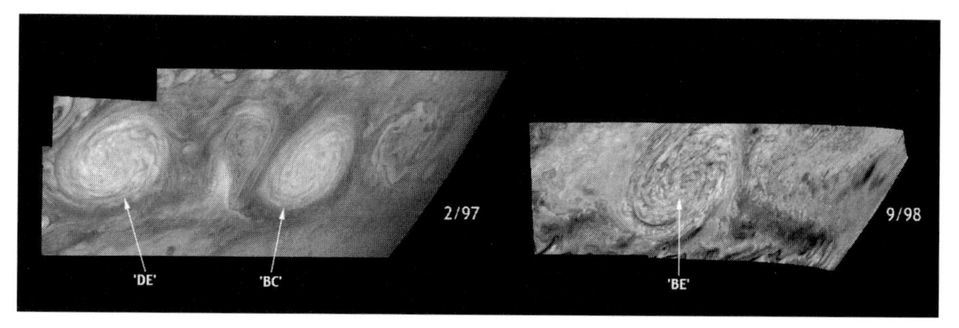

The 'traffic jam' of two 'white ovals' squeezing a cyclonic system, which Galileo had documented on its sixth orbit, merged in February 1998. Unfortunately, the planet was on the far side of the Sun at the time, so the merger was not actually witnessed, but the aftermath was recorded by Galileo on its seventeenth orbit. The two views present the same viewing geometry. The earlier view (left) combines imagery at 0.756, 0.727, and 0.889 micron to display variations in cloud height and thickness. The white ovals rise far into the stratosphere, but the lack of haze over the pear-shaped cyclone shows that it is deeper. The post-merger image (right) is at 0.757 micron, with negligible methane absorption. The cyclonic storm was evidently consumed by the process of merger. The appearance of the merged white oval is anomalous, in that it is capped by an unusually thick layer of high cold ammonia cloud. This may be a transitionary feature, which will dissipate over time.

upwelling in the centre, which in turn leads to downwelling around it," Orton explained. As the white ovals merged, the sudden increase in pressure may have driven the column even higher, making its top even colder. In fact, the centre of the merged oval was determined to be a degree or so cooler than its surroundings. This infrared anomaly could be explained if the new oval was capped by a thick layer of high-level cloud that was temporarily obscuring the peripheral downwelling. Observations over subsequent orbits would document how the new oval evolved.

SUBSTORMS

In mid-August 1998, as Galileo neared apojove, it collected particles and fields data on the 'deep tail' (defined to start at a distance of 100 Rj) to further investigate how the plasma in the inner magnetosphere escapes down the tail and leaks into the solar wind.

In the case of the Earth, high-energy particles from the solar wind enter the magnetosphere on the sunward side. Some particles head for the magnetic poles and trigger aurora, but most are trapped inside the magnetosphere. They cannot remain trapped for ever, otherwise the magnetosphere would fill up with plasma. During geomagnetic storms, these trapped particles are ejected into space in the form of vast blobs of plasma called 'plasmoids'. "It took 25 years of study to figure that out for Earth – we're just starting on that journey of understanding for Jupiter," noted Claudia Alexander, a JPL plasma physicist. This second 'mini tour' was particularly favourable for investigating magnetic substorms because its apojove was directly down-Sun, where such activity is most prevalent.

DARK SPOT

The seventeenth orbit presented a brand new target for the integrated photo-polarimeter and radiometer, in the form of a 'dark spot'. Dark spots are rare in the Jovian atmosphere and this one was the darkest ever seen. A swath spanning 60 degrees of longitude and extending from the equator down to 40 degrees south was scanned in order to document both the spot and its context.

A map of Jovian temperatures near the 0.25-bar pressure level established that the spot was warmer than its environment. The warm cloud-free conditions indicated that it was a region where dry upper-atmospheric gas flow had converged, made a hole in the cloud as it was forced to descend, and warmed as its density increased. The temperatures corresponded to those at the top of the troposphere. The 0.25-bar map also indicated 'temperature waves' in a warm region (this was the first time that atmospheric temperatures had been scanned by the radiometer at a spatial resolution better than about 2,000 kilometres, allowing such fine structures to be detected). Thermal waves had been seen before which were independent of the cloud structure, but those had been much larger in size. Interestingly, although the waves were clearly 'channelled' within the warm band, they produced no counterpart in the visible cloud structure.

On its seventeenth orbit, Galileo investigated a strangely dark spot which had appeared recently. It was suspected of being a downwelling making a hole in the ammonia cloud deck, exposing a deeper layer beneath. The integrated photopolarimeter and radiometer made a temperature map of the 0.250 bar pressure level (left). The same area is shown in visible light (middle) and in terms of cloud-top temperatures (right). Unlike 'infrared hot-spots', this dark spot was truly warmer than its surroundings in the upper troposphere. The documented area spans from the equator to 40 degrees south.

The dark spot was unlike the dry and relatively cloud-free hot spot that the probe sampled, however. "Both appear warm at 5 microns," explained Glenn Orton, "but the dark spot is truly warmer than its surroundings, whereas the '5-micron hot spots' appear to be the same temperature as their surroundings. The typical visual colour of 5-micron features is a dark blue-grey, instead of black. The colour and depth of the cloud tops revealed by these features is one of the most enduring mysteries of Jupiter – what makes Jupiter's clouds at various depths coloured the way they are?"

AURORAL ELECTRO-JETS

Meanwhile, ground observers had been busy. A study[12] conducted using NASA's Infrared Telescope Facility in Hawaii established that violent winds race around Jupiter's polar regions at supersonic speeds. Since the energy for these jetstreams is derived from the magnetosphere, and they are associated with aurorae, they were named 'auroral electro-jets'. This investigation established that the auroral regions are linked through the planet's magnetic field to the plasma sheet, which derives its material from Io's volcanoes.

"You need a lot of energy to keep that plasma sheet rotating," explained Steve Miller of the Department of Physics and Astronomy at University College London, who led the team which made the observations. "At the rate that Io is pumping out gas and dust – a rate of 1 tonne per second – we estimate 10 million megawatts of power is required. What is happening, is that the plasma sheet is siphoning off some of the reservoir of rotational energy that is stored up in Jupiter. Our discovery of the auroral electro-jet shows how the plasma sheet 'couples' to the planet by a sort of electromagnetic friction. This involves electric currents flowing through the plasma sheet, along the magnetic field, and then closing the switch across the aurorae." This discovery did not come as a surprise, however. "We have had a model that predicted this for some while, but now we really know its true."

Above the 'homopause' at 1 to 2 microbars, the individual atomic and molecular species of the Jovian atmosphere settle out according their mass, forming layers of H, H_2, and He with the ions H^+ and H_3^+. The region above the homopause also corresponds to the thermosphere. So the ionosphere composed of H_3^+ and (higher up) H^+ co-exists with the thermosphere's H, H_2 and He.

The wind speed was determined by the doppler shift of spectral features due to molecular hydrogen H_3^+ ions flowing around the ribbons of the auroral activity at a pressure of 0.3 microbar, where charged particles interact with the ionosphere. The electrically charged molecules are accelerated by the ambient electromagnetic forces to speeds of 2.8 kilometres per second.

The friction between the electro-jet and Jupiter's atmosphere also produces a great deal of energy. This helped to explain why the temperature near the top of the Jovian atmosphere is about 1,000 K, which is several hundred degrees warmer than could be accounted for by insolation alone. As Miller explained, "As the ionospheric ions rush around the electro-jet they drag – as a result of collisions – the neutral thermosphere with them. This stores up 10^{15} to 10^{16} joules of energy in the motion of the neutrals. In the steady-state, this energy must be dissipated somehow. We've calculated that when you 'switch off' the electro-jet you find the energy dissipates in about 1,000 seconds. This means that the forces of friction between that part of the thermosphere which is forced to rotate along with the ions and the the rest of the thermosphere outside of the electro-jet dissipates 10^{12} to 10^{13} watts, and eventually this will warm up the upper atmosphere (the thermosphere and the ionosphere) all over the planet."

CONSOLIDATION

Galileo had been suffering a spate of glitches while it was deep within the magnetosphere, pre-empting the perijove and exit sections of its encounter sequences. The eighteenth orbit was frustrated. The ultraviolet spectrometer was to have picked up a long-term study of the amount of hydrogen in the Jovian atmosphere. These observations had to be performed after perijove, on the night-side – in the absence of sunlight, the variation in hydrogen loss is due to vertical atmospheric mixing and by interaction with charged particles. This would provide information on the interactions between the magnetosphere, the ionosphere and the cloud tops. However, Galileo safed itself before this observation could be made. Luckily, the infrared spectrometer had already made an observation of the merged white oval.

Galileo 'safed' itself just after perijove on its nineteenth orbit. The ultraviolet spectrometer was to have observed hydrocarbons in the sunlit part of the atmosphere to study how these chemicals were related to the dynamics of the clouds in the upper atmosphere. The 'safing' also pre-empted a campaign to collect data at seven specific longitudes in order to build up a picture spanning 100 degrees of longitude. The solid-state imaging system was to take three images at each longitude. The first was to have included the bright limb, the second was to have been near the centre of the disk, and the third was to have sought lightning on the night-side (that is, for a total of 21 images). At the same time, the ultraviolet spectrometer was to have made 15 observations to provide information on the role and abundance of water in the atmosphere.

On the next perijove passage, the twentieth, the integrated photopolarimeter and radiometer was scheduled to measure the fine-scale temperature field of a broad swath of the atmosphere, but this was frustrated. As part of the cautious strategy for using this instrument after its filter wheel had been reinstated, it had been left running in polarimetry mode for two orbits. It was switched back to radiometry for this study, but it suffered a fault.

"When we examined the very first data," explained Larry Travis of the Goddard Institute for Space Studies, "it appeared that the radiometry detector, or one of the amplification stages for that detector signal was dead, because the data readouts were essentially flat – that is, no discernible signal." This malfunction was even more mysterious than the filter-wheel problem. "All we had were some speculations, one being that age and radiation effects may have led to a cracking or debonding of the conducting epoxy which connects the pyroelectric detector to the rest of the detector circuit." This was because it is tricky to solder wires to such a detector. "If this speculation had any basis, then we reasoned that heating the entire instrument might hold the slim chance of closing the cracks between the epoxy and the detector element." Much to the astonishment of Richard Chandos, the lead electronics engineer who developed the instrument – and who had put the odds as less than 1 in 10 – this worked, and the radiometry function was restored for future use.

It was otherwise a productive encounter, however. Following perijove, the solid-state imaging system and the ultraviolet spectrometer made a series of observations of a number of features in the Jovian atmosphere, including 'equatorial waves', high-speed jets, clouds in the north and south equatorial belts, and white ovals. Up to half a dozen images were taken of each feature in order to track it across the daylight hemisphere and, in some cases, on into darkness, with the objective of determining the three-dimensional dynamics of cloud motion with both high spatial and temporal resolutions to investigate the wind and storm patterns in detail.

DUSK-SIDE PLASMA SHEET

As Galileo left, the particles and fields instruments made 8 days of real-time observations. This was the spacecraft's first opportunity to map the dusk-side portion of the magnetosphere. The impingement by the solar wind creates very different environments on the dawn-side and dusk-side of the magnetosphere. Because the solar wind compresses the sunward side of the magnetosphere, the plasma trapped in the magnetosphere is squeezed into a smaller region of space as it rotates from the night-side to the day-side of the planet. A few hours later, at dusk, the pressure from the solar wind eases, and the trapped plasma is able to expand. Some of the plasma is accelerated to a velocity fast enough to escape from Jupiter.

The objective of the survey was to observe the characteristics of the plasma and the electric and magnetic fields in this region so as to figure out where the plasma acceleration actually takes place. The *in situ* measurements of the dusk-side would hopefully indicate how the plasma accelerates down the magnetotail and leaks out into the solar wind. In effect, Io's volcanoes were 'polluting' the outer regions of the Solar System.

WINDING DOWN

The twentieth orbit marked a milestone in the extended mission, because although Galileo had started to reduce its perijove in order to be able to return to Io, this also reduced the orbital period, which in turn reduced the time available to replay taped data. Therefore, as the focus shifted to Io and its torus, the giant planet received progressively less attention.

13

Moonlets and rings

NEW MOONS

It took almost two centuries after Galileo Galilei discovered Jupiter's four large satellites for another to be found. In 1892, Edward Barnard of the Lick Observatory above San Jose in California spotted a faint speck of light close alongside the planet. Once he had verified that it was a moon that circled the planet every 12 hours, he called it Amalthea after a nymph who is reputed to have nursed Zeus as an infant. Clearly smaller than any of Galileo's moons, it was thought likely to be a captured asteroid.

Other satellites were subsequently discovered telescopically (the most recent being Leda, discovered by Charles Kowal of the California Institute of Technology in 1974) all of which orbit beyond Callisto.

This was the situation when the Voyagers ventured into the Jovian system in 1979. Both were programmed to image Amalthea. In addition to discovering that Jupiter had a system of rings made of fine dust, they discovered Thebe orbiting just outside Amalthea, and Metis and Adrastea orbiting within it.

An icy body orbiting closer than about 2.5 Rj would be broken up by tidal forces. A more metallic body would be able to survive to about 1.5 Rj. These ranges are derived from a study by the nineteenth-century French mathematician Edward Roche of how moons might be built up by accretion, but they also define the 'limits' within which a moon cannot stray and survive. The four inner moonlets orbit between these two limits.

Although Amalthea is clearly irregular in shape, and its 270-kilometre long primary axis is tidally locked and aimed at Jupiter, it is too small for much surface detail to be resolved in the best Voyager image, even at its 420,000-kilometre point of closest approach.

The fact that all of the inner moonlets were found to be reddish supported the argument that they have been coated with sulphur from Io's volcanic plumes, ions of which circulate in the Jovian magnetosphere.

On its second orbit of Jupiter, Galileo made its first observations of the ring system and the related moonlets. The next time around, while it was deep within

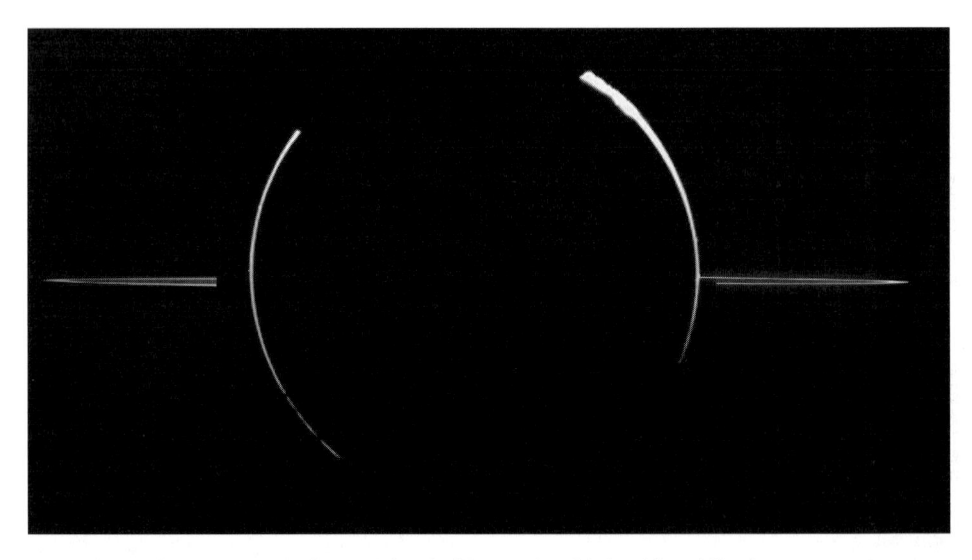

Jupiter's ring system was imaged by Galileo on its third orbit, while the spacecraft was in the planet's shadow. The nearly edge-on ring is visible because its fine dust particles are forward-scattering sunlight. The system is composed of a flat main ring, a toroidal halo interior to the main ring, and a gossamer ring exterior to the main ring. Only the main ring is visible in this short-duration exposure. In fact, this mosaic is a composite, with the arcs of the planet's limb being from a similar image taken by a Voyager.

Jupiter's shadow, the spacecraft observed the main ring as it forward-scattered sunlight. It made another observation on its sixth orbit, and then the infrared spectrometer observed Amalthea and Thebe to determine their surface compositions.

On the eighth orbit, Galileo's ultraviolet spectrometer observed Elara. Discovered in 1905 by Charles Perrine at Lick Observatory, Elara was one of a quartet of moonlets[1]* that orbit out beyond Callisto in a plane inclined at almost 30 degrees to the main system. Although clearly captured asteroids, the fact that their orbits are so similar suggests that they share a common origin. On the tenth orbit, Galileo observed Himalia, another member of the quartet, which was also discovered by Perrine.

After reducing its perijove to facilitate a return to Io, Galileo was able to inspect the inner moonlets in greater detail. The close-ups of Amalthea yielded the clearest look yet of a feature known as Ida. What previously had been resolved only as a blotchy bright spot was seen to be a 50-kilometre long streak. Although this might be the crest of a ridge, it may also be a ray of ejecta thrown out by an impact. Other patches of relatively bright material can be seen elsewhere but none has Ida's linear shape. The nature of the bright patch in Gaea, the crater near Amalthea's south pole, is unknown.

Thebe and Metis were recorded with a resolution of about 2 kilometres per pixel.

* For Notes and References, see pages 413–424.

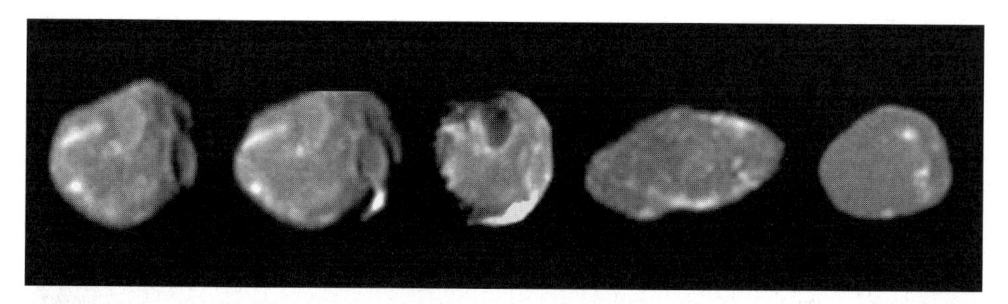

Five views of Amalthea.

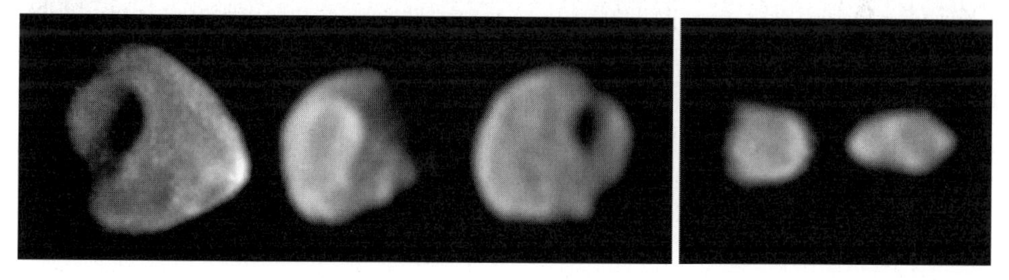

Three views of Thebe (left) and two of Metis (right).

"For the first time, we can start to resolve surface features on them," reported Duane Bindschadler, the science planning and operations chief. "We see things that are distinctly impact craters."

On Thebe, which is only about 60 kilometres across, there is a massive 40-kilometre wide crater, tentatively named Zethus.

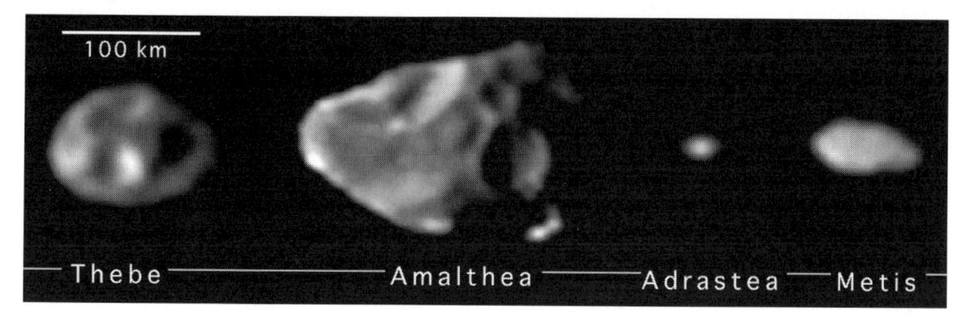

This montage shows the four inner moonlets to scale and in order (Metis orbits closest to Jupiter).

THE RING SYSTEM

The ring system starts at about 92,000 kilometres from the planet's centre, and extends out to about 250,000 kilometres. As Jupiter's radius is 70,000 kilometres, the

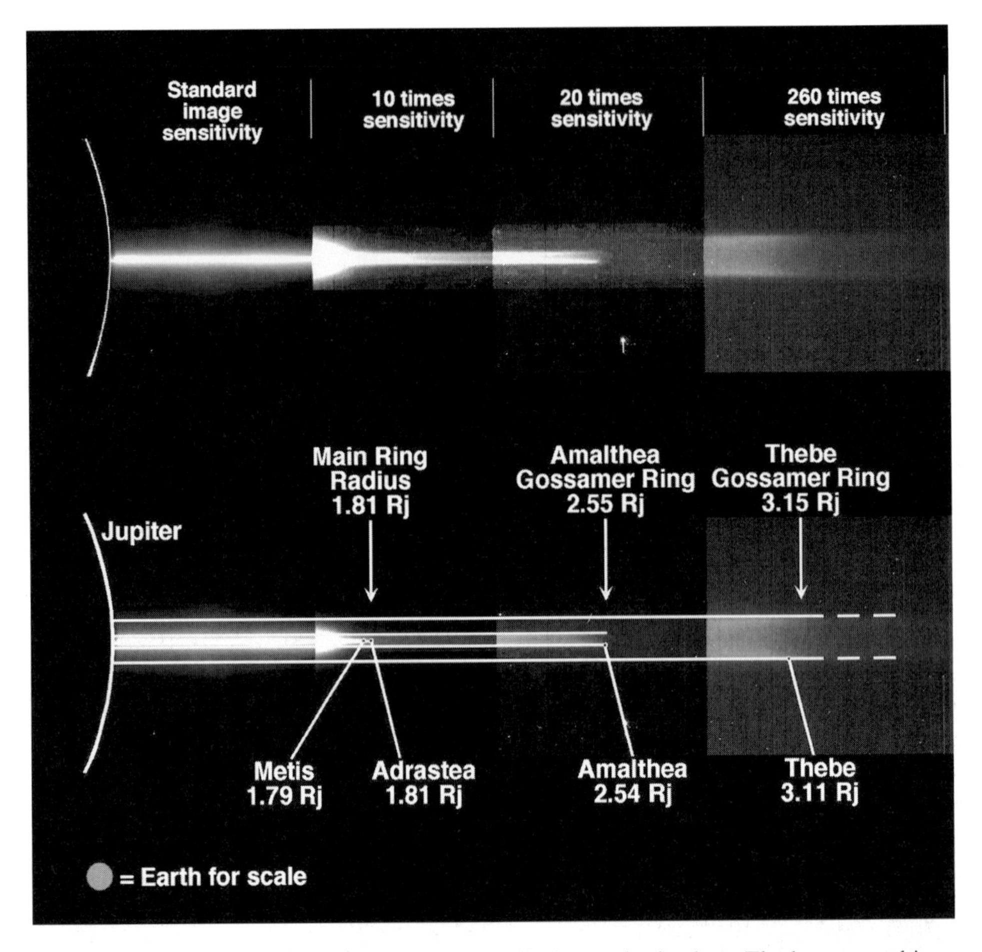

The mosaic is shown twice. The upper panel displays only the data. The lower panel is annotated to show the locations of the ring moonlets, and to show a match between the image and a simple geometrical model of the two gossamer rings. Images at increasing distance from the planet required progressively longer exposure times.

rings are really close in. The Voyagers discovered a flattened main ring, an inner cloud-like ring (dubbed the 'halo') and a hint of a very faint outer ring.

Galileo confirmed the existence of this outer ring – which has been called the 'gossamer' ring due to its translucency. Furthermore, it established this faint outer structure to comprise a pair of rings, one embedded within the other. "The structure of the gossamer ring was totally unexpected," noted Michael Belton, the leader of the solid-state imaging system team, after he had studied the ring system and its relationship to the inner moonlets.

The inner moonlets have "bizarre surfaces of undetermined composition that appears dark and red," said Joseph Veverka of Cornell University, Ithaca, New York, and a member of the imaging team. They are heavily cratered by meteoroid impacts. It is these impacts that sustain the rings. "We can see the gossamer-bound dust coming off Amalthea and Thebe," explained Joseph Burns, also of Cornell.

The vital clue was that the orbits of the outermost pair, Amalthea and Thebe, are slightly inclined to Jupiter's equatorial plane – as is the gossamer ring. Furthermore, not only are the components of the gossamer ring tilted at an angle to one another, their planes precess on a period of months.

The rings are dust kicked up by interplanetary meteoroids which are drawn in by Jupiter's gravitational attraction and accelerated until they are travelling so fast that when they strike one of the small inner moons they penetrate deep into the surface before vaporising. This has the effect of blasting debris away with sufficient energy to escape from the moonlet. If it were not for these small rocky bodies orbiting so close to Jupiter, there would be no rings.

"We now believe it is likely that the main ring comes from Adrastea and Metis," explained Burns. With a diameter of just 8 kilometres, and an orbit which lies just outside the main ring, Adrastea is "most perfectly suited for the job". Both Metis and Adrastea are close to Jupiter's equatorial plane, and, hence, so too is the main ring.

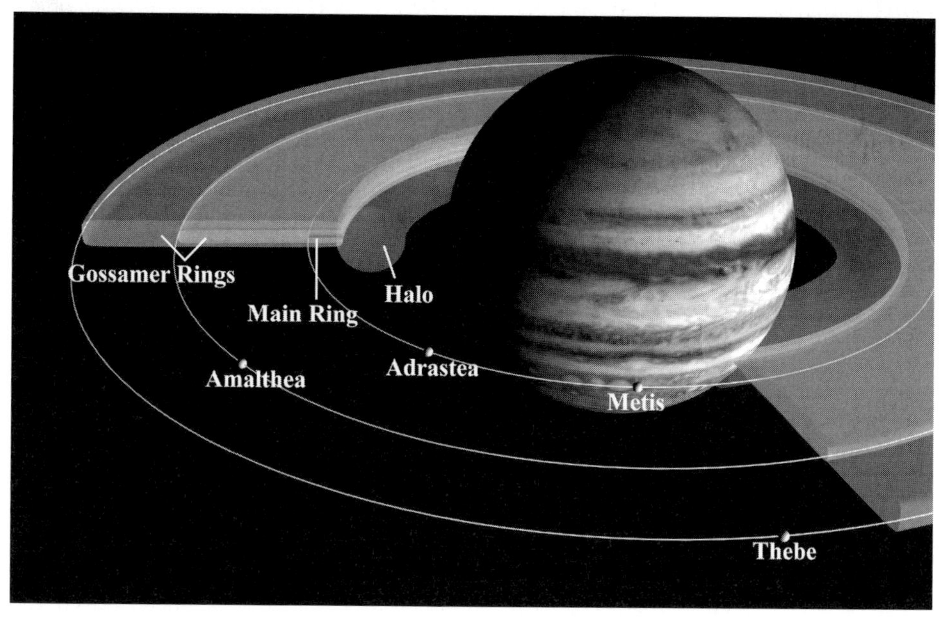

This schematic cutaway shows the geometry of Jupiter's rings in relation to the planet and to the inner moonlets, which are the source of the dust which forms the rings. The innermost and thickest ring is the halo. The thin main ring is bounded by Adrastea and shows a marked decrease in brightness near the orbit of the innermost moon, Metis (a thousand kilometres closer to Jupiter than Adrastea; but this difference is too small to depict). The main ring is composed of fine particles knocked from Adrastea and Metis. Similarly, Amalthea and Thebe supply the dust for the outer gossamer rings (which are thicker than the main ring because the orbits of their source moonlets are inclined to the planet's equator).

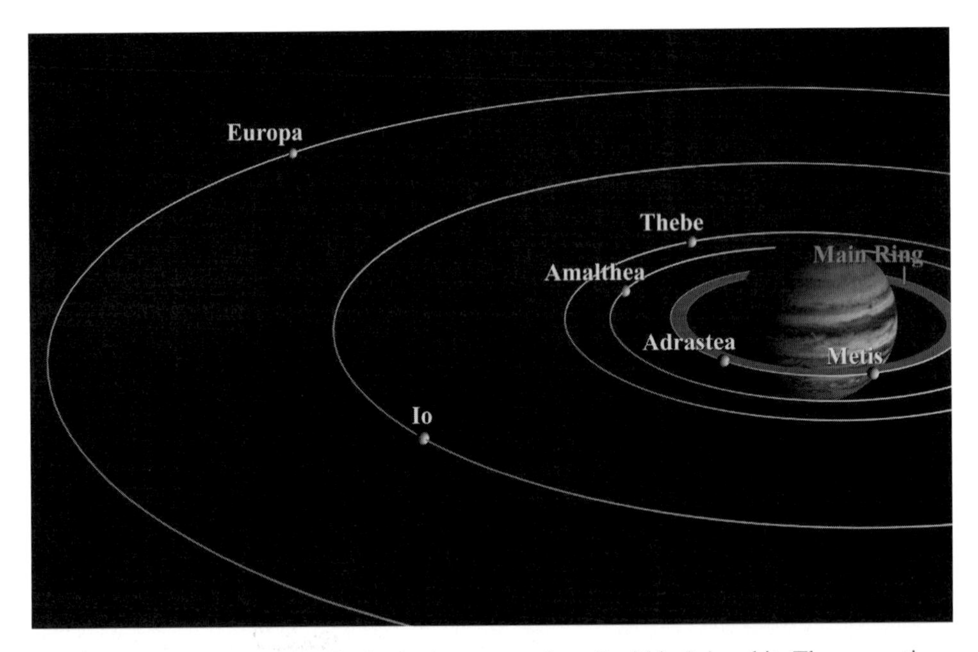

This schematic shows that Jupiter's ring system is well within Io's orbit. The energetic particle instrument on Galileo's amospheric probe measured the radiation in this zone as it fell towards the planet on 7 December 1995. The environment is too harsh for Galileo to venture, so it can study the inner moonlets only from a distance.

Therefore, Amalthea and Thebe produce the gossamer ring, and Adrastea and Metis produce the main ring. The halo is different, however. The main ring is only 30 kilometres thick, but the halo within it extends for thousands of kilometres above and below the equatorial plane. It is believed that the magnetic field in close to Jupiter draws the electrically charged dust out of the disk. The halo, therefore, comprises particles which have been drawn out from the main ring at its inner edge. "We now know the source of Jupiter's ring system," Burns summed up, "and how it works."

'OUTER' RINGS

Are the interplanetary meteoroids accelerated by Jupiter's gravitational field blasting dust off the larger moons? Galileo's dust detector monitored each encounter sequence. "Our *in situ* measurements go one step further than optical studies of the moonlets," explained Harald Krüger of the Max Planck Institute in Heidelberg, "because we can directly study the physical parameters of the ejected particles." It could measure the speed, trajectory and mass of each individual grain. It saw a cloud of dust grains surrounding each fly-by target,[2] with a sharp peak in submicron-sized particles within half an hour of closest approach. The velocity of the dust was only 8 kilometres per second, which was the relative speed of the spacecraft during an encounter, so the dust was travelling with the moon. The gravitational attraction of even a large moon was too weak to have captured the accelerated interplanetary dust, so these grains must have come from the moons, blasted off by impacts.

A detailed analysis established that most of the ejected material pursued ballistic trajectories, and so would fall back within a matter of hours, or days. The rain of meteoroids is ongoing, so this cloud is in a steady-state, finely balanced between the creation and depletion of dust.[3]

The dust detector's data was significant because it provided the first direct measurement of the process inferred to be active on the inner moonlets, which are far beyond Galileo's reach. The projectile and target materials, and the projectile speeds are all relevant parameters in making a model of the process, so the Galileo data provided a significant improvements over laboratory experiments.

Some of the dust from Ganymede has actually made a ring around Jupiter. This is much too tenuous to be detected optically, however. In fact, even the dust cloud surrounding the moon had a density of just one mote within the volume of a cube 20 metres on a side.

The dust detector also found evidence for a faint ring of interplanetary and interstellar dust at a range of 600,000 kilometres. The grains, which were in the size range 0.6 to 1.4 microns – about the size of smoke particles – had been passing through the Jovian system, but were then captured by the planet's magnetic field. "If these particles are just the right size," noted Joshua Colwell of the Laboratory for Atmospheric and Space Physics at the University of Colorado at Boulder, "they lose energy to the magnetosphere and are captured in the ring. Smaller grains are deflected away from the magnetosphere, and larger grains retain enough of their energy to avoid capture." The fact that the dust was in a retrograde orbit indicated that it originated from outside the Jovian system.

Mihaly Horanyi, also of the Laboratory for Atmospheric and Space Physics, has computed that Jupiter's break-up of comet Shoemaker Levy 9 in 1992 should have left behind a thin ring too.

Uranus and Neptune both have systems of thin rings or short ring-arcs, so the knowledge gained of Jupiter's ring system by Galileo will provide insight for comparative studies.

"Rings are important dynamical laboratories to look at the processes which probably went on billions of years ago," Joseph Burns pointed out, "when the Solar System was forming from a flattened disk of dust and gas."

The scientists eager to study how moonlets create and 'shepherd' rings are in for a real treat once the Cassini spacecraft reaches Saturn.

14

Fiery Io

VOYAGER'S IO

The Voyagers' discovery of nine simultaneously active volcanic sites confirmed the study by Stan Peale, Pat Cassen and Ray Reynolds which predicted that Io's interior was heated by extreme tidal stress.[1]* The innermost of the four large satellites, Io is so close in to Jupiter that differential gravity has distorted the moon's shape. The surface of the hemisphere which faces Jupiter forms a 100-metre tall bulge. Cyclical perturbations by Europa (and to some extent by Ganymede) make Io's orbit slightly elliptical. As Io orbits, the fluctuation in the gravitational field induces tidal stresses upon the bulge, and indeed upon the interior of the moon, and this physical stress manifests itself as heat. Such a high degree of thermal differentiation would create a distinct core which, if it was metallic, may well generate a dipole magnetic field. However, the Voyagers did not fly close enough to detect this – if indeed it exists. Certainly, the severe tidal heating has given rise to rampant volcanism.

Brad Smith posited that Io's silicate lithosphere has been transformed into the floor of a global ocean of molten sulphur several kilometres deep, and that the visible surface is actually a layer of frozen sulphur several hundred metres thick capping this ocean.[2] Volcanism is the result of a localised build-up of sulphur dioxide gas within this ocean, which opens a vent in the crust. A plume is powered by explosive venting. In some cases, there is an accompanying extrusion of fluid sulphur. Once the gas pressure is relieved, the vent seals itself. This model explained the sulphurous ocean as a natural result of the extreme thermal differentiation.

There was a clear precedent for this. When the Earth's Moon was accreting, the infall of vast planetesimals kept its outer few hundred kilometres so hot that it formed a magma ocean. As a result of thermal differentiation, lightweight aluminium-based silicates rose to the surface and crystallised to form a crust of anorthosite. Once this had thickened sufficiently to resist further impacts, the tail-end of the Great Bombardment formed the large basins, some of which were subsequently flooded by upwellings of dark lavas.[3] Although the details were

* For Notes and References, see pages 413–424.

different in Io's case, the result was strikingly similar. All of this was disputed by other researchers, of course, but at last there was real data on Io to work with.

IO'S VOLATILES

By radiating energy to the vacuum of space, the intense volcanic activity serves to cool Io from the relentless tidal heating. The rate at which energy is transferred through the crust is called the 'heat flow'. In 1981, Dennis Matson[4] used observations by an infrared telescope to calculate this flow, and found it to be 30 times greater than that of the Earth. By being so very hot, Io is a useful point of contrast for comparative studies of the 'terrestrial' planets.

As far as we can tell, Io is desiccated. Any water that it may have possessed (and being so far from the Sun, it must have originally accreted a fair proportion of ice) has long since been lost to space. The water molecules would have been dissociated by solar ultraviolet, hydrogen would have been swept away, and the oxygen would have been partly recovered by the crust and undergone chemical reactions.

In fact, earlier in its development, while Io still possessed hydrated minerals, volcanic plumes would have been driven by water steam. Geyser activity would have been global in extent, continuous and ongoing for millions of years. When the supply of water was finally exhausted, this activity would have ceased and, denied one means of cooling, the crust would have heated up until the temperature attained the point at which the next most readily vented gas was boiled off. Io has evidently lost its water, nitrogen, carbon dioxide and neon. Io it were not for tidal heating, Io would undoubtedly have retained these volatiles and may well have resembled ice-encrusted Europa. Alternatively, if Europa was more intensely heated, it would boil off its ice and become like Io.[5]

SULPHUR CHEMISTRY

At the current time, Io is boiling off its sulphur dioxide. As sulphur dioxide in a volcanic plume condenses and falls back, it 'paints' a frosty halo on the surface around the vent. Other volatiles, including gaseous sulphur, also condense out and create a chemically diverse blanket of pyroclastic.

Any sulphur atoms will join with other sulphur to form molecules of varying numbers of atoms; that is, 'allotropes' of sulphur. The length that a chain of sulphur atoms can achieve is temperature dependent. As temperature is a measure of kinetic energy, an allotrope of a specific length will be literally shaken apart above a given temperature.

At 'room temperature' on Earth, sulphur is a pale yellow solid. If heated to 400 K, it melts and turns orangy. Also, as the temperature is increased, the viscosity and colour of the liquid change. At 435 K it turns clear pink, and at 465 K it becomes viscous and turns red. A transition in the molecular structure occurs at 500 K, and it develops a black tarry constituency. It begins to lose its viscosity at 600 K, and at

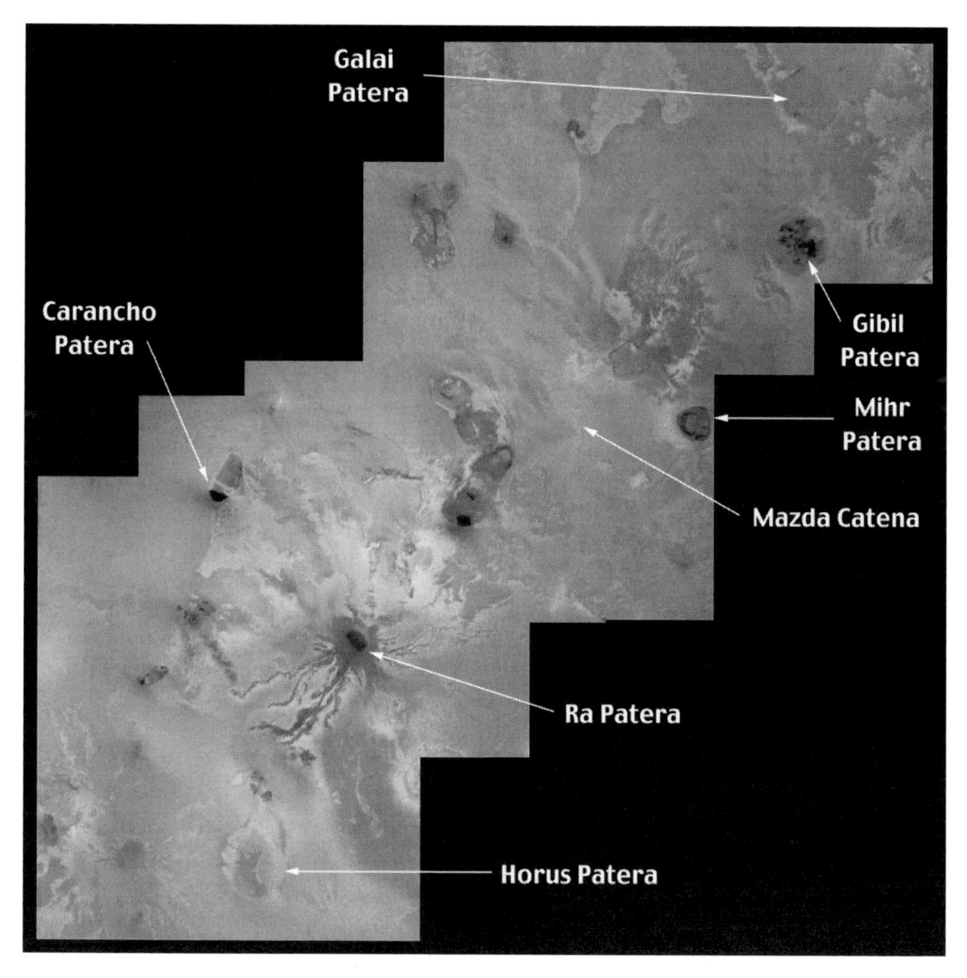

This Voyager mosaic shows Ra Patera – one of Io's largest shield volcanoes – at the centre of radiating lava flows.

650 K it is a dark runny fluid. It reaches its minimum viscosity before it vaporises. The boiling point is dependent on pressure – it will boil off at 715 K at 1 bar, but it will do so at 450 K at 0.001 bar. This, of course, applies to pure sulphur; the presence of impurities will raise the boiling point at any pressure.

As the colour and viscosity of sulphur evolve as it is heated, they also do so as it cools. Evidently, the 'pizza-like' surface represents subtle sulphur chemistry. As a hot dark flow cools, the colour changes, and it becomes viscous and slows, which correlates with the fact that the material in the calderas and in close to the vents is dark, and the variation in hues further out corresponds to material of different compositions, ages and temperatures. In effect, Io is being continuously resurfaced with fluid sulphurous flows and sulphur dioxide pyroclastic.

An analysis of Io by Alfred McEwen and Laurence Soderblom[6] concluded that the plumes are driven by explosive gas venting, and that fluidic 'lava' extrusions are less violent events. This combination of processes appeared to be sufficient to explain the observed structure of a large vent like Ra Patera.

With such clear evidence of sulphurous volcanism, geologists began to doubt whether Io underwent terrestrial-style basaltic volcanism, as the temperatures needed to melt silicate rock are significantly higher than will boil sulphur.

However, as photo-geologists studied the fine detail of the Voyager imagery, they found contradictory evidence. Michael Carr,[7] for example, noted that limb views showed mountains. Haemus Mons, prominent in the south polar region, spans 200 kilometres across its base, and rises about 10 kilometres above the surrounding plain. How was this constructed? It does not seem to be a volcanic edifice. It displays the appearance of being a fractured crustal block that has been shoved up by tectonic forces – a massif. If the observable surface is just the frozen cap on a sulphurous ocean, are the peaks deep-rooted structures protruding through it? Or are the sulphurous flows only a veneer on a silicate lithosphere which has undergone fracturing and uplift? Furthermore, the large vents have steep cliff-like walls plunging several kilometres to caldera floors. If these volcanic complexes were sulphur constructs, the caldera walls would have slumped.

The issue of silicate volcanism remained unresolved, however, since the Voyager infrared spectrometers had found no direct evidence of the temperatures required to melt rock. The Pele and Loki flows were measured to be 425 K warmer than their immediate – freezing – surroundings. However, the spatial resolution of the instrument was low, and the readings were averaged over wide areas. Although the derived temperature was consistent with sulphur chemistry, it was not possible to rule out the possibility that there was molten rock in localised areas within the instrument's field of view. In 1986, infrared telescope observations by a team led by Torrence Johnson increased the peak temperature to 900 K, which was sufficiently hot to demonstrate that silicate volcanism was active at least on an occasional basis. Hopefully, Galileo's higher resolution instruments would determine whether Io's surface was primarily basaltic in nature.

PHYSIOGRAPHY

Between them, the Voyagers recorded about 35 percent of Io's surface, covering the south polar and equatorial zones at a resolution of at least 5 kilometres; the best coverage was on the sub-Jovian hemisphere between 275 and 360 degrees longitude at a resolution of 500 metres per pixel.

Utilising the principle of superposition, Gerry Schaber, of the US Geological Survey in Flagstaff, Arizona, defined three main physiographic units[8, 9]:

- mountain materials
- various plains materials
- vent-related materials.

The art of photo-geology involves four stages: identify the various geological units

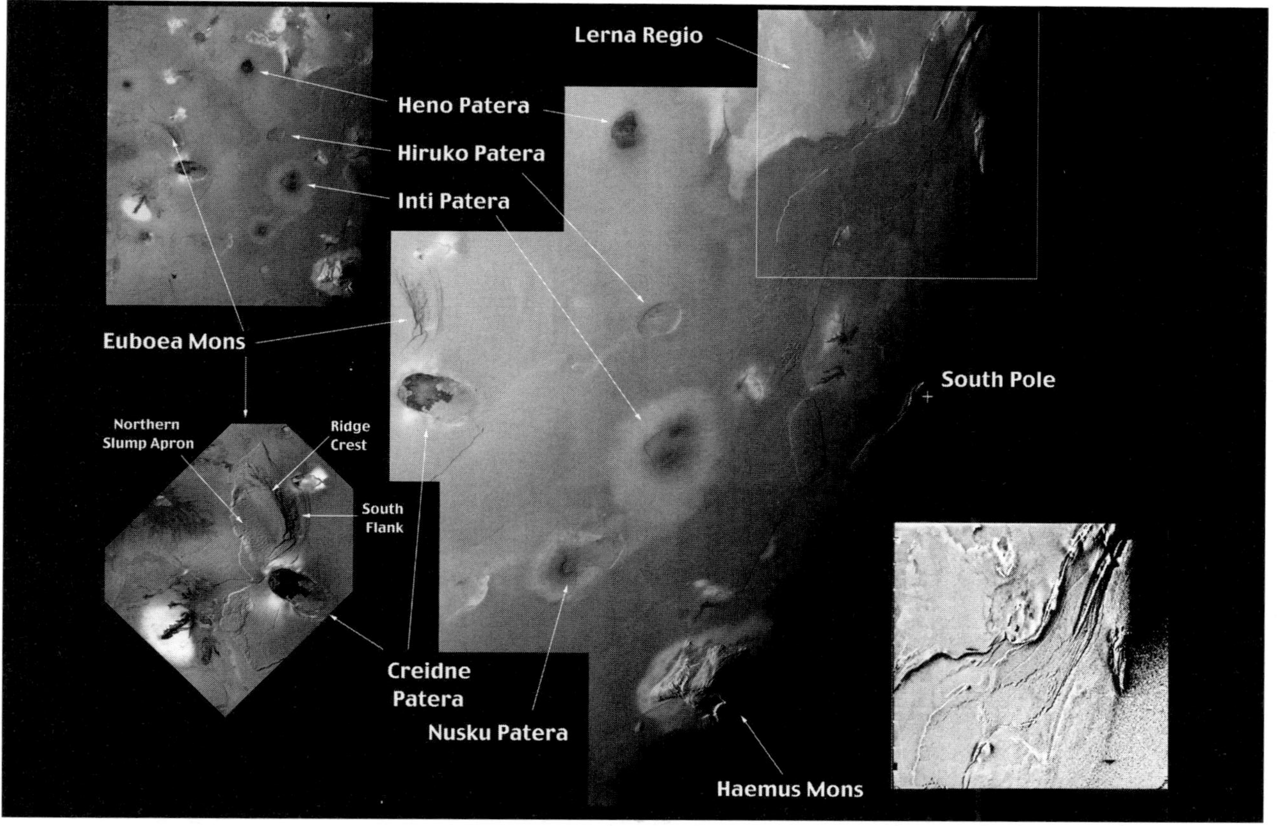

This montage shows the terrain near Io's south pole. The sub-Jovian hemisphere is lit, the anti-Jovian hemisphere is in darkness. In addition to volcanic calderas, this region includes several mountain blocks. Haemus Mons is well illuminated on the terminator. An analysis of Euboea Mons under more favourable lighting (lower left) shows that a linear ridge divides a steep slope on one side and a gentle incline on the other, and that there are terraces suggestive of slumping. The flat-layered plain of Lerna Regio is shown in the computer-processed insert (lower right).

in terms of their surface characteristics; chart the distribution of each unit; determine the relative sequencing; and then formulate a model to explain their origins. While mapping can be done fairly objectively, the act of interpretation is subjective, and thus contentious, so there can be many interpretations of the same raw data – witness, for example, the debate over silicate versus sulphurous volcanism.

Impact craters provide an excellent means of determining the relative sequence of deposits, but craters on Io are notable for their complete absence. Torrence Johnson[10] calculated that Io's volcanoes eject 10 billion tonnes of material per year, and if this was averaged over the whole moon, it would produce a 100-metre thick blanket every million years. This rate of resurfacing is more than sufficient to smother the impact craters which must be made from time to time. Nevertheless, localised superposition relationships can be inferred from cross-cutting features and embayments of the mountains.

The 'mountain material' is surrounded by, and stands above, the material that surrounds the lower slopes. Individual mountains are 100 to 200 kilometres across at the base. Euboea Mons in the southern hemisphere – near Creidne Patera and Haemus Mons – is one of the tallest examples; it rises 13 kilometres above the plain.[11] Although the absolute ages of the mountains are unknown, the superposition relationships indicate that they are the oldest exposed features.

The term 'plains material' was introduced as a generic name for several types of flat terrain which is collectively the most widespread of the three physiographic units. Plains material in the polar areas tends to be dark brown to black, but that at lower latitudes is predominantly red-yellow. The 'inter-vent plains' are characterised by smooth surfaces of intermediate albedo and in some cases are crossed by low scarps. They were interpreted as plume fall-out interbedded locally with flows and fumerolic materials. The 'layered plains' are broad flat surfaces crossed by grabens and prominent scarps. The most conspicuous layered plains were observed in the southern polar region, but the northern polar zone was not well observed by the Voyagers. Schaber interpreted the multi-coloured surficial patterns as anhydrous mixtures of sulphur allotropes, sulphur dioxide frost and sulphurous salts of sodium and potassium.

A study of the scarps and other relief features by Jack McCauley[12] found evidence of local undercutting, slumping, and etching due to venting of sulphur dioxide from a shallow depth. If there are fractures, liquid sulphur dioxide is able to rise. As it approaches the surface, and the pressure decreases, it will sublimate to gas while still underground and then vent explosively, thereby 'eroding' the surface manifestation of the fracture. The sulphur dioxide will promptly condense, and fall nearby as white frost (this would account for numerous bright patches seemingly not associated with large central vents). An undercut scarp would then slump. A sudden release of gas could even produce a collapse structure resembling a crater, as occurs on Earth when explosive venting of steam produces a marr.

Schaber concluded that the transection by faults, overlapping deposition and 'erosional' features indicated a complex geological history. The vertical scale of the scarps – some were almost 2 kilometres high – implied that there is an underlying silicate lithosphere at a shallow depth, as Carr had proposed.

The 'vent-related materials' are directly derived from the primary fissure, or caldera, of a volcanic site. A comprehensive analysis of the Voyager imagery identified at least 300 vents. Most of the calderas are about 40 kilometres across, but Loki, one of the largest, is just over 200 kilometres wide. The most prominent volcano, by virtue of its large plume halo, is Pele. The depressed floors of the calderas are dark, suggesting that they contain 'lakes' of molten sulphur – which would be black. In the four months between the two Voyagers, parts of the 'lava lake' in Loki appeared to have crusted over, and other areas seemed to have opened new vents. Numerous older flows are in evidence, ranging from vast coalescing units up to 700 kilometres long to sinuous flows radiating from vents for up to 300 kilometres. And, as Carr had pointed out, some caldera walls are cliffs almost 2 kilometres high. This structural integrity implies that the underlying vents are based in a silicate lithosphere. However, vents initially opened by silicate volcanism may well subsequently be exploited by sulphurous volcanism. A number of kinds of volcano were distinguished, including 'shields', 'discoids', and large central features which had given rise to multiple – 'digitate' – flows.

Although the vents appeared to be the result of 'hot spots', their random distribution did not suggest the presence of large-scale convection cells in Io's

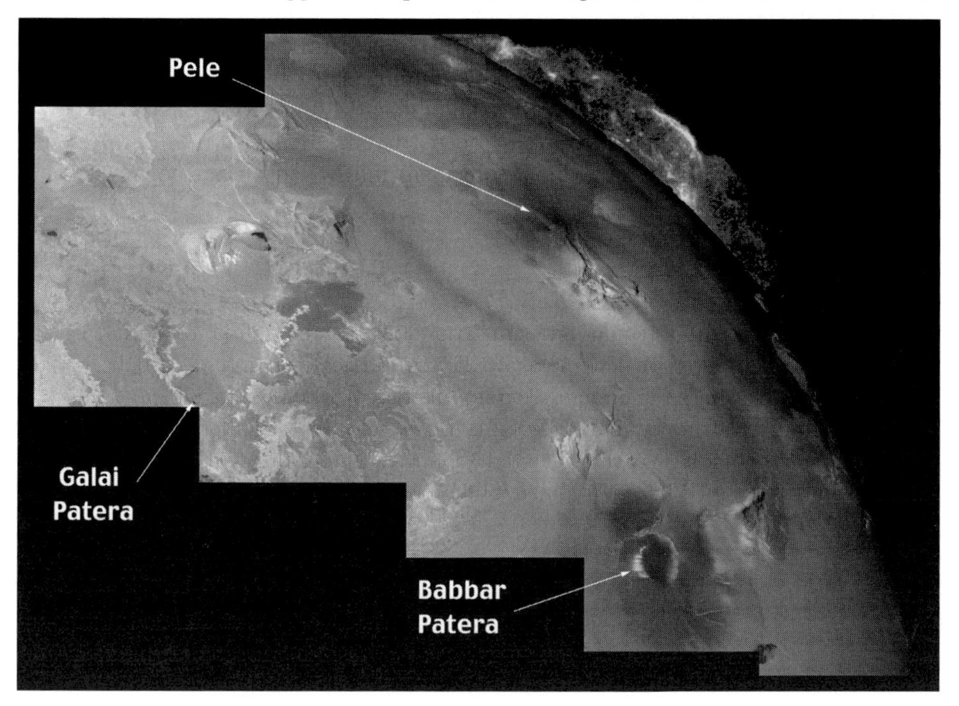

This Voyager mosaic shows the volcano Pele in eruption. Although the plume appears to be over a point nearer the horizon, it is actually centred over the vent (arrowed). It is just more difficult to see when projected against the surface. The plume rises several hundred kilometres. Notice that the pyroclastic fall-out from the plume has 'painted' a dark ring on the surface forming a 1,000-kilometre diameter halo.

deep interior. Nor were there any patterns of activity suggestive of plate tectonic processes at work.

A comprehensive analysis by Susan Kieffer[13] determined that Io's plumes are very similar to terrestrial water geysers. Her model showed that the sulphur dioxide in an energetic plume derived from a depth of several kilometres, where it would be heated to 1,400 K upon coming into direct contact with silicate lava. Once boiled under pressure, the gaseous sulphur dioxide would force open pre-existing cracks in the brittle crust and rise without transferring significant heat to its conduit – that is, adiabatically – to vent explosively upon reaching the surface.

Despite the supposition that the volcanic activity is driven by the build up of gas pressure, no explosive 'Plinian'-style enormous ash plumes have been observed on Io.

As Galileo approached Jupiter, therefore, the popular mood was that the plumes were well understood, so the spacecraft's task would be to determine whether the volcanic flows were primarily of a silicate or sulphurous nature.

GALILEO'S FLY-BY

Prior to Voyager, it was expected that the solar wind would penetrate the 'polar cusps' of the Jovian magnetosphere – as happens in the Earth's magnetosphere – and give rise to auroral displays, and indeed, Voyager 1 detected an aurora which extended for 30,000 kilometres in an arc around Jupiter's north pole. During this time, the plasma wave instrument noted radio emissions from Io's torus, and the 'whistling' resembled that created by terrestrial aurora. It was soon realised, however, that the Jovian aurorae are not produced by the solar wind, but by the presence of volcanic Io deep within the magnetosphere. Furthermore, the moon is electrically linked to the planet by a pair of 'flux tubes' which run down the magnetic field lines to Jupiter's polar regions.

The 3 May 1996 issue of the journal *Science* contained papers reporting upon Galileo's fly-by of Io during its approach to Jupiter on 7 December 1995. Although the scan platform's imaging instruments had been inert for this encounter, the particles and fields instruments had taken data. Also, the Deep Space Network had measured the doppler on the spacecraft's radio signal, both to track the spacecraft and to survey the moon's gravitational field.

The energetic particle detector, which was part of the particles and fields suite, sampled the charged particles in the spacecraft's environment. Passing Io, it detected bidirectional electron flows aligned with the Jovian magnetic field lines, making the first *in situ* observation of the flux tubes. This flow of particles constitutes an electrical current of several million amps with an overall energy deposition into the Jovian atmosphere of a trillion watts – this makes it the most powerful 'direct current' in the Solar System. "These 'beams' are a remarkable particle accelerator," reflected the instrument team's chief, Donald Williams of the Applied Physics Laboratory at Johns Hopkins University, Baltimore, Maryland. The flux tubes carry sufficient energy to generate visible auroral emissions at the 'footprints' (one in each polar zone) where they impinged upon the atmosphere.

While surveying the inner region of the Jovian magnetosphere and Io's plasma torus, the magnetometer made a significant discovery concerning the moon itself. "Instead of increasing continuously as the spacecraft approached Jupiter," Torrence Johnson explained, as it passed Io, "the magnetic field strength took a sudden drop of about 30 percent." It was as if Io resides inside a 'bubble' within the Jovian magnetosphere. A hole in the magnetosphere surrounding Io strongly suggested that the moon has an intrinsic magnetic field strong enough to shield the moon from the intense Jovian field. However, it was not immediately evident how Io came to generate such a strong magnetic field.

"It's an astonishing result," reflected Margaret Kivelson of the University of California at Los Angeles, and the chief of the magnetometer team. "Completely unexpected." The result had to be 'preliminary', however, because it was based on the low time-resolution data transmitted in near-real time. The high time-resolution version – together with that from all the other physics instruments – was not due for replay until early June 1996. Because the Io fly-by had been a fairly remote 1,000-kilometre range, it would be difficult to isolate the moon's intrinsic field – if it had one – from fields generated by the torus and the flux tubes; but if it could be confirmed it would make Io the first planetary satellite to be proved to possess an intrinsic magnetic field.

As it passed by Io on the way in, Galileo's plasma wave spectrometer found "a very dense cloud of ionised oxygen, sulphur and sulphur dioxide which must have been pumped into that region by Io's relentless volcanic activity", noted Lou Frank, chairman of the Magnetosphere Working Group. The fact that these ions had not been swept up by the Jovian magnetosphere supported the case for Io having a magnetosphere of its own. The Pioneer 10 occultation had indicated the presence of an ionosphere ranging out to about 100 kilometres. "Passage of the spacecraft through an ionosphere was unexpected." Images from the Voyagers indicated that the plumes rose only to a few hundred kilometres. However, the observable structure was the sulphur dioxide condensation; gases that did not condense would rise higher. The environment around Io was evidently time-variable and it would be one of Galileo's tasks to characterise this variability.

As Torrence Johnson summarised at 'The Three Galileos' conference in January 1997, in presenting the early results: "There's still debate over whether the observed magnetic signature is due entirely to the interaction of Io with the plasma flow in the magnetosphere, or whether there is some contribution from an intrinsic field from Io itself." The only way to clarify this situation would be to return to Io and make a much closer fly-by. With a little luck – if Galileo survived its primary mission – this would be possible.

The Jovian magnetic field turns with the planet every 10 hours. If Io really does possess a dipole field strong enough to ward off the Jovian magnetosphere, its surface will be shielded from the charged particles circulating in the magnetosphere. Furthermore, the dust ejected into space by the volcanic plumes will not be caught by the Jovian field unless it emerges from the 'bubble'. But the Io fly-by provided evidence that Io is indeed the source of the high-velocity dust streams which both Ulysses and Galileo had detected millions of kilometres from Jupiter. "These dust impacts continued up to the time of Galileo's Io fly-by and then ceased," reported Eberhard Grün of the Max Planck Institute for Astrophysics in Heidelberg, Germany, and the leader of the dust science team. "My preliminary interpretation is that Io is – in some way – the source of the dust streams." As Galileo performed its orbital tour, the dust detector was to study the characteristics of the dust within the Jovian system, and so reveal the mechanism by which it was accelerated to escape velocity. A study by Amara Graps, also of the Max Planck Institute, identified a 42-hour periodicity in the density of the streams. This was the period of Io's orbit, and was therefore proof that the dust was of volcanic origin.

Monitoring the doppler on the radio signal throughout the Io fly-by measured the moon's gravitational field, which in turn meant that the moment of inertia (a key dynamical parameter) could be computed. The low value indicated that Io is not a homogeneous sphere.[14] It must possess a two-layer structure, with a dense metallic core (which is most likely iron and iron sulphide) englobed by a partially molten silicate mantle over which there is a volcanically active crust. The data showed that the core would be 52 percent of the moon's radius if it was iron mixed with iron sulphide, and only 36 percent of the radius if it was iron alone; but even this would represent a "giant iron core". Such a core could not have formed as Io was accreted from the solar nebula, it must represent billions of years of thermal differentiation sustained by the tidal heating.

The first Io images were taken on 25 June 1996 from a range of 2.25 million kilometres, while Galileo was inbound to Ganymede. The resolution was only 25 kilometres per pixel at the 972,000-kilometre closest approach, so Galileo would not be able to rival the imagery from its predecessors. Accumulating observations over a few years would hopefully identify patterns of activity, however. One advantage that Galileo had over the Voyagers was that its solid-state detector technology was sufficiently sensitive to measure the temperature of Io's surface in Jupiter's shadow, when it was radiating thermal emission rather than reflecting sunlight. This was also the first opportunity to observe Io in the near-infrared to chart the composition of its surface.[15]

LOKI AND PELE

Galileo's first task was to catalogue the extent to which the volcanically active moon had changed since 1979. "The changes we're seeing on Io are dramatic," reported Michael Belton of the National Optical Astronomical Observatories in Tucson, the leader of the imaging team. "The colours of material on the ground, and their distribution, has changed substantially."

However, the Loki caldera had barely changed. This was a surprise because Loki is one of the most persistently active sites. According to Alfred McEwen of the University of Arizona in Tucson, and a member of the imaging team, Loki accounts about 25 percent of Io's global energy flow. In fact, its energy output corresponds to 100,000 times that of Yellowstone, Wyoming, which is one of the largest terrestrial resurgent calderas. Diana Blaney of JPL had utilised the Infrared Telescope Facility

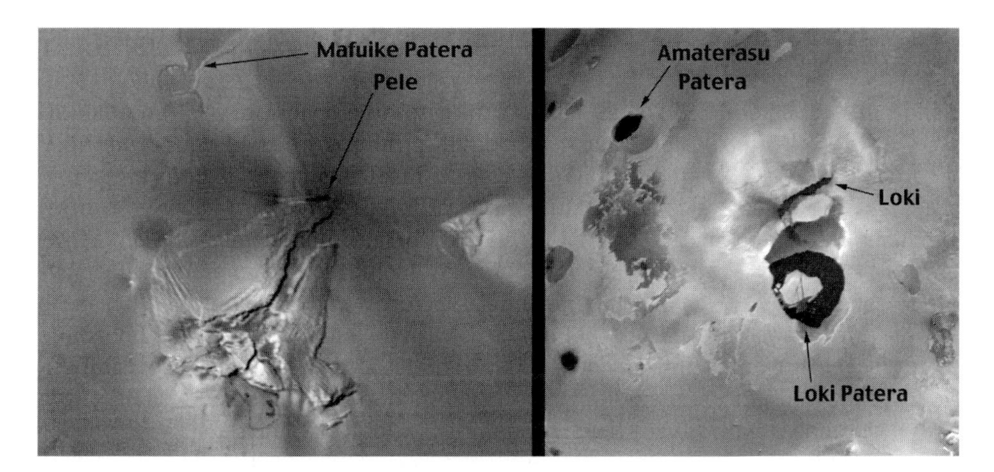

These Voyager images show Pele (left) and Loki (right). Notice that Pele's plume has made a distinctive pattern, and that the mountain to the southwest is similar to Euboea Mons (see earlier figure) with steep slopes separated from gentle slopes by ridges, and terraces that suggest slumping. At 200 kilometres in diameter, Loki's caldera is one of Io's largest. The white 'island' within the caldera is an elevated plateau-like block. The dark linear feature to the north of the caldera was the source of the plumes observed by the Voyagers (although it had switched from one end to the other in the few months between their fly-bys). This is evident from the fact that the white plume pyroclastic is centred around this feature rather than the caldera.

in Hawaii to monitor Io in the decade leading up to Galileo's arrival, and established that both Loki and Pele undergo periods of intense activity, interspersed by periods of relative quiescence. The debate concerning whether Loki produced sulphurous or silicate lava had yet to be resolved.[16, 17, 18, 19]

Pele was of historical significance because it was the volcano whose plume Linda Morabito had spotted in a Voyager 1 image in March 1979, and thereby discovered that Io is volcanically active. When Voyager 2 had flown by several months later, however, Pele had been inactive. When Galileo made its first observation, it saw an intense hot spot in Pele's caldera, but no evidence of a plume.

RA PATERA

Ra Patera is a shield volcano near the sub-Jovian point, with colourful lava flows ranging up to 300 kilometres from its dark central vent. It had not been active in 1979. When the Hubble Space Telescope inspected Io in July 1995, it found that Ra Patera had laid down a 350-kilometre wide area of yellowish-white pyroclastic since its last observation, in March 1994,[20] but its resolution was insufficient to study this deposit in detail. Galileo was to have observed Ra during its approach to Jupiter in December 1995 but the problem with the tape recorder had precluded this, so Ra was a high priority for the spacecraft's first orbit.

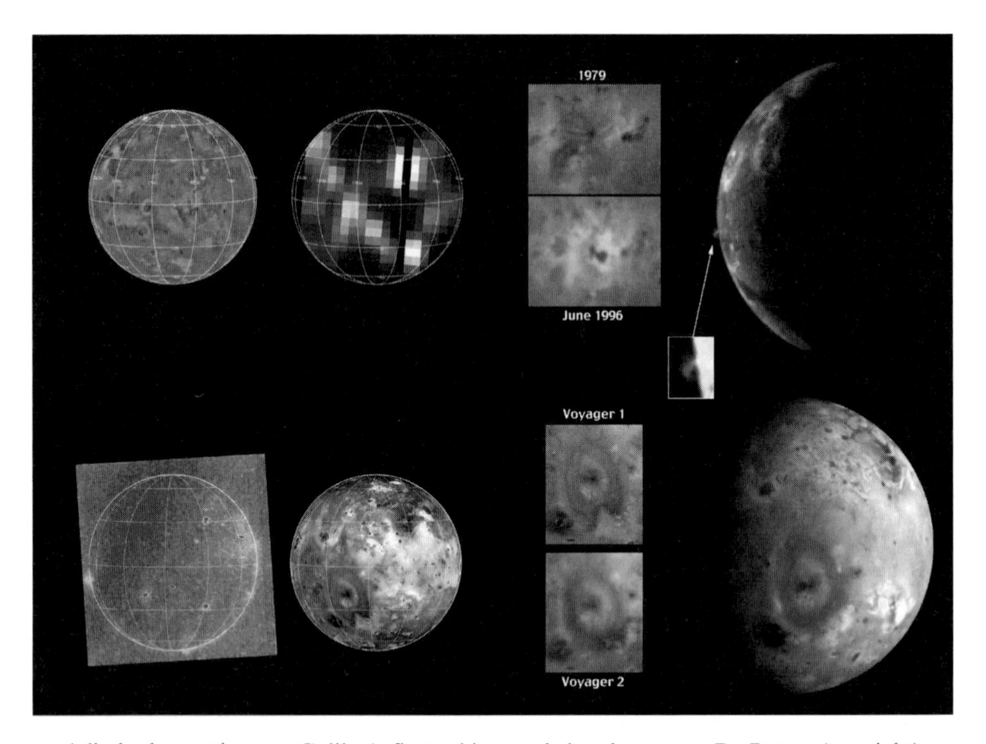

A limb observations on Galileo's first orbit revealed a plume over Ra Patera (top right). The vent may well have been active since its eruption in mid-1995. Galileo imaged this site from a distance to determine how it had changed since 1979 (centre right). It had a new lava flow (dark) arcing around to the southeast. Galileo also documented changes to Pele (bottom right). Between the Voyager fly-bys, the distal (that is, remote from the vent) plume deposits changed markedly, restoring the southern section of the halo. The site had not changed much since, however. The solid-state imaging system was able to image Io in Jupiter's shadow (bottom left) and observed volcanic hot spots and auroral glows. Although the infrared spectrometer also observed Io in eclipse (top left) it had a lower spatial resolution. Even so, it identified several new hot spots. A corresponding hemisphere from the Voyager era is shown for each eclipse observation for reference.

Galileo saw a 100-kilometre tall plume from Ra Patera poking over the limb. A comparison of Voyager and Galileo imagery revealed that this had resurfaced an area of 40,000 square kilometres. The ultraviolet spectrometer observed the plume glowing in the dark (possibly the result of fluorescence of sulphur and oxygen ions; the energetic particles circulating in the Jovian magnetosphere first dissociate the molecules, and then induce them to fluorescence).

"This is very different from what we see with volcanic eruptions on Earth," pointed out Torrence Johnson. "Terrestrial eruptions cannot throw material to such high altitudes." The plumes are able to rise so high, and to paint such a wide area, because the moon has a weak gravitational field. In fact, Io is only a little larger than

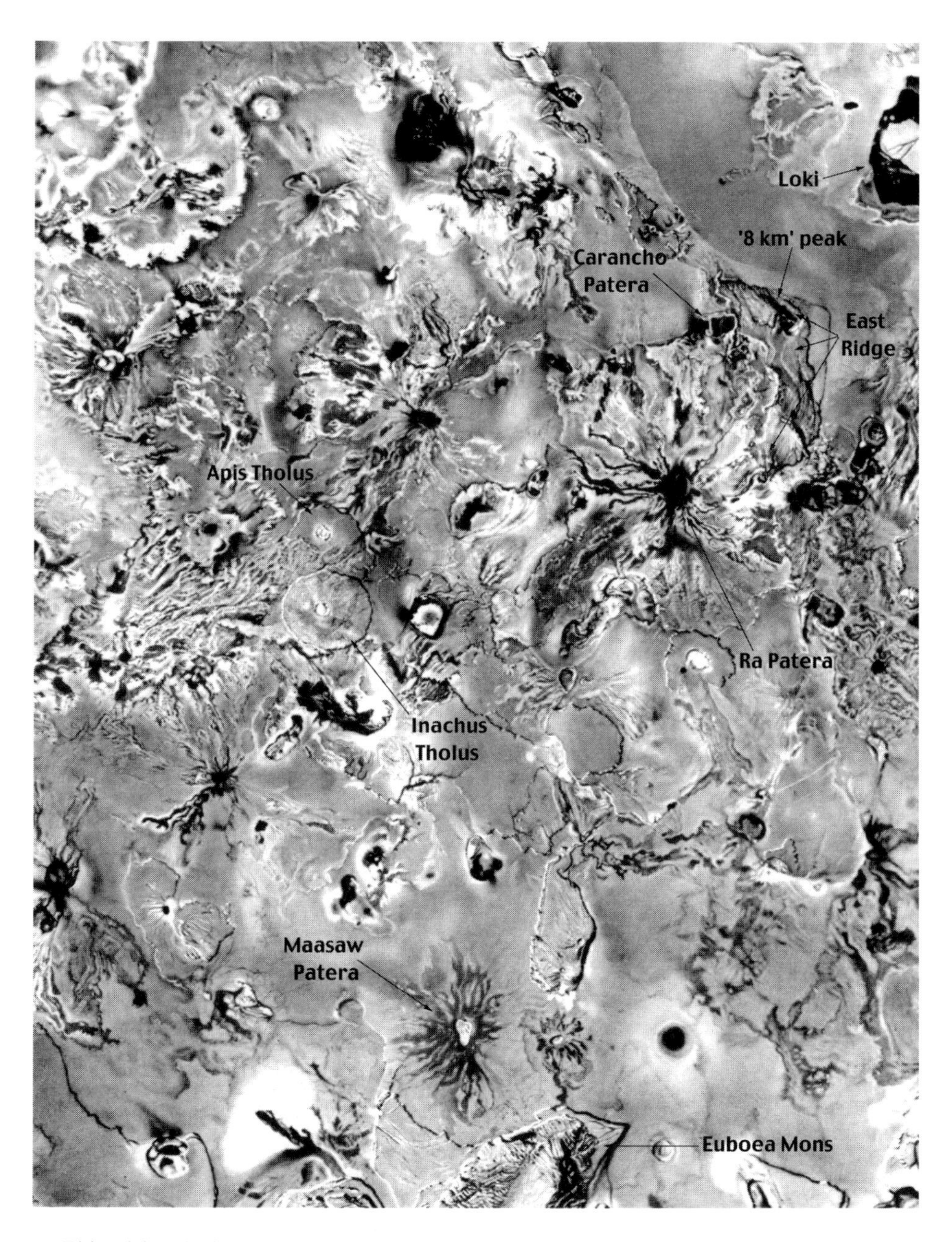

This airbrushed map based on a Voyager mosaic highlights the intensely volcanic landscape southwest of Loki. Many of the vents – most notably Ra Patera – show radial lava flows, but other calderas appear 'clean'. The 'domes' of Apis and Inachus Tholus are the nearest thing to terrestrial-style volcanic mountains. (Courtesy USGS Astrogeology Branch)

the Earth's Moon, on which there are patches of dark pyroclastic from ancient 'fire fountains', but such activity is localised on the Earth due to its stronger gravity and the resistance of its dense atmosphere.

The yellow surface flows visible in the Voyager imagery indicated that Ra Patera extruded sulphurous lava.[21, 22, 23, 24] The new flow extending off to the southeast of the caldera was dark,[25] so if it was sulphurous it had to be still warm. However, although the plume indicated that the site was still active, Galileo did not detect a hot spot on the surface.

This eruption was studied in detail by processing Voyager imagery to create a topographic map of the area.[26] At 450 kilometres across, Ra Patera is Io's largest 'shield' volcano, but its summit rises less than 1 kilometre above the surrounding plain. As a result, its lavas have run down shallow gradients ranging from 0.3 to 0.1 degrees. To flow so readily, the lava must have been of extremely low viscosity. The low shields and overlapping flows of the Snake River Volcanic Plain in Idaho may be a close terrestrial analogue for this area of Io.[27, 28]

The stereoscopic analysis identified a 600-kilometre long mountainous feature starting east of Ra Patera and arcing around to the northeast. About 50 kilometres east of Ra, this forms a 60 × 90-kilometre plateau some 500 metres high, but as it runs north it progressively rises to the 8-kilometre tall peak situated just east of the Carancho Patera. The most recent lava flow followed this plateau's southern periphery.

VOLCANO MONITORING

The lack of plumes was surprising. There were only two, possibly three. The observations of Io in eclipse showed seven hot spots and diffuse glows from plumes, so it was decided that this aspect of the volcano-monitoring campaign should be increased on subsequent orbits. The plume deposits at Aten and Surt, which had been emplaced in the brief interval between the two Voyager fly-bys, had faded somewhat, but the correlation between sites of 'recent' activity and the dark and red materials was confirmed.

As Galileo made its way back in on its second orbit, it resumed Io volcano monitoring. At its closest point of approach – 440,000 kilometres – it took a succession of limb images as its viewing angle advanced by 10 degrees of longitude in order to conduct a 'plume inventory'. It saw plumes over from Prometheus and, possibly, nearby Culann Patera, but that was all.

A regional image of the anti-Jovian hemisphere confirmed that Prometheus had changed since 1979; in fact, it had extruded a long dark lava flow. Furthermore, the plume was rising from the 'flow front' and so was about 100 kilometres west of where it had been previously.[29] The migration of the plume site would have to be accounted for by the model which explained plumes in terms of explosive venting of volatiles boiled deep in the crust. Had the gas found a more convenient route through the crust?

Pillan Patera's caldera had darkened since the first orbit, and the infrared

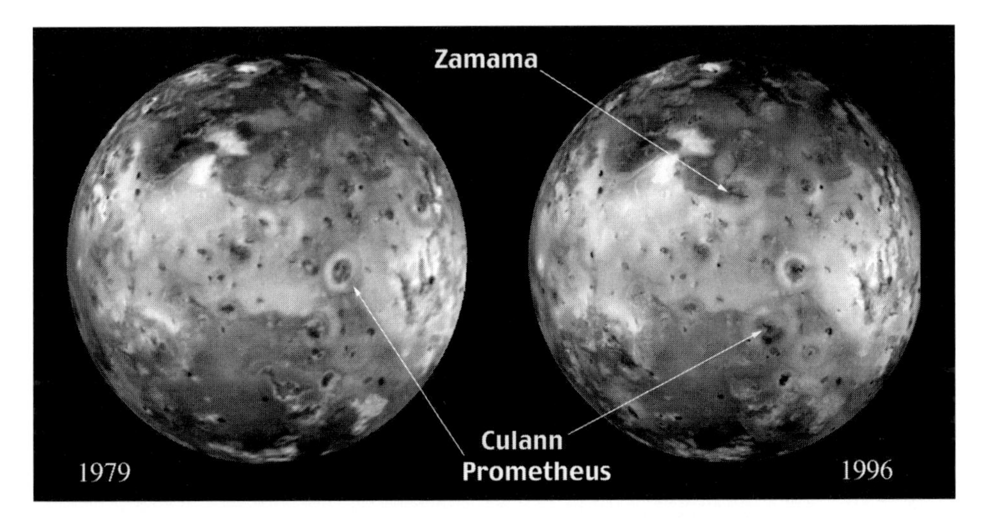

On its second orbit, Galileo documented the anti-Jovian hemisphere to identify changes since 1979 – there were marked changes at Prometheus and Culann, and the Zamama flow had appeared.

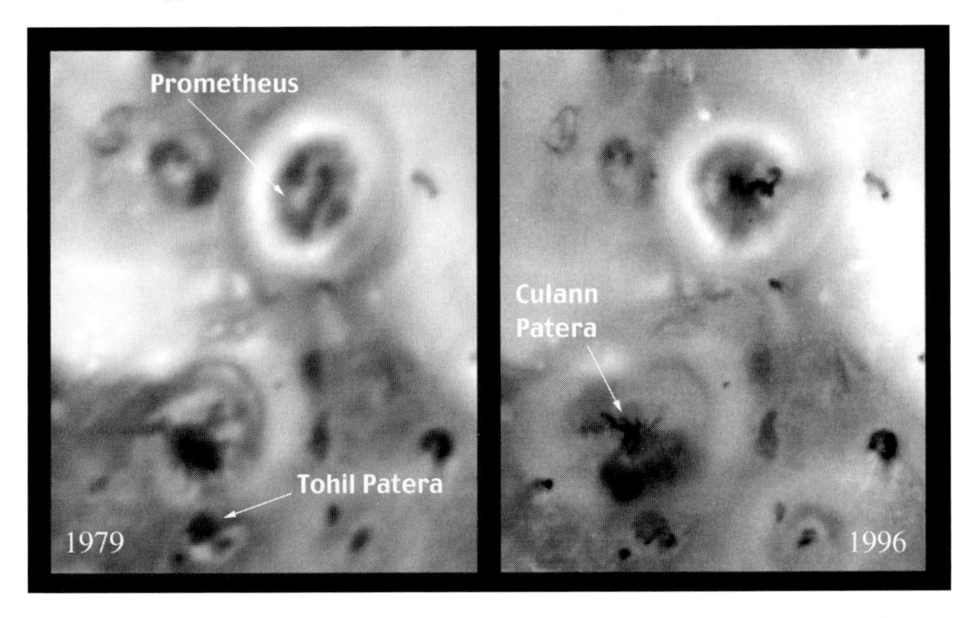

The changes to Prometheus and Culann since 1979 are shown in greater detail in these images (the Galileo image was taken on the second orbit).

spectrometer saw it as a hot spot.[30] Zamama showed considerable surface changes since 1979, having produced many intricate flows. It had been detected as a hot spot during the previous orbit. The infrared spectrometer found the background temperature on the dark hemisphere to be 80 to 85 K, but there were localised hot

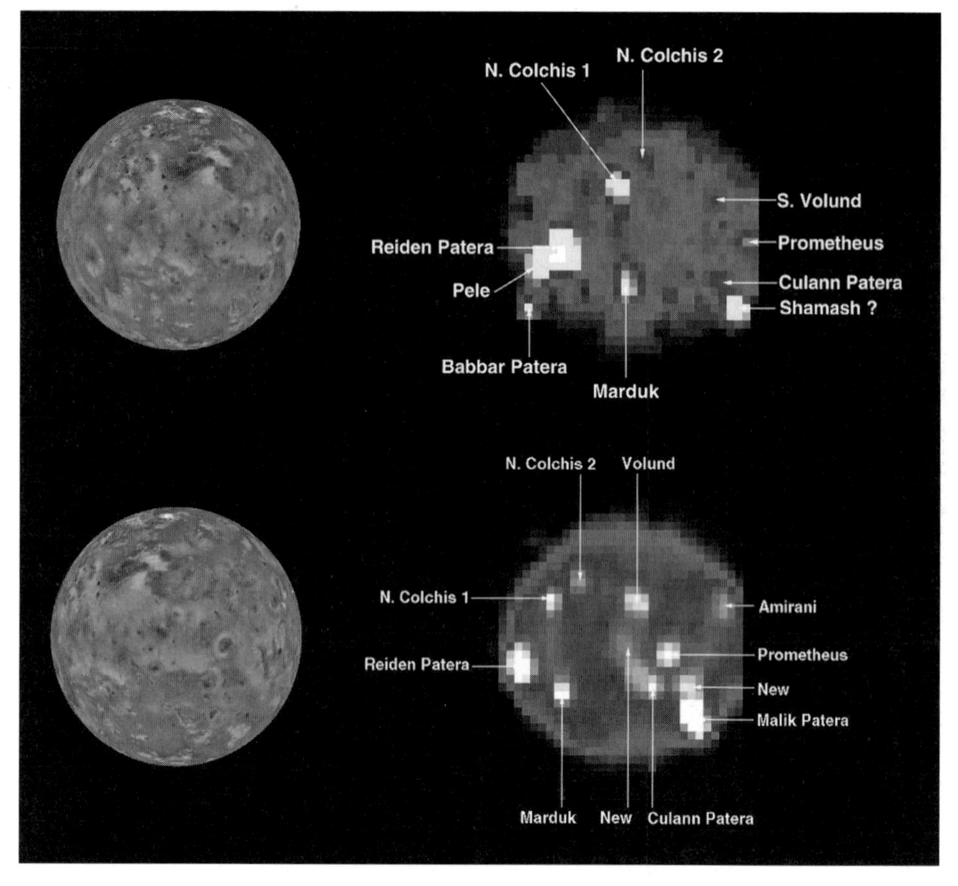

On Galileo's second orbit, the infrared spectrometer continued to monitor volcanic hot spots. The corresponding Voyager hemispheres are provided for reference.

spots in the range 420 to 620 K near volcanically active sites, which was consistent with sulphurous lava.

At 244,000 kilometres, Galileo's third orbit yielded the closest Io encounter of the primary mission. The infrared spectrometer documented Io's anti-Jovian hemisphere at a resolution of 135 kilometres per pixel, recording each pixel at 408 wavelengths. The data was processed[31] to isolate infrared excesses, in order to survey the population of volcanic sites. The previously known emission sources were readily identified, and many additional small dark calderas were located, as were warm plains. In fact, there were many more hot volcanic centres than had been expected. Most sources were highly localised, implying that they are small calderas. A number of mountains were also identified on the anti-Jovian hemisphere (these would later be named Dorian, Rata, Tohil, Euxine, Skythia, and Gish-Bar Mons), and if Galileo proved able to return to make a close fly-by of Io, this survey would assist in planning the observations.

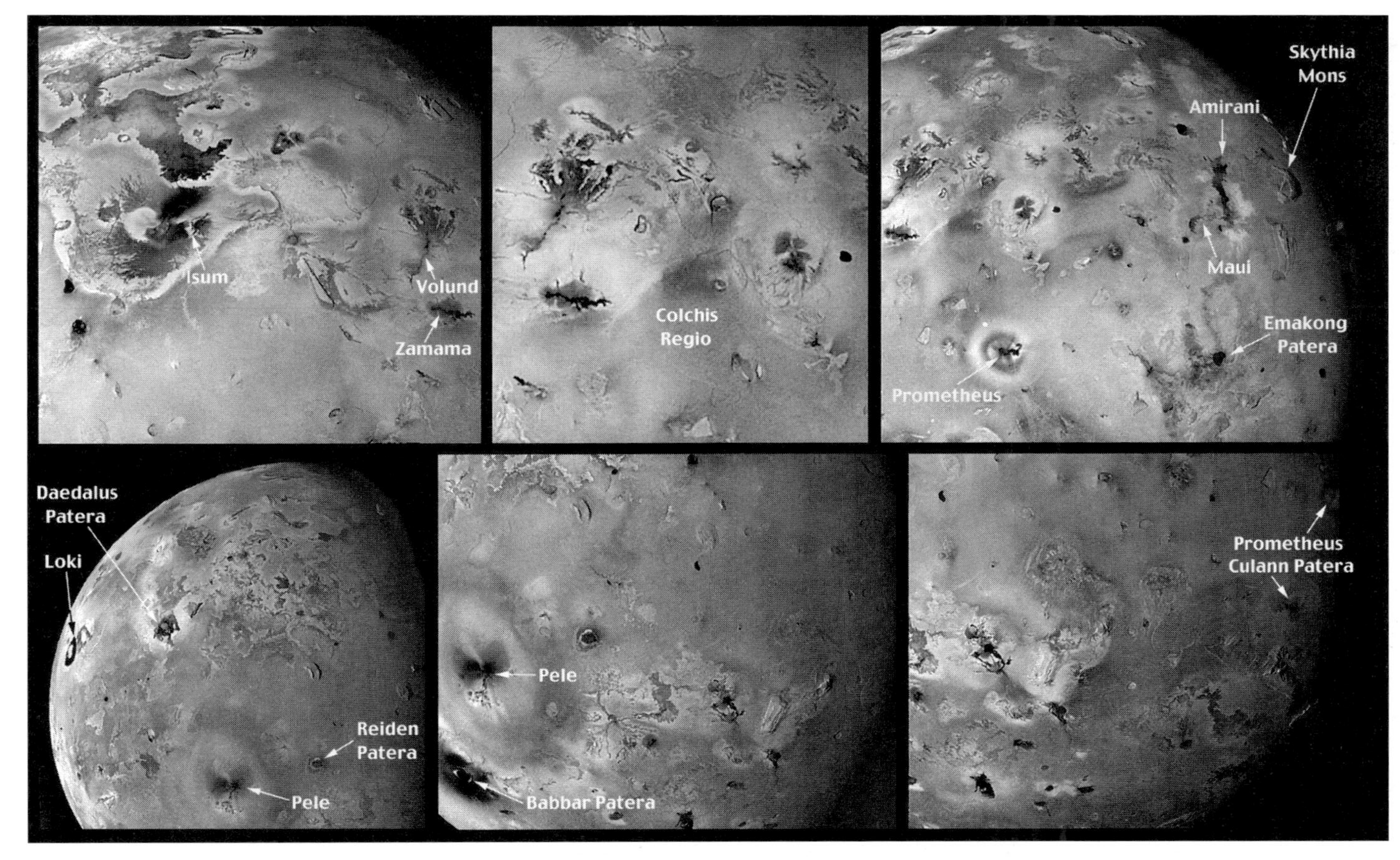

At 244,000 kilometres, Galileo's third orbit provided the closest approach to Io of the primary mission, so the anti-Jovian hemisphere was mapped in detail.

Pele had resumed plume activity since Galileo had noted its hot spot in June 1996, and its fall-out had masked the recent 'darkening' of the Pillan Patera caldera. When the Hubble Space Telescope saw Io transiting Jupiter's disk in July, it caught the plume from Pele on the limb. Up to this time, plumes had been observed only by spacecraft in Jovian space, so this distant detection opened up new opportunities for long-term monitoring.[32] With image processing, the Hubble's imagery provided a resolution of 150 kilometres. This was sufficient to show Jupiter in detail that was comparable to that of a naked-eye observer standing on Io.[33] In fact, this was the largest plume yet measured. It was calculated that material emerged from the vent at a speed in excess of 3,000 kilometres per hour and rose to an altitude of 400 kilometres.

The combination of high spatial and spectral resolution infrared spectrometer data enabled the distribution of sulphur dioxide frost on Io's surface to be charted for the first time in order to reveal the deposition patterns and the relation of the frost distribution to volcanic activity. It utilised sulphur dioxide's characteristic near-infrared absorption of sunlight. The false-colour rendition of the data represented the maximum concentration as white, and little or no material as black. The surveyed area spanned the longitude range 120 to 270 degrees at a resolution of 120 kilometres per pixel.

The strongest concentration of sulphur dioxide is on the Colchis Regio. Although many of the known hot spots are in areas having a low concentration, the infrared spectrometer had not identified any significant hot spots on the plain to the south of Colchis which is poor in sulphur dioxide.

The 1,000-kilometre wide red halo around Pele showed dark at this wavelength, indicating that its pyroclastic is not sulphur dioxide (it is primarily sulphurous in

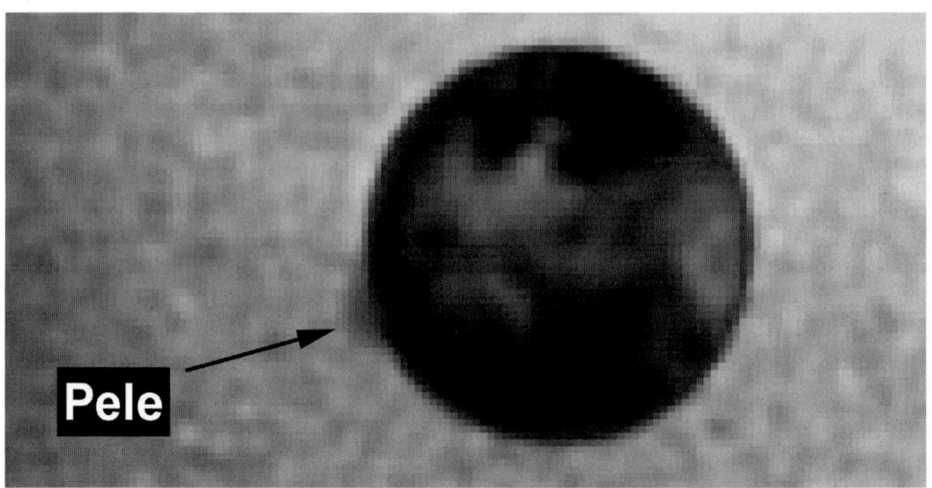

When the Hubble Space Telescope took this image of Io in transit across Jupiter's disk in July 1996, it caught the plume from Pele on the limb. The spatial resolution was not as good as Galileo's, but this observation demonstrated that the Hubble would be able to undertake long-term monitoring of plumes. (Courtesy STScI)

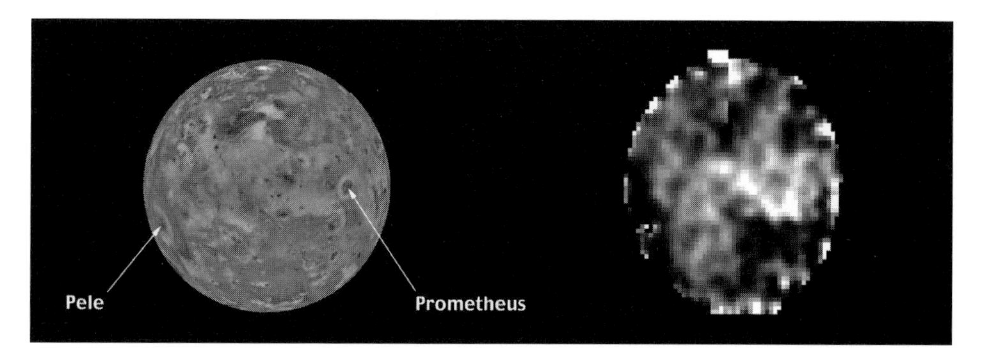

Although the Voyagers revealed that sulphur dioxide condensed from volcanic plumes, the deposition pattern was unknown. On the particularly close fly-by on its third orbit, Galileo's infrared spectrometer measured the solar reflection spectrum in order to chart the distribution across the anti-Jovian hemisphere so that this could be correlated to the locations of vents. The map represents maximum concentration as white, and areas with little or no sulphur dioxide as black. The spatial resolution is about 120 kilometres per pixel. The corresponding Voyager hemisphere is provided for reference. The strongest concentration is on Colchis Regio. Hot spots like Pele are depleted in sulphur dioxide.

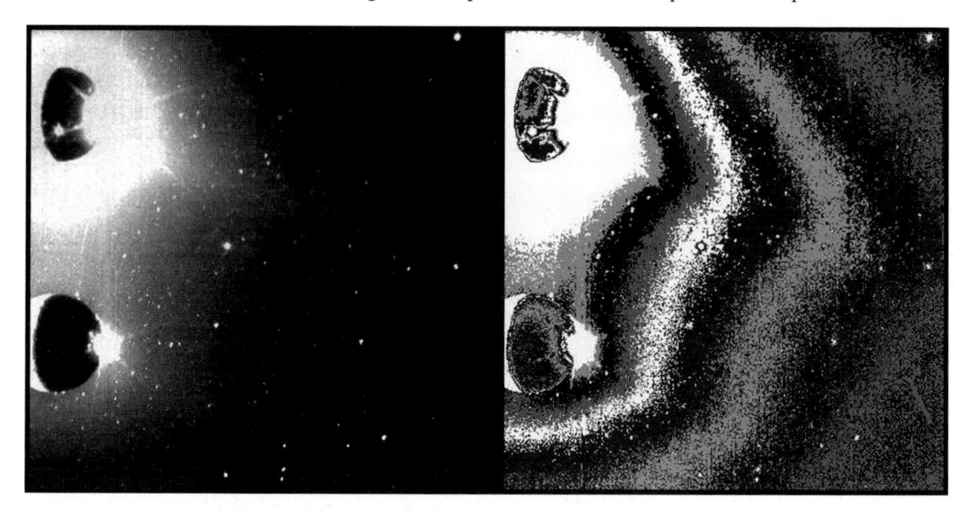

On its third orbit, Galileo took extended exposures to record the glowing sodium cloud around Io. The solid-state imaging system used the 'on-chip mosaicking' technique, in which the same pixel array is repeatedly exposed and the aim is adjusted in between. In this case a 'clear' filter (upper left) and a 0.589-micron filter (lower left) were swapped. Most of Io's surface is in shadow, with only a crescent illuminated. The 'star burst' on the limb is sunlight scattered by the plume from Prometheus. The bright spot over by the terminator is thermal emission from Pele's hot spot. Part of the diffuse sky glow is scattered light from Prometheus's plume, but much of it comes from the sodium cloud, and the isophotes document its tremendous extent.

form). The smaller ring at Prometheus, on the other hand, was prominent, so its plume is depositing sulphur dioxide.

Long-exposure images were taken through a number of filters to study Io's sodium cloud. Most of the observed hemisphere was in shadow, so Io presented a crescent with the Pele plume near the terminator and the plume from Prometheus scattering sunlight on the limb. The isophotes documented the diffuse glow of sunlight scattered by sodium atoms in the volcanic moon's extensive envelope. Being electrically neutral the sodium atoms cannot be caught and accelerated by the magnetosphere, so they linger around the moon.

Even though Pele was almost 30 degrees beyond the limb when Galileo inspected Io on its fourth orbit, the top of the plume was visible. This indicated that the material was still rising 460 kilometres above the surface. Amirani also had a plume; and there were intense hot spots at Kanehekili and Gish-Bar Patera. By the sixth orbit, the bright materials around Ra Patera had faded significantly. Despite a profusion of discrete hot spots on the surface, and comprehensive limb surveys, it was not clear why there should be so few plumes.

Diffuse glows observed near the limb in eclipse were probably either emissions by neutral oxygen or sulphur in volcanic plumes, or auroral glows induced in Io's patchy atmosphere by Jupiter's magnetosphere.

The infrared spectrometer's multispectral study had yet to find direct evidence of any iron-bearing minerals, so it was still not possible to establish whether Io's volcanism was primarily sulphurous or silicate. However, Galileo had had several opportunities to study Io in Jupiter's shadow and had measured the temperatures of

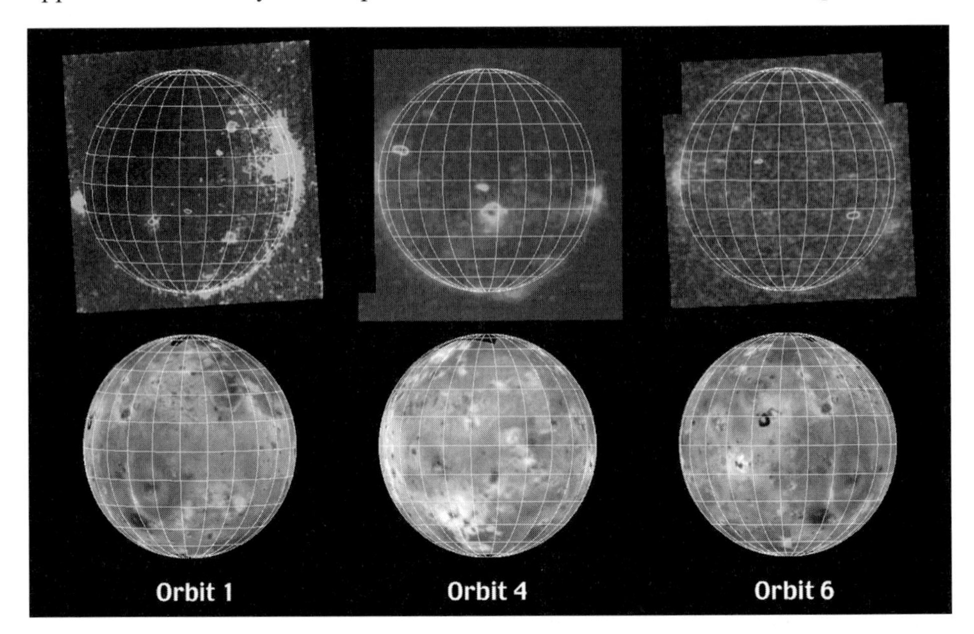

The initial observations of Io in eclipse had proved so fruitful that this type of imaging was rescheduled.

the hot spots by their radiated energy; the fact that many exceeded 700 K strongly implied the presence of silicate lava.

OCCULTATION

A mission objective was achieved on 26 February 1997, when Galileo was occulted by Io and the radio science team was able to gather valuable data on Io's ionosphere. Of course, the spacecraft was nowhere near the moon at the time – it was 3 million kilometres from it. This was only the second Io occultation. The first had been when Pioneer 10 had flown behind the moon in 1973. Galileo's orbital tour would eventually yield three occultations (each of which provided two opportunities to sample refraction data), so it would be able to measure both the temporal and the geographical variation of Io's environment for correlation with the extent of volcanic activity.

The previous Europa fly-by had been arranged to adjust the inclination of the spacecraft's orbit around Jupiter by just the right angle to set up this occultation. Of course, this tilt would subsequently have to be cancelled out. Whereas Europa had aligned the spacecraft's trajectory for 'free', by virtue of the gravitational slingshot effect, the spacecraft would have to reverse this plane change on its own. The most cost-effective point at which to execute the manoeuvre was at apojove, on 14 March. The 15.8 metres per second out-of-plane adjustment was made by pulsing the thrusters 2,800 times over a period of 12 hours – in fact, this was the biggest 'trim' burn yet.

"Galileo was a true trailblazer," enthused Todd Barber, JPL's Galileo propulsion engineer. "No mission has ever asked so much of a bipropellant propulsion system, and Galileo rose to the challenge marvellously, although not without some 'hiccups' [most notably malfunctioning valves] along the way to keep things interesting."

BACK TO THE ROUTINE

Shortly after Galileo had finished its sixth encounter sequence, infrared telescopes on the Earth detected the onset of a major thermal brightening at Loki,[34] so the absence of significant surface modification upon Galileo's return in April was something of a surprise. A plume observed in eclipse was initially suspected of being from Loki, but further analysis showed that its source was further west, most probably in the vast complex of light-toned deposits of Acala Fluctus.

In fact, there is surprisingly little topographical relief around Loki, it is simply a caldera on the open plain, and all the recent activity was confined to its floor. Old flows extend across the plain to the northeast, to a dark linear feature where a pair of plumes had been observed by the Voyagers, one at either end (which is why Loki was counted twice in the 1979 plume count).

The major surprise of the eighth orbit was the number of plumes. The limb views showed plumes at Prometheus (as expected), and also at Kanehekili, Zamama,

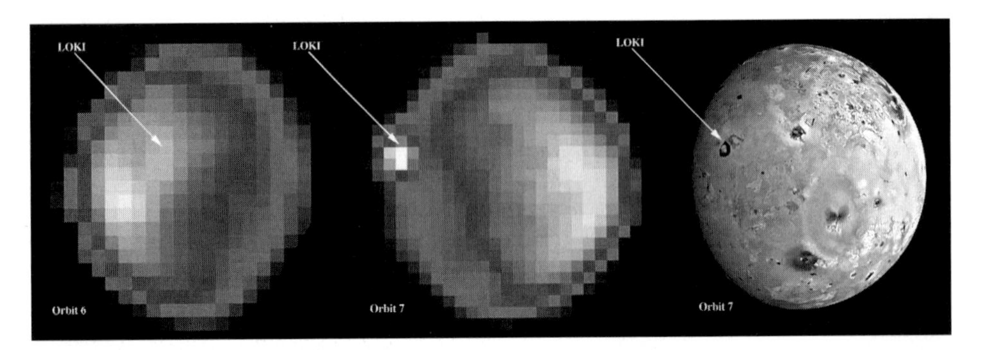

Galileo's infrared spectrometer had observed Loki several times. It had been quiescent early in the mission, but it was progressively becoming more active. It was imaged by Galileo in daylight at 2.95 microns on 21 February 1997 (left). A few weeks later, on 12 March, the Infrared Telescope Facility in Hawaii observed Io in eclipse and detected an intense glow, indicating that a major eruption was in progress. Galileo imaged Loki in darkness on 4 April (centre). The visual image (right) provides the same hemisphere as the infrared observation on the seventh orbit.

Marduk and Amirani. Observations in eclipse confirmed that the plume west of Loki was rising from Acala Fluctus, and revealed a remarkable 'field' of 14 intense hot spots near the equator in a region spanning 30 degrees either side of the sub-Jovian meridian.

The eclipse imagery on the ninth orbit showed that the hot spots at Pele and Loki were still active, and new ones had developed at Svarog and Marduk. There was a diffuse glow over the Acala Fluctus hot spot.

These observations strongly supported Andrew Ingersoll's proposal[35] that the density and composition of Io's envelope were correlated with volcanic outgassing.

The ultraviolet spectrometer had detected sulphur dioxide emissions from Io during recent orbits, documenting the variability of the volcanic activity. An analysis by Amanda Hendrix of the University of Colorado at Boulder determined that Io's gaseous envelope is derived from both outgassing from volcanoes and sublimation of surface frosts. "The thickness of Io's sulphur dioxide atmosphere varied with both time and location over the past year," said Hendrix. This data also confirmed Io to be dry. Although it would have accreted as much water as its neighbour, Europa, "any water present on Io probably disappeared billions of years ago, when volcanic activity commenced".

There had been spectroscopic reports of water on Io. Farid Salama and Jesse Bregman of the Ames Research Center had announced at the American Astronomical Society in June 1993 that NASA's Kuiper Airborne Observatory had obtained infrared reflection spectra indicating strong absorption of the hydroxyl free-radical, suggesting water-ice on the surface. The exact wavelength absorbed by water-ice is very sensitive to its physical state. The 2.79-micron band suggested that the individual water molecules were 'caged' in the crystalline structure of solid sulphur dioxide. "We have finally seen the spectral signature of something for which we have been looking for years – water on Io," Bregman had reported.

A single image on the ninth orbit recorded two plumes. The one on the limb was rising from Pillan Patera, a newly active caldera just northeast of Pele. The other plume was over Prometheus. This one was casting a shadow in the late-afternoon illumination and Galileo viewed it from directly overhead. The vent is sited near the centre of the bright and dark rings, and the shadow of the mushroom plume can clearly be seen.

Although Io has undoubtedly long-since vented its endogenic water, the *surficial* materials may still be hydrated by incorporating water produced by oxygen dissociated from volcanic gases combining with protons (hydrogen nuclei) in the Jovian magnetosphere.[36] Galileo's ultraviolet spectrometer established that if water-ice has indeed been bound in by this process, it must be a very minor constituent.

PILLAN PATERA ERUPTS!

On 5 July 1997, the Hubble Space Telescope discovered a 200-kilometre tall plume rising from Pillan Patera.

It was calculated that the vent was ejecting material at 2,880 kilometres per hour. Although the gas was very hot, it rapidly condensed. "Other observations have inferred sulphur dioxide 'snow' in Io's plumes," said John Spencer of Lowell Observatory in Flagstaff, Arizona, who had found Pillan's plume while checking for activity at nearby Pele, which was active as well, "but this image offers direct observational evidence."

A week earlier, Galileo had recorded an intense high-temperature hot spot in Pillan Patera's caldera, but its imagery was not replayed until after the Hubble team had reported detecting the plume.

Remarkably, the same Galileo image which showed Pele's plume on the limb had captured a plume rising over Prometheus, but in this case the viewing angle was vertical and because the volcano was near the dusk terminator the shadow of the plume was visible extending across the surface far to the east. The mushroom shape of the structure was readily apparent. The sulphur dioxide plumes preferentially scatter blue sunlight, so Prometheus's shadow had a reddish hue.

The magnitude of the eruption at Pillan Patera was revealed upon Galileo's return – on its tenth orbit. In the Voyager era, Pillan had appeared as a simple caldera, devoid of surrounding deposits. Galileo had found it unchanged. Although there had initially been no sign of thermal activity, albedo variations during the first few orbits had indicated a growing degree of activity within the caldera. Following the Hubble Space Telescope's detection of the plume, terrestrial infrared telescopes had identified a hot spot at this location in July.[37, 38] In terms of raw thermal emission the caldera had faded by Galileo's return, but the plume was still present and there

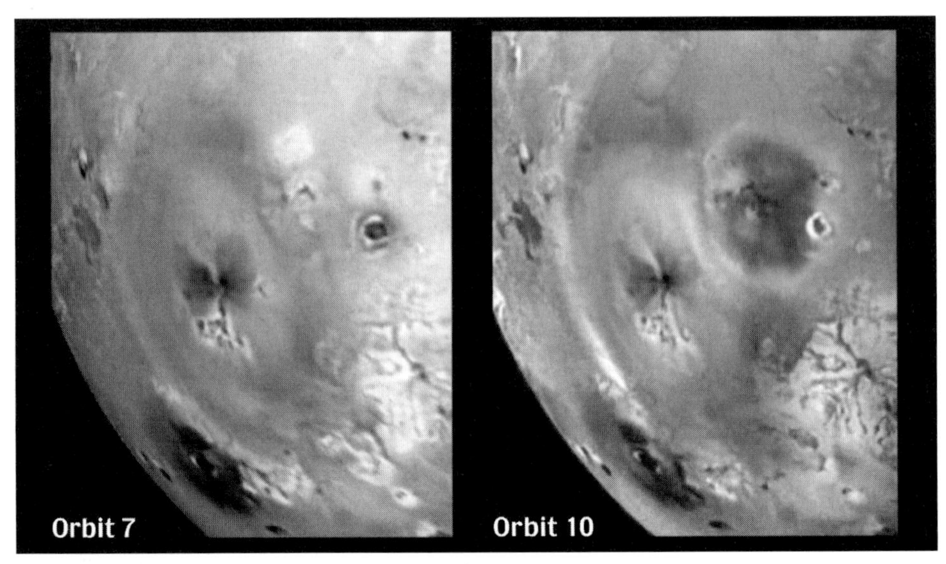

These images were taken five months apart. The eruption of Pillan Patera, just northeast of Pele, had laid down a dark pyroclastic blanket about 400 kilometres across, masking part of Pele's halo. There are several dark lava flows at the heart of the pyroclastic. The dark material is similar to that emitted by Babbar Patera, on the far side of Pele's halo.

was a dark circular feature about 400 kilometres in diameter. The dark pyroclastic had disfigured Pele's reddish halo.

"This is the largest surface change Galileo has observed on Io," reported Alfred McEwen of the University of Arizona, Tucson. "Most of the volcanic plume deposits on Io show up as white, yellow, or red due to sulphur compounds. This new deposit is grey. This tells us it has a different composition – probably richer in silicates." In this respect, Pillan Patera resembled Babbar Patera, whose pyroclastics had impinged upon the southwestern arc of Pele's halo. In addition to spewing out a great deal of pyroclastic, Pillan's eruption had also extruded a number of dark lava flows extending up to 75 kilometres northeast from the vent. Two hot spots noted in the eclipse view corresponded to the vent and to the tip of the flow feature. Silicate volcanism had long been inferred, but this was the first example documented by before-and-after views.

The eclipse views showed that the lava at Pillan Patera exceeded 1,700 K.[39] In fact, it was so hot that it saturated the detector, so it could have been about 2,000 K. This was conclusive proof that Io undergoes silicate volcanism. The hottest terrestrial lava nowadays is only about 1,500 K.[40] "These very hot lavas erupting on Io are hotter than anything which has erupted on Earth for billions of years," explained McEwen.

MYSTERIES

Early on, Galileo had established that Acala Fluctus had acquired a new bright patch since 1979. It detected a faint hot spot on the sixth orbit, a plume was detected during eclipse on the next three orbits, and the hot spot was bright on the ninth orbit, but it had now faded and there was no evidence of any plume on the limb even though the viewing geometry was favourable. Perhaps it was a 'stealthy' plume, such as had been suggested by Torrence Johnson.[41] When viewed on the limb, plumes condensing sulphur dioxide were visible because they were scattering sunlight. Because stealthy plumes were composed of extremely hot gas and did not carry significant particulates, they could not be seen visually; they could be inferred only from their infrared emissions when viewed in eclipse.

In addition to identifying sites of 'hot' activity during its volcano-monitoring, Galileo was providing data to characterise the vast relatively cool plains so as to determine the thermal inertia of the surficial material and hence infer the nature of the subsurface. When Io was in eclipse, the infrared spectrometer was able to observe the compositional variation across its surface and to measure the rate at which the surface cooled.

The eclipse observations also revealed that the number of discrete hot spots near the sub-Jovian point had increased to 26, and a diffuse glow now covered the area. The Voyagers had imaged this area, which contained many shield volcanoes.[42] Galileo's orbital tour did not favour imaging the sub-Jovian hemisphere in daylight, and the spatial resolution of the long-range imagery was too low to correlate each hot spot with an individual vent, so this mystery would be difficult to resolve.

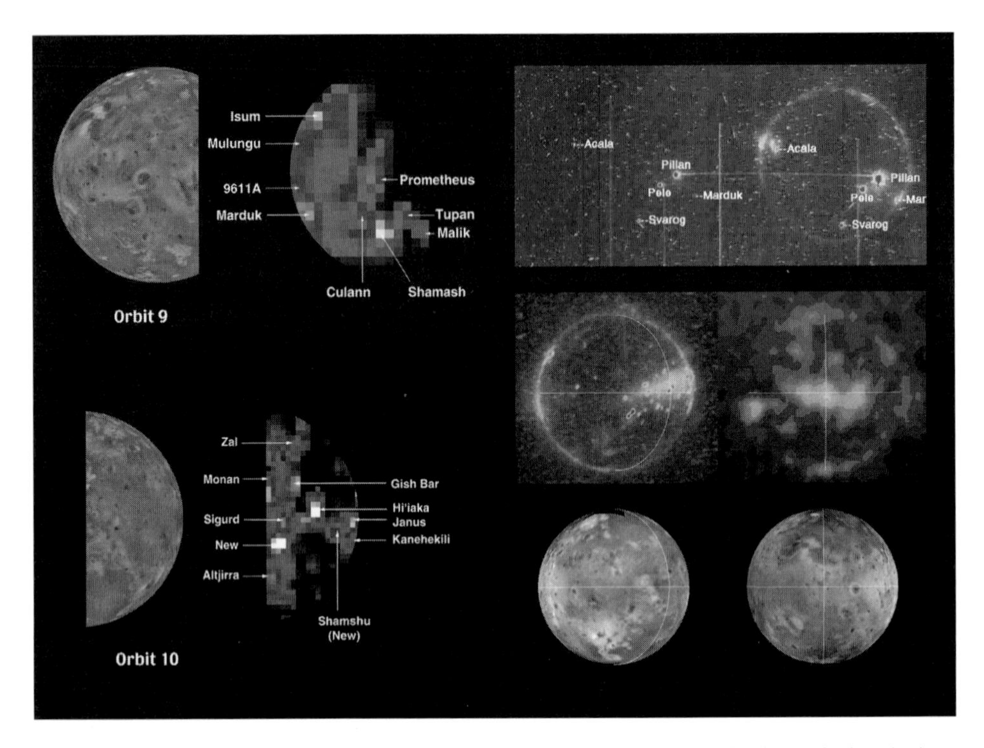

Galileo's infrared spectrometer's volcano-monitoring continued (left) through the ninth and tenth orbits, and the solid-state imaging system observed Io while it was in eclipse (right), using different filters in order to measure the temperatures of the hot spots. On the ninth orbit (top right), the lava in Pillan Patera was measured at 1,700 K, but it saturated the detector and so may have been several hundred degrees hotter. The tenth orbit's imagery (middle right) showed enhanced concentrations of volcanic gases (most of it sulphur dioxide) near the sub-Jovian and anti-Jovian regions, with context images (bottom right) for reference.

A month later, in October 1997, the Imaging Spectrograph on the Hubble Space Telescope discovered that Io had 'polar caps' of glowing hydrogen gas.

"This was a surprise," noted Frederick Roesler of the University of Wisconsin at Madison, who had studied Io for many years before using the new instrument. "What is the hydrogen's origin? And why is it glowing? We're not sure." This prompted speculation that Io's poles are coated in a frost of molecular hydrogen sulphide, or some other hydrogen-bearing frost. If such a frost was dissociated and released hydrogen, this could glow by re-radiation of absorbed solar ultraviolet. The discovery was interesting, because the hydrogen glow seemed to bear no relationship to the glows from oxygen and sulphur atoms that had already been observed in the equatorial regions. The hydrogen glow "behaves completely differently", Roesler noted. Of course, it was always possible that the glowing hydrogen was stimulated by the electrical currents flowing in the flux tubes. If accelerated protons – hydrogen ions – were interacting with the surface at the pole to make neutral hydrogen atoms, the glow could be re-radiation of absorbed solar ultraviolet. If this was the case, then the flux tubes imposed a delightful symmetry, with auroral displays on both Jupiter and Io.

The eclipse observations showed evidence for enhanced concentrations of sulphur dioxide in the sub-Jovian and anti-Jovian areas. A strong concentration on Colchis Regio had already been noted. This antipodal symmetry could indicate that outgassing is more substantial due to enhanced heat flow, possibly as a consequence of greater gravitational tidal stresses?

As Galileo's two-year primary mission drew to a conclusion, the fact that the spacecraft's basic systems and instruments were very healthy prompted an extension. Although this would initially focus on Europa, the prospects for eventually making a return to Io looked promising.

EXTENDED MISSION

From the viewpoint of Galileo's observations of Io, the transition from the primary to the extended missions was seamless. The routine involved the solid-state imaging system making limb observations to survey plumes and, together with the infrared spectrometer, monitoring hot spots and mapping the composition of the surface, while the integrated photopolarimeter and radiometer measured the thermo-physical properties of the pyroclastic deposits and the inter-vent plains, and the ultraviolet spectrometer studied the interaction between the volcanic plumes, the moon's tenuous atmosphere and the Jovian magnetosphere. The plasma torus was studied by the particles and fields suite and observed from afar by the extreme-ultraviolet spectrometer.

On 28 March 1998, a few hours before its fourteenth orbit's perijove, Galileo flew within 250,000 kilometres of Io, which was almost as close as the encounter on the third orbit, when the spacecraft had received its first good look at the volcanic landscapes. At that time, monochrome images at a resolution of a few kilometres per pixel had been obtained, this time colour imagery was taken because it is considerably easier to identify surface materials in imagery taken through filters. A number of small greenish patches – immediately dubbed 'golf courses' – were discovered, as were subtle violet hues at the cores and margins of bright sulphur dioxide-rich regions. Many of the small dark calderas[43] were ringed by bright red pyroclastic, believed to be deposits of sulphur allotropes. These would be the best views of Io until the spacecraft began to reduce its perijove.

A GLOBAL PERSPECTIVE

During the final week of May, shortly before Galileo encountered Io on its fifteenth orbit, Loki began a progressive infrared brightening which continued through to early January 1999. Such brightenings occur every few years, and the 5-micron flux from Loki typically reaches a level of 25 to 50 percent of that from the sunlit disk.

This rise was one of the longest, and also one of the best recorded. The Wyoming Infrared Observatory's observations were particularly welcome, because at 10.3-microns they were close to the peak of the hot spot's emission. This put constraints upon the total volume of material erupted and the total heat flow. A preliminary analysis showed that Loki accounted for about 60 percent of the total 10.3 micron flux from Io when the moon was within Jupiter's shadow. The remainder of the flux was distributed across the rest of the disk and it was presumably from multiple faint hot spots.

Loki's volcanic eruption rate exceeded the most effusive terrestrial outpourings of lava – the 'flood basalts' that occur when a deep mantle convection plume melts right through the continental lithosphere and accumulates a volcanic plain, with India's Deccan and Washington State's Columbia River Plateaux (which were formed 65 and 15 million years ago respectively) being 'recent' examples.

Three of Galileo's instruments could measure temperature. The infrared spectrometer and the integrated photopolarimeter and radiometer – acting as a radiometer – were optimised to measure chilly inter-vent plains and sulphur dioxide frosts. They could measure temperatures typical of sulphurous volcanism, but as their detectors saturated at 700 K they could not accurately measure the temperature of silicate volcanism. However, these higher temperatures could be inferred from solid-state imaging system observations at different wavelengths and it had observed dozens of calderas in which the lava was literally sizzling. With the exception of these localised hot spots, however, most of Io's surface was far below freezing (it was about 120 K) because Io is so far from the Sun.

Galileo was best able to measure lava temperatures when Io was in Jupiter's shadow, when the vents could be seen glowing in the dark without the surrounding terrain reflecting sunlight. Anything hotter than 700 K indicated silicate volcanism. As evidence accumulated during the primary mission, Galileo found temperatures exceeding 1,500 K at a dozen sites.

The temperatures were measured by fitting the infrared signatures to computer models of the lava production process. One of the variables was the composition of the magma. A high concentration of magnesium and iron-rich minerals (prompting a lava to be described as being 'mafic') was the most likely reason for the extremely high temperatures.

"We've tentatively identified magnesium-rich orthopyroxene in lava flows around the hot spots," confirmed Alfred McEwen.

This raised questions about Io's composition and evolution. Mafic magma is dense, so it should sink rather than rise within a thermally differentiated planet. The lithosphere should be capped by a crust of low-density minerals rich in silica, sodium and potassium. In the case of Io, however, the high-temperature volcanism suggests that the crust itself is composed of this dense material. "How do we explain that?" McEwen wondered. "Maybe some process mixes the crust back into Io's interior, so that the crust has a higher density?" The dark maria on the Earth's Moon were formed when dense lava erupted from deep fractures in the floors of large impact basins. Io, however, no longer possesses such basins; its volcanism is driven by the extreme heat derived from gravitational tidal action. Io is so hot inside that the lithosphere is very thin, the magma is never far from the surface, and the lithosphere is primarily extruded lava which has cooled sufficiently to crystallise. However, there is a vertical 'cycle', so as it is buried by subsequent flows, the older material sinks and melts.

The Earth's lithosphere is subjected to plate tectonics. At a spreading margin on the ocean floor, the fault is continuously sealed by mafic magma which rises and crystallises. An oceanic plate literally floats on the mantle from which it is derived. The main difference is that the mantle is warmer. As a oceanic plate cools, its density increases, it becomes less buoyant, and it tends to sink. If the ocean is wide, by the time an oceanic plate is subducted under a lightweight continental plate it will have accumulated a thick layer of sediment from continental run-off. Much of this sediment is drawn down with the diving plate and when it is heated in the upper mantle the water in hydrated minerals is released. The presence of this water significantly lowers the temperature at which minerals melt. Magma formation is a process of partial melting, and if water is present the earliest melt will be composed of the minerals whose melting point is lowest. This melt will rise through the crust to drive volcanism with high-silica andesitic and rhyolitic lavas. Without water, it would require a higher temperature to induce melting, and because this would contain a higher proportion of refractory minerals the lava would be more mafic. By transporting water into the melting zones beneath oceanic trenches, plate tectonics has promoted the formation of buoyant continental crust. After the water within the lava has been released as steam, it makes its way back into the ocean, thereby completing the 'water cycle'.

As Io is desiccated there is no water cycle to induce partial melting and thus moderate the production of mafic lavas. However, Io's crust is rich with sulphur. Could this play a similar role? "To the best of our knowledge," said Laszlo Keszthelyi, a geologist from the Hawaiian Volcano Observatory working at the University of Arizona, "and it's actually pretty good on this point, sulphur and sulphur compounds do not act like water when mixed in a silicate melt. Some modification of the silica chains is inevitable, but sulphur often produces an immiscible liquid and gets out of the silicate system. When it does, it usually drags with it specific metal atoms, usually chalcophiles – copper, iron, etc. – which will modify the magma chemistry, but sulphur is not expected to play the role that water does in the Earth's plate tectonic cycle. We believe that the bulk of the sulphur in Io is in the iron-rich core, and the rest has escaped from the silicate magma in either the gaseous phase or in an immiscible liquid, and so is in the crust." This is why Io undergoes distinct sulphurous volcanism. The sulphur in the crust evidently has its own complex 'sulphur cycle', and apart from being driven by the transfer of heat from the lava, this is largely independent of the 'silicate cycle'.

TORUS INTERACTIONS

The first visible-light colour images of Io while in Jupiter's shadow were obtained in May 1998. It was observed glowing in a variety of vivid red, green and blue hues.

These emissions were produced by the charged particles flowing in Jupiter's magnetosphere interacting with the moon's volcanically driven envelope – the gases were fluorescing. The various hues resulted from different states of excitation, and probably also due to compositional differences between the plumes.

A green glow near the middle of the hemisphere facing the camera (which was the leading hemisphere) indicated a concentration of oxygen which almost certainly derived from dissociated sulphur dioxide. This was within the moon's magnetospheric wake. The bright glow running along the polar limbs could have been due to oxygen in a higher state of excitation, although it may well have been sodium or hydrogen (without knowing the frequency of the emission line, it was impossible to ascertain). "The most interesting of the aurorae," said Paul Geissler of the University of Arizona's Lunar and Planetary Laboratory, "are the bright blue glows at the sides of the satellite." In the orientation that the moon was viewed, these were the sub-Jovian and anti-Jovian regions. The increased concentration of sulphur dioxide from the plumes in these regions was evidently interacting with the electrons flowing in the flux tubes, and it was glowing. The Io glows had not been detected previously, because they were so faint that they could be observed only in the darkness of Jupiter's shadow. "This is our first detailed look at visible aurorae on a Solar System satellite," Geissler reported.[44] "It helps us to understand Io's atmosphere, and the processes that generate the emissions."

Io's disk-averaged emissions diminished over time upon entering eclipse but, surprisingly, the localised blue glows brightened as the eclipse proceeded, possibly indicating that some of the electrical current flow between Io and Jupiter is conducted through the interior of the moon.

While Io was within Jupiter's shadow, Galileo made two observations an hour or so apart. The first, just after the moon entered eclipse, was to identify any short-term thermal effects as the surface reacted to the shade. The second, towards the end of the eclipse, recorded the extent to which the surface had cooled in order to measure the thermal conductivity of the surficial material, which is a function of both composition and physical structure.

For its next few orbits, Galileo approached Io no closer than about 750,000 kilometres, so it focused on routine volcano monitoring. Jupiter was studied in late October 1998 by a pair of telescopes mounted in the payload bay of STS-95/Discovery, the Shuttle flight more popularly noted for the fact that its crew included John Glenn.

The Ultraviolet Spectrograph Telescope for Astronomical Research (UVSTAR) had been developed by the University of Arizona's Lunar and Planetary Laboratory and the Centre for Advanced Research in Space Optics of the University of Trieste in Italy. The team was led by Lyle Broadfoot, the principal investigator for the Voyager ultraviolet spectrometer. UVSTAR was a pair of telescopes with imaging spectrographs that covered overlapping spectral ranges from 500 to 1,250 Ångströms, spanning the far-ultraviolet and on into the extreme-ultraviolet wavelengths – the range which includes most of the energy radiated by the Io plasma torus. It was uniquely capable for studying Io's torus, because its high spectral resolution enabled the torus to be imaged in each of its emission lines in order to chart the distribution

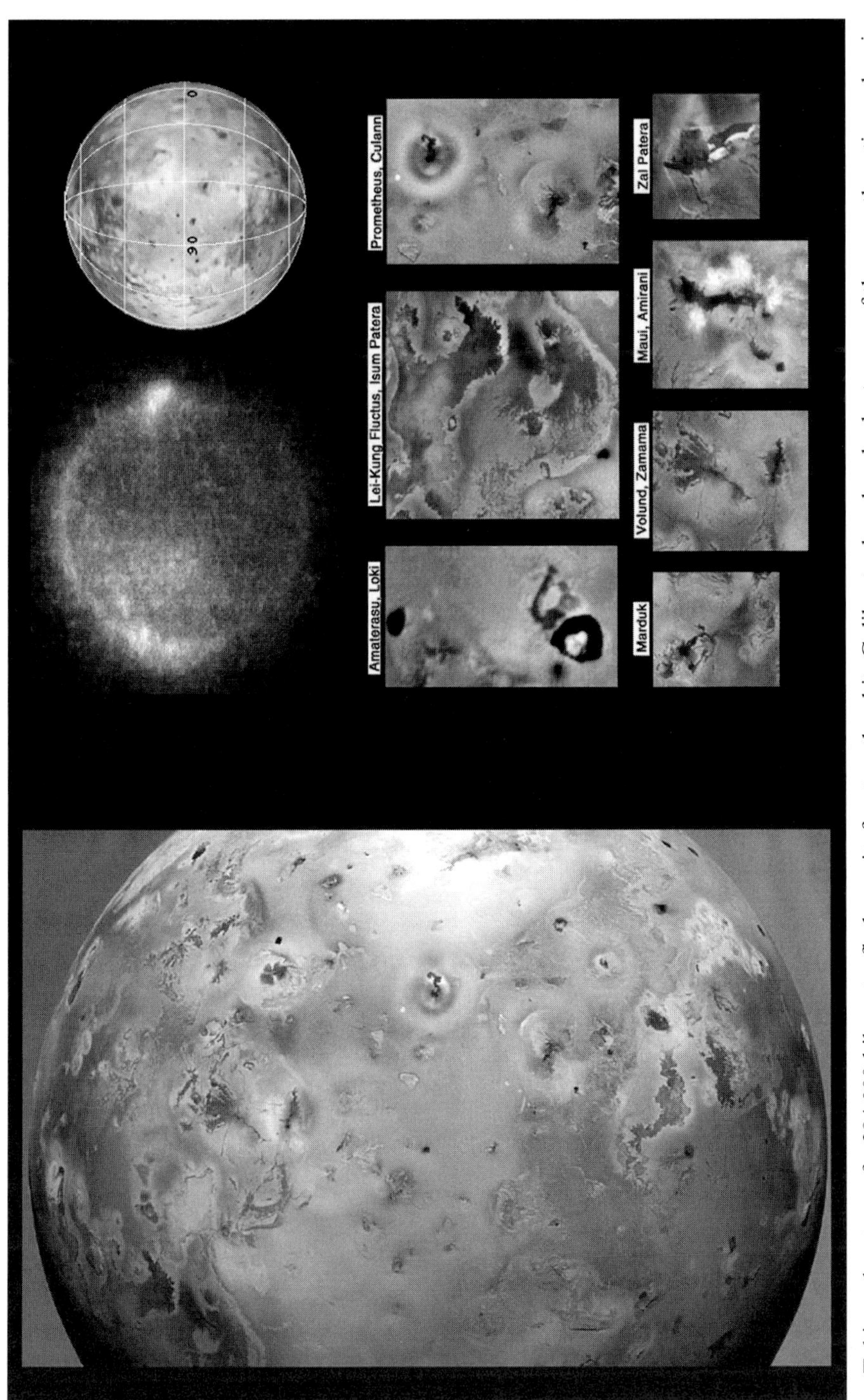

Taking advantage of a 294,000-kilometre fly-by on its fourteenth orbit, Galileo took a close look at some of the recently active volcanic vents on the anti-Jovian hemisphere with a resolution of about 3 kilometres per pixel. And the eclipse observations during the fifteenth orbit recorded intense equatorial glows at the sub-Jovian and anti-Jovian regions.

of each of the ionic species. UVSTAR made 15 torus observations over the nine-day mission.[45] It had the advantage over the Hubble Space Telescope in sensing further into the ultraviolet. Galileo had just passed apojove, and was replaying the tape of its seventeenth sequence, so it was not in a position to make coordinated observations.

On its way back up the Jovian magnetotail, Galileo was to pass through the plasma sheet. "The plasma sheet is not very massive but it is big," noted Claudia Alexander, one of JPL's plasma physicists. At Galileo's distance of 90 Rj, it should have taken hours to cross it. "We got ready to take the measurements, but when the spacecraft arrived, the plasma sheet wasn't there! What a surprise. Imagine missing something so large." Of course, the spacecraft could not collect data all the time. It was a programmed observation. "There was probably a coronal mass ejection that hammered Jupiter and it changed the configuration of the magnetosphere," Alexander speculated. "We had a lot of 'missed' predictions in the beginning of this mission, but Galileo is really improving our understanding of Jupiter's magnetic environment."

When Galileo 'safed' itself approaching perijove on its sixteenth orbit it had already passed Io, but when it was disabled two orbits later most of the Io observations were lost because the encounter was after perijove.

At the American Geophysical Union's Spring 1999 meeting in Boston, astronomers from the Kitt Peak National Observatory in Arizona reported that Io's torus contained chlorine. This was in addition to the already-known sulphur, oxygen, sodium and potassium.

The two most common inorganic compounds of chlorine are hydrogen chloride (which is a colourless gas emitted by terrestrial volcanoes) and sodium chloride (which is more commonly referred to as 'table salt'). The presence of chlorine in the torus implies that Io's volcanoes are venting hydrogen chloride and that the moon's surface may be encrusted with sodium chloride. "Io seems to have a higher proportion of chlorine in its atmosphere than any object in the Solar System," pointed out Nick Schneider of Colorado University at Boulder. Actually, the ratio of chlorine on Io was a billion times greater than on Earth. Io is desiccated now, but if salt was formed early on – perhaps by hydrothermal activity producing mineral-rich water – a deposit of salt would have accumulated when the volatiles were boiled off. Considering the amount of water on Europa, if Io had formed in a similar state, it may well have left behind a vast residue of salt.

PERIJOVE REDUCTION

Galileo had made a fly-by of Io on the way in to Jupiter in December 1995, but the planned observations had been cancelled due to the faulty tape recorder. During its primary mission, the spacecraft had maintained its perijove above 9 Rj in order to avoid the worst of the radiation in the inner magnetosphere, and the Io specialists had had to monitor the activity on the volcanic moon from afar. When the mission was extended, Europa had been made the primary focus, but now – as the mission drew towards its conclusion – it was time to initiate the eagerly awaited return to Io.

As Io orbited at only 5.5 Rj, Galileo's perijove would have to be reduced, but this could only be achieved by a series of gravitational slingshots because there was insufficient propellant for such a major manoeuvre. Although the perijove could be reduced using any moon, it was most efficiently done using the outermost one. One of the constraints on setting up the final Europa fly-by had been to create a Callisto encounter on the next orbit.

The objective of the Perijove Reduction Campaign[46] was to reduce the spacecraft's perijove to 393,000 kilometres, so that it would fly just within Io's orbit at closest approach. Lowering the orbit, however, would also reduce the orbital period, so the amount of data that would be able to be taken on each encounter sequence would be limited to that which could be replayed during the cruise through apojove.

An important theme for the Io observations during the perijove reduction campaign was to secure *in situ* data on how the moon interacted with its torus, and how the torus interacted with the Jovian magnetosphere. The objective was to figure out how the material carried away from Io by its plumes was distributed throughout the rest of the system. Just as Io would not be the same if it were not where it was (in close to Jupiter, and locked into an orbital resonance with Europa) the Jovian magnetosphere would not be so active if it were not for Io pumping up the torus.

The radiation in close to Io was intense and Galileo's systems had already suffered glitches near perijove – a situation that could only get worse. "There's a good chance that Galileo will not survive," warned Ron Baalke. "In fact, since it completed its primary mission at the end of 1997, the spacecraft has been operating beyond its design in terms of its accumulated radiation exposure."

Galileo's twentieth orbit – the first of the perijove reduction campaign – focused upon high-resolution imaging of Callisto's surface in order to resolve issues raised by the previous investigation. The spacecraft approached Io no closer than 800,000 kilometres, but at that time the infrared spectrometer made multispectral observations of Colchis Regio in general, and of Prometheus in particular. The study of Callisto continued on the following orbit, but the main attraction was Io because the reduced perijove took it within 124,000 kilometres. This was not just Galileo's closest since entering Jovian orbit, it was also better than three times closer than the closest Voyager. Also, since the encounter was only seven minutes after perijove, the balky gyroscopes were commanded off to ensure that they could not induce the spacecraft to 'safe' itself and abort the programme. "We'll have the first real passage since 1995 through the outer edge of the Io torus," explained Torrence Johnson. "Galileo will spend nearly a day inside the edge of the torus, in a region of space between Io and Europa. We know, generally, what the torus is, and what it's made of; now we're going to look at its detailed structure." High time-resolution observations by the particles and fields instruments were undertaken for 2 hours while in Io's vicinity, at the deepest penetration of the torus.

The ultraviolet spectrometer observed Io while it was in Jupiter's shadow to continue the study of its aurora-like atmospheric glow. Just before Io re-emerged, the solid-state imaging system observed its sodium cloud. Over successive orbits, the extent of plume activity was to be correlated with the density of the sodium cloud to

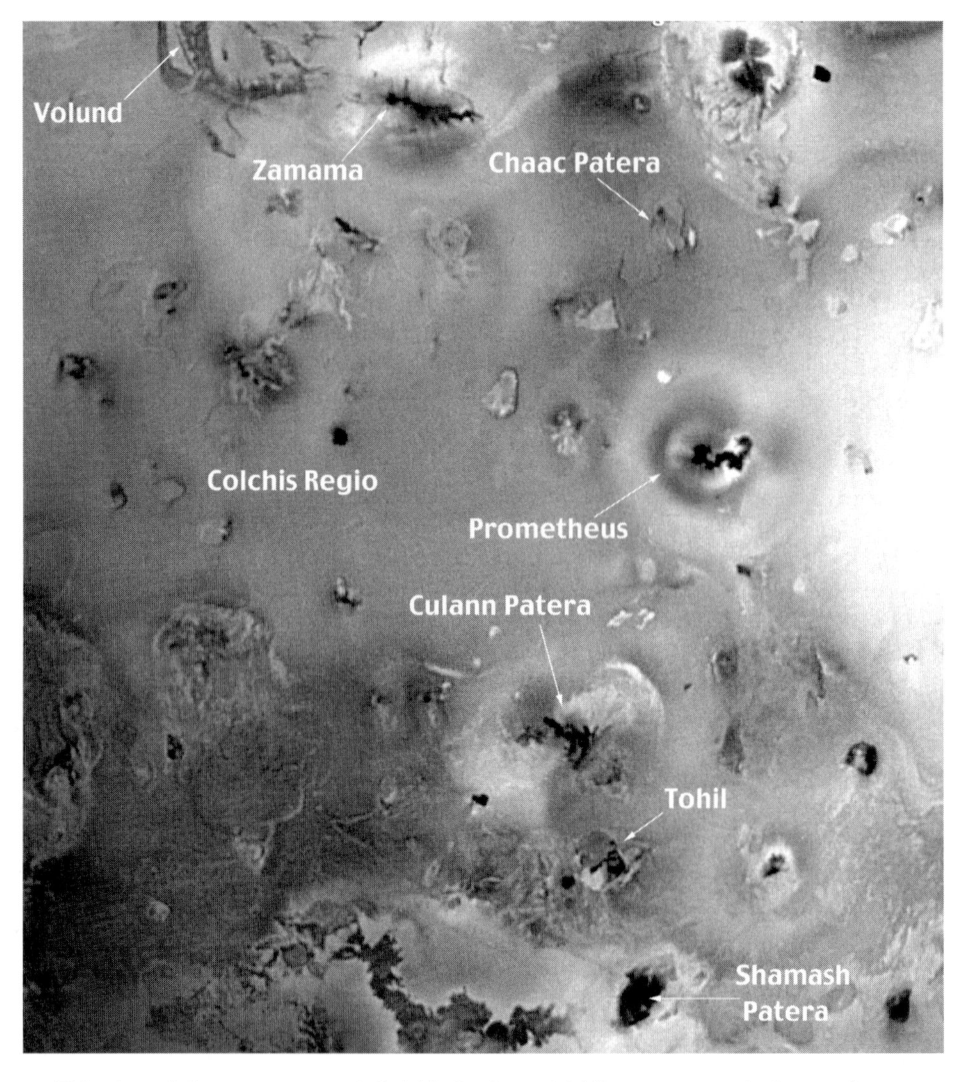

This view of the eastern part of Colchis Regio at 1.3 kilometres per pixel was taken on Galileo's twenty-first orbit, during a 124,000-kilometre fly-by.

identify its sources and, if possible, the processes by which the neutral atoms were removed from the moon. It would also shed light on the process which maintained the plasma in the torus.

With every reduction in perijove, the spatial resolution of the infrared spectrometer's high spectral resolution imagery would improve. Even at 124,000 kilometres, with 60 kilometres per pixel, it was able to map the surface's composition in unprecedented detail.

In addition, the solid-state imaging system was finally able to begin regional surveying of the anti-Jovian hemisphere with a resolution of 1 to 2 kilometres, which was comparable with the best Voyager coverage. This facilitated direct comparisons of the lava flows around the major volcanic centres over the 20-year interval. Pairs of images were made of selected features for stereoscopic analysis.

Telescopic observers had maintained that Io, although distinctly reddish, is not really as striking as depicted in the Voyager imagery. Those colours were derived from the filters used to make the individual images for integration as a 'colour' rendition. As a result, Io had entered the history books with its volcanic features resembling slices of Italian sausage scattered on a mozzarella pizza. Galileo's filters were better able to render Io as the human eye would see it, which showed that the sulphurous surface is a tapestry of pastel green-tinged hues with blacks, reds and oranges representing the complex flows and pyroclastic deposits around the vents.

On its twenty-first orbit, Galileo saw a 100-kilometre plume rising from near the site of a plume at Masubi that had been seen by the Voyagers – in fact, Galileo had seen plumes at several locations within this general area.

Improved imagery of Pillan Patera further documented the dramatic changes over the past three years. Pele's red pyroclastic had started to mask Pillan Patera's dark material, but it had not yet obscured it. A small unnamed vent to the east had also erupted recently, and deposited dark material surrounded by a yellow ring, which was most visible where it covered the dark feature from Pillan's 1997 eruption.

The current state of Prometheus was revealed in unprecedented detail, too. The high solar illumination was ideal for infrared spectrometry to study the composition of the surface but it washed out topographical relief. The long-lived plume had created a ring of bright white and yellow pyroclastic, most likely rich in sulphur dioxide frost. The dark spoke-like pattern had evidently been created by distinct jets within the plume, although the precise mechanism was not understood.

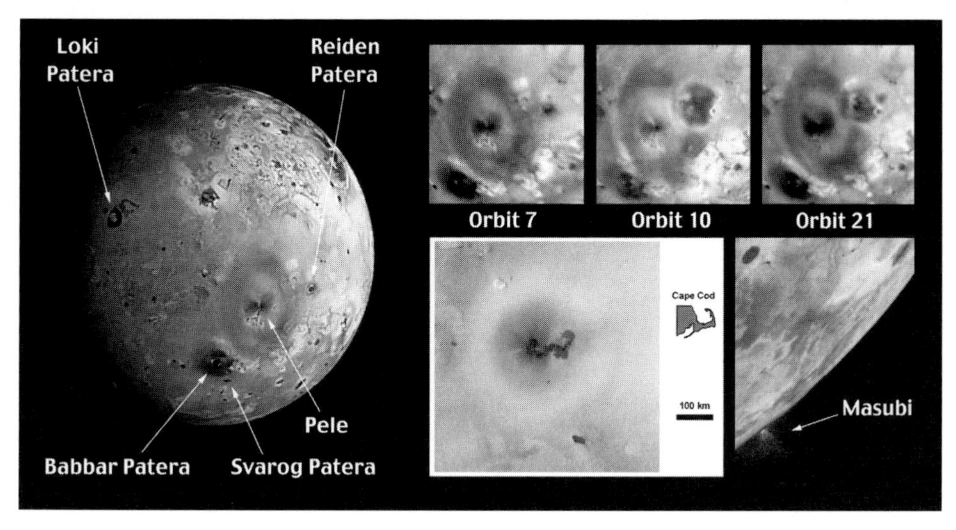

On its twenty-first orbit, Galileo caught a plume from Masubi on the limb and it further documented the evolution of Pele and Pillan Patera, and Prometheus.

Galileo's twenty-second orbit produced only a 730,000-kilometre Io fly-by, but the lower perijove facilitated *in situ* measurements even deeper within the torus, and so the particles and fields instruments sampled at high time-resolution for six hours. In addition to magnetic fields and particle interactions, the plasma wave spectrometer observed electromagnetic effects in order to study the dynamical processes within the torus.

A particular objective was to detect emissions produced as plasma was accelerated by an efficient wave-particle interaction believed to be primarily responsible for transferring energy from Jupiter's magnetic field to plasma in the torus, and thence into the outer portions of the magnetosphere. However, two hours after perijove, Galileo suffered a spurious reset signal. Although the computer successfully recovered, the glitch interrupted the tape recorder, and the observations were terminated early.

In fact, the spacecraft had suffered the most intense exposure to radiation since it had made its deep penetration of the magnetosphere in December 1995. "We anticipated the spacecraft's star scanner would detect about 300 to 400 pulse counts of radiation," admitted Jim Erickson, "so imagine our surprise when the instrument showed Galileo had flown through 1,400 pulse counts!" This was even more than it was expected to endure on the coming Io fly-by. "Then again, that's why we're exploring Jupiter and its moons: to discover unusual phenomena." At that time, the spacecraft was at its deepest penetration of the torus, and about to cross through the plasma sheet, but these factors had been taken into account when making the prediction of the likely radiation dosage.

Two weeks previously, Robert Howell at the Wyoming Infrared Observatory, a regular Io observer, had announced the onset of the largest heat pulse seen since 1986. The volcanic activity had evidently pumped up the torus with charged particles, and Galileo had passed right through it. "This was a great dress rehearsal for the Io encounters," Erickson reflected. "We have been wondering how the spacecraft might hold up when it gets in close to Io – this latest brush with radiation makes us believe that the odds of survival may be fairly good!" Galileo demonstrated its resilience again the following day, when radiation induced the back-up sensor which monitored the spacecraft's spin rate to send false readings, but the computer isolated the sensor. It went on to survive two spurious resets before it emerged from the torus. However, when the ultraviolet spectrometer was observing the torus as the spacecraft pulled away from Jupiter, the instrument's microprocessor suffered a glitch and switched itself off. Its observations were to have been correlated with the *in situ* particles and fields data and ongoing volcano-monitoring, to investigate how the torus was replenished by volcanic activity.

The twenty-third orbit, in September 1999, was the final encounter with Callisto to reduce Galileo's perijove. Six hours of high time-resolution particles and fields data was taken while within the torus, and the plasma wave spectrometer made another attempt to record emissions from plasma in the process of being accelerated.

"We haven't been this far into the torus since 1995," pointed out Duane Bindschadler, the science planning and operations chief. "The radiation levels we saw back then were relatively low, but apparently that's not the case now." The planners were torn between seeking a close look at an active volcano, possibly even flying through a plume, and the desire for the spacecraft to survive long enough to report its observations.

So how long would it take for the torus to settle down from the recent infusion? "That's an interesting question," admitted Claudia Alexander, a plasma physicist at JPL. "We don't know how long it takes for energetic particles in Jupiter's magnetosphere to diffuse." Well, now they would have an opportunity to find out!

Galileo reached apojove on 27 September – less than a fortnight after its perijove passage. All the way out, the particles and fields instruments had been reporting in real-time, in order to monitor the dusk-side of the magnetosphere and on down the magnetotail. As it made its way back in, the Outreach Program sponsored a three-day workshop for educators to put the Io fly-by into context. This drew an analogy between Io's volcanoes and the resurgent caldera at the Yellowstone National Park in Wyoming. This spread a thick blanket of pyroclastic for thousands of kilometres when it last erupted, some 600,000 years ago. "It's almost like being there," said chief outreach coordinator, Leslie Lowes. "Yellowstone is the closest we can come to taking teachers to Io – without actually putting them on a spacecraft."

"Io has lots of thermal areas, just like Yellowstone," confirmed Bill Smythe, a member of the infrared spectrometer team. "The volcanic plumes get most of the attention, but there are probably also things like fumeroles and geysers." Yellowstone's 'Old Faithful' is probably a close terrestrial analogy for a plume on Io. In fact, if this water geyser was transplanted to Io, and the weak gravity and the absence of air resistance were taken into account, it would reach to a height of about 37 kilometres on steam power alone!

THE PLUMES

It was hoped that Galileo would pass through a plume on its coming fly-by, and so make unprecedented observations.

A study[47] of the plumes observed by the Voyagers suggested that there might be two types. The Pele-style was large, faint and short-lived. The Prometheus-style was smaller, brighter and longer-lived. Whereas the Pele-style plumes produced red pyroclastic, most of the fall-out from the Prometheus-style was white – although in some cases they produced small patches of red material in close to the active centre. The Voyager infrared spectrometer had not had the spatial resolution to distinguish the temperature of a small hot spot from its surroundings but, as the hot spot associated with Pele's plume was very distinct, it was proposed that this plume was sulphur which had been vaporised upon coming into contact with molten silicate, while the Prometheus plume seemed to be sulphur dioxide vaporised by cooler sulphurous magma. The red deposits near the vents was possibly magmatic gases emerging directly from the deep lava source rather than from upper crustal volatiles. The faded red deposits surrounding Surt and Aten – both of which were active between the two Voyager fly-bys – were considered to be similar. The hot particulate-free 'stealthy' plumes were proposed as a refinement of this basic classification scheme. In fact, Galileo detected only one Pele-style plume – Pele itself – and it was detected only intermittently. However, if it was stealthy it may have been continuously active.

Galileo found that many of the sources of the Prometheus-style plumes were hot spots in excess of 1,000 K, and hence these too were the result of silicate rather than sulphurous volcanism. The ongoing volcano-monitoring campaign observed such plumes at Kanehekili, Amirani, Maui, Zamama, Volund and Marduk, and established that they can remain active for years. Significantly, there was evidence of lateral 'migration' of the sites of these plumes. The Prometheus plume was now rising from a site almost 100 kilometres west of where it had been in 1979. There was some evidence of similar migration at Amirani, Masubi and Marduk. The fact that the Prometheus-style plumes were associated with high-temperature hot spots on dark flows on the plain prompted the suggestion that they mark where crustal sulphur dioxide is being evaporated by silicate magma on or very close to the surface.[48]

Although the heights of the plumes observed by the Voyagers could be directly measured, the speed at which the material left the vents had to be inferred, and a variety of figures were proposed. Galileo had extensively observed plumes, but "when we look at a photograph of a plume, we really aren't certain what we're seeing," said David Goldstein of the University of Texas at Austin. "It might be gas. It might be dust entrained in the gas. Or it might be gas that has condensed out to form ice crystals. We can't be sure." Goldstein was heading up a team which was developing a computer model of a plume vent on Io. This work built upon that by Sue Kieffer in 1982. Their Monte Carlo direct simulation traced the particles as they were blasted out under pressure from the vent, and took into account energy lost to thermal emission, collisions between particles, and the energy drawn from Io's torus. The key model parameters included the size of the vent, the temperature of the surrounding terrain, and the temperature and velocity of the gas that was carrying the particles. A comparison of the simulation with photographs of plumes provided insight into the physics of the volcanoes. If Galileo observed a plume close-in, this would help to refine the model by providing better boundary conditions. "We need to know the particle velocities, their temperatures coming out of the vent, and how they interact with the surface."

"By looking at the polarisation of sunlight passing through the plume," Bill Smythe noted, "using the photopolarimeter, we should get some valuable information about the temperature and density of particles coming out of the vents."

"One of the most important features of this model", said Victor Austin, another member of the team, "is the 'canopy shock', where the gas rises to its apex and then falls back on itself. One way to think of the shock [wave] is a running water hose held straight up in the air. The water decelerates by gravity, and then falls back down. The same thing happens to a volcanic plume, but the rising gaseous fluid is supersonic and so it forms a shock near the turnaround point." If Galileo could make direct observations of this feature, it would refine the model. The shock region was predicted to be very thin. As rising gas passed through the shock, it would rapidly cool, and produce a thin layer of sulphur dioxide crystals in the immediate post-shock region. Although this general model was well established, the devil was in the detail.

Some of Io's vents seem to have a dark ring about 150 kilometres out. Although this could be due to the deposition of a different coloured material, it may be due to something more interesting. "Many of our simulations show that warm material from the plume crashes down about 150 kilometres from the vent," Goldstein pointed out. "We think this might explain the dark rings. These could be 'scouring rings', places where the sulphur dioxide frost is worn down to dirt and rock." The model predicted specific characteristics. "If we see sharp edges in the close-up images, it would help confirm our model."

THE FLOWS

Volcanism is the process by which Io sheds the heat induced by tidal stresses, so the way to understand Io is to determine its heat flow. To do this, it is necessary to estimate the ratio of silicate to sulphurous lava, because the very hot silicate will transport far more energy from the interior onto the surface.

"The biggest mystery about Io's volcanoes is why they're so hot," Bill Smythe noted. "At 1,800 Kelvin, the vents are about one-third the temperature of the surface of the Sun!" Early in the Earth's history, ultramafic lavas were of such temperatures, but its lithosphere is now dominated by plate tectonics, and the melting point of the hydrated lava produced when an oceanic plate is subducted by a continental margin is about 300 K cooler. "It's very surprising to see lava flows on Io as hot as the ancient flows on Earth." After Voyager, "we thought that *all* the lava flows were sulphurous, but sulphur vaporises at about 700 Kelvin. These hotter lavas have to be basaltic. Now the question is – are *any* of the lava flows sulphurous!? Galileo has detected areas on Io with temperatures between 300 Kelvin and 600 Kelvin. That's about right for sulphur, but these could also be places where tiny volcanic vents at around 1,800 Kelvin are surrounded by cold ground." From the distance that Galileo observed earlier, the average temperature would appear to be cooler. "We need higher resolution data to figure out what's going on. If we're lucky, Galileo will fly right over one of these spots, and we'll have the answer."

The geological objectives of this Io fly-by therefore included determining the distribution of heat across the moon, the surface composition, the layering of deposits, the caldera walls, and the shapes of lava flows.

"We really want to know what the shapes and edges of the lava flows look like," Smythe said, "because that can tell us a lot about the properties of the lava. On Earth, lava flows form little side lobes that range in size from a few metres to centimetres." The shape and size of the lobes on Earth can be correlated with a lava's viscosity and other physical properties. "That's what we want to do on Io, and when we start seeing how the lobes formed, we'll know what kind of flows these are."

The front of this fan-shaped lava flow in Pu'u O'o on Kilauea in Hawaii is producing finger-like ('digitate') flows. (Courtesy JD Griggs, USGS)

CALCULATED RISK

As Galileo approached Io's trailing hemisphere on 11 October 1999, the moon was almost in line between the Sun and Jupiter. Having just passed perijove, the spacecraft was within the moon's orbit. Pele was in the centre of the disk in darkness on the way in. The spacecraft flew low over the dawn terminator on the equator as it crossed Io's orbit and Prometheus was in the centre of the disk in daylight as it withdrew. Most of the activity was concentrated into several hours around perijove and the Io fly-by.

"Galileo has already survived more than twice the radiation it was designed to withstand," noted Jim Erickson, "so we're keeping our fingers crossed that it will complete this encounter with flying colours."

"We expect the spacecraft to survive the fly-by," assured Wayne Sible, the deputy project manager, "although the radiation may cause its computers to reset. Radiation might even cause irreversible damage to critical electronic components." If the spacecraft was crippled within the torus, it would never be able to replay its taped observations of Io. "Galileo has more than fulfilled its mission objectives. It seems reasonable to take a calculated risk for a much closer look at a scientifically rich target."

"If there's an unexpected radiation storm, there's not much we can do," observed Duane Bindschadler philosophically.

IO, AT LAST!

The Io fly-by sequence began with the integrated photopolarimeter and radiometer making remote observations of Io to chart the thermal characteristics of the approach-hemisphere. By locating hot spots and determining the temperature of the nearby area, this data would provide context for the higher resolution views to follow. Observing at night permitted the instrument to measure thermal emissions on the surface, free of solar reflection, and since it was coming up to dawn the surface had had plenty of time to cool down. Because it was sensitive farther into the infrared than the infrared spectrometer, the radiometer was better able to chart the vast areas which – although cool – nevertheless radiate a substantial fraction of the overall heat flow.

During the primary mission, when Galileo had been restricted to remote observations, the radiometer had produced day/night disk-averaged data for various phase angles. On the closer passes of the perijove reduction campaign, the improved spatial resolution had permitted it to characterise various terrain types. This data was to be used to analyse the high-resolution map which would be produced during this fly-by. One study[49] constructed a thermo-physical model of these cooler sites, taking into account both thermal inertia derived from earlier observations and, because the heat radiated from the surface depends upon whether the Sun is heating it up, the effect of insolation.

The particles and fields instruments were to have begun five hours of high time-resolution measurements of the structure and dynamics of the plasma, dust and electromagnetic fields in the torus about three and a half hours before perijove, as the spacecraft crossed 6.4 Rj inbound, ran through perijove at 5.5 Rj, and climbed back out to 5.9 Rj. However, the first half of this programme was lost when one of the computers suffered a radiation-induced fault in its solid-state memory and 'safed' itself. The engineers were soon able to confirm that one 'bit' at a specific memory address was now permanently damaged. As the priority task was to pick up the encounter sequence, the software was rewritten to avoid using this address, which was in a part of memory employed by both the integrated photopolarimeter and radiometer and the infrared spectrometer. Galileo was deep within the magnetosphere, nearing perijove, when the revised sequence was uplinked. There was a real chance that radio interference would cause it to be incorrectly received, in which case the spacecraft would reject the commands and remain 'safed', but the transmission was accepted. It was fortunate that the Io encounter was two and a half hours after perijove, rather than before it! "It was a heroic effort to pull this off," reported Erickson. "Our team diagnosed and corrected a problem that we had never come across before, and they put things back on track."

The particles and fields investigation at perijove had been lost, but the Io fly-by was back on. "We waited for years for this encounter, and we did everything in our power to make it happen," assured Eilene Theilig, the spacecraft's sequence chief.

"Before every encounter, we go through various contingency scenarios," explained Nagin Cox, one of the systems engineers, "including a possible safing. That preparation paid off and the anomaly resolution team swung into action quickly."

"It was poetry in motion," observed Olen Adams, working on the Galileo's C&D subsystem. "People were moving around these aisles like it was a relay race. Every single person had to perform perfectly. We could not afford a single 'gotcha'. If one person got sick, or one PC crashed, or a command didn't make it to the spacecraft, it wouldn't have worked."

"I knew that if the radiation had triggered one memory fault, there was a good chance it'd trigger another," reflected Tal Brady, the C&D subsystem manager, "I was very relieved, first when we got the spacecraft out of safing, and then again – later – when the fly-by data was recorded successfully."

Fifty minutes out from Io, the particles and fields instruments began an hour's sampling of the moon's environment to study the way in which the Jovian magnetic field first snatches and then progressively accelerates ionised material ejected from Io.

The magnetometer and plasma wave spectrometer teams hoped that the close fly-by would penetrate into the electromagnetic 'bubble' that the spacecraft had detected in December 1995, and confirm that Io had a magnetic field. The dust detector team was hoping to characterise the particulate debris ejected by the volcanic plumes. And the Deep Space Network was recording doppler data to probe the moon's internal mass distribution.

LOW PASS

"We'll be targeting four major volcanoes," explained Duane Bindschadler, "Pillan Patera, Prometheus, Loki, and Pele. The pictures will be great!" The programme called for over 100 images of ten targets, all with a resolution better than the very best achieved by Voyager and a few images – at closest approach – with a resolution as fine as 7 metres. "We want to learn more about the differences and similarities between volcanoes on Io and volcanoes on Earth."

Although the infrared spectrometer images would have a resolution of only a few hundred metres, this was two orders of magnitude better than that on the 127,000-kilometre pass on the twenty-first orbit. "You can hide a lot in 60 kilometres," noted Bill Smythe. The resolution of the integrated photopolarimeter and radiometer would vary from 56 kilometres per pixel when distant, to 13 kilometres at closest approach. There was therefore scope for real discoveries by three of the imaging instruments.[50]

As Io loomed, the infrared spectrometer examined Loki's caldera in darkness in search of hot spots. It advanced to Pele, which was near the terminator, and nearby Pillan Patera where the early morning Sun offered ideal conditions to study the topography of this recently active caldera. At this time the camera joined in, taking high-resolution images of each volcanic complex in order to document the site, and the infrared spectrometer peered into the caldera for signs of ongoing activity.

Soon after the spacecraft's closest approach to Io, attention turned to Ot Mons on Colchis Regio. The origin and structure of Io's isolated mountains was still debated, so it was hoped that high-resolution imagery would help to decide between the competing theories.

Next on the schedule was a high-resolution observation of Zamama, where lava seemed to be erupting from a fissure rather than from a central vent. Determining

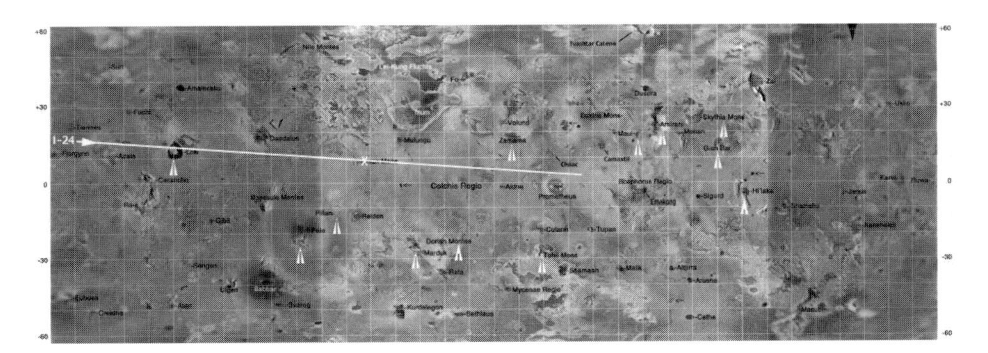

This Cartesian projection of Io traces Galileo's ground track during its fly-by in October 1999. The spacecraft approached the trailing hemisphere in darkness, crossed the dawn terminator, and then withdrew in daylight over the anti-Jovian hemisphere.

the temperatures of the lava flows around Prometheus was next, followed by a regional image of Colchis Regio in order to provide context for the earlier high-resolution imagery, some context for Zamama, and then further imagery of Prometheus for stereoscopic analysis. The infrared spectrometer observed a green 'golf course' on Dorian Mons, and another at Tohil, where a mountain and a caldera are in close association.

As it climbed over the anti-Jovian hemisphere, Galileo took medium-resolution images of the Amirani-Maui complex (which contains the longest lava flow in the Solar System), Skythia Mons, and Gish Bar Patera, then an oblique view of the topography of the Hi'iaka caldera and mountain unit on the evening terminator. By this time, Pillan Patera was on the far limb, perfectly positioned to display its plume, if this was still active.

FAST CAMERA

As Galileo switched to playback, the engineers relaxed. The "calculated risk" had paid off. "Everything is looking pretty good," reported Jim Erickson. "None of the normal things that would have indicated problems have shown up." There had been the 'safing' event earlier on, of course. "We are thrilled that the spacecraft handled this fly-by so well, particularly because it endured a strong dose of radiation." The spacecraft had survived its deep torus passage, so the prospects for another encounter on the next orbit were excellent. However, as the data trickled in it was discovered that the radiation had interfered with some of the observations.

Most of the imagery had been shot in 'fast camera' mode, in which the solid-state imaging system processed an image by averaging the brightness, to compress it, and then flushed it to tape as soon as possible in order to minimise the time that the image was resident in the CCD array. However, it was found that radiation had caused the process to become desynchronised, and this type of imagery was seriously degraded. It was decided that the camera's electronics had suffered sufficient radiation to prevent this readout mode from functioning properly and instead of summing 2×2 blocks of adjacent pixels, the faulty algorithm added pixels from opposite sides of the image.

"When we're flying the spacecraft through the high-radiation zone near Io's orbit, we have to plan for the likely radiation, and figure out how to deal with it," Erickson pointed out. "We used several different modes, to see how each would work. Now that we know this particular camera mode didn't work well amidst the radiation, for the next Io fly-by we'll use another of our six modes [which was not degraded]."

As soon as it was realised that the 'fast camera' imagery was corrupted, the spacecraft was told to replay the data from the other instruments. However, once the rest of the data had been recovered, the flawed images were downloaded and then examined by Greg Levanas and Kenneth Klaasen, imaging specialists at JPL's Measurement Technology Center, who first figured out how the summation process had malfunctioned and then devised a procedure using LabVIEW software supplied by National Instruments in Austin, Texas, to 'repair' the images.

"It was like watching a scrambled cable signal on television," explained Torrence Johnson. Their task was "to unscramble the signal – to break the code that was inadvertently introduced by the radiation near Io."

Laszlo Keszthelyi, a Galileo research associate at the University of Arizona in Tucson, was impressed. "They only had one-fourth of the data needed to reconstruct the images." Once they had inspected the corrupted images, they "found a way to intelligently guess the missing bits; it seemed to be mathematically impossible, but they pulled it off."

The most intense radiation had also induced a fault with the infrared spectrometer. It turned out that the motion of the diffraction grating was anomalous. High-spatial resolution imagery had been obtained, but it was not across the full wavelength range. It was sufficient to permit temperatures to be measured, but not to derive surface composition. "Despite several attempts, the grating is still stuck," reported Robert Carlson later, "probably due to a failure in the drive circuitry. It's not as bad as it might sound, we can still collect multispectral images at fifteen quite useful wavelengths."

On 19 November, a 'Space Sciences Update' in the James Webb Auditorium at NASA's Washington Headquarters presented the preliminary findings from the Io fly-by. A few hours later, the University of Arizona's Gerard Kuiper Space Sciences Building hosted an amplified briefing by the Lunar and Planetary Laboratory's staff.

With at least 100 active sites, Io had been rather more active than expected – with Loki, Pele and Prometheus being the most active.

Io has some low shield volcanoes – notably Ra Patera – but most of its volcanic sites are sunk into the plains. A caldera is a collapse structure which forms after a major eruption when magma drains from a shallow reservoir and the overriding crust, lacking support, collapses into the cavity. The terrestrial equivalent is a resurgent caldera. Yellowstone, one of the largest, appears to be located over a hot mantle plume. Nevertheless, it erupts only every half million years or so. When it erupts, it is enormously violent and blasts out vast quantities of pyroclastic.[51] Because Io is so hot, its calderas are active virtually continuously and so they do not build up pressure for explosive eruptions.

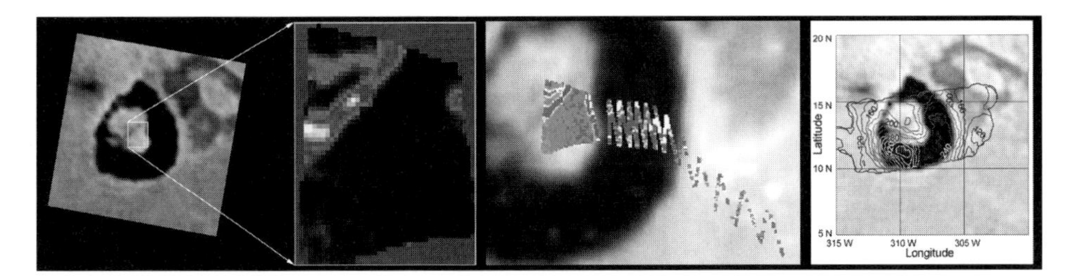

As Galileo approached Io for its first close fly-by, the infrared spectrometer determined the thermal characteristics of the southeastern part of the 'island' within Loki's caldera (second left, with an earlier visual view for context, left). As the instrument was slewed towards Pele, it continued to take data (centre), to extend its survey over the dark floor of the caldera and across the plain. The integrated photopolarimeter and radiometer was able to chart the temperature field across the caldera (right).

Most of Io's calderas are basically circular, but some of them have a linear side, which suggests that the location and the shape of a caldera is partly determined by faults in the silicate lithosphere immediately below the blanket of flows and pyroclastic. Some of these faults may be related to those associated with mountain-building. About half of Io's calderas have dark floors, and thermal studies[52] show that these have recently been active.

LOKI

The most 'powerful' volcanic site in the Solar System, Loki produces more heat than all of the Earth's active volcanoes combined. "Loki has an enormous caldera," said Rosaly Lopes-Gautier, the infrared spectrometer's science coordinator, "which is repeatedly flooded by lava over an area larger than the state of Maryland."

Long-term ground-based infrared monitoring of Io since the 1980s had revealed that Loki underwent periodic 'brightenings', typically lasting several months, about once per year.

Galileo had observed Loki in darkness on orbits 7, 9, 16 and 22. A major brightening was detected on 12 March 1997 by the Infrared Telescope Facility on Mauna Kea in Hawaii. When Galileo observed on orbit 7, Loki's eruption was at its peak, but the activity had subsided by orbit 9. The evolution of the temperature distribution implies a spreading and cooling lava flow.[53] The fact that the solid-state imaging system (which is better suited to higher temperature sites than is the infrared spectrometer) saw Pele and Pillan Patera glowing in the dark, but did not see Loki on that occasion was consistent with its lava flow having cooled. Loki brightened again in May 1998, and was observed by Galileo a month later, on orbit 16.

By August 1999, Loki was at a low point in its eruptive cycle, and the Infrared Telescope Facility observed it glowing faintly on 9 August 1999. The next day, however, roughly 20 hours before Galileo's fly-by on orbit 22, it had 'bloomed' by a factor of ten. The Wyoming Infrared Observatory monitored the brightening develop through September. As Galileo made its close fly-by in October it passed directly over Loki.

The infrared spectrometer's thermal scan across Loki's caldera achieved a real milestone in the mission,[54] because it was able to map the *inside* of a caldera with resolutions varying from about 25 kilometres per pixel for general context, down to 500 metres per pixel for the detailed study. The only fault was the failure of the diffraction grating's motor, which limited it to 15 wavelengths. However, the accumulation of so much data across so few channels provided a thermal map with a very high signal-to-noise ratio, in which fine-scale temperature variations could be discerned with confidence.

This investigation focused on the southeastern quadrant of the 'island' within the caldera. This is elevated above its surroundings, and is remarkably flat. Despite being within an active caldera, the white material is frozen, implying that it is a thick deposit of sulphur dioxide and that the underlying structure had not been active recently. However, the centre is warmer than the periphery, indicating that the heat flow from the centre is somewhat greater. Significantly, the warmer part is also marked by albedo variations.

The infrared spectrometer continued to record data as the scan platform swung southeast towards Pele, sampling a zig-zag pattern across the floor of the caldera and over its rim onto the frozen surface outside. The heat flowing from the dark caldera material is about ten times greater than that from the island, which is in turn ten times that from the surrounding plain.[55]

The integrated photopolarimeter and radiometer's coordinated study of the temperatures in and around Loki were presented as a contour map, which showed that most of the lava in the vast horseshoe of the caldera is at a remarkably uniform 250 K. In fact, enormous amounts of volcanic heat are needed to maintain such a large area at that temperature. At only 125 K, the surrounding terrain is typical of sulphur dioxide frost on an inter-vent plain.

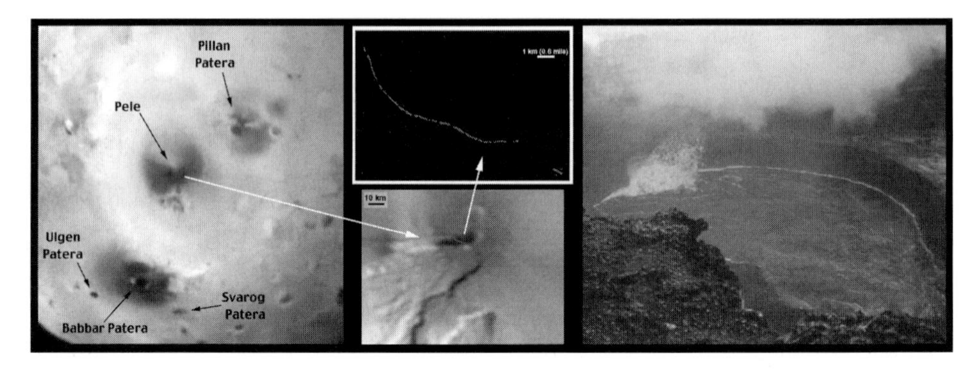

During its close fly-by in October 1999, Galileo peered down into Pele's caldera. The thin curving line, which is 10 kilometres long and 50 metres wide, is believed to be hot lava exposed where a lava lake makes contact with the caldera's rim, such as occurs on a far smaller scale in the lava lake in Pu'u O'o on Kilauea in Hawaii (right). (Courtesy Laszlo Keszthelyi, Hawaiian Volcano Observatory and USGS)

It is likely that the site of the recent eruption is the southwestern corner of the caldera, as this is much hotter – at least 400 K. The otherwise uniform temperature of the caldera floor implies that there is a uniformly thick frozen crust covering the lava lake. The thermal brightenings evidently correspond to when lava broaches the surface. It was remarkable that even though these eruptions release so much energy, the caldera is so vast that such events cause so little change. What must conditions on Loki be when it is really active?

PELE

As Galileo neared the terminator, it was able to look straight down at Pele, and saw lava glowing brightly. This finally pin-pointed the eruptive centre. An image taken through a near-infrared filter that viewed only material hotter than 1,000 K observed only a thin curving line some 10 kilometres long, but only 50 metres across. Although the maximum temperature measured by the infrared spectrometer was 900 K, its high spatial resolution scan did not actually cross the most active area. Temperatures of 1,300 K had been measured in Pele previously, but as such lava would cool and become invisible to this filter in a few minutes, this image records a part of the volcano which is at most *several minutes old.* Superimposing the thermal data on the best visual image indicates that the hot lava is on the margin of the caldera, which suggests that the caldera contains a fluid lava lake encrusted with a cooling scum that is repeatedly broken where it is in contact with the cliff-like walls. Variations in brightness along the line can be interpreted as being due to differing amounts of hot lava exposed at the surface. In contrast to volcanoes which erupt in pulses, and flood large areas with lava that cools over time, Pele's intensely hot vent is remarkably stable, which suggests that this extremely active lava lake is constantly exposing fresh lava. It is strikingly similar in form to that of Pu'u O'o on Kilauea in Hawaii, except that its area is thousands of times larger, measuring fully 10×15 kilometres.[56]

The lava lake in Pu'u O'o on Kilauea in Hawaii is only 100 metres across. It is feeding lava through a deep channel into a lava tube. (Courtesy JD Griggs, USGS)

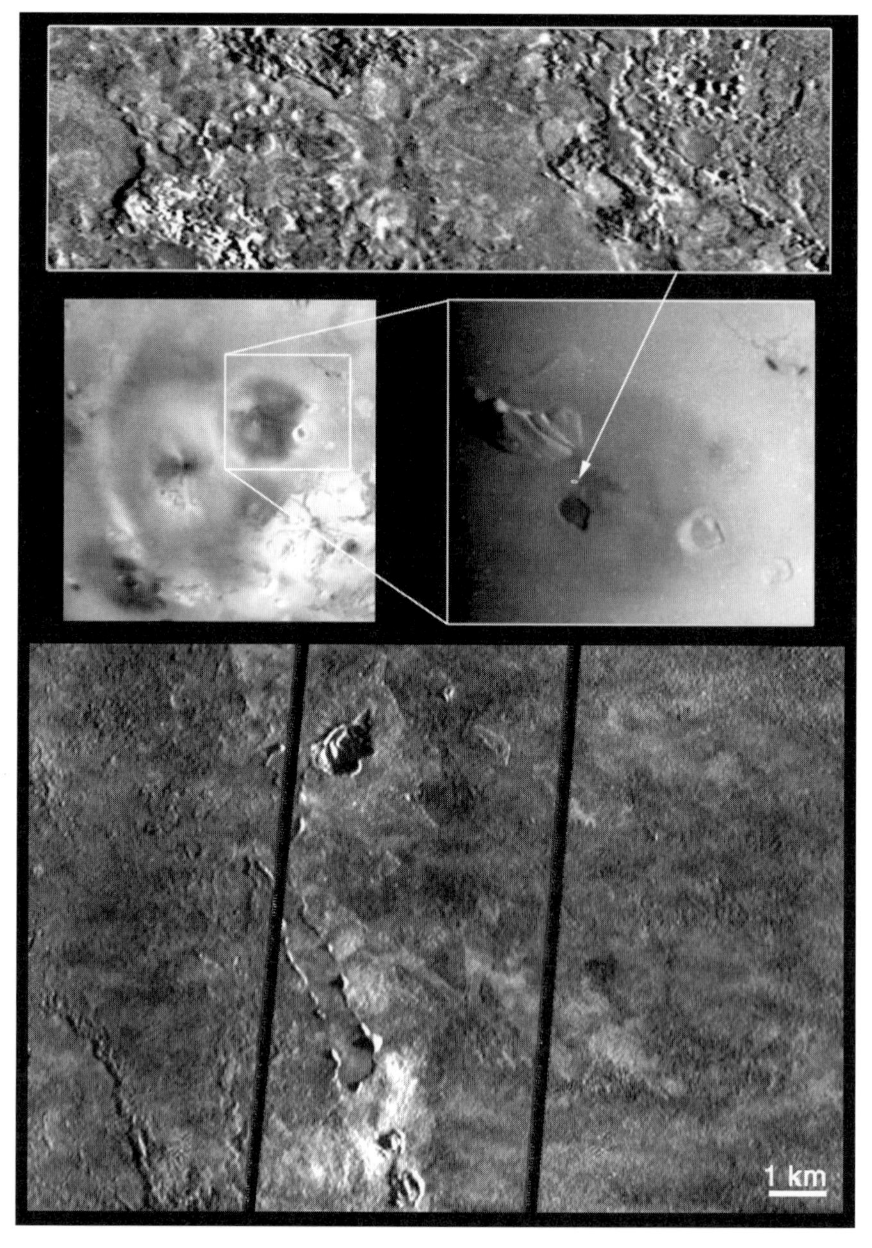

During its close fly-by in October 1999, Galileo inspected a small section of the recent lava flow extruded by Pillan Patera. Taken from a height of 620 kilometres, this image (top) spans an area 2.2 × 7.2 kilometres and shows features as small as 9 metres. The complex mix of smooth and rough areas, and clusters of pits and domes are similar to features found on Earth, but this variety of different types of lava flows within such a small area is remarkable.

The Voyagers identified sulphur dioxide in plumes, and when the Hubble Space Telescope started observing Io, it was able to identify sulphur monoxide and monatomic sulphur vapour. As Galileo made its fly-by, the Hubble detected molecular sulphur vapour in Pele's plume.[57] Once the data had been combined, a model based on plume chemistry[58] suggested that Pele's lava is 1,450 Kelvin, in good agreement with Galileo's measurements of 1,200 to 1,470 K.

It would appear that Pele's lava is a magnesium-rich orthopyroxene, and that the plume is derived from a fire fountain in the caldera's lava lake rather than by explosive degassing by the lava as it rises towards the vent.

PILLAN PATERA

When Pillan Patera made its dramatic appearance in 1997, it laid a 400-kilometre diameter deposit of dark pyroclastic on the northeastern arc of Pele's plume halo. Galileo observed it from time to time over the following years, and documented the changes. The vent is a classic example of a caldera sunk into the plain. In addition to the dark pyroclastic, it has sent a flow of dark lava north across the plain.

As Galileo passed 600 kilometres overhead, it took a high-resolution strip covering an area of about 2.2 × 7.2 kilometres. The Sun had only just risen, so the long shadows emphasised topographic relief. This revealed a complex mix of smooth and rough terrain, with clusters of pits and domes; some house-sized. The cliff to the west of the flow varies in height between 3 and 10 metres. Although similar volcanic terrains are found on Earth and Mars, such a variety of different types of lava flows had not been seen before within a small area, so Io's surface is evidently being modified on a local scale by a wide variety of volcanic processes.

Galileo's 25-metre resolution imagery crosses the margin of the new lava flow. A channel 140 metres wide and 3 kilometres long[59] has fed the flow, which is significantly lower than the adjacent plain. Although the lava flow may have pooled in a pre-existing shallow depression, it is possible that the superheated lava eroded the plain and formed its own cavity.

Up to this time, it had not been possible to inspect Pillan's dark pyroclastic in detail to rule out the possibility that it was black sulphur. It was confirmed to be silicate-rich ash. The Pillan data is the most convincing evidence so far that Io produces ultramafic lavas. The difficulty in proving this is that surface deposits are rapidly contaminated by sulphurous compounds, so it was necessary to study a very young uncontaminated flow.

ZAMAMA

Zamama is believed to have formed during the time period between the Voyager visits in 1979 and Galileo's first look at Io in the summer of 1996. While volcano-monitoring, Galileo imaged Zamama on its fourteenth orbit from 250,000 kilometres, when a plume was present, and again from half that distance on its

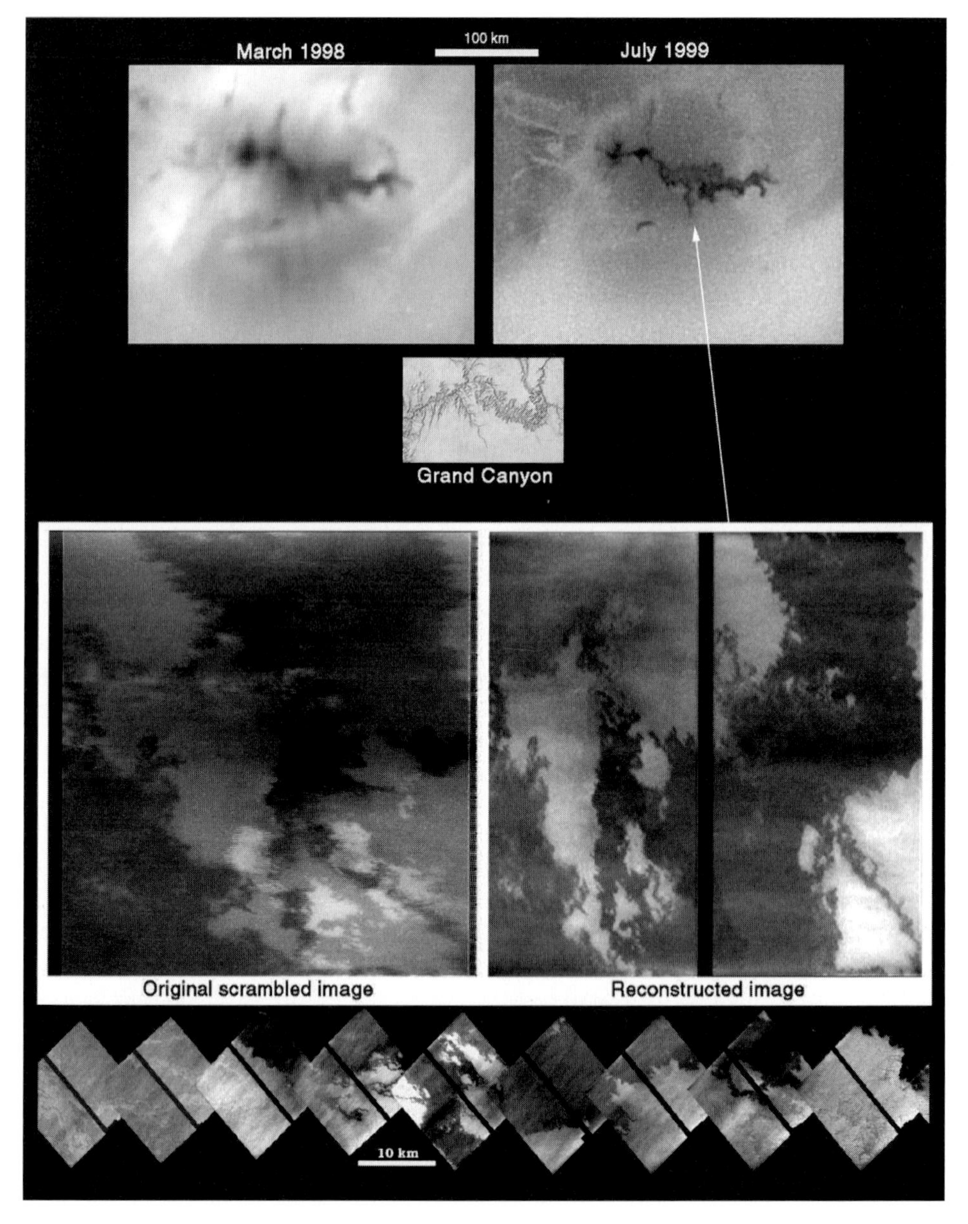

During its close fly-by in October 1999, Galileo imaged part of the Zamama lava flow, but the solid-state imaging system's 'fast camera' mode was disrupted by the radiation so close to Jupiter. However, clever processing 'reconstructed' the image.

twenty-first orbit, by which time the plume had ceased. Combining data from the infrared spectrometer and the solid-state imaging system showed that its lava is hotter than 1,100 K, which is too hot for sulphur. Its generally linear character prompted initial speculation that Zamama is a fissure eruption, but this turned out not to be the case.

A sequence of images across Zamama with a resolution of about 50 metres per pixel were scheduled for the October fly-by to determine whether the lava was flowing in open channels or in well-insulated lava tubes, but the 'fast camera' mode images were corrupted by radiation interference.[60] However, they were able to be 'repaired'. The convoluted margins of the flows suggest a smooth ropey lava known by its Hawaiian name of 'pahoehoe'. The vent is actually at the western end of the long linear feature. The lava appears to emerge from the vent by way of a series of narrow channels (less than 1 kilometre wide), then merge and flow far to the east. Overall, Zamama strongly resembles a terrestrial basalt flow.

PROMETHEUS

Remote observations of Prometheus by Galileo's infrared spectrometer during its volcano-monitoring campaign confirmed that Prometheus is extremely active. In fact, it may have been continuously active over the past 20 years because its plume has been in evidence every time it has been sought since being discovered by Voyager, prompting the name 'Old Faithful of the Outer Solar System'.

Prometheus was in the centre of Io's disk as Galileo drew away from the moon. In earlier, lower resolution imagery, it had appeared that all of the dark material comprised a single long lava flow. Galileo's 120-metre resolution was ten times better than the previous best view of the volcano and its immediate surroundings. The northeastern end of the long dark feature is actually a D-shaped caldera, 28 kilometres long and 14 kilometres wide. The caldera's floor is coated by plume fall-out, so it isn't hot. An infrared wavelength sensitive to thermal emission rather than reflected insolation revealed the hottest areas. It identified two major hot spots: the cooler one is located at the southern rim of the caldera; the hotter one is at the western end of the lava flow. Galileo had already measured the lava temperature at 1,100 K, so this is silicate-rich.

The sulphur-rich snowfield around the caldera is intriguingly hummocky, possibly due to the supersonic blasts from the vent 'plastering' viscous material onto one side of pre-existing mounds, and progressively building them up.

The high-resolution imagery revealed the lava flows in unprecedented detail, and produced several insights. "It appears that Prometheus has characteristics remarkably similar to those of the Kilauea in Hawaii," said Laszlo Keszthelyi, "with flows that travel through lava tubes and produce plumes when they interact with cooler materials."

Kilauea has been continuously active for more than 16 years so it, too, is long-lived. Like Prometheus, Kilauea's lava tubes transport the lava from the vent to the front of the flow, but they are only 10 kilometres long. The lava then disgorges into the Pacific Ocean and generates plumes of steam.

The model derived from all the latest data proposed that the deep reservoir is beneath the caldera, but it has broached the surface 15 kilometres south of the rim;

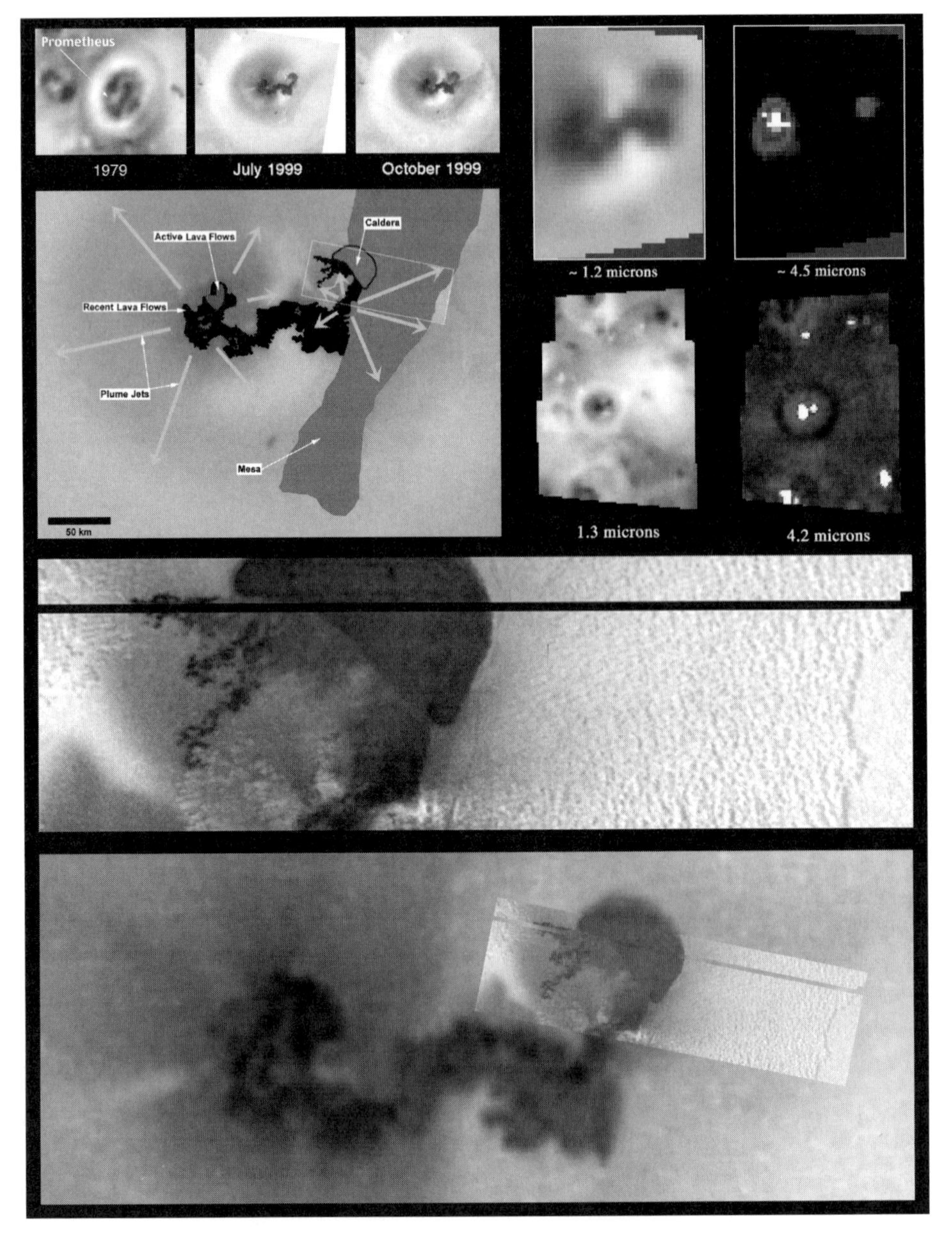

During its close fly-by in October 1999, Galileo discovered that the D-shaped feature at the eastern end of Prometheus's dark flow is actually a caldera. In addition to extruding several short sinuous flows, this has given rise to a vast flow extending 100 kilometres to the west. Most of this flow has been extruded since 1979. Back then, the plume rose from just south of the caldera, but it now rises from the front of this flow. In fact, when the contextual image (bottom) was taken, the spacecraft was peering straight down into the plume.

this corresponds to the eastern hot spot. The lava then travels 100 kilometres west through lava tubes to the front of the flow, where it disgorges onto the surface and produces the plume by vaporising sulphur-dioxide frost; this is the more intense hot spot. A detailed analysis of the thermal data found a third, somewhat fainter hot spot near the kink in the dark flow (about a third of the way from the source) where there seems to be a smaller break-out of lava.

A comparison of the images taken by Galileo in July and October documented the pace of activity. The break-out from the kink in the lava flow appears to have taken place during this period, extruding a new dark deposit to the north of the older lava flow. It was also apparent that the gas discharge from the vent at the eastern end of the flow has increased, because there is a new fan of dark material streaming out from this location. A recent bright crescent-shaped deposit across the middle of Prometheus suggests that the main plume – to the west – has yielded to the increased gas emission to the east. The 1979 plume rose from the southerly hot spot near the caldera.

This fly-by provided the first opportunity to study the pyroclastic in detail. A 4.1-micron image readily shows Prometheus's fall-out. Surprisingly, there is little correlation between the concentration of sulphur dioxide and albedo. The ring is asymmetrical and the sulphur dioxide is concentrated towards the outer edge of the ring which is prominent in visible-light imagery and at short infrared wavelengths. Possibly the white ring is contaminated by a short-chained sulphur allotrope.

"I expect we'll be changing some of our modelling," reflected David Goldstein, who had been studying plumes in terms of venting under pressure from deep fractures. "At least some of the plume material does not come from nice neat volcanic vents; it seems to come from lava overflowing ice/frost, or perhaps slumping of cliff faces which exposes fresh sulphur dioxide ice. Either way, the source flow is more complicated."

Galileo discovered that the plume from Prometheus is produced when hot lava crossing the frozen plain sublimates sulphur dioxide pyroclastic frost. The manner in which lava from Kilauea produces a steam cloud when it pours from a lava tube into the ocean may be a close terrestrial analogy. (Courtesy Laszlo Keszthelyi, Hawaiian Volcano Observatory and USGS)

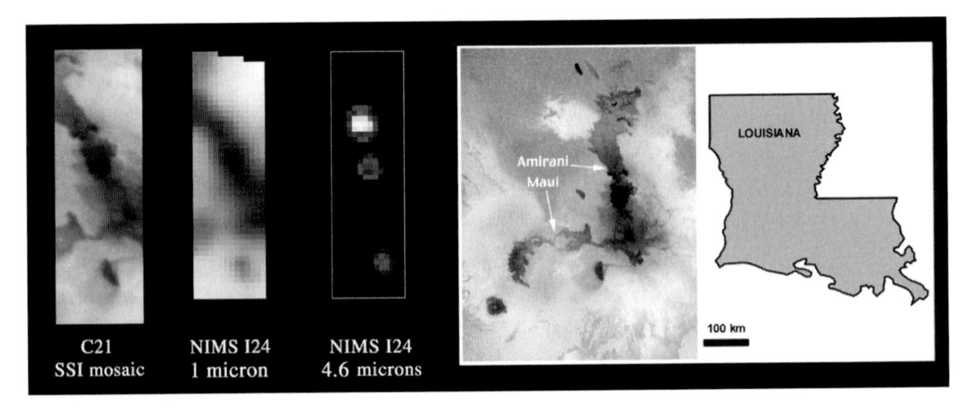

The infrared spectrometer observed the Amirani-Maui volcanic complex during its close fly-by in October 1999, pin-pointing a number of hot spots.

This general view shows the tremendous scope of the Amirani-Maui volcanic complex and a number of nearby mountain units.

AMIRANI-MAUI

The Amirani-Maui volcanic complex had been imaged by Galileo on its 124,000-kilometre encounter on orbit 21. They had been thought to be two separate features. However,[61] Galileo found Maui to be the active front of a lava flow from a vent at

Amirani extending westward for more than 250 kilometres. The infrared spectrometer saw a hot spot at Maui, so this enormous flow is evidently still active. Other flows were seen extending northward from the Amirani vent.

During Galileo's Io fly-by, the infrared spectrometer scanned the Amirani complex from a distance of 25,000 kilometres, yielding images with a resolution of about 6.5 kilometres per pixel. The 1-micron image shows light and dark areas on the surface which served to line up the spectrometer data with the camera's image. The 4.6-micron image documented the thermal emission from three separate hot spots whose locations corresponded to the darkest features in the visual image. This reinforced the belief that there is a correlation between hot spots and dark material. Interestingly, although two of the hotspots are on the Amirani flows, the third is at the eastern edge of the distinctive caldera to the south, which may be working up to an eruption.

MOUNTAINS

Io's mountains stand in isolation, they do not form ranges – which indicates that they are not the result of large-scale tectonism, as are the Earth's 'mobile belts'. The fact that they rise directly off the plain, without foothills, makes them all the more impressive. When it had been thought that Io's volcanism was primarily sulphurous, it had been obvious that the mountains could not be volcanic constructs because sulphur does not have the structural strength to form a peak. So what are they?

Even as Galileo was exploring the Jovian system, the origin of Io's mountains remained a topic of research. In a recent paper,[62] Paul Schenk of Lunar and Planetary Institute in Houston and Mark Bulmer of the Center for Earth and Planetary Science at the National Air and Space Museum analysed the shape of Euboea Mons to assess a possible landslide. Galileo had yet to improve on Voyager's coverage of this feature, so they produced a topographical map using a pair of stereo images from 1979. Located about 40 kilometres east of Creidne Patera far in the southern hemisphere, Euboea Mons is 175 × 240 kilometres, and has a prominent curving ridge crest peaking at about 10.5 kilometres high. This divides the steep southern flank from the much smoother and shallower northern slope leading to a thick and heavily ridged deposit possessing a scalloped outer margin. It seemed as if the entire north face had slumped, and the northern deposit was an apron of rock and debris from a major avalanche – if so, it would be the largest debris apron in the Solar System.[63] If Euboea Mons was ancient, its steep southern flank suggested structural strength, yet the collapsed northern flank indicated structural failure. Because Euboea Mons was relatively intact, its polygonal shape lent support to the proposition that such mountains are individually faulted and uplifted blocks of crustal material. Schenk and Bulmer argued that Euboea Mons rests on a shallow, overthrusting fault dipping to the north. This model avoided the contradiction of a single edifice displaying both structural strength and structural collapse by positing that the massif is basically rigid; the failure was triggered by the act of uplift, and the debris is the volcanic material which had been deposited on the block. The fact that

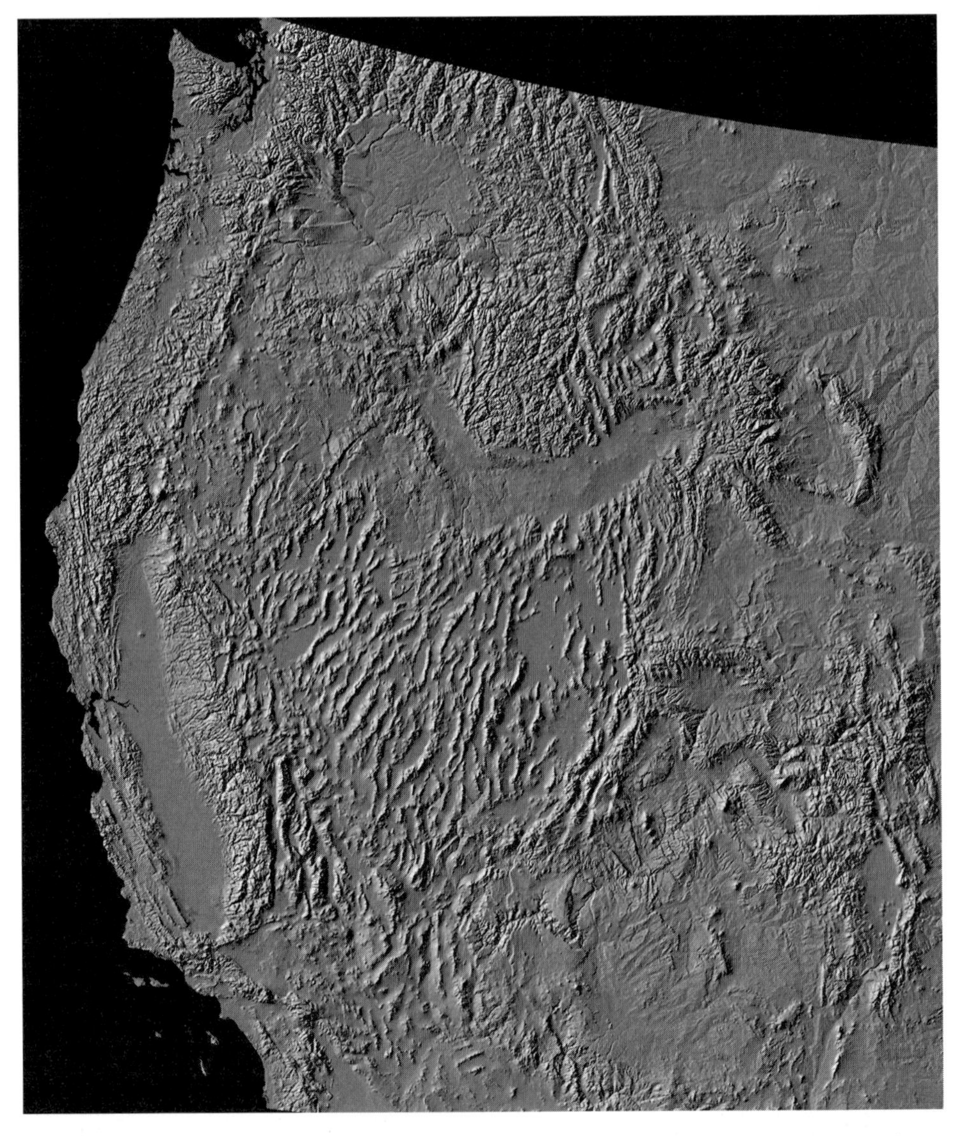

The western region of the United States has been shaped by tectonic forces. Unlike the isolated peaks on Io, terrestrial mountains form 'mobile belts'. (Courtesy USGS)

the northern flank is smooth implies that it is a formerly horizontal surface. The 6-degree slope presumably matches that of the fault. The southern flank is therefore a heavily eroded, but still 10-kilometre tall exposure – a magnificent 'window' into the structure of the crust for a future field geologist.

So what caused the fault and the uplift? Io is so volcanically active that it is resurfaced at an average rate of 1 centimetre per year – at this rate even the most deeply buried volcanic layers will be repeatedly recycled. Schenk and Bulmer argued that, as layer after layer of new volcanic material is erupted, older buried material will be forced to subside. As it does so, it will suffer lateral compression, which may induce thrust-faulting such as appears to have created Euboea Mons. This is appealing, because it draws together several lines of evidence. The mountains are massifs and they could have been upthrusted recently – they are not *necessarily* ancient. Despite extensive slumping, the underlying structures are rigid silicate blocks; it is the weaker sulphurous lava and the pyroclastics deposits (laid down prior to the uplift) that have slumped.

This suggests that Io's mountains will tend to be polygonal and combine steep slopes with shallower slopes showing signs of collapse. A major theme for Galileo's fly-by was therefore to investigate the shapes, slopes and scarps of the mountains.

Significantly, Schenk and Bulmer cited no evidence of lava flows, vents, calderas or any other volcanic features on Euboea Mons. This indicates that the deep fracture which prompted the uplift did not provide a route for lava to reach the surface to modify the massif. Evidently, although Io's volcanism has given rise to vast calderas, it has not produced edifices similar to Earth's stratovolcanoes – which is another fact for comparative planetologists to ponder. Io's calderas have several unique characteristics.[64] In many cases, calderas abut mountains,[65] which suggests a deep structural relationship.[66, 67] The calderas would appear to have formed more recently than the nearby and evidently structurally related uplifted massifs.

When Skythia Mons and nearby mountains were imaged at resolutions of 500 meters per pixel, they revealed topography ranging from angular peaks to gentler plateaux surrounded by gently sloping debris aprons. The angular mountains appear to be younger than the rounded ones. The ridges parallel to the margins imply progressive slumping under the force of the moon's weak gravity.[68] Evidently, the mountains are in various states of deterioration. Their absolute ages are not known,

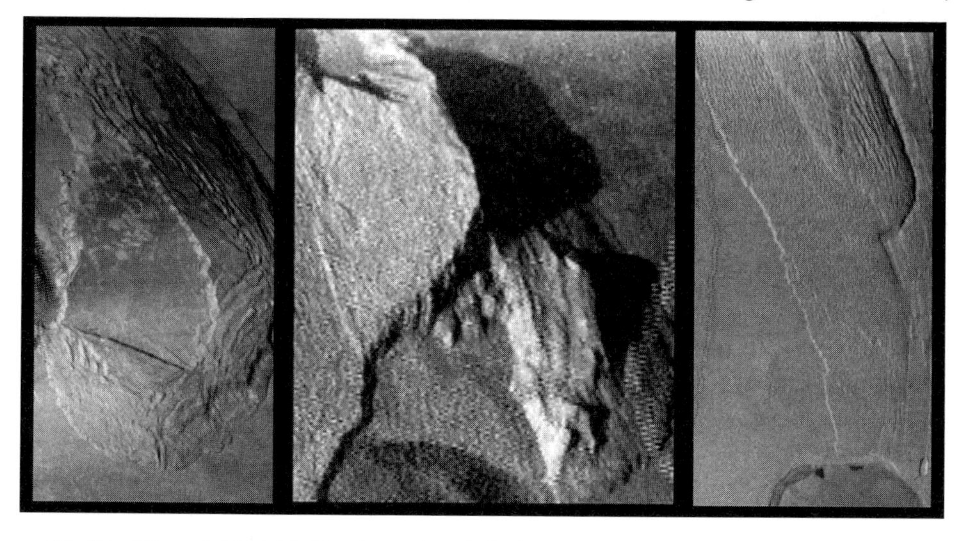

Skythia Mons (left), Gish Bar (centre) and the low terrain unit south of Monan Patera, all east of the Amirani-Maui complex (see earlier figure).

Ot Mons, on Colchis Regio, was inspected at high resolution during Galileo's close fly-by in October 1999. It shows a rather lumpy landscape strikingly different to other areas of Io observed at comparable resolution.

but the fact that they seemed to be collapsing under their own weight indicates that they are fairly recent creations. Similar ridges are present at the base of Olympus Mons on Mars (which *is* a volcanic edifice) so a comparative analysis may provide insight into the active processes on both planets.

One of the images shows a small fraction of the north face of Ot Mons in Colchis Regio at the very high resolution of 9 metres per pixel.[69] The lumpy form of the "relatively nondescript low mountain" (as Laszlo Keszthelyi expressed it) indicates that this particular massif is in a highly degraded state. Intriguingly, the albedo variation on this mountain slope is the greatest seen to date on Io.

APOJOVE MANOEUVRE

"The Io fly-by has shown us gigantic lava flows, and lava lakes, and towering collapsing mountains," summed up Alfred McEwen. "Io makes Dante's '*Inferno*' seem like another day in paradise".

"We've been having a feast looking at the material from Io," noted Rosaly Lopes-Gautier. "We've been waiting for such high-resolution images of Io for more than ten years."

When the engineers checked out Galileo after it emerged from the torus and established it to be in very good shape, the second Io fly-by was authorised. On 2 November, the day before the spacecraft passed apojove and started back in, it made a trim burn, so that it would pass close over the moon's south polar zone. This trajectory offered a Europa fly-by on the way in, with the 300-kilometre Io fly-by following two hours after perijove.

POLAR PASSAGE

For the engineers, the Io fly-by on 26 November 1999 became a real "white knuckle ride". Galileo's C&D Subsystem 'safed' itself two hours before perijove. The signal

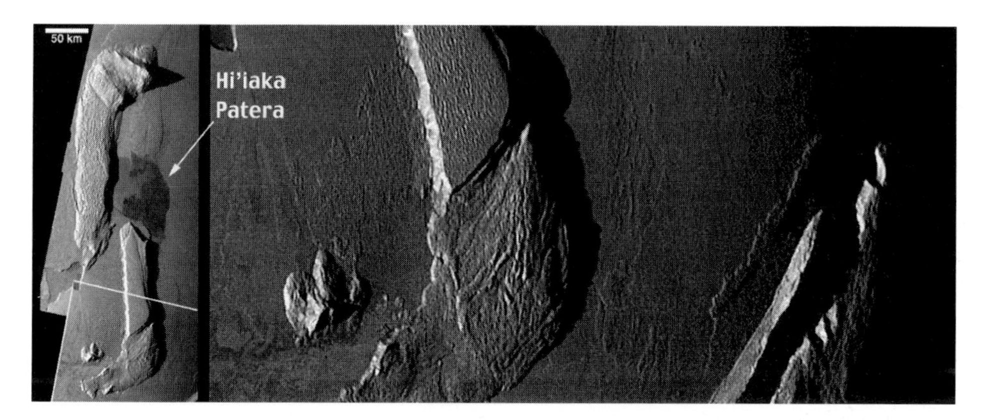

During its close fly-by in October 1999, Galileo documented the large mountain blocks near Hi'iaka Patera on the evening terminator with a resolution of 267 metres per pixel.

took 35 minutes to reach Earth. "With so little time to spare," recalled Jim Erickson, "it would have been easy to think 'no way can we recover this'. But our team members rose to this challenge, in some cases leaving behind half-eaten Thanksgiving dinners." A range of contingency plans had been drawn up following the previous fly-by. "This planning paid off in a big way." They managed to resume the encounter sequence 4 minutes after the closest approach to Io, to make the outbound observations of the daylit hemisphere, but the night-time observations scheduled for the approach had been lost, as had the best opportunity for imaging the south polar zone and the high-resolution studies of the Emakong, Tupan and Shamshu calderas. However, the medium resolution imagery of these targets (which had been intended to provide context) was taken on the way out, together with Malik, Zal, Culann, Prometheus and a caldera complex at high northern latitude.

Galileo achieved apojove on 15 December, so the short orbit gave little time to transmit the taped imagery.

EMAKONG

Emakong Patera, a few hundred kilometres east of Prometheus, is essentially surrounded by lava flows which are notable for being yellow-white (most flows – both on Io and other planets – are dark). This prompted speculation that Emakong's caldera spilled moderately hot sulphurous lava rather than very hot silicate lava. There are dark sinuous channels in the light flows which so resemble lava channels winding their way down shallow slopes on terrestrial volcanoes that they have been similarly interpreted on Io.[70] Their serrated margins imply that a fluid lava has embayed all the peripheral crevices as it ran along the channel. While some of the channels feed a darker lava flow, the largest of them, about 100 kilometres long and half a kilometre wide, feeds a light-toned flow. Evidently, the dark material is black sulphur which evolves to yellow as it cools. Emakong is one of the best cases for the sulphurous volcanism that was once believed to be predominant on Io. The fact that

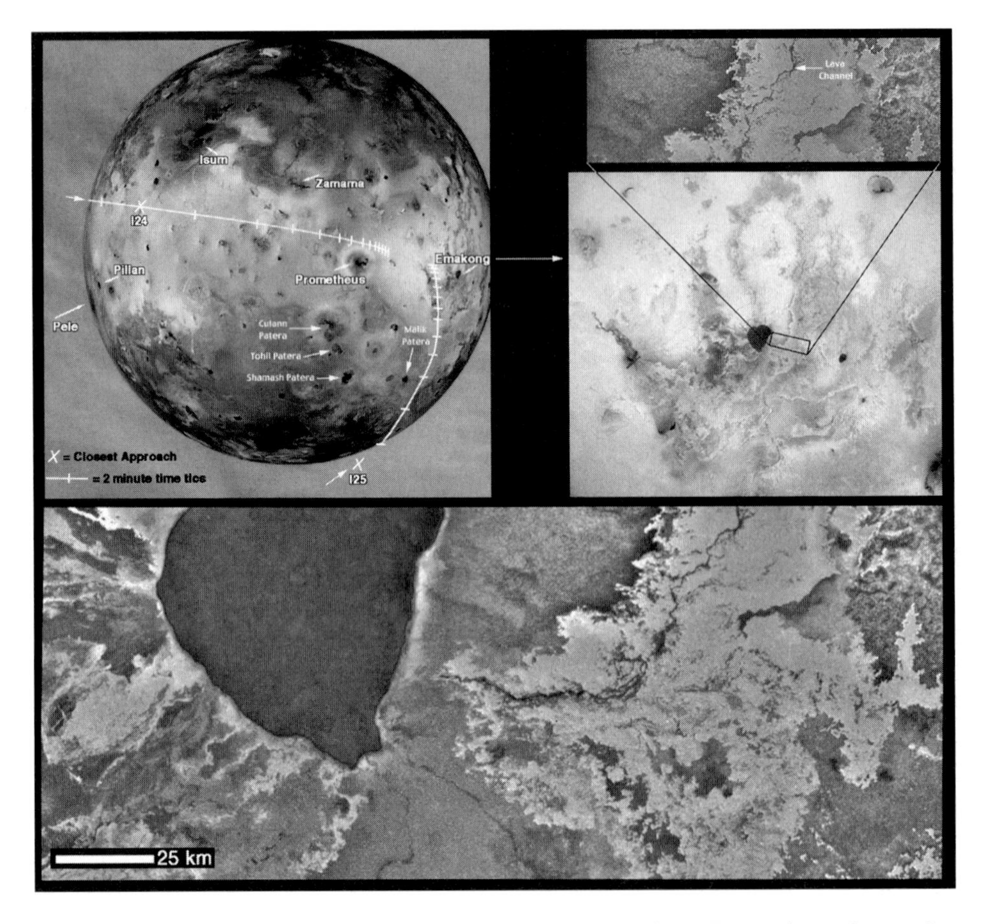

During its fly-by in November 1999, Galileo inspected Emakong Patera, located east of Prometheus, and found that its light-toned lava flows are sustained by dark sinuous lava channels.

the infrared spectrometer did not observe any hot spots implies that the caldera has not been active for some time.

TVASHTAR

As Galileo withdrew, it gained an oblique view of the north polar region. This was its first opportunity to take a close look at the recently named Tvashtar Catena – a chain of some of Io's largest calderas to the north of the Amirani-Maui region. This area had been observed from afar on the fourteenth and twenty-first orbits. Tvashtar had not previously been listed as being active, but Galileo found a curtain of lava rising almost 2 kilometres into the sky!

The active lava was hot enough to induce what the solid-state imaging team

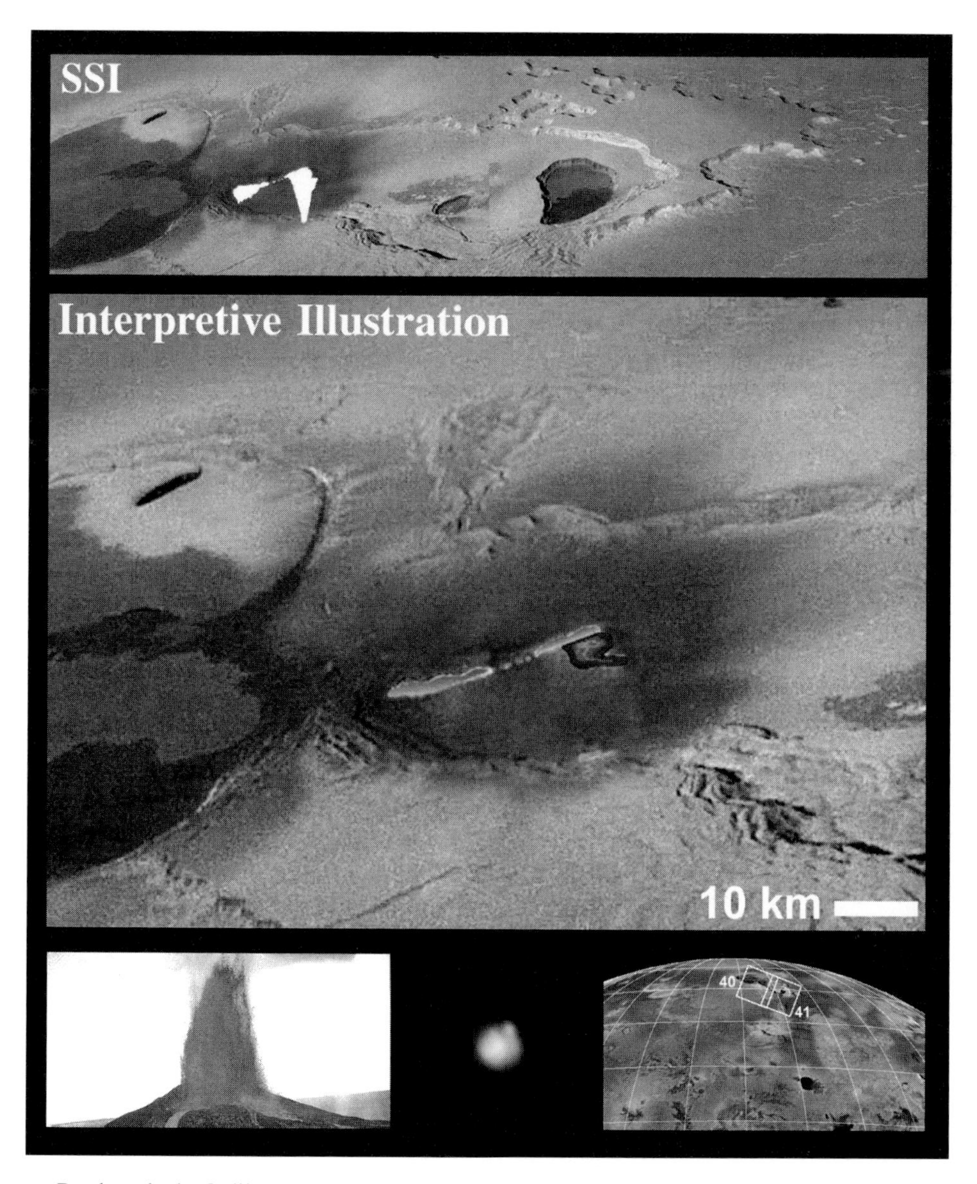

By sheer luck, Galileo caught a 'fire fountain' in one of the calderas of Tvashtar Catena during its fly-by in November 1999. It rose nearly 2 kilometres, whereas terrestrial fire fountains (bottom left) rarely rise higher than a few hundred metres. The eruption was also observed by the Infrared Telescope Facility (bottom centre).

referred to as 'bleeding', which occurred when the source was so bright that electrons leaked onto adjacent elements of the CCD detector array yielding a 'white out'. Nevertheless, the general shape of the feature could be inferred. The wavy line was interpreted as being a fire fountain from a fissure. The hot material 'below' the line

The fire fountain from this fissure eruption in the caldera of Pu'u O'o on Kilauea rises only 30 metres. (Courtesy JD Griggs, USGS)

was interpreted as hot lava flowing away from the fissure. A fire fountain in a Hawaiian volcano rarely rises more than a few hundred metres. The prospect of fire fountains on Io had been suggested by Laszlo Keszthelyi[71] to explain observations of Loki, but it had not been possible to verify this.

A few hours after the fly-by, Glenn Orton at the Infrared Telescope Facility at Mauna Kea in Hawaii made an infrared observation. From this perspective, the active centre was near the limb. The image revealed a bright spot (at the 1-o'clock position) that corresponded to the lava fountain. The line-of-sight would have been just 5.5 degrees above the local horizon. The fact that the hot spot was so bright confirmed that it was a fire fountain rising into the sky (in such an oblique view, surface lava would have been difficult to see). Co-observer John Spencer of the Lowell Observatory at Flagstaff in Arizona (who had seen this type of eruption previously) was delighted. "We thought that some of these eruptions might be due to lava fountains, but it is incredible to see that idea confirmed so spectacularly." The nearby 10-metre Keck telescope was also observing Io, using a newly installed 'adaptive optics' system, and it documented the event at a number of infrared wavelengths. A combination of ground-based and Galileo observations indicated that the lava was hot – possibly as hot as 1,800 K.

"Catching one of these fountains was a 1-in-500-chance observation," estimated Alfred McEwen. The detection was all the more remarkable because it was not as if the spacecraft had 'noticed' it and fired off a picture as a target of opportunity. The observation had been planned well in advance and inserted into the observational sequence for this orbit because, for the first time, the spacecraft's trajectory would

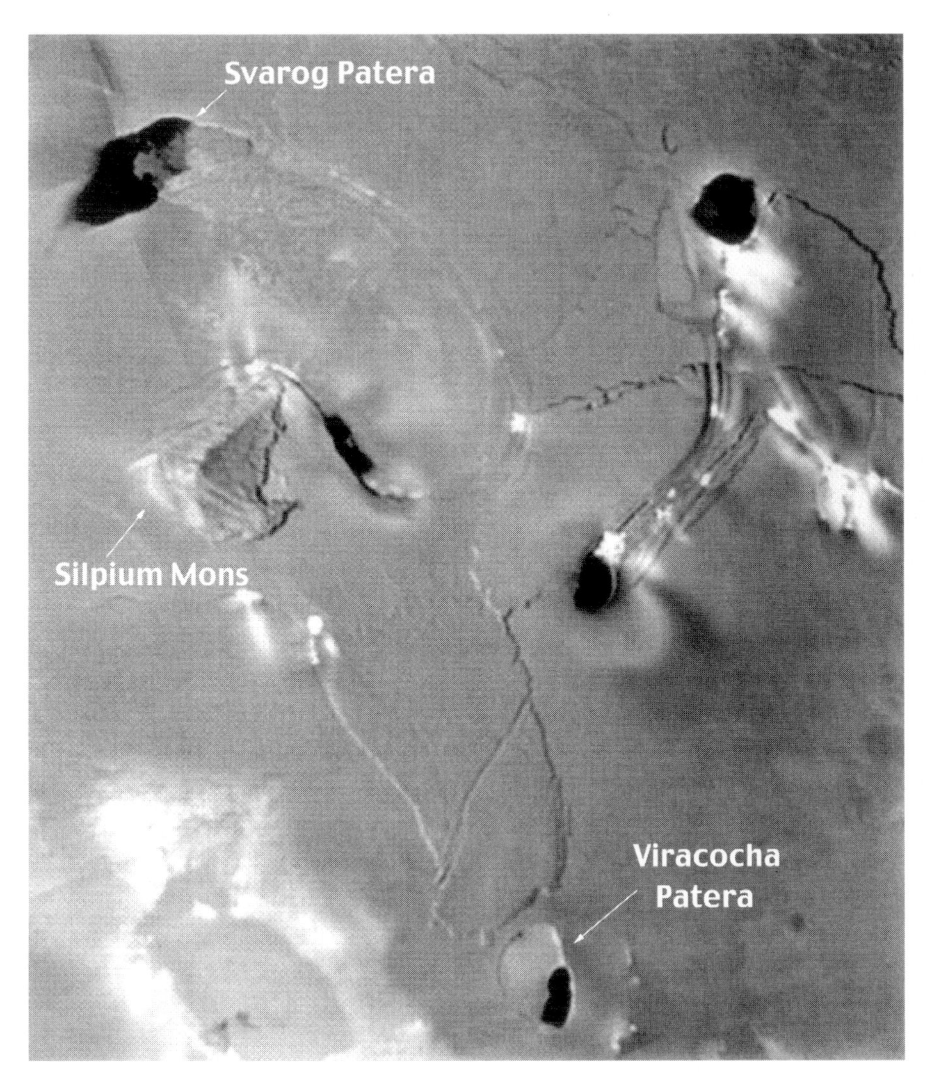

This Voyager mosaic of the Silpium Mons area hints at a structural relationship between mountains and calderas on Io.

accommodate such a northerly target. If Galileo had not been told to take a picture, the nature of this eruption would have remained unidentified. What else had we been missing?

MUSHY MANTLE?

As a planetary interior melts, the amount of iron and magnesium in the melt increases with the extent of melting. Depending upon the pressure, a lava temperature of 1,800 K would indicate up to 30 percent melting. This prompted Laszlo Keszthelyi and Alfred McEwen, with Jeffrey Taylor of Hawaii's Institute of Geophysics and Planetology, to reconsider a discarded idea that the moon's interior is a partially-molten 'mush' of crystals in magma.[72, 73]

When Stan Peale, Pat Cassen and Ray Reynolds predicted that Io would be volcanically active, they said that it would have an ocean of magma beneath a thin volcanic crust. There were problems with this, however. One argument was that a magma ocean would soon lose heat due to rapid convection and would solidify. Another problem was that mountains would be eroded at their bases faster than they could be built up. The idea was discarded as inappropriate when it was concluded that Io's lava was primarily sulphurous.

Analysis of the reflection spectra imdicates that the lavas at all of Io's active volcanoes are similar in composition. They contain a significant proportion of pyroxene. This is consistent with high amounts of iron and magnesium. Galileo's long-term volcano-monitoring established that the geographical distribution of volcanic activity is uniform, which indicates that the source of this mafic magma is global. Keszthelyi and colleagues therefore suggested that Io's interior is a sticky slurry of molten rock and suspended crystals beneath the solid lithosphere some 100 kilometres thick. A mushy mantle would not convect so vigorously, and therefore would not cool rapidly, but it would ensure continuous mixing and maintain the high proportion of mafic minerals in the zone feeding the volcanism.

A mushy mantle would tend to eat away at the lower crust and induce partial melting. This would help to break the crust into large blocks, some of which, as suggested by Paul Schenk and Mark Bulmer, form massifs when tilted and upthrusted.

A recent analysis[74] of the composition of the materials around Io's vents prompted a model in which the crust is a thick blanket of ultramafic lava, sulphurous lava, and various types of pyroclastic, with a reservoir of liquid sulphur sandwiched between this volcanic crust and the underlying lithosphere of magnesium-rich komatiite-type basalt.[75, 76] The composition of a lava flow depends upon the depth from which its magma is derived, with sulphurous flows being derived from individual shallow reservoirs.

If Io turns out to have an intrinsic magnetic field, this would argue in favour of a solidified interior. The lack of such a field would support a mushy interior because the magnetic field would be generated by electrical currents circulating within the convecting metallic core. If Io's core is englobed by a magma ocean, the mantle would be hotter than the core, and this would inhibit motions in the core. On the other hand, if the mantle were solid, the core would convect and generate a magnetic field. Unfortunately, the attempt to resolve this issue was frustrated because Galileo was not coaxed back to life until just after closest approach.

Galileo's context view documented the fact that Tvashtar Catena is a complex of calderas within calderas. The largest structure is about 100 $\times$ 290 kilometres in size, which rivals the complex on the summit of Olympus Mons on Mars. There is one of the 'golf courses' in the northernmost of the calderas. The close-up image established that although this corresponds to a distinctive albedo change, there is no associated topographic variation, which implies that it is a pyroclastic feature. Also of great interest is the large irregular flat mesa. Its steep scalloped margins indicate that the base of the 1-kilometre tall cliff has been eroded by a process called 'sapping'. On Earth, this arises when ground water springs undermine a cliff. The presence of such features on Mars was considered to be proof that water was once exposed on the planet's surface. In Io's case, the agent is presumably pressurised sulphur dioxide which sublimates to gas upon reaching the vacuum at the surface, and the rapid gaseous expansion breaks open the base of the cliff. The gas subsequently condenses and settles onto the surface as frost. As the frost becomes buried by a succession of deposits, it will be heated, and eventually resume its liquid state underground, thereby completing the cycle.

ZAL

As Galileo withdrew, it took an image of Zal Patera in the far north. The dark western rim of the caldera abuts a triangular plateau. A patch of reddish pyroclastic extending to the south could indicate that sulphurous gases are escaping.

MILLENNIUM MISSION

Contrary to concern that the intense radiation within Io's torus would disable Galileo when it went to inspect the volcanic moon, the spacecraft had come through unscathed, and this had facilitated the hoped-for follow-up. Although the November 1999 fly-by concluded the mission from an administrative point of view, the spacecraft was healthy, and had sufficient propellant and electrical power to sustain another year of orbital operations. Funding was therefore approved, on an incremental basis, for a 'Millennium Mission' to continue the orbital tour in order to pursue joint magnetospheric observations with the Cassini spacecraft (on its way to Saturn) when it flew through the Jovian system in December 2000 to shed light on how the solar wind billows around Jupiter and propagates on out into the outer Solar System.

Extending Galileo's mission into the twenty-first century was possible primarily due to excellent navigation during the orbital tour. No matter how healthy the spacecraft might be, it would be obliged to cease operations when it ran out of propellant, but precise navigation had minimised the need to fire the thrusters to 'tweek' the trajectory, and this had allowed the mission to be repeatedly extended. The spacecraft's consumables were not infinite, however. "I've seen a prediction that just 9 kg of propellant will remain at the end of the Millennium Mission," said Todd Barber, the former Galileo propulsion engineer who had transferred to the Cassini team, "which is only 1 percent of the launch load!"

Although the extended-extended mission began with a fly-by of Europa on 3 January 2000, the trajectory took Galileo within 212,000 kilometres of Io, and its imagery of Culann Patera established that a lava tube runs from the western rim of

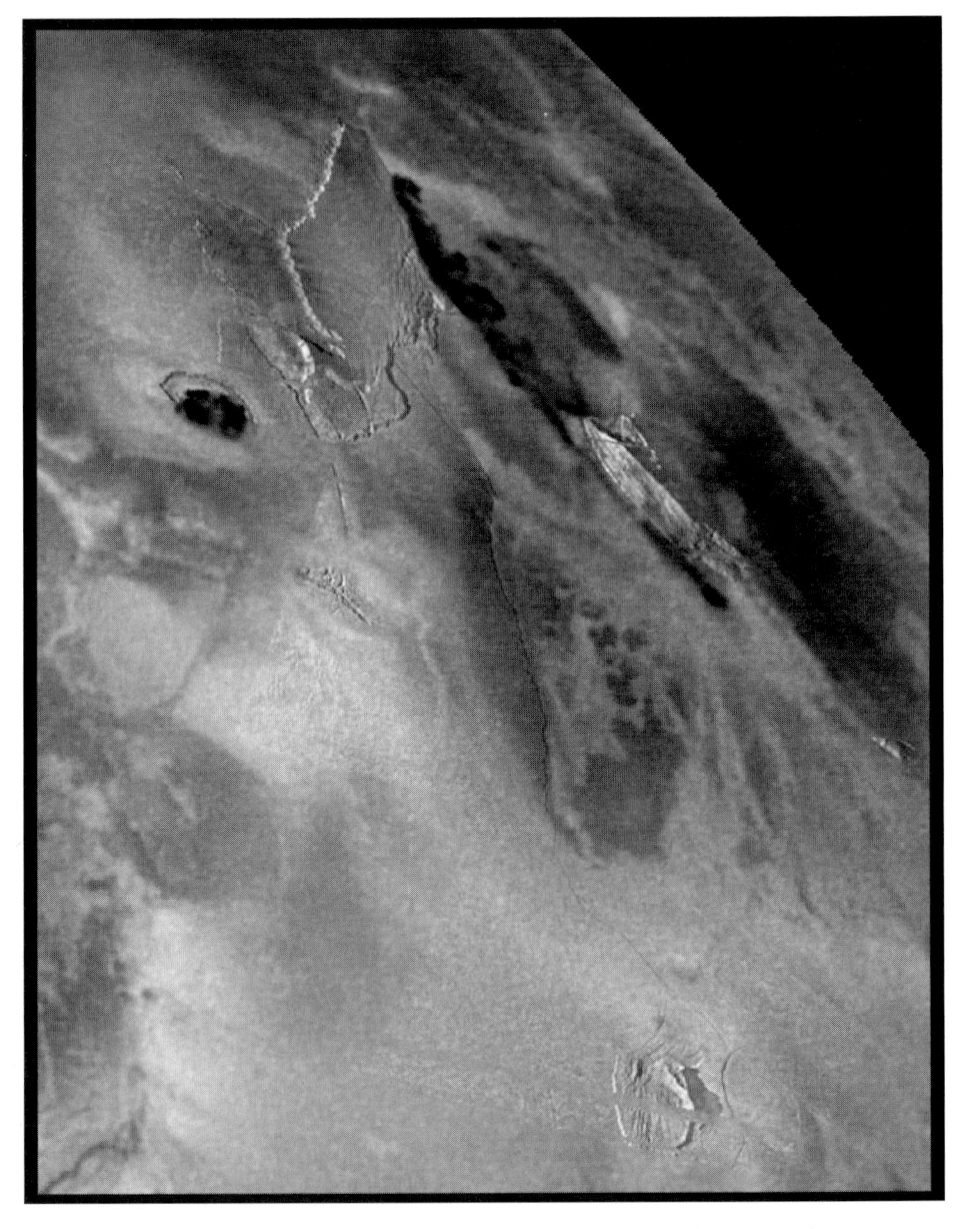

Galileo inspected the Zal Patera area during its fly-by in November 1999. The western side of the caldera is marked by dark lava. The shadows indicate that the northernmost part of the adjacent plateau sits about 2 kilometres above the plain.

the caldera to feed the lava flows to the northwest.[77] The infrared spectrometer investigated the distinctive 'golf course' in the caldera.

The objectives for the 200-kilometre Io fly-by on 22 February were chosen to pick up sites which had previously been frustrated by spacecraft glitches, and to take a second look at the some of the most interesting phenomena.

"Io's volcanoes are so active that the surface is always changing, and with each fly-by we get new and different observations," noted Torrence Johnson. "This time we expect to be able to observe the effects of the eruptions we saw in the October and November Io fly-bys."

Galileo ignored two spurious resets during the approach to Io, which was just over an hour after perijove. The first observation – the darkened trailing hemisphere – was made by the integrated photopolarimeter and radiometer, which also made a thermal scan of Loki comparing its caldera and the surrounding terrain in order to derive the heat flowing from the active centre. This instrument then made a thermal map of the sulphur frosts on Daedalus Patera to the east. The infrared spectrometer thermally mapped the lava lake in Pele's caldera at a resolution of 1 kilometre per pixel – which was very high spatial resolution for this instrument. Once across the terminator, the integrated photopolarimeter and radiometer switched to polarimetry in order to characterise the physical properties of the sulphur frosts of the Mulungu Patera region.

An image taken on the twenty-first orbit – which had provided an opportunity to view Io from the then-unprecedentedly close range of 124,000 kilometres – had revealed a cliff on the eastern tip of a plateau near Isum Patera which seemed to have been eroded by sapping, so this had been scheduled for a close inspection this time around.

A double strip of six images at 185 metres per pixel built an east–west mosaic north of Prometheus taking in Chaac Patera, a hot spot located in one of the 'golf courses' and Camaxtli Patera. Prometheus, Culann and Camaxtli had all turned out to be complex thermal structures, with several hot spots.[78] While over Chaac, Galileo took a series of swaths at 7 metres per pixel showing the floor of the caldera and its wall in great detail.

Another follow-up target was the source of Prometheus's plume, which had been found to be a cloud of sulphur dioxide rising from where a lava flow encroached on a field of sulphur dioxide frost. This time, Galileo took high-resolution imagery to study this process in detail.

The Tohil area was imaged again for a stereoscopic study of the relationship of the mountain to the caldera. Switching its attention to the north, Galileo recorded the Amirani-Maui region, and further north still, took a full-colour image of the lava fountain which had been discovered in Tvashtar Catena. It followed up with some oblique views looking east towards the terminator in order to document the topography of Zal Patera and Shamshu Patera.

Remarkably, considering past experiences, Galileo was able to complete its observational programme without incident. "We're thrilled," Jim Erickson reported, as the spacecraft began to replay the taped data.

Galileo's mission would continue, but it would not return to Io. Its three close passes by the volcanic moon had more than made up for the disappointment of having to cancel the imaging which had been planned for the encounter in December 1995, when the spacecraft had arrived at Jupiter. The long-term monitoring of the degree of volcanic activity, and the close-up views from the final phase of the mission had shed considerable light upon the nature of this remarkable moon. Perhaps one of the most profound insights was that Io has a great deal in common with the ancient Earth so, in studying Io, we are discovering the Earth.

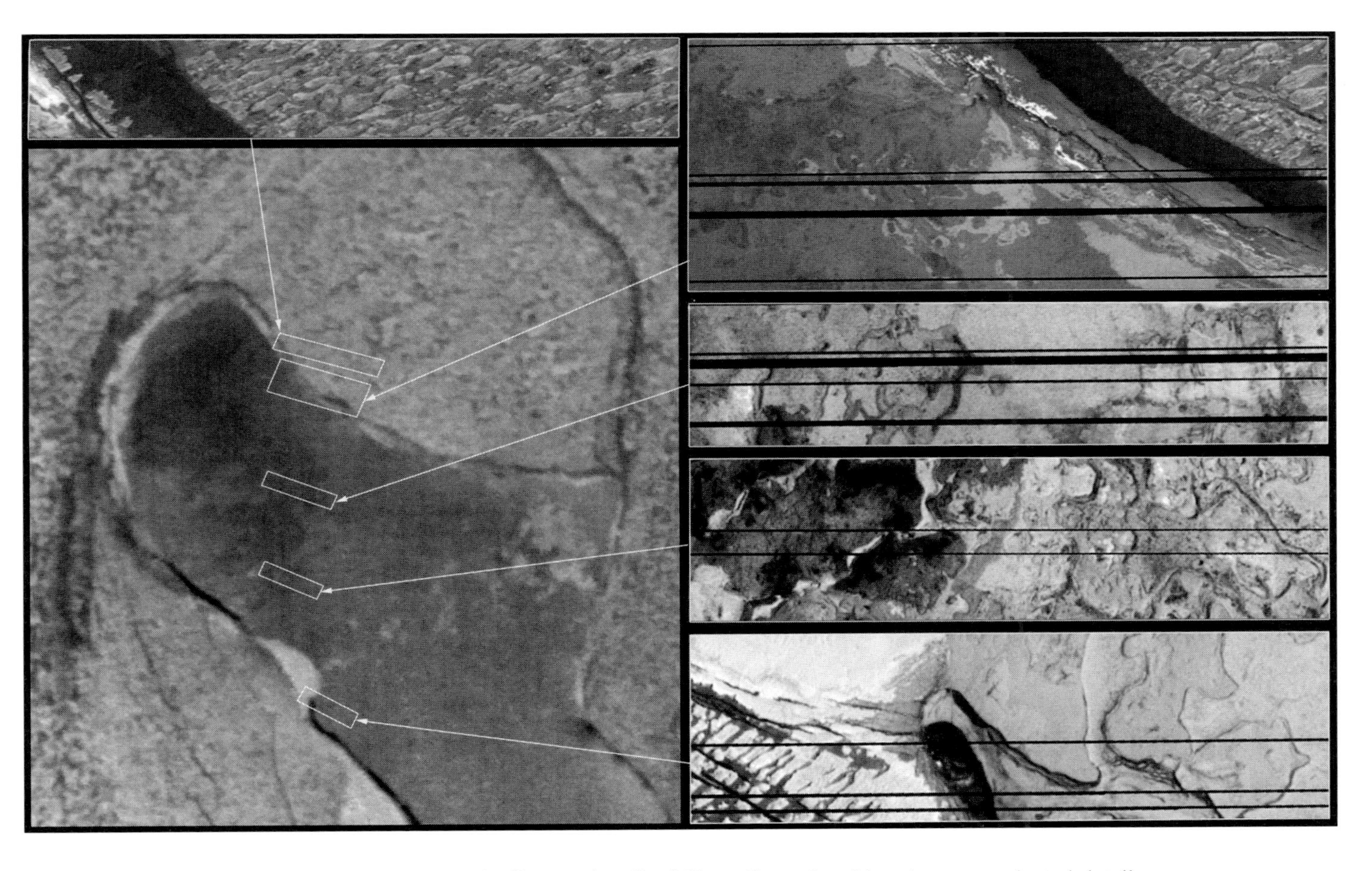

This series of 7-metre per pixel swaths show the floor and wall of Chaac Patera's caldera in unprecedented detail.

The plume from Prometheus is produced when the lava flow (dark) encroaches upon, and sublimates to gas, the frozen plain of sulphur dioxide pyroclastic (light-toned). Notice the 'rays'.

15

Passing the torch

A BONUS

Galileo's 'Millennium Mission' was to conclude with a series of joint observations of the Jovian magnetosphere in conjunction with the Cassini spacecraft, which was to pass through the Jovian system in December 2000 *en route* to Saturn.

"One spacecraft will be inside Jupiter's magnetic envelope, with the other outside where it can observe the powerful solar wind compressing that envelope," explained Dennis Matson of the Cassini science team. "From the two vantage points, we will watch cause-and-effect as the wind changes the magnetic properties around Jupiter." In fact, the joint observations started as early as February, when the plasma wave spectrometers monitored radio-frequency emissions emanating from the magnetosphere. "It's a real bonus for both missions," confirmed Torrence Johnson, the Galileo project scientist.

SATURN

Shortly after discovering that Jupiter had satellites, Galileo Galilei also made pioneering observations of Saturn. Although he saw what appeared to be an enormous satellite on either side, they seemed to hold fixed positions rather than orbit the planet. Over the next two years, they seemed to fade away, then reappear; very mysterious.

In 1655, the Dutch astronomer Christiaan Huygens discovered that Saturn has a 'normal' satellite, which he named Titan. The following year, he realised that the planet's mysterious appendages are flat rings which reflect the Sun, and when they had seemed to disappear they were edge-on. In 1671, the Italian astronomer Giovanni Domenico Cassini,[1]* in Paris, spotted Iapetus. He added Rhea in 1672 and Tethys and Dione in 1684, and in 1675 he discerned a thin dark gap in the rings. This

* For Notes and References, see pages 413–424.

This mosaic of Saturn and its major moons is a composition of Voyager images.

had first been reported by William Ball a decade earlier, but it has been called Cassini's Division.[2]

CASSINI

Although by no means a clone of Galileo, the Cassini spacecraft is certainly a close relative. It has been designed in the same spirit, as a multi-sensor vehicle to follow up on the Pioneer and Voyager fly-bys by making an *in-depth* study of the Saturnian system – the planet, its magnetosphere, its ring system, and its satellites – during a four-year-long primary mission in much the same way as Galileo has done at Jupiter.[3]

Cassini carries an atmospheric probe, but this is not intended for the planet itself, it is for Titan, which is distinguished by being the only planetary satellite to have a dense atmosphere. It is composed primarily of nitrogen, but is laced with methane and ethane. As methane in the atmosphere is destroyed by solar ultraviolet, the hydrocarbon derivatives create a smog, or haze, which can be seen above the limb. At Saturn's distance from the Sun – 9.5 AU – the methane condenses and falls as rain, which may pool in depressions on the surface and form oceans. Water-ice on the surface will be as tough as granite. To the Voyager cameras, Titan's atmosphere formed a featureless orange cloak.

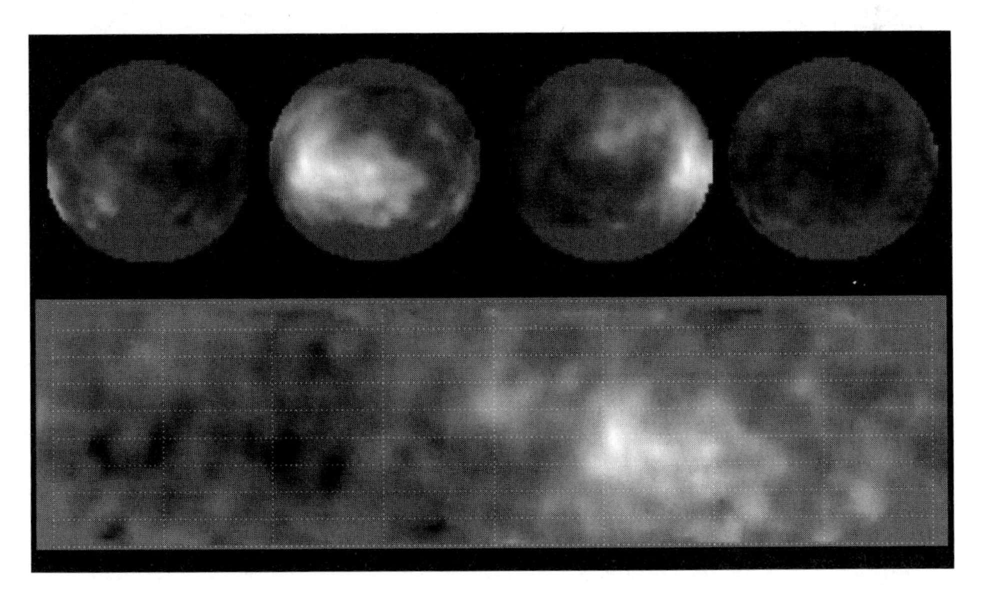

By repeatedly imaging Titan over its 16-day axial rotation at near-infrared wavelengths (0.85 and 1.05 microns) the Hubble Space Telescope was able to sense the reflectivity of the moon's surface (which is otherwise masked by dense methane clouds) and make an albedo map. One prominent bright area is about the same size as Australia. The low-albedo areas may represent oceans of ethane. (Courtesy STScI and Peter Smith, University of Arizona)

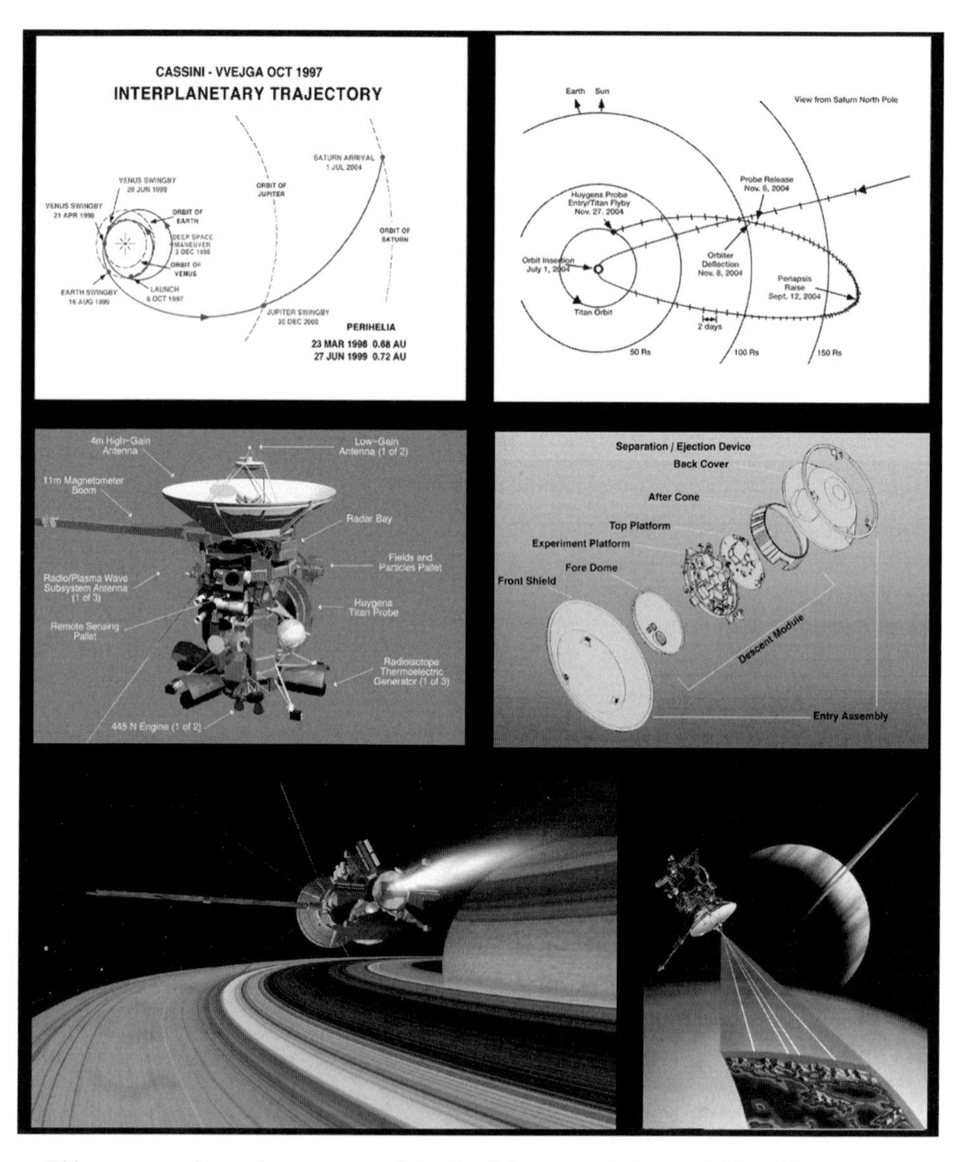

This montage shows the structure of the Cassini spacecraft (centre left) and its Huygens probe (centre right), the interplanetary cruise (top left), the capture orbit and the probe's trajectory to Titan (top right), artists' impressions of orbital insertion (bottom left), and mapping Titan's surface by synthetic aperture radar (bottom right).

In October 1994, however, a team led by Peter Smith of the University of Arizona's Lunar and Planetary Laboratory used the Hubble Space Telescope to observe Titan in a near-infrared band in which the haze is sufficiently transparent for the surface to be characterised in terms of reflectivity.[4] Only the equatorial and mid-latitudes could be studied in this way and the surface resolution was just 500 kilometres, but it was sufficient to make the first crude map of Titan's surface in terms of albedo variations. The most prominent bright feature is comparable in size to the continent of Australia, and the dark areas may be oceans.

Cassini's probe, which was supplied by the European Space Agency and is appropriately named for Christiaan Huygens, will study the atmosphere during a 2.5-hour descent; it will then touchdown – or perhaps splashdown – and report on surface conditions for up to half an hour.

Over successive fly-bys during its orbital tour, Cassini will map the entire surface of Titan through its atmospheric blanket utilising a synthetic aperture radar – in the same way as the Magellan spacecraft mapped Venus in the early 1990s.

The Cassini spacecraft does not have Galileo's spin-bearing assembly, so it does not have spinning and despun parts. The entire spacecraft is to be spun up to make particles and fields measurements and be stabilised for imaging. Nor does Cassini use a deployable antenna, it has a rigid structure; and instead of a reel-to-reel tape recorder, Cassini has a large solid-state mass data storage system. For power in the frozen reaches of the outer Solar System, Cassini has a trio of RTGs, one of which was Galileo surplus. The atmospheric probe is carried on the side of the bus, rather than tucked in under the propulsive unit, which in Cassini's case has a pair of co-aligned engines for redundancy.

Cassini's mission in Saturnian space is essentially to mirror Galileo's in Jovian space, so many of its instruments are similar and some are direct derivatives. The detector technologies have been improved, of course, and the names have been revised, but the lineage is clear. The solid-state imaging system is now the imaging science system, the ultraviolet spectrometer is the ultraviolet imaging spectrograph, and there is a new magnetometer and a radio and plasma wave spectrometer. In fact, the engineering model of the near-infrared mapping spectrometer has been rebuilt as Cassini's visual and infrared mapping spectrometer. For the same reasons, the Huygens probe will use an atmospheric structure instrument, and the main spacecraft will monitor the probe's signal for the doppler wind experiment. However, the fact that Cassini's probe is intended to land on Titan means that it carries a science package designed to identify the physical characteristics of the surface environment.

So why not sample Saturn's atmosphere, too? In fact, this had been included in the initial mission plan, but mass limitations prompted the deletion of the planetary probe early in the development process, and, in light of Galileo's success with Jupiter, an *in situ* study of Titan was considered to be more important than sampling Saturn.

Unlike Galileo, the Cassini spacecraft was to be launched on a rocket rather than within the Shuttle's payload bay. However, the spacecraft was so heavy that it needed the most powerful rocket in the inventory – the Titan 4. Furthermore, it required the latest version of this vehicle, which had flown only once before, so the countdown for Cassini's launch on 15 October 1997 was a tense affair. In the event, the Titan functioned flawlessly, the Centaur upper stage boosted the spacecraft to escape velocity, and Cassini proved to be healthy. Even with such a powerful escape stage, Cassini required a roundabout trajectory involving two fly-bys of Venus, one

On 15 October 1997, the Cassini-Huygens spacecraft was successfully dispatched by a Titan 4B-Centaur.

of the Earth, and one of Jupiter – with the rather less readily pronounced acronym VVEJGA – and a total flight time of over six years, with arrival in 2004.

Cassini's passage through the Jovian system in December 2000 will mark a 'handing over' of the torch of planetary exploration from Galileo to Cassini and a switch of focus from Jupiter to Saturn.

UPON REFLECTION

Galileo has set a high standard, and it is going to be a very hard act to follow. Despite the failure of its high-gain antenna to deploy and the problems with its tape recorder, it has proved itself to be admirably suited to its mission. But even as it had been pursuing its interplanetary cruise to Jupiter – taking in the sights along the way – and Cassini was being developed, the philosophy underlying their design was under question.

Although pairs of Pioneers and Voyagers had been sent out for early reconnaissance, the cost of developing the Space Shuttle severely limited NASA's planetary exploration budget. Few proposals were funded, and many of those that were started were either trimmed back or cancelled to ease Shuttle overruns. Only single spacecraft were ordered in each case – one Magellan, one Ulysses, one Galileo, and one Mars Observer. The fiasco over the selection of rocket escape stages, and then the delay imposed by the loss of the Challenger, further inflated the cost of such missions. This led to a strategy of assigning each planetary objective a single mission, and with the prospect of this being the only opportunity to a generation of scientists, each spacecraft was equipped with comprehensive suites of instruments designed to pursue a number of themes, and mission timescales were designed to provide *in-depth* studies. The original plan for Magellan had been a multi-sensor platform, but the continuing Shuttle overruns led to this being cancelled. Ulysses was to have involved a pair of spacecraft, one supplied by the United States and one by Europe, but NASA's was cancelled. Galileo and Mars Observer survived. The resurrected 'low cost' Magellan carried just one instrument, a synthetic aperture radar, and it served as a new role model.

When Dan Goldin took over as the agency's administrator in 1992, he initiated a change in thinking for space exploration with the mantra 'faster, cheaper, better'. This antipathy towards all-in-one missions was reinforced by the loss of the Mars Observer in 1993 as it was preparing itself to brake into orbit around its objective. Instead, small spacecraft would be designed for a rapid turnaround, either to pursue focused science operations or to test technologies for future missions. Although this strategy, which endorsed 'risk', has had notable successes, it has also produced lamentable failures. Upon Cassini's launch in 1997, it was derided in the Press as the last of the 'dinosaurs'.

WHAT OF THE FUTURE?

There is talk of developing 'low-cost' missions which would dispatch 'lightweight' spacecraft to orbit Uranus and Neptune in order to follow up the Voyager 2 fly-by, and to reconnoitre the Pluto-Charon system. To follow up Galileo, there is a plan for a Europan orbiter, and even a lander. The design of these spacecraft will have to balance science capability against cost, but it is impracticable to develop a Europan mission as if it was a Mars probe – the issue is mass.

Scientific instrumentation accounted for only 118 kg of the Galileo orbiter's 2,660 kg. This was, in fact, the same payload mass as the Voyagers. At 935 kg, Galileo's propellant load was heavier than a complete Voyager spacecraft (815 kg). At 340 kg, Galileo's atmospheric probe was heavier than a Pioneer (260 kg), but it carried a similar payload (28 kg). In fact, at 216 kg, Galileo's propulsive module was almost as massive as a Pioneer. Hence, most of Galileo's 3-tonne mass derived from its mission requirements. Although Ulysses did not have to enter Jovian orbit, only 55 kg of its 370 kg was science instruments. Nevertheless, it required an IUS augmented by a kick-motor to dispatch it on the direct route to Jupiter.

Operational considerations will make spacecraft intended for the outer planets heavier than those intended for the inner Solar System. Firstly, they cannot use solar transducers for power; they have to carry RTGs, which, in addition to being heavy, impose complex systems. Powerful communications systems with larger antennas are required, and the light-travel time involved in sending spacecraft to such distances means that they have to be able to operate autonomously for extended periods. All this requires sophisticated computers, and a high degree of redundancy is needed to guard against 'show stopping' failures. With such overheads, it is amazing that Galileo was able to accommodate as much scientific instrumentation as it carried.

The mass of a spacecraft intended for Jovian orbit will be dominated by the engine and the propellant needed to brake into Jovian orbit. This 'overhead' is worse if there is then a requirement to enter orbit around one of the moons. In the case of Europa, there is no option of using aerobraking in order to save on propellant, as is being done at Mars. The case is even worse if the probe has to soft-land on the moon. Any such mission will turn into a heavyweight simply because of the propulsive requirements, and this mass will translate into a heavy launcher, and thence into launch cost. There is also little point in cutting back on instrumentation in order to reduce this lifting requirement. Galileo's scientific potential derived from the fact that it had a suite of complementary instruments.

In other words, despite the 'faster, cheaper, better' mantra that is currently in vogue, and can deliver benefit in the inner Solar System, Galileo and Cassini are not 'dinosaurs', they are extremely well suited to the task of undertaking *in-depth* exploration of the outer planets on missions involving long travel times and multiple objectives in complex environments.

Congress authorised the development of the Galileo mission in 1977. After much delay, it was launched in 1989 on a six-year interplanetary cruise, and it has been exploring the Jovian system since 1995. Since its conception, therefore, the project has spanned over a quarter of a century. By any measure, it has been a major undertaking. Next time you look up at Jupiter in the sky and test your eyesight by trying to resolve the Galilean satellites, save a thought for the small robotic exploring machine which has endured so much to return the illustrations used in this book.

Facts and figures

1. JUPITER

Insolation (Earth = 1)	0.04
Mean distance from Sun (AU)	5.2028039
Mean distance from Sun (million km)	778.34
Orbital eccentricity	0.0485
Orbital inclination to ecliptic (degrees)	1.3
Mass (Earth = 1)	318
Mass (Sun = 1)	1/1047355
Mass (kg)	1.9×10^{27}
Volume (Earth = 1)	1,325
Density (g/cm^3)	1.33
Surface gravity (Earth = 1)	2.69
Escape velocity (km/s)	60
Radius (1 Rj) (km)	71492
Diameter: polar (km)	134,200
equatorial (km)	142,800
Axial obliquity to ecliptic (degrees)	3.1
Magnetic axis offset to spin axis (degrees)	9.5
Orbital period around Sun (years)	11.86
Rotational period (magnetic field)	9 h 55 min

2. THE GALILEAN MOONS

	Io	Europa	Ganymede	Callisto
Radius (km)	1,817	1,568	2,634	2,403
Diameter (km)	3,634	3,136	5,268	4,806
Mass (kg)	8.9316×10^{22}	4.79982×10^{22}	1.48186×10^{23}	1.07593×10^{23}
Mass (Earth = 1)	1.4960×10^{-2}	8.3021×10^{-3}	2.47×10^{-2}	1.807×10^{-2}
Surface gravity (Earth = 1)	0.183	0.135	0.145	0.127
Mean distance from Jupiter (km)	421,600	670,900	1,070,000	1,883,000
Mean distance from Jupiter's centre (Rj)	5.905	9.5	15.1	26.6
Orbital period (days)	1.769137786	3.551181041	7.15455296	16.6890184
Density (g/cm^3)	3.57	3.01	1.94	1.86
Orbital eccentricity	0.0041	0.009	0.002	0.007
Orbital inclination (degrees)	0.040	0.470	0.2	0.35
Orbital speed (km/s)	17.34	13.74	10.9	8.21
Escape velocity (km/s)	2.56	2.02	2.74	–
Visual albedo	0.61	0.64	0.43	–

3. INNER MOONLETS

	Metis	Adrastea	Amalthea	Thebe
Mean radius (km)	30	$13 \times 10 \times 8$	$131 \times 73 \times 67$	55×45
Distance from Jupiter (km)	128,000	129,000	181,000	222,000
Orbital period (days)	0.294780	0.29826	0.49817905	0.6745
Orbital eccentricity	–	–	0.003	0.015
Orbital inclination (degrees)	–	–	0.40	0.8
Mass (kg)	9×10^{16}	2×10^{16}	7.2×10^{21}	8×10^{17}
Rotational period	–	–	sync	sync

4. NEXT FOUR MOONLETS

	Leda	Himalia	Lysithea	Elara
Mean radius (km)	5	85	12	40
Distance from Jupiter (km)	11,094,000	11,480,000	11,720,000	11,737,000
Orbital period (days)	238.72	250.5662	259.22	259.6528
Orbital eccentricity	0.14762	0.15798	0.107	0.20719
Orbital inclination (degrees)	26.07	27.63	29.02	24.77
Mass (kg)	6×10^{15}	9.5×10^{18}	8×10^{16}	8×10^{17}
Rotational period (days)	–	0.4	–	0.5

5. OUTER FOUR MOONLETS

	Ananke	Carme	Pasiphae	Sinope
Mean radius (km)	10	15	18	14
Distance from Jupiter (km)	21,200,000	22,600,000	23,500,000	23,700,000
Orbital period (days)	631	692	735	758
Orbital eccentricity	0.16870	0.20678	0.378	0.275
Orbital inclination (degrees)	147	164	145	153
Mass (kg)	4×10^{16}	9×10^{16}	2×10^{17}	8×10^{16}

6. ASTEROIDS

	951 Gaspra	243 Ida
Mean radius (km)	$5 \times 5.5 \times 9$	11×26
Mean distance from Sun (AU)	1.82	2.74
Orbital period around Sun (years)	3.28	4.84
Orbital eccentricity	0.173	0.042
Orbital inclination (degrees)	4.10	1.14
Rotational period (h:min)	7:03	4:38
Semimajor axis of orbit (AU)	2.21	2.86

7. SPACECRAFT

Spacecraft	Launch	Mission
Pioneer 10	2 March 1972	First Jupiter fly-by on 4 December 1973
Pioneer 11	5 April 1973	A Jupiter slingshot on 3 December 1974 led to the first Saturn flyby on 1 September 1970
Voyager 1*	5 September 1977	A Jupiter slingshot on 5 March 1979 led to a Saturn fly-by on 12 November 1980
Voyager 2*	20 August 1977	A series of slingshots produced a 'grand tour' taking in Jupiter on 9 July 1979, Saturn on 26 August 1981, Uranus on 24 January 1986 and Neptune on 25 August 1989
Galileo	18 October 1989	After a lengthy interplanetary cruise, it entered Jovian orbit on 7 December 1995
Ulysses	6 October 1990	A Jupiter slingshot on 8 February 1992 placed it in a high-inclination solar orbit
Cassini	15 October 1997	A Jovian slingshot on 30 December 2000 sent it towards Saturn, to enter orbit in 2004

* Although Voyager 1 was launched after Voyager 2, it flew a faster trajectory and promptly overtook its mate.

8. GALILEO ENCOUNTER SEQUENCE

No.	Focus	Closest satellite		Perijove		Encounter sequence	Designation
		Date	km	Date	Rj		
0	Io	7 Dec 1995	892	7 Dec 1995	4.0	7–8 Dec	J0
1	Ganymede	27 Jun 1996	844	28 Jun 1996	11.0	23–30 Jun	G1
2	Ganymede	6 Sep 1996	250	7 Sep 1996	10.7	1–8 Sep	G2
3	Callisto	4 Nov 1996	1,104	6 Nov 1996	9.2	2–11 Nov	C3
4	Europa	19 Dec 1996	692	19 Dec 1996	9.2	15–22 Dec	E4
5	–	–	–	19 Jan 1997	?	–	J5
6	Europa	20 Feb 1997	587	20 Feb 1997	9.1	17–23 Feb	E6
7	Ganymede	5 Apr 1997	3,059	4 Apr 1997	9.1	30 Mar–6 Apr	G7
8	Ganymede	7 May 1997	1,585	8 May 1997	9.3	4–11 May	G8
9	Callisto	25 Jun 1997	416	27 June 1997	10.8	22–29 Jun	C9
10	Callisto	17 Sep 1997	524	18 Sep 1997	9.2	14–20 Sep	C10
11	Europa	6 Nov 1997	2,690	6 Nov 1997	–	2–8 Nov	E11
12	Europa	16 Dec 1997	200	16 Dec 1997	8.8	15–17 Dec	E12
13	Europa	–	–	10 Feb 1998	8.9	–	J13
14	Europa	29 Mar 1998	1,649	29 Mar 1998	8.8	28–31 Mar	E14
15	Europa	31 May 1998	2,521	1 Jun 1998	8.8	30 May–2 Jun	E15
16	Europa	21 Jul 1998	1,837	21 Jul 1998	8.8	20–22 Jul	E16
17	Europa	26 Sep 1998	3,598	26 Sep 1998	8.9	25–28 Sep	E17
18	Europa	22 Nov 1998	2,281	22 Nov 1998	8.9	21–23 Nov	E18
19	Europa	1 Feb 1999	1,495	1 Feb 1999	9.1	31 Jan–2 Feb	E19
20	Callisto	5 May 1999	1,311	3 May 1999	9.4	2–5 May	C20
21	Callisto	30 Jun 1999	1,050	2 Jul 1999	7.3	30 Jun–3 Jul	C21
22	Callisto	14 Aug 1999	2,288	12 Aug 1999	7.3	11–14 Aug	C22
23	Callisto	16 Sep 1999	1,053	14 Sep 1999	7.3	13–17 Sep	C23
24	Io	11 Oct 1999	500	11 Oct 1999	5.5	10–12 Oct	I24
25	Io	26 Nov 1999	300	26 Nov 1999	5.7	25–27 Nov	I25
26	Europa	3 Jan 2000	343	–	–	–	E26
27	Io	22 Feb 2000	200	–	–	–	I27
28	Ganymede	20 May 2000	808	–	–	–	G28
29	Ganymede	28 Dec 2000	–	–	–	–	G29

IAU nomenclature

FEATURE TYPES

Feature	Description
Albedo feature	–
Catena, catenae	Chain of craters
Cavus, cavi	Hollows, irregular steep-sided depressions usually in arrays or clusters
Chaos	Distinctive area of broken terrain
Chasma, chasmata	A deep, elongated, steep-sided depression
Colles	Small hills or knobs
Corona, coronae	Ovoid-shaped feature
Crater, craters	A circular depression
Dorsum, dorsa	Ridge
Eruptive centre	Active volcanic centers on Io
Facula, faculae	Bright spot
Farrum, farra	Pancake-like structure, or a row of such structures
Flexus, flexus	A very low curvilinear ridge with a scalloped pattern
Fluctus, fluctus	Flow terrain
Fossa, fossae	Long, narrow, shallow depression
Labes, labes	Landslide
Labyrinthus	Complex of intersecting valleys
Lacus*	'Lake'; small plain
Large ringed feature	Cryptic ringed features
Lenticula, lenticulae	Small dark spots on Europa
Linea, lineae	A dark or bright elongate marking, may be curved or straight
Macula, maculae	Dark spot, may be irregular
Mare*	'Sea'; large circular plain
Mensa, mensae	A flat-topped prominence with cliff-like edges
Mons, montes	Mountain
Oceanus*	A very large dark area on the moon
Palus*	'Swamp'; small plain
Patera, paterae	An irregular crater, or a complex one with scalloped edges

Planitia	Low plain
Planum	Plateau or high plain
Plume	–
Promontorium*	'Cape'; headland
Regio, regiones	A large area marked by reflectivity or color distinctions from adjacent areas
Reticulum, reticula	Reticular (net-like) pattern on Venus
Rima, rimae*	Fissure
Rupes, rupes	Scarp
Scopulus	Lobate or irregular scarp
Sinus	'Bay'; small plain
Sulcus, sulci	Subparallel furrows and ridges
Terra	Extensive land mass
Tessera, tesserae	Tile-like, polygonal terrain
Tholus, tholi	Small domical mountain or hill
Undae	Dunes
Vallis, valles	Valley
Vastitas	Extensive plain

* Used only on the Earth's Moon.

IAU SCHEMES FOR NAMING FEATURES ON JUPITER'S SATELLITES

Amalthea	People and places associated with the Amalthea myth
Thebe	People and places associated with the Thebe myth
Io	
Active eruptive centres	Fire, sun, thunder gods and heroes
Catenae	Sun gods
Fluctus	Name derived from nearby named feature, or fire, sun, thunder, volcano gods, goddesses and heroes, mythical blacksmiths
Mensae	People associated with Io myth, derived from nearby feature, or from Dante's *Inferno*
Montes	Places associated with Io myth, derived from nearby feature, or from Dante's *Inferno*
Paterae	Fire, sun, thunder, volcano gods, heroes, goddesses, mythical blacksmiths
Plana	Places associated with Io myth, derived from nearby feature, or from Dante's *Inferno*
Regiones	Places associated with Io myth, derived from nearby feature, or from Dante's *Inferno*
Tholi	Places associated with Io myth, derived from nearby feature, or from Dante's *Inferno*

Europa	
Chaos	Places associated with Celtic myths
Craters	Celtic gods and heroes
Flexus	Places associated with the Europa myth
Large ringed features	Celtic stone circles
Lenticulae	Celtic gods and heroes
Lineae	People associated with the Europa myth
Maculae	Places associated with the Europa myth
Regiones	Places associated with Celtic myths
Ganymede	
Catenae	Gods and heroes of ancient Fertile Crescent people
Craters	Gods and heroes of ancient Fertile Crescent people
Faculae	Places associated with Egyptian myths
Fossae	Gods (or principals) of ancient Fertile Crescent people
Regiones	Astronomers who discovered Jovian satellites
Sulci	Places associated with myths of ancient people
Callisto	
Large ringed features	Homes of the gods and of heroes
Craters	Heroes and heroines from northern myths
Catenae	Mythological places in high latitudes

JUPITER'S SATELLITES

NAME	LAT	LONG	DIAM	MAP	AD	ORIGIN
Amalthea						
Crater						
Gaea	80.0S	90.0W	0.0	–	1979	Greek mother earth goddess who brought Zeus to Crete
Pan	55.0N	35.0W	0.0	–	1979	Greek; goat-god, son of Amalthea and Hermes in some legends, also Zeus' foster brother
Facula						
Ida Facula	20.0N	175.0W	0.0	–	1979	Greek; mountain where Zeus played as a child
Lyctos Facula	20.0S	120.0W	0.0	–	1979	Greek; area in Crete where Zeus was raised

Io

Regio						
Bactria Regio	45.8S	123.4W	0.0	I1550	1979	Io passed through this area of ancient Iran in her wanderings
Bosphorus Regio	0.0N	120.0W	1200.0	–	1997	'Ford of the Cow'; Io wandered through here while trying to escape from the gadfly
Chalybes Regio	45.5N	83.2W	0.0	I1713	1979	Greek; Io passed through here in her wanderings
Colchis Regio	5.3N	199.8W	0.0	I1550	1979	Greek; Io passed through this part of Asia Minor in her wanderings
Illyrikon Regio	72.0S	160.0W	0.0	–	1997	Io passed by here in her wanderings
Lerna Regio	64.0S	292.6W	0.0	I1549	1979	Greek; meadows of Lyrcea
Media Regio	4.6N	58.8W	0.0	I1491	1979	Greek; Io passed through this part of Iran in her wanderings
Mycenae Regio	37.3S	165.9W	0.0	I1550	1979	Greek; in some legends, Io was transformed there
Tarsus Regio	43.7S	61.4W	0.0	I1491	1979	Io passed through here in her wanderings
Planum						
Argos Planum	47.0S	318.2W	140.0	I1491	1985	Where Io was captured by Zeus
Danube Planum	20.9S	258.7W	150.0	I1550	1985	Where Io passed by in her wanderings
Dodona Planum	56.8S	352.9W	390.0	I1549	1979	Greek; where Io went after the death of Argus
Ethiopia Planum	44.9S	27.0W	105.0	I1491	1985	Where Io passed by in her wanderings
Hybristes Planum	54.0S	21.1W	150.0	I1549	1985	Where Io passed by in her wanderings
Iopolis Planum	34.5S	333.5W	125.0	I1491	1985	Town where Io was wor shipped as moon goddess (present-day Antioch)
Lyrcea Planum	40.3S	269.3W	310.0	I1550	1985	Plain where Io was born
Nemea Planum	73.3S	275.5W	500.0	I1549	1979	Greek; where Io was turned into a cow by Zeus and given to Hera
Eruptive centre						
Amirani	25.9N	114.5W	0.0	I1550	1979	Georgian god of fire
Kanehekili	18.0S	40.0W	0.0	–	1997	Hawaiian thunder god
Loki	17.9N	302.6W	0.0	I1491	1979	Norse blacksmith, trickster god

Marduk	27.1S	207.5W	0.0	I1550	1979	Sumero-Akkadian fire god
Masubi	43.6S	54.7W	0.0	I1491	1979	Japanese fire god
Maui	20.0N	122.0W	0.0	I1550	1979	Hawaiian demigod who sought fire from Mafuike
Pele	18.6S	257.8W	0.0	I1550	1979	Hawaiian goddess of the volcano
Prometheus	1.6S	153.0W	0.0	I1550	1979	Greek fire god
Surt	45.5N	337.9W	0.0	I1491	1979	Icelandic volcano god
Volund	25.0N	184.2W	0.0	I1550	1979	Germanic supreme smith of the gods
Zamama	18.0N	174.0W	0.0	–	1997	Babylonian sun, corn, and war god
Patera						
Agni Patera	40.5S	334.2W	20.0	I1491	1985	Hindu god of fire
Aidne Patera	2.0S	178.0W	50.0	–	1997	Irish creator of fire
Altjirra Patera	34.0S	108.0W	50.0	–	1997	Australian sky god whose voice is thunder
Amaterasu Patera	37.7N	306.6W	100.0	I1491	1979	Japanese sun goddess
Angpetu Patera	21.2S	10.6W	45.0	I1491	1985	Dakota name meaning the sun
Aramazd Patera	73.4S	337.7W	38.0	I1549	1985	Armenian thunder god
Arusha Patera	38.0S	101.0W	60.0	–	1997	Hindu god of the rising sun.
Asha Patera	8.6S	225.8W	90.0	I1550	1979	Persian spirit of fire
Atar Patera	30.2N	278.9W	125.0	I1491	1979	Iranian personification of fire
Aten Patera	47.9S	310.1W	40.0	I1491	1979	Egyptian sun god
Babbar Patera	39.5S	272.1W	95.0	I1491	1979	Sumerian sun god
Bochica Patera	61.0S	20.6W	50.0	I1549	1979	Chibcha sky god
Carancho Patera	1.5N	317.2W	30.0	I1491	1985	Bolivian legendary hero who received fire from an owl
Cataquil Patera	24.2S	18.7W	125.0	I1491	1985	Inca god of thunder and lightning
Catha Patera	53.0S	100.0W	60.0	–	1997	Etruscan sun god
Creidne Patera	52.4S	343.5W	125.0	I1549	1979	Celtic smith god
Culann Patera	19.9S	158.7W	100.0	I1550	1979	Celtic smith god
Daedalus Patera	19.1N	274.3W	40.0	I1491	1979	Greek hero, smith; father of Icarus
Dazhbog Patera	54.0N	301.6W	0.0	I1713	1979	Slavonic sun god
Dingir Patera	4.0S	342.1W	40.0	I1491	1985	Sumerian sun god; means 'shining'
Dusura Patera	37.0N	119.0W	70.0	–	1997	Nabataean sun god
Emakong Patera	3.2S	119.1W	80.0	I1550	1979	Sulca (New Britain) man who brought fire
Fo Patera	40.5N	192.0W	50.0	–	1997	Chinese fire and sun god
Fuchi Patera	28.4N	327.9W	45.0	I1491	1979	Ainu fire goddess
Galai Patera	10.7S	288.3W	90.0	I1491	1979	Mongol fire god
Gibil Patera	14.9S	294.9W	95.0	I1491	1979	Sumerian fire god
Gish Bar Patera	17.0N	90.0W	150.0	–	1997	Babylonian sun god
Hatchawa Patera	58.2S	35.1W	75.0	I1549	1985	Yaroro (Slavic) god who, in form of a boy, gave fire to mankind

Heiseb Patera	29.7N	244.8W	60.0	I1550	1985	Bushman devil who represents fire
Heno Patera	56.5S	312.1W	65.0	I1549	1979	Iroquois god of thunder
Hephaestus Patera	2.0N	289.5W	50.0	I1491	1979	Greek smith god
Hi'iaka Patera	3.0S	80.0W	80.0	–	1997	Sister of Pele
Hiruko Patera	65.1S	329.3W	80.0	I1549	1979	Japanese sun god
Horus Patera	9.6S	338.6W	125.0	I1491	1979	Egyptian falcon-headed solar god
Huo Shen Patera	15.1S	329.3W	90.0	I1491	1982	Chinese god of fire
Ilmarinen Patera	14.0S	2.8W	40.0	I1491	1985	Finnish blacksmith with supernatural creative powers
Inti Patera	68.1S	348.6W	75.0	I1549	1979	Inca sun god
Isum Patera	29.0N	208.0W	100.0	–	1997	Assyrian fire god
Janus Patera	3.0S	42.5W	60.0	–	1997	Italian sun god
Kane Patera	47.8S	13.4W	115.0	I1491	1979	Hawaiian god of sunlight
Karei Patera	2.0N	16.0W	100.0	–	1997	Semangan (Malay Penninsula) thunder god
Kava Patera	16.5S	342.1W	75.0	I1491	1985	Persian blacksmith
Khalla Patera	6.0N	303.4W	80.0	I1491	1985	Bushman sun in form of man often referred to as the hunter
Kibero Patera	11.6S	305.5W	60.0	I1491	1982	Yaroro toad who lives in underworld giving mankind fire
Kurdalagon Patera	49.5S	218.0W	50.0	–	1997	Ossetian celestial smith
Laki-oi Patera	37.5S	62.5W	130.0	–	1997	Bornean hero who invented fire
Loki Patera	12.6N	308.8W	250.0	I1491	1979	Norse blacksmith, trickster god
Lu Huo Patera	38.4S	354.1W	90.0	I1491	1985	Stove fire associated with Chinese god of the hearth fire
Maasaw Patera	40.0S	340.4W	30.0	I1491	1979	Hopi (USA) god of fire and death
Mafuike Patera	13.9S	260.0W	110.0	I1550	1979	Hawaiian demigoddess whose fingers held fire
Malik Patera	34.2S	128.5W	85.0	I1550	1979	Babylonian, Caananite sun god
Mama Patera	10.6S	356.5W	30.0	I1491	1985	Chagaba (Chibcha, Colombia) word for sun
Manua Patera	35.5N	322.0W	110.0	I1491	1979	Hawaiian sun god
Masaya Patera	22.5S	348.1W	125.0	I1491	1979	Nicaraguan smith god
Maui Patera	16.5N	124.0W	45.0	I1550	1979	Hawaiian demigod who sought fire from Mafuike
Mbali Patera	31.5S	7.1W	60.0	I1491	1982	Pygmy word representing fire itself
Menahka Patera	30.7S	346.2W	40.0	I1491	1985	Mandan (USA) name for the sun
Mihr Patera	16.4S	305.6W	60.0	I1491	1979	Armenian fire god
Mithra Patera	58.8S	267.9W	25.0	I1549	1985	Persian god of light

Monan Patera	19.0N	106.0W	50.0	–	1997	Brazilian god who destroyed the world with fire and flood
Mulungu Patera	17.0N	218.0W	50.0	–	1997	African thunder god
Nina Patera	38.3S	164.2W	425.0	I1550	1979	Inca fire god
Ninurta Patera	16.5S	315.3W	75.0	I1491	1985	Babylonian god of the spring sun
Nusku Patera	64.7S	4.6W	90.0	I1549	1979	Assyrian fire god
Nyambe Patera	0.6N	343.9W	50.0	I1491	1979	Zambezi sun god
Paive Patera	45.5S	0.0W	69.0	I1491	1985	Saami-Lapp sun god
Pautiwa Patera	32.8S	347.9W	40.0	I1491	1985	Hopi (USA) name for the sun
Pillan Patera	12.0S	244.0W	80.0	–	1997	South American thunder, fire, and volcano god
Podja Patera	18.2S	304.9W	60.0	I1491	1985	Tungu spirit who keeps the fire
Purgine Patera	2.6S	297.7W	20.0	I1491	1985	Mordvinian (Russia) thunder god
Pyerun Patera	56.0S	252.2W	40.0	I1549	1985	Slavonic god of thunder
Ra Patera	8.6S	325.3W	30.0	I1491	1979	Egyptian sun god
Rata Patera	35.5S	199.5W	30.0	–	1997	Maori sun hero
Reiden Patera	13.4S	235.7W	70.0	I1549	1979	Japanese thunder god
Ruwa Patera	0.4N	3.0W	50.0	I1491	1979	African sun god associated Mt. Kilimanjaro
Sed Patera	2.8S	303.8W	70.0	I1491	1985	Phoenician chariot rider of the Sun
Sengen Patera	32.8S	304.3W	55.0	I1491	1979	Japanese; deity of Mt. Fugiyama
Sethlaus Patera	52.0S	194.0W	80.0	–	1997	Etruscan celestial smith
Shakuru Patera	23.6N	266.4W	70.0	I1550	1979	Pawnee (USA) sun god of the East; gives light and heat
Shamash Patera	33.7S	152.1W	110.0	I1550	1979	Assyro-Babylonian sun god
Shamshu Patera	8.0S	64.0W	90.0	–	1997	Arabian sun goddess
Shoshu Patera	19.6S	324.1W	50.0	I1491	1982	Caucasian patron of fire
Sigurd Patera	5.0S	98.0W	60.0	–	1997	Norse sun hero
Siun Patera	49.8S	1.4W	50.0	I1491	1985	Nanai (Siberia) sun god
Sui Jen Patera	19.2S	4.3W	40.0	I1491	1985	Chinese hero who discovered fire
Svarog Patera	48.3S	267.5W	70.0	I1549	1979	Russian smith god
Talos Patera	26.1S	356.5W	20.0	I1491	1982	Nephew of Daedalus; also a blacksmith
Taranis Patera	70.8S	28.6W	105.0	I1549	1985	Celtic thunder god
Taw Patera	33.3S	0.0W	15.0	I1491	1982	Monguor word for fire or hearth
Tiermes Patera	22.5N	351.5W	50.0	–	1997	Lapp thunder god
Tohil Patera	26.3S	156.5W	20.0	I1550	1979	Central American god who gave fire to man
Tol-Ava Patera	1.7N	322.0W	70.0	I1491	1985	Mordvinian (Russia) goddess of fire
Tung Yo Patera	18.7S	2.5W	20.0	I1491	1985	Chinese fire god

Tupan Patera	18.0S	141.0W	50.0	–	1997	Thunder god of the Tupi-Guarani Indians of Brazil
Ukko Patera	33.0N	20.0W	40.0	–	1997	Finnish thunder god
Ulgen Patera	40.4S	288.0W	49.0	I1491	1979	Siberian progenitor god who struck first fire
Uta Patera	35.3S	24.9W	30.0	I1491	1979	Sumerian sun god
Vahagn Patera	23.8S	351.7W	70.0	I1491	1979	Armenian fire god
Viracocha Patera	61.2S	281.7W	55.0	I1549	1979	Qechua sun god
Zal Patera	42.0N	76.0W	130.0	–	1997	Iranian sun god
Tholus						
Apis Tholus	11.2S	348.8W	0.0	I1491	1979	Greek; name for Epaphus, son of Io and Zeus
Inachus Tholus	15.9S	348.9W	0.0	I1491	1979	Greek; river god, father of Io
Mons						
Boosaule Montes	4.4S	270.1W	590.0	I1550	1985	Cave where Io bore Epaphus
Caucasus Mons	33.0S	239.0W	150.0	–	1997	Io passed by these mountains while trying to escape from the gadfly
Crimea Mons	76.1S	244.8W	0.0	I1549	1985	Where Io passed by in her wanderings
Dorian Montes	24.0S	198.0W	450.0	–	1997	Region in ancient Greece
Egypt Mons	41.0S	257.0W	300.0	–	1997	Io ended her wanderings here
Euboea Montes	46.3S	339.9W	0.0	I1491	1985	Where Io passed by in her wanderings
Euxine Mons	27.0N	126.0W	200.0	–	1997	Io passed by here in her wanderings
Haemus Montes	68.9S	46.6W	0.0	I1549	1979	Where Io passed by in her wanderings
Ionian Mons	9.0N	236.0W	150.0	–	1997	Io crossed this sea in her wanderings
Nile Montes	52.0N	253.0W	450.0	–	1997	Where Zeus restored Io to her human form
Ot Mons	4N	218W	–	–	1999	–
Rata Mons	35.0S	201.0W	200.0	–	1997	Maori sun hero
Silpium Mons	52.6S	272.9W	0.0	I1549	1979	Greek; where Io died of grief in some legends
Skythia Mons	26.0N	98.0W	200.0	–	1997	Io passed by here in her wanderings
Tohil Mons	29.0S	157.0W	300.0	–	1997	Central American god who gave fire to man
Mensa						
Echo Mensa	79.6S	357.4W	0.0	I1549	1985	Mother of Iynx
Epaphus Mensa	53.5S	241.3W	0.0	I1549	1985	'Child of touch', son of Io and Zeus
Hermes Mensa	43.0S	247.0W	130.0	–	1997	Freed Io from Argus

Iynx Mensa	61.1S	304.6W	0.0	I1549	1985	Cast a spell on Zeus so he fell in love with Io
Pan Mensa	49.5S	35.4W	0.0	I1549	1985	Father of Iynx
Catena						
Mazda Catena	8.6S	313.5W	0.0	I1491	1979	Babylonian sun god
Reshet Catena	0.8N	305.6W	0.0	I1491	1985	Aramaic sun god
Fluctus						
Acala Fluctus	11.0N	337.0W	300.0	–	1997	Japanese fire god
Euboea Fluctus	45.1S	351.3W	0.0	I1491	1985	Where Io passed by in her wanderings
Fjorgynn Fluctus	11.5N	358.0W	300.0	–	1997	Norse thunder god
Kanehekili Fluctus	16.0S	38.0W	250.0	–	1997	Hawaiian thunder god
Lei-Kung Fluctus	38.0N	204.0W	400.0	–	1997	Chinese thunder god
Marduk Fluctus	27.0S	209.0W	150.0	–	1997	Sumero-Akkadian fire god
Masubi Fluctus	48.0S	60.0W	800.0	–	1997	Japanese fire god
Tung Yo Fluctus	16.4S	357.8W	0.0	I1491	1985	Chinese fire god
Uta Fluctus	32.6S	19.2W	0.0	I1491	1985	Sumerian sun god

Europa						
Large ring feature						
Callanish	16.0S	333.4W	100.0	–	1997	Stone circle in the Outer Hebrides, Scotland
Tyre	31.7N	147.0W	148.0	–	1997	Greek; the seashore from which Zeus abducted Europa. Changed from Tyre Macula
Chaos						
Conamara Chaos	9.5N	273.3W	127.0	–	1997	Rugged part of western Ireland named for Conmac, son of the Queen of Connacht
Crater						
Cilix	1.2N	181.9W	23.0	I1493	1985	Brother of Europa; Rand control-point crater
Govannan	37.5S	302.6W	10.0	–	1997	One of the Children of Don, a smith and brewer
Mael Duin	17.0S	197.9W	2.0	–	-	Celtic hero
Mannann'an	2.0N	240.0W	30.0	–	1997	Irish sea and fertility god
Morvran	5.7S	152.2W	25.0	I1493	1985	Celtic; ugly son of Tegid
Pwyll	26.0S	271.0W	38.0	–	1997	Celtic god of the underworld
Rhiannon	81.8S	199.7W	25.0	I1499	1985	Celtic heroine
Taliesin	23.2S	137.4W	48.0	I1493	1985	Celtic, son of Bran; magician
Tegid	0.6S	164.0W	29.0	I1493	1985	Celtic hero who lived in Bula Lake
Flexus						
Cilicia Flexus	47.6S	142.6W	639.0	I1241	1979	Land named for Cilix on his search for Europa

Delphi Flexus	69.7S	172.3W	1125.0	I1499	1985	Where the cow led Cadmus before it stopped at the site of Thebes
Gortyna Flexus	42.4S	144.6W	1261.0	I1241	1979	Place on Crete where Zeus brought Europa
Phocis Flexus	48.6S	197.2W	298.0	I1499	1985	Where the cow lead Cadmus before it stopped at the site of Thebes
Sidon Flexus	64.5S	170.4W	1216.0	I1241	1979	Another name for Tyre; where Europa was born
Linea						
Adonis Linea	51.8S	113.2W	758.0	I1241	1979	Greek; son of Phoenix, nephew of Europa
Agave Linea	12.6N	273.0W	1250.0	–	1997	Daughter of Harmonia and Cadmus
Agenor Linea	43.6S	208.2W	1326.0	I1241	1979	Greek; Europa's father
Alphesiboea Linea	28.0S	182.6W	1642.0	I1493	1985	Son of Phoenix, nephew of Europa
Androgeos Linea	12.8N	278.8W	645.0	–	–	Son of Minos
Argiope Linea	8.2S	202.2W	934.0	I1241	1979	Greek; another name for Telephassa
Asterius Linea	17.7N	265.6W	2753.0	I1241	1979	Greek; Europa's husband after Zeus
Astypalaea Linea	76.5S	220.3W	1030.0	–	1997	Sister of Europa
Autonoe Linea	17.0N	164.5W	1340.0	–	–	Daughter of Harmonia and Cadmus
Belus Linea	11.8N	228.3W	2580.0	I1241	1979	Greek; Agenor's twin brother
Cadmus Linea	27.8N	173.1W	1212.0	I1241	1979	Greek; brother of Europa
Chthonius Linea	0.1N	311.3W	1850.0	–	1997	Survivor of the men Cadmus sowed with dragon's teeth, a founder of Thebes
Echion Linea	13.1S	184.3W	1217.0	I1493	1985	Survivor of the men Cadmus sowed with the dragon's teeth; a founder of Thebes
Harmonia Linea	27.0N	168.0W	925.0	–	1997	Wife of Cadmus
Hyperenor Linea	3.1S	314.7W	2200.0	–	1997	Survivor of the men Cadmus sowed with dragon's teeth, a founder of Thebes
Ino Linea	5.0S	163.0W	1400.0	–	1997	Daughter of Harmonia and Cadmus
Katreus Linea	39.5S	215.5W	245.0	–	–	Son of Minos
Libya Linea	56.2S	183.3W	452.0	I1241	1979	Greek; Agenor's mother
Minos Linea	45.3N	195.7W	2134.0	I1241	1979	Greek; son of Europa and Zeus
Onga Linea	38.8S	209.6W	806.0	–	–	Phoenician name for Athene
Pelagon Linea	34.0N	170.0W	800.0	–	1997	King who sold Cadmus the cow with a white full moon on each flank

Pelorus Linea	17.1S	175.9W	1770.0	I1241	1979	Greek; survivor of the men Cadmus sowed with the dragon's teeth; a founder of Thebes
Phineus Linea	33.0S	269.2W	1984.0	I1241	1979	Greek; brother of Europa
Phoenix Linea	14.5N	184.7W	732.0	I1493	1985	Brother of Europa
Rhadamanthys Linea	18.5N	200.8W	1780.0	I1493	1985	Son of Europa and Zeus
Sarpedon Linea	42.2S	89.4W	940.0	I1241	1979	Greek; son of Europa and Zeus
Tectamus Linea	17.9N	181.9W	719.0	I1493	1985	Father of Asterius
Telephassa Linea	2.8S	178.8W	800.0	I1493	1985	Europa's mother
Thasus Linea	68.7S	187.4W	1027.0	I1241	1979	Greek; brother of Europa
Thynia Linea	57.9S	148.6W	398.0	I1499	1985	Peninsula between Black and Marmara Seas, where Phineus sought Europa
Macula						
Boeotia Macula	54.0S	166.0W	22.0	–	1997	Place where Cadmus led cow before it stopped at site of Thebes
Cyclades Macula	64.0S	192.0W	105.0	–	1997	Islands where Rhadamanthys reigned
Thera Macula	47.7S	180.9W	78.0	I1241	1979	Greek; place where Cadmus stopped in his search for Europa
Thrace Macula	46.6S	171.2W	173.0	I1241	1979	Place in northern Greece where Cadmus stopped in his search for Europa
[Tyre Macula]	31.7N	147.0W	148.0	I1241	1979	Greek; the seashore from which Zeus abducted Europa. Changed to Tyre (large ring feature)

Ganymede

Regio						
Barnard Regio	0.8N	1.0W	2547.0	I1242	1979	Edward E.; American astronomer (1857–1923)
Galileo Regio	35.7N	137.6W	3142.0	I1242	1979	Italian astronomer (1564–1642)
Marius Regio	12.1N	199.3W	3572.0	I1242	1979	Simon; German astronomer (1570–1624)
Nicholson Regio	34.0S	356.7W	3719.0	I1242	1979	Seth B.; American astronomer (1891–1963)
Perrine Regio	38.8N	30.0W	2145.0	I1242	1979	Charles D.; American astronomer (1867–1951)

Sulcus						
Akitu Sulcus	39.0N	197.0W	380.0	–	1997	Where Marduk's statue was carried each year
Anshar Sulcus	21.5N	202.9W	1181.0	I1242	1979	Assyro-Babylonian; celestial-world home of Lakhmu and Lakhamu
Apsu Sulci	34.8S	235.5W	1281.0	I1817	1979	Sumero-Akkadian; primordial ocean
Aquarius Sulcus	50.0N	11.5W	1341.0	I1242	1979	Greek; Zeus set Ganymede among the stars as the constellation of Aquarius, the water carrier
Arbela Sulcus	22.3S	353.6W	1896.0	I1650	1985	Assyrian town where Ishtar was worshipped
Bubastis Sulci	79.8S	263.1W	2197.0	I1860	1988	Town in Egypt where Bast was worshipped
Byblus Sulcus	38.0N	202.0W	600.0	–	1997	Ancient Phoenecian city where Adonis was worshipped
Dardanus Sulcus	39.3S	20.2W	2559.0	I1242	1979	Greek; where Ganymede was abducted by Zeus disguised as an eagle
Dukug Sulcus	81.3N	352.7W	467.0	I1810	1985	Sumerian holy cosmic chamber of the gods
Elam Sulci	57.4N	205.5W	1866.0	I1499	1985	Ancient Babylonian seat of sun worship, in present-day Iran
Erech Sulcus	5.4S	175.7W	998.0	I1536	1985	Akkadian town that was built by Marduk
Harpagia Sulcus	14.9S	319.1W	1398.0	I1242	1985	Greek; where Ganymede was abducted an eagle
Hursag Sulcus	10.7S	234.5W	928.0	I1548	1985	Sumerian mountain where winds dwell
Kishar Sulcus	8.3S	218.7W	1213.0	I1242	1979	Assyro-Babylonian; terrestrial-world home of Lakhmu and Lakhamu
Lagash Sulcus	10.4S	162.8W	1407.0	I1536	1985	Early Babylonian town
Mashu Sulcus	31.1N	209.2W	3030.0	I1242	1979	Assyro-Babylonian; mountain with twin peaks where sun rose and set
Mysia Sulci	9.6S	28.6W	4221.0	I1242	1979	Greek; where Ganymede was abducted by an eagle
Nineveh Sulcus	26.0N	60.0W	1000.0	–	1997	City where Ishtar was worshipped
Nippur Sulcus	40.9N	191.5W	2158.0	I1499	1985	Sumerian city

Nun Sulci	49.3N	318.8W	1090.0	I1242	1979	Egyptian; chaos; primordial ocean; held germ of all things
Philae Sulcus	68.5N	175.0W	900.0	–	1997	Temple that was the chief sanctuary of Isis
Philus Sulcus	44.0N	212.0W	473.0	I1242	1979	Greek; where Ganymede and Hebe were worshipped as rain-givers
Phrygia Sulcus	12.4N	19.3W	3205.0	I1242	1979	Greek; kingdom in Asia Minor where Ganymede was born
Sicyon Sulcus	36.5N	12.0W	1146.0	I1242	1979	Greek; where Ganymede and Hebe were worshipped as rain-givers
Sippar Sulcus	15.8S	191.0W	1539.0	I1536	1985	Ancient Babylonian town
Tiamat Sulcus	3.2N	209.2W	1310.0	I1242	1979	Assyro-Babylonian; tumultuous sea from which everything was generated
Ur Sulcus	48.0N	179.0W	950.0	I2331	1985	Ancient Sumerian seat of moon worship
Uruk Sulcus	8.4N	169.0W	2456.0	I1242	1979	Babylonian city ruled by Gilgamesh
Xibalba Sulcus	35.0N	80.0W	2000.0	–	1997	Mayan 'place of fright'; destination of those escaped violent death
Catena						
Enki Catena	39.5N	13.2W	151.0	–	1997	Principal water god of the Apsu
Khnum Catena	32.1N	350.9W	59.0	–	1997	Egyptian creation god
Nanshe Catena	14.7N	355.0W	59.0	–	1997	Goddess of springs and canals, daughter of Enki
Crater						
Achelous	60.3N	13.5W	51.0	I1242	1979	Greek river god; father of Callirrhoe, Ganymede's mother
Adad	55.9N	180.0W	59.0	I1242	1979	Assyro-Babylonian god of thunder
Adapa	71.3N	30.2W	54.0	I1242	1979	Assyro-Babylonian; lost im mortality when, at Ea's advice, he refused food of life
Agreus	15.2N	235.4W	72.0	I1548	1985	Hunter god in Tyre
Agrotes	62.5N	199.5W	61.0	I1565	1985	Tyre; greatest god of Gebal; farmer god
Aleyin	16.3N	134.7W	12.0	–	1997	Son of Ba'al, spirit of springs
Ammura	30.9N	344.2W	61.0	I1242	1979	Phoenician; god of the west
Amon	33.4N	223.3W	102.0	I1565	1985	Theban king of gods
Amset	14.5S	178.2W	10.0	–	1997	One of the four gods of the dead, son of Horus
Anat	3.1S	127.9W	28.0	I1498	1985	Assyro-Babylonian goddess of dew; Also used as Rand control-point crater

Andjeti	52.4S	159.1W	65.0	I1769	1985	Egyptian; first god of Busirus
Anhur	32.6N	193.9W	25.0	–	1997	Egyptian warrior god
Antum	4.4N	220.0W	21.0	I1548	1985	Babylonian; wife of Anu
Anu	63.6N	346.3W	57.0	I1242	1979	Sumerian-Akkadian god of power, of heavens
Anubis	82.7S	118.5W	97.0	I1860	1988	Egyptian jackal-headed god who opened the underworld to the dead
Ashima	37.7S	122.4W	82.0	I1769	1985	Semitic-Arab god of fate
Asshur	53.0N	335.7W	23.0	I1242	1979	Assyro-Babylonian warrior god
Aya	66.1N	326.2W	42.0	I1242	1979	Assyro-Babylonian; wife of Shamash
Ba'al	24.0N	331.6W	52.0	I1242	1979	Phoenician; Canaanite god
Bau	24.1N	53.3W	81.0	I1890	1988	Goddess who breathed into men the breath of life; daughter of Anu and patroness of Lagash
Bes	25.8S	180.4W	0.0	I1817	1985	Egyptian god of marriage
Chrysor	16.5N	134.9W	6.0	–	1997	Phoenecian god; inventor of bait, fishing hooks and line, first to sail
Cisti	32.0S	65.0W	65.0	–	1997	Iranian healing god
Danel	4.3S	25.2W	54.0	I1242	1979	Phoenician; mythical hero versed in art of divination
Diment	22.4N	353.6W	47.0	I1242	1979	Egyptian goddess of the dwelling place of the dead
Ea	18.7N	149.2W	20.0	–	1997	Assyro-babylonian god of water, wisdom, and the earth
El	1.9N	151.2W	50.0	–	1997	'Father of Men', existed before the birth of gods
Enkidu	27.9S	328.4W	121.0	I1242	1982	Friend of Gilgamesh
Enlil	53.9N	314.8W	43.0	I1242	1979	Assyro-Babylonian; nature god of the air, hurricanes, and nature
En-zu	12.2N	168.6W	7.0	–	1997	Babylonian moon god
Epigeus	24.0N	181.0W	320.0	–	1997	Phoenecian god
Erichthonius	15.5S	174.6W	35.0	–	1997	Possible father of Ganymede
Eshmun	17.8S	191.5W	99.0	I1242	1979	Phoenician; divinity of Sidon
Etana	72.5N	344.1W	49.0	I1242	1979	Assyro-Babylonian; asked the eagle for an herb to give him an heir
Gad	12.4S	137.5W	68.0	I1498	1985	Semitic god of fate or good fortune
Geb	58.3N	187.2W	62.0	I1565	1985	Heliopolis Earth god
Geinos	18.0N	221.0W	45.0	I1548	1985	Tyre; god of brick making
Gilgamesh	61.7S	123.9W	145.0	I1242	1979	Assyro-Babylonian; sought immortality after Enkidu died

Gir	35.2N	146.6W	77.0	I1649	1985	Sumerian god of summer heat
Gula	62.7N	13.9W	38.0	I1242	1979	Assyro-Babylonian; health god
Halieus	35.2N	168.0W	90.0	I1649	1985	Tyre; fisherman god
Hapi	31.3S	212.4W	85.0	I1817	1988	Egyptian god of the Nile
Hathor	70.4S	268.1W	59.0	I1242	1979	Egyptian goddess of joy and love
Hay-tau	15.8N	133.6W	28.0	–	1997	Nega god, spirit of forest vegetation
Ilah	22.8N	161.0W	71.0	I1493	1985	First Sumerian sky god
Ilus	11.5S	110.8W	41.0	I1498	1985	Ganymede's brother
Irkalla	31.1S	114.7W	116.0	I1769	1985	Sumerian goddess of under world, seen by Enkidu in a dream
Ishkur	0.1N	11.5W	83.0	I1808	1985	Sumerian god of rain
Isimu	8.1N	2.5W	90.0	I1808	1985	Sumerian god of vegetation
Isis	67.9S	197.2W	68.0	I1242	1979	Egyptian goddess; wife of Osiris
Kadi	48.8N	181.0W	94.0	I1565	1985	Babylonian goddess of justice
Keret	16.0N	35.2W	36.0	I1242	1979	Phoenician hero
Khensu	1.8N	152.8W	15.0	–	1997	Egyptian moon god
Khepri	21.5N	148.1W	50.0	–	1997	God of transformations for the Heliopitans
Khonsu	38.0S	189.7W	86.0	I1817	1988	Egyptian moon god
Khumbam creator god	25.3S	338.7W	60.0	I1242	1979	Assyro-Babylonian; Elamite
Kingu	35.7S	227.4W	91.0	I1817	1988	Assyro-Babylonian; conquered leader of Tiamat's forces whose blood was used to create man
Kishar	70.7N	352.5W	79.0	I1242	1979	Assyro-Babylonian; terrestrial progenitor goddess
Kittu	0.5S	336.6W	33.0	I1550	1985	Assyro-Babylonian god of justice
Kulla	34.8N	115.0W	82.0	I1649	1985	Sumerian god of brick making
Latpon	61.0N	175.0W	45.0	–	1997	One of the sons of El
Lugalmeslam	23.5N	195.0W	70.0	–	1997	Sumerian god of the under world
Lumha	37.3N	155.2W	71.0	I1649	1985	Title of Enki as patron of singers; also Babylonian priest
Maa	1.0N	203.8W	30.0	–	1997	Egyptian god of the sense of sight
Mehit	29.7N	165.1W	50.0	I1649	1985	Egyptian lion-headed goddess; Anhur's wife
Melkart	10.0S	185.8W	111.0	I1242	1979	Phoenician; divinity of Tyre
Min	28.7N	3.2W	35.0	I1890	1988	Egyptian fertility god
Mir	4.0S	231.4W	22.0	I1548	1985	West Semitic god of wind
Misharu	5.3S	338.3W	95.0	I1650	1985	Assyro-Babylonian god of law

Mont	43.5N	314.0W	10.0	–	1997	Theban war god
Mor	29.5N	329.3W	40.0	I1242	1979	Phoenician; spirit of the harvest
Mot	10.5N	166.2W	25.0	–	1997	Spirit of the harvest, one of the sons of El
Mush	13.5S	115.0W	97.0	I1498	1985	Sumerian male deity; upper parts are human, lower parts a serpent
Nabu	47.3S	10.1W	44.0	I1242	1979	Sumerian god of intellectual activity
Namtar	62.4S	351.6W	58.0	I1242	1979	Assyro-Babylonian plague demon
Nanna	18.5S	244.4W	48.0	I1548	1985	Sumerian moon god; god of wisdom
Nefertum	43.0N	323.0W	30.0	–	1997	Original divine son of the Memphis triad, son of Ptah
Neheh	70.8N	57.3W	57.0	I1810	1985	Egyptian god of eternity
Neith	28.9N	9.0W	93.0	I1890	1988	Egyptian warrior goddess; goddess of domestic arts
Nergal	38.5N	202.3W	8.0	–	1997	Assyro-babylonian king of the underworld
Nidaba	19.0N	123.8W	188.0	I1498	1985	Sumerian grain goddess
Nigirsu	61.2S	327.4W	70.0	I1242	1979	Assyro-Babylonian; god of the fields, war god
Ningishzida	14.0N	190.5W	30.0	–	1997	Sumerian vegetation god
Ninkasi	56.8N	53.8W	75.0	I1980	1988	Sumerian goddess of brewing
Ninki	6.6S	120.9W	170.0	I1498	1985	Consort to Ea, Babylonian god of water
Ninlil	7.6N	118.7W	91.0	I1498	1985	Chief Assyrian goddess; As shur's consort
Ninsum	13.3S	140.1W	91.0	I1498	1985	Minor Babylonian goddess of wisdom; Gilgamesh's mother
Nut	60.1S	268.0W	93.0	I1242	1979	Egyptian goddess of the sky
Osiris	37.8S	165.2W	109.0	I1242	1979	Egyptian god of the dead
Ptah	67.0S	214.3W	45.0	I1860	1988	Sovereign god of Memphis; patron of artisans
Punt	26.1S	242.2W	228.0	–	1997	Land east of Egypt where Bes originated. Changed from Punt Facula
Ruti	11.9N	310.7W	35.0	I1242	1979	Phoenician; Byblos god
Sapas	56.5N	37.9W	65.0	I1242	1979	Assyro-Babylonian; torch of the gods
Sati	30.5N	14.9W	98.0	I1890	1988	Wife of Khnum, Egyptian god of the Cataracts
Sebek	59.5N	178.9W	70.0	I1818	1979	Egyptian crocodile god
Seima	16.6N	217.4W	33.0	I1548	1985	Mother goddess of the Arameans
Seker	40.8S	351.0W	117.0	I1871	1988	Egyptian god of the dead at Memphis

Selket	16.7N	107.4W	140.0	I1498	1985	Tutelary goddess who guarded intestines of the dead
Serapis	10.0S	50.0W	155.0	–	1997	Egyptian healing god
Shu	42.5N	357.9W	54.0	I1818	1988	Egyptian god of air
Sin	51.9N	359.2W	30.0	I1242	1979	Babylonian moon god
Tammuz	12.9N	232.7W	44.0	I1548	1985	Akkadian youthful god of vegetation; Ishtar's son
Tanit	56.7N	40.7W	45.0	I1242	1979	Assyro-Babylonian; Carthaginian goddess
Ta-urt	26.5N	306.5W	85.0	I1818	1988	Egyptian childbirth goddess
Teshub	72.3S	281.0W	60.0	–	1994	Elamite god of the tempest
Thoth	42.4S	146.0W	107.0	I1769	1985	Egyptian moon god; invented all arts and sciences
Tros	11.0N	31.1W	109.0	I1242	1979	Greek; father of Ganymede
Upuant	45.0N	321.5W	15.0	–	1997	Jackal-headed warrior god, god of the dead
Zakar	30.5N	335.2W	150.0	–	1997	Assyrian supreme diety
Zaqar	57.5N	41.3W	52.0	I1242	1979	Assyro-Babylonian; Sin's messenger who brought dreams to men
Facula						
Abydos Facula	34.1N	154.0W	165.0	I1493	1985	Egyptian town where Osiris was worshipped
Akhmin Facula	28.0N	191.0W	225.0	–	1997	Egyptian town where Min was worshipped
Busiris Facula	14.9N	216.1W	348.0	I1536	1985	Town in lower Egypt where Osiris was first installed as local god
Buto Facula	12.6N	204.3W	236.0	I1536	1985	Swamp where Isis hid Osiris' body
Coptos Facula	9.4N	209.8W	332.0	I1536	1985	Early town from which caravans departed
Dendera Facula	0.0N	257.0W	114.0	I1548	1985	Town where Hathor was chief goddess
Edfu Facula	26.8N	147.7W	187.0	I1493	1985	Egyptian town where Horus was worshipped
Heliopolis Facula	19.5N	147.6W	50.0	–	1997	Sacred Egyptian city of the sun
Hermopolis Facula	22.0N	196.0W	200.0	–	1997	Place where Unut was worshipped
Memphis Facula	15.4N	132.5W	344.0	I1498	1985	Ancient capitol of lower kingdom
Ombos Facula	3.8N	238.6W	90.0	I1548	1985	Egyptian town where Sebek's triad worshipped; present Kom Ombo
[Punt Facula]	26.1S	242.2W	228.0	I1817	1985	Land east of Egypt where Bes originated. Feature reclassified as Punt Cater

[Sais Facula]	37.9N	14.2W	137.0	I1890	1988	Capital of Egypt in mid–7th century B.C. Name no longer valid
Siwah Facula	7.5N	143.2W	220.0	I1498	1985	Oasis oracle of Zeus-Ammon; visited by Alexander
Tettu Facula	38.6N	160.9W	86.0	I1493	1985	Egyptian town where Hatmenit and Osiris were worshipped
Thebes Facula	6.0N	202.4W	475.0	I1536	1985	Ancient capitol of upper kingdom
Fossa						
Lakhamu Fossa	12.5S	228.3W	392.0	I1548	1985	Dragon monster, or divine natural force produced by Apsu and Tiamat
Lakhmu Fossae	30.3N	142.3W	2871.0	I1498	1985	Dragon monster, or divine natural force produced by Apsu and Tiamat
Zu Fossae	53.0N	129.4W	1386.0	I1493	1985	Dragon of chaos slain by Marduk

Callisto

Large ring feature						
Adlinda	46.0S	33.0W	600.0	I1239	1979	Eskimo; ocean depths where souls are imprisoned after death
Asgard	32.0N	139.8W	1347.0	I1888	1979	Norse; the home of the gods
Valhalla	15.9N	56.6W	2748.0	I1239	1979	Norse; Odin's hall, where he received the souls of slain warriors
Catena						
Eikin Catena	8.5S	15.9W	191.0	–	1997	Norse river
Fimbulthul Catena	8.4N	65.4W	378.0	–	1997	Norse river
Geirvimul Catena	49.0N	347.1W	90.0	–	1997	Norse river
Gipul Catena	70.2N	48.2W	588.0	I1239	1979	Norse river
Gomul Catena	35.4N	48.0W	324.0	–	1997	Norse river
Gunntro Catena	19.3S	343.3W	136.0	–	1997	Norse river
Sid Catena	48.7N	105.4W	78.0	–	1997	Norse river
Svol Catena	11.0N	37.1W	140.0	–	1997	Norse river
Crater						
Adal	75.4N	80.8W	40.0	I1239	1979	Norse; son of Karl and Erna
Aegir	45.9S	104.4W	46.0	–	1997	Norse sea god
Agloolik	47.9S	82.9W	49.0	–	1997	Eskimo spirit of the seal caves
Agroi	43.3N	11.0W	55.0	I1239	1979	Finno-Ugric god of twins
Ahti	41.8N	103.1W	52.0	I1888	1987	Finnish god of water; sends fish to the fisherman
Ajleke	22.4N	101.3W	46.0	I1888	1987	Saami god of holidays

Akycha	72.5N	318.6W	67.0	I1239	1979	Alaskan name of the sun
Alfr	9.7S	222.8W	60.0	I1239	1979	Norse dwarf
Ali	59.3N	56.2W	61.0	I1239	1979	Norse; strongest of men
Anarr	44.1N	0.6W	47.0	I1239	1979	Norse dwarf
Aningan	50.5N	8.2W	287.0	I1239	1979	Moon god of Greenland Eskimos
Arcas	85.0S	66.8W	41.0	–	1997	Callisto's child by Zeus
Askr	51.7N	324.1W	64.0	I1239	1979	Norse; first man, created from a log drifted ashore on a beach
Audr	31.0S	81.2W	70.0	–	1997	Ottar's ancestor
Austri	81.3S	64.1W	9.0	–	1997	Norse dwarf
Aziren	35.4N	178.3W	64.0	I1888	1987	Estonian spirit of death
Balkr	29.1N	11.9W	64.0	I1239	1979	Norse; Ottar's ancestor
Barri	31.6S	71.0W	83.0	–	1997	Ottar's ancestor
Bavorr	49.2N	20.3W	84.0	I1239	1979	Norse dwarf
Beli	62.6N	81.7W	50.0	I1239	1979	Celtic; father of Caswallawn
Biflindi	53.8S	74.3W	57.0	–	1997	Another name for Odinn
Bragi	75.7N	61.7W	65.0	I1239	1979	Skaldic; god of poetry
Brami	28.9N	19.2W	67.0	I1239	1979	Norse; Ottar's ancestor
Bran	24.3S	207.7W	89.0	I1239	1979	Celtic; omnipotent god who watched over people
Buga	22.2N	323.9W	54.0	I1239	1979	Tungu heaven god
Buri	38.7S	46.2W	98.0	I1239	1979	Norse dwarf
Burr	42.5N	135.5W	74.0	I1239	1979	Norse giant; his sons raised up heaven's vault and shaped the Earth
Dag	58.6N	74.2W	40.0	I1239	1979	Norse; Ottar's ancestor
Danr	62.5N	77.8W	48.0	I1239	1979	Norse; king against whom Konr marched
Dia	73.0N	50.4W	35.0	I1239	1979	Greek; Callisto's sister
Doh	30.4N	142.1W	55.0	–	1997	Ketian shaman who created the earth
Dryops	77.6N	21.3W	42.0	I1239	1979	Greek; son of Dia by Apollo
Durinn	67.0N	90.1W	49.0	I1239	1979	Norse dwarf
Egdir	33.9N	35.9W	58.0	I1239	1979	Norse; shepherd for the giants
Egres	42.5N	176.6W	38.0	I1888	1987	Karelian deity of the harvest of beans
Erlik	66.8N	1.3W	39.0	I1239	1979	Russian first man who became a devil
Fadir	56.4N	12.7W	81.0	I1239	1979	Norse farmer
Fili	64.3N	349.5W	42.0	I1239	1979	Norse dwarf
Finnr	15.5N	4.3W	65.0	–	1979	Norse dwarf
Freki	79.9N	352.0W	48.0	I1239	1979	Norse; wolf's name meaning 'unsatiable'
Frodi	68.3N	139.1W	44.0	I1239	1979	Norse; Hledis' father
Fulla	73.5N	103.7W	45.0	I1239	1979	Norse; maid to Frigg, queen of the gods

Fulnir	60.3N	35.5W	46.0	I1239	1979	Norse; son of Thrael and Thyr
Gandalfr	81.0S	63.3W	12.0	–	1997	Norse dwarf
Geri	66.7N	354.2W	38.0	I1239	1979	Norse; wolf's name meaning 'greedy.'
Ginandi	85.7S	50.0W	30.0	–	1997	Ottar's ancestor
Gisl	57.2N	34.8W	39.0	I1239	1979	Norse; steed ridden by Aesir
Gloi	49.0N	245.7W	112.0	I1239	1979	Norse dwarf
Goll	57.5N	319.5W	55.0	I1239	1979	Norse; servant to the gods
Gondul	59.9N	115.5W	50.0	I1888	1979	Norse; a Valkyrie
Grimr	41.6N	215.2W	90.0	I1239	1979	Norse; a name for Odin
Gunnr	64.7N	106.2W	58.0	I1888	1979	Norse; a Valkyrie
Gymir	63.8N	49.2W	40.0	I1239	1979	Norse; another name for the sea-god, Legir
Habrok	74.8N	129.3W	40.0	I1239	1979	Norse; a hawk
Haki	24.9N	315.1W	69.0	I1239	1979	Norse giant
Har	3.6S	357.9W	44.0	I1239	1979	Norse; a name for Odin
Hepti	64.5N	23.9W	41.0	I1239	1979	Norse dwarf
Hijsi	61.6N	169.3W	51.0	I1888	1987	Karelian deity of hunting
Hodr	69.0N	91.0W	76.0	I1239	1979	Norse; Baldr's blind brother who shot Baldr unknowingly
Hoenir	33.9S	261.2W	84.0	I1239	1979	Norse; god who gave souls to first humans
Hogni	13.5S	4.5W	65.0	I1239	1979	Norse; Ottar's ancestor
Holdr	44.1N	109.2W	61.0	I1888	1988	Son of Karl and Snor in Rigdismal
Igaluk	5.6N	315.9W	105.0	I1239	1979	Alaskan name of the Moon
Ilma	30.0S	167.2W	53.0	–	1988	A celestial divinity of air
Ivarr	6.1S	321.5W	68.0	I1239	1979	Norse; Ottar's ancestor
Jalkr	38.6S	83.2W	74.0	–	1997	Another name for Odin
Jumal	58.8N	119.5W	60.0	I1888	1987	Estonian sky god
Jumo	65.9N	12.3W	37.0	I1239	1979	Finno-Ugric heaven god
Kari	48.2N	117.9W	32.0	I1888	1979	Ottar's ancestor
Karl	56.4N	330.7W	42.0	I1239	1979	Norse; Rigr's son with Amma
Keelut	77.3S	92.1W	47.0	–	1997	Eskimo evil spirit who resembles a hairless dog
Kul	62.7N	123.5W	43.0	I1888	1987	Komi wood spirit
Lempo	25.6S	319.5W	46.0	–	1988	Finno-Ugric evil spirit
Ljekio	47.9N	161.4W	36.0	I1888	1987	Finnish god of grass, roots of trees
Lodurr	51.2S	270.8W	76.0	I1239	1979	Norse; god who gave first humans goodly colour
Lofn	57.0S	24.0W	200.0	–	1997	Norse goddess of marriage
Loni	3.6S	214.9W	86.0	I1239	1979	Norse dwarf
Losy	65.3N	323.1W	62.0	I1239	1979	Mongolian; Mongol evil snake; tried to kill all living things
Lycaon	45.2S	5.8W	55.0	–	1997	Callisto's father

Maderatcha	30.5N	95.9W	57.0	I1888	1987	Saami sky god
Mera	64.1N	75.8W	36.0	I1239	1979	Greek; another nymph of Artemis seduced by Zeus
Mimir	32.6N	53.2W	51.0	I1239	1979	Norse giant
Mitsina	57.5N	104.7W	43.0	I1888	1979	Alaskan old man who perished while hunting on ice
Modi	66.4N	120.9W	43.0	I1239	1979	Norse; son of Thor and Sif
Nakki	56.6S	69.9W	65.0	–	1997	Finnish water god
Nama	57.2N	331.3W	61.0	I1239	1979	Altaic hero who built ark to save his family from the flood
Nar	1.7S	46.4W	63.0	I1239	1979	Norse dwarf
Nerrivik	17.1S	57.1W	37.0	I1239	1979	Alaskan name of Sedna
Nidi	66.7N	96.5W	43.0	I1239	1979	Norse dwarf
Nirkes	29.8N	163.6W	48.0	I1888	1988	Karelian patron of squirrel hunting
Njord	16.5N	132.8W	34.0	–	1988	Nordic gods called the Vanir; pacific, benevolent, guardians of man
Nori	45.4N	343.5W	86.0	I1239	1979	Norse dwarf
Norov-Ava	54.6N	113.7W	47.0	I1888	1987	Mordvinian mistress of the field
Nuada	62.1N	273.2W	66.0	I1239	1979	Irish chieftan god
Numi-Torum	50.2S	93.4W	65.0	–	1997	Mansi creator god
Nyctimus	62.7S	3.8W	29.0	–	1997	Brother of Callisto
Oluksak	47.9S	63.9W	74.0	–	1997	Eskimo god of lakes
Omol	42.2N	118.4W	60.0	I1888	1987	Komi wood spirit
Orestheus	46.7S	47.5W	30.0	–	1997	Brother of Callisto
Oski	57.2N	269.3W	57.0	I1239	1979	Norse; a name for Odin
Ottar	61.5N	104.8W	50.0	I1888	1979	Innsteinn's son and Freyja's favorite
Pekko	18.3N	5.4W	61.0	I1239	1979	Finno-Ugric god of barley
Randver	72.3S	53.6W	21.0	–	1997	Ottar's ancestor
Reginleif	66.2S	97.2W	32.0	–	1997	Servant of the gods
Reginn	39.7N	90.8W	51.0	I1888	1979	Norse dwarf
Reifnir	50.8S	63.8W	39.0	–	1997	Ottar's ancestor
Rigr	70.9N	245.0W	54.0	I1239	1979	Norse; another name for the god Heimdall
Rongoteus	53.5N	106.8W	35.0	I1888	1987	Karelian deity of the harvest of rye
Rota	27.8N	109.8W	55.0	I1888	1987	Deity of the underground world
Saga	0.0N	326.0W	0.0	I1239	1979	Scandinavian goddess, wife of Odin; Rand control point crater
Sarakka	3.7S	53.7W	56.0	I1239	1979	Finno-Ugric goddess of child-birth
Seqinek	55.5N	25.5W	80.0	I1239	1979	Eskimo; the sun
Sholmo	53.9N	16.4W	58.0	I1239	1979	Finno-Ugric heaven god
Sigyn	35.8N	29.2W	44.0	I123	1979	Norse; Loki's wife

Skeggold	49.6S	31.9W	39.0	–	1997	Servant of the gods
Skoll	55.6N	315.3W	55.0	I1239	1979	Norse wolf
Skuld	10.1N	37.7W	81.0	I1239	1979	Norse; maiden living near Yggdrasill who governed the fate of humans
Sudri	55.4N	137.1W	69.0	I1239	1979	Norse dwarf
Sumbur	67.1N	324.8W	38.0	I1239	1979	Russian (Buriat) world mountain
Tapio	30.6N	109.7W	56.0	I1888	1987	Finnish deity of the wood who sends game to the hunter
Thekkr	80.8S	61.6W	10.0	–	1997	Norse dwarf
Thorir	32.0S	67.3W	43.0	–	1997	Ottar's ancestor
Tindr	2.5S	355.5W	64.0	I1239	1979	Norse; Ottar's ancestor
Tontu	27.6N	100.3W	47.0	I1888	1987	Finnish god of housekeeping
Tornarsuk	28.7N	128.6W	104.0	I1888	1979	Greenland legendary hero
Tyll	43.3N	165.4W	65.0	I1888	1987	Estonian epic hero; struggled with a giant
Tyn	70.8N	233.6W	60.0	I1239	1979	Great god of Germanic peoples
Uksakka	49.5S	42.3W	22.0	–	1997	Lapp protector goddess
Valfodr	1.2S	247.8W	81.0	I1239	1979	Norse; a name for Odin, god of wisdom
Vali	9.8N	325.2W	46.0	I1239	1979	Norse; Ottar's ancestor
Vanapagan	38.1N	158.0W	62.0	I1888	1987	Estonian, a wicked giant
Veralden	33.2N	96.1W	75.0	I1888	1987	Saami god of fertility
Vestri	45.3N	52.8W	75.0	I1239	1979	Norse dwarf
Vidarr	11.9N	193.6W	84.0	–	1988	Norse god
Vitr	22.4S	349.3W	76.0	I1239	1979	Norse dwarf
Vu-Murt	22.9N	170.9W	79.0	I1888	1987	Estonian spirit of water
Vutash	31.9N	102.9W	55.0	I1888	1987	Estonian spirit of water
Ymir	51.4N	101.3W	77.0	I1239	1979	Norse; giant from whom Earth was created
Yuryung	54.9S	86.1W	74.0	–	1997	Yakutian heaven god

ASTEROIDS

Gaspra

Regio

Dunne Regio	15.0N	15.0W	–		1994	James; early Galileo Project planner.
Neujmin Regio	2.0N	80.0W	–		1994	Grigorij N.; Russian astronomer; discoverer of Gaspra (1885–1946)
Yeates Regio	65.0N	75.0W	–		1994	Clayne; early Galileo Project manager

Craters					
Aix	47.9N	160.3W	6.0	1994	Spa in France
Alupka	65.0N	65.0W	0.0	1994	Spa in Crimea, Ukraine
Baden-Baden	46.0N	55.0W	0.0	1994	Spa in Germany
Badgastein	25.0N	3.0W	0.0	1994	Spa in Austria
Bagnoles	55.0N	122.0W	0.0	1994	Spa in France
Bath	13.4N	9.7W	10.0	1994	Spa in England
Beppu	3.9N	58.4W	5.0	1994	Spa on Kyushu, Japan
Brookton	27.7N	103.3W	6.0	1994	Spa in New York, USA
Calistoga	30.0N	2.0W	0.0	1994	Resort in California, USA
Carlsbad	29.7N	88.8W	5.0	1994	Spa in Czech Republic
Charax	8.6N	0.0W	10.7		
Helwan	22.4N	118.9W	6.0	1994	Spa in Egypt
Ixtapan	11.9N	86.9W	5.0	1994	Spa in Mexico
Katsiveli	55.0N	65.0W	0.0	1994	Spa in Crimea, Ukraine
Krynica	49.0N	35.0W	0.0	1994	Health resort in Poland
Lisdoonvarna	16.5N	358.1W	10.0	1994	Spa in Ireland
Loutraki	42.0N	140.0W	0.0	1994	Spa in Greece
Mandal	23.5N	46.5W	5.0	1994	Spa in Norway
Manikaran	62.0N	155.0W	0.0	1994	Spa in India
Marienbad	35.4N	81.8W	5.0	1994	Spa in Czech Republic
Miskhor	15.0N	65.9W	5.0	1994	Spa in Crimea, Ukraine
Moree	15.1N	164.4W	6.0	1994	Spa in Australia
Ramlosa	15.0N	4.9W	10.0	1994	Spa in Sweden
Rio Hondo	31.7N	20.7W	7.0	1994	Spa in Argentina
Rotorua	18.8N	30.7W	6.0	1994	Spa in New Zealand
Saratoga	50.0N	270.0W	0.0	1994	Spa in New York, USA
Spa	51.5N	152.0W	6.0	1994	Health resort in Belgium
Tang-Shan	59.0N	256.0W	0.0	1994	Spa in China
Yalova	29.0N	10.0W	0.0	1994	Health resort in Turkey
Yalta	57.6N	261.3W	5.0	1994	Spa in Crimea, Ukraine
Zohar	23.0N	118.0W	0.0	1994	Spa in Israel

Ida					
Regio					
Palisa Regio	23.0S	34.0E	23.0	1997	Johann; Austrian astronomer, discovered Ida (1848–1925)
Pola Regio	11.0S	184.0E	8.0	1997	Place where Palisa (discoverer of Ida) observed
Vienna Regio	8.0N	2.0E	13.0	1997	Where Palisa discovered Ida
Dorsum					
Townsend Dorsum	25.0N	30.0E	40.0	1997	Tim E.; Galileo imaging team member (d. 1989)
Craters					
Afon	6.5S	0.0E	25.0	1994	Cave in Russia
Atea	5.7S	18.9E	2.0	1997	Cave in the Muller Range of Papua New Guinea

These annotated images show the locations of various features on asteroids Gaspra, Ida and Dactyl. (Courtesy Jennifer Blue, USGS)

Azzurra	30.5N	217.2E	9.6	1997	Flooded cave (known as the Blue Grotto) on the island of Capri in southern Italy
Bilemot	27.8S	29.2E	1.8	1997	Lava tube in Korea
Castellana	13.4S	335.2E	5.2	1997	Cave in Puglia region, Italy
Choukoutien	12.8N	23.6E	1.1	1997	Site where Peking Man was discovered
Fingal	13.2S	39.9E	1.5	1997	Cave in the Hebrides
Kartchner	7.0S	179.0E	0.9	1997	Cave in Arizona
Kazumura	32.0S	41.1E	2.1	1997	Lava tube in Hawaii
Lascaux	0.8N	161.2E	11.8	1997	Cave in France noted for its prehistoric paintings
Lechuguilla	7.9N	357.1E	1.5	1997	Cave in Carlsbad National Park, New Mexico
Mammoth	18.3S	180.3E	10.2	1997	Longest limestone cavern known on Earth
Manjang	28.3S	90.5E	1.0	1997	Lava tube in Korea
Orgnac	6.3S	202.7E	10.6	1997	Cave in France
Padirac	4.3S	5.2E	1.9	1997	Cave with underground river in France
Peacock	2.0S	52.0E	0.2	1997	Cave in Florida, USA
Postojna	42.9S	359.9E	6.0	1997	Large cave in Slovenia
Sterkfontein	4.1S	54.1E	4.7	1997	Cave in South Africa
Stiffe	27.9S	126.5E	1.5	1997	Karst cave in Sulmona, Italy
Undara	2.0N	113.8E	8.5	1997	Lava tube from Undara Volcano, North Queensland, Australia
Viento	12.2N	343.9E	1.6	1997	Lava tube in Spain
Dactyl					
Craters					
Acmon	39.0S	138.0E	0.3	1997	One of the original three Dactyls
Celmis	46.0S	220.0E	0.2	1997	One of the original three Dactyls

Glossary

aa A basaltic lava with a rough, jagged surface.

achondrite A stony meteorite, coarsely crystallised, with sizable fragments of various minerals visible to the naked eye.

adiabatic cooling This takes place in a gas when it does not undergo a net loss of heat energy, but decreases in temperature only as a result of expansion as its pressure drops. It takes place in a well-mixed atmosphere.

Adrastea This Jovian moon is the smallest of the four inner satellites. It was discovered by Dave Jewitt and Ed Danielson of the Voyager team, in 1979. Its longest dimension is 20 kilometres across but details of its actual shape are not available. It skims along the main ring's outer edge. It was named after a nymph of Crete to whose care Rhea entrusted the infant Zeus.

aerosol A sol (homogeneous suspension, more fluid than a gel) in which the dispersion medium is a gas containing solid particles or liquid droplets.

agents of change The constructive and destructive processes that reshape a planetary surface.

albedo Reflectivity of an object, expressed as the ratio or percentage of reflected to incident (incoming) light or electromagnetic radiation. It is the ratio of the amount of solar radiation reflected from an object to the total amount incident upon it.

Alfvén wing Electromagnetic waves that are generated when plasma flows past an electrically conducting body such as Jupiter's moon Io.

altimetry The measurement of elevation or altitude.

Amalthea This Jovian moon is the largest of the four inner moons. It was discovered by Edward Barnard in 1892, who eventually chose a name suggested by Flammarian for the satellite. Amalthea (a goat in some accounts, a princess of Crete in others) suckled Zeus as a young child. It lies at the outer periphery of the inner gossamer ring. The largest crater, Pan, appears to be bowl-shaped and 90 kilometres wide. A bright spot at the south pole is associated with a smaller crater named Gaea. Amalthea's longest axis is 247 kilometres.

ammonia NH_3

ammonium hydrosuphide NH_4SH.

Ananke [a-NANG-kee] A small moonlet of Jupiter. It was discovered by Seth Nicholson in 1951. It was named for the goddess of necessity, the daughter of Zeus and Themis.

anorthosite A type of igneous rock composed almost entirely of feldspar.

Anticyclone A high pressure atmospheric circulation whose relative direction of rotation is clockwise in the northern hemisphere, anticlockwise in the southern hemisphere, and undefined at the equator.

anti-Jovian Used to refer to the side of a moon that is facing away from Jupiter.
antipodal point The opposite point with respect to any given point, through the centre of a body.
apojove The furthest point from Jupiter in an orbit about that planet.
aphelion The point in the path of a planet, asteroid, comet, or other body that is furthest from the Sun.
asteroid A small rocky celestial body in orbit around the Sun which shows no evidence of an atmosphere or other types of activity associated with comets. A concentration of such bodies makes up a belt in the space between the orbits of Mars and Jupiter.
aurora A glow from a planet's atmosphere produced by the impact and interaction of charged particles in a planet's magnetosphere with the atmospheric atoms and molecules.

basalt Fine-grained igneous rock (rich in mafic minerals) that has erupted onto the surface.
base A substance which reacts with acid to make a salt. A substance with this property is said to be basic.
basin A very large impact structure.
bedrock Continuous solid rock that underlies regolith and is exposed at outcrops.
bolide Projectile; a meteor or meteorite.
breccia Coarse-grained rock composed of angular fragments of pre-existing rock, bound within a finer grained matrix.
bright terrain A type of surface on Ganymede which is lighter than other regions on this moon which is often associated with grooves (*see also* dark terrain).

caldera A large volcanic crater on the summit of a volcano produced by the collapse of the underground lava reservoir, or on a plain caused by a massive explosion.
Callisto [kah-LISS-toe] One of the large moons of Jupiter. It was discovered by Galileo Galilei in 1610. In Greek mythology, Callisto was the daughter of Lycaon. Zeus wanted to woo her, and so disguised himself as Artemis and seduced her. He changed her into a bear in order to hide her from his jealous wife Hera.
Carme [KAR-mee] A small moonlet of Jupiter. It was discovered by Seth Nicholson in 1938. It was named for a nymph and attendant of Artemis; mother, by Zeus, of Britomartis.
Centaur In Greek mythology, a being with the head, arms, and torso of a man, and the body and legs of a horse. The personification of wisdom and beastliness: the two natures of humankind.
charge-coupled device (CCD) An electronic device that consists of a regular array of light sensitive elements that emit electrons when exposed to light.
Chicxulub crater A very large impact crater near the Yucatan Peninsula in Mexico. The effects of this particular impact may have been responsible for the extinction of the dinosaurs 65 million years ago.
chondrite A stony meteorite, composed of finely crystallised material.
clathrates When water ices contain small amounts of 'impurities' (e.g. ammonia) they are known as clathrates, and have a lower melting temperature than pure water-ice.
coma A roughly spherical region of diffuse gas which surrounds the nucleus of a comet. Together, the coma and the nucleus form the comet's head.
comet A comet is a planetesimal in orbit around the Sun. Comets are believed to be composed of dust and volatile ices. When close to the Sun, comets become heated enough to produce a coma of gas and dust. As this gas and dust moves outward from the comet, it is 'blown' away by the solar wind and forms the comet's tail. Comets either orbit the Sun, or pass through the Solar System on hyperbolic orbital paths.

compression A 'packing' or reduction in the amount of data being transmitted, using a mathematical formula similar to averaging, in order to optimise the transmission of data from the spacecraft. Once received, this data is 'unpacked' (decompressed) to reconstruct the full image. However, if noise (*see* radiation) is present it may cause incorrect values for the picture elements it was averaged with, which also affects the resulting image.

convection The transfer of heat from a region of high temperature to a region of lower temperature by the displacement of the cooler molecules by the warmer molecules.

coordinates The large Galilean satellites of Jupiter rotate on axes perpendicular to their orbits, and their periods of rotation are synchronised with their orbital periods, so they maintain one hemisphere facing Jupiter. Cartographers took advantage of this to define the meridian as passing through the 'sub-Jovian' point. Accordingly, the 'anti-Jovian' point is in the centre of the hemisphere which permanently faces away from the planet. By convention, longitude is measured exclusively west of the meridian. The 'leading' hemisphere is centred at 90 degrees longitude, the anti-Jovian point is at 180 degrees, and the 'trailing' hemisphere is centred at 270 degrees.

Coriolis effect Also known as the Coriolis deflection, it is the deflection relative to the Earth's surface of any object moving above the Earth resulting from the velocity-dependent pseudo-force in a reference frame which is rotating with respect to an inertial reference frame. An object moving horizontally is deflected to the right in the northern hemisphere and to the left in the southern hemisphere.

crater A typically bowl-shaped pit, depression, cavity, or hole, generally of considerable size and with steep slopes, sometimes surrounded by a raised rim, typically formed by explosion during meteorite impact.

crater density The number of craters on a surface per unit area.

crater size distribution The relative numbers of craters of given sizes on a surface.

cross-cut (principle of cross-cutting relationships) An interruption of a geologic feature by another, which can give an indication of the relative ages of these geologic features/events. (e.g. a fault cutting across an impact crater would be younger than the crater.)

crust The outermost layer of the lithosphere, composed of relatively low-density materials.

cyclone A low-pressure atmospheric circulation whose relative direction of rotation is anticlockwise in the northern hemisphere, clockwise in the southern hermisphere, and undefined at the equator.

Dactyl In Greek mythology, a legendary being that lived on Mount Ida.

dark terrain A type of surface on Ganymede and Callisto which is darker than other regions on these moons (*see also* bright terrain). On Ganymede dark terrain is often associated with furrows.

Deimos In Greek mythology, a son of Ares (Mars) who, with brother Phobos, was a constant companion to his father.

Doppler effect This was first described by the Austrian physicist Christian Johann Doppler in 1842, while writing about the colours of binary stars. As a source of radiation, either sound or light, is moving towards or away from an observer, the apparent frequency of the radiation is shifted. If the source is approaching, then the shift is towards higher frequencies. If the source is receding the shift is to lower frequencies.

ejecta The material thrown from an impact crater during its formation.

Elara [EE-lar-uh] A small moonlet of Jupiter. It was discovered by Charles Perrine in 1904. It was named for a paramour of Zeus; mother of the giant Tityus.

ellipse A set of points, the sum of whose distance from two fixed points (the foci) is constant.

An ellipse is essentially a circle that has been stretched out of shape. When describing ellipses, the eccentricity defines how 'stretched out' it is.

erosion The process of physically removing weathered materials; the wearing away of soil and rock by weathering, mass wasting, and the action of streams, glaciers, waves, wind and groundwater.

ERT Earth-received-time, is the soonest that a signal indiating an event experienced by the spacecraft would reach the Earth (taking into account the OWLT).

escarpment A long, more or less continuous cliff or relatively steep slope facing in one general direction, produced by erosion or faulting.

Europa [yur-ROH-pah] One of the large moons of Jupiter. It was discovered by Galileo Galilei in 1610. In Greek mythology, Europa, the beautiful daughter of Agenor, king of Tyre, was seduced by Zeus, to whom he appeared as a gentle white cow; he assumed the shape of a white bull and persuaded her to take a ride on his back, and then he carried her away across the sea to Crete, where she bore several children, including Minos.

exosphere This is the extreme outer fringe of a planetary atmosphere in which the gas density is so low that the mean free path of particles depends upon their direction with respect to the local vertical (being greatest when travelling upwards). In the case of a small moon possessing a tenuous gaseous envelope, the term is employed to indicate that this 'atmosphere' is exceedingly thin.

facula A circular to subcircular region which is brighter than the surrounding area.

fault A fracture or zone of fractures in a planet's crust, accompanied by displacement of the opposing sides.

feldspar A group of rock-forming minerals that make up about 60 percent of the Earth's crust.

footprint The point on Jupiter where a flux tube from Io makes contact with the planet's upper atmosphere. There is one near each magnetic pole, in line with Io's position in orbit.

flux tube The electrical current formed as charged particles from Io flow back and forth along Jupiter's magnetic field lines.

fracture A break or crack.

friction The rubbing of one surface or thing against another.

furrow In the case of Ganymede furrows refer to depressions which occur in the dark terrain.

Galilean satellites Io, Europa, Ganymede, Callisto, the four largest satellites of Jupiter which were discovered in 1610 by Galileo Galilei.

Ganymede [GAN-ee-meed] One of the large moons of Jupiter. It was discovered by Galileo Galilei in 1610. In Greek mythology, Ganymede was a young Trojan boy, son of Tros and Calirrhoe, who was carried to Olympus by Zeus disguised as an eagle, whereupon he was made cupbearer to the Olympian gods.

gas giants The outer Solar System planets – Jupiter, Saturn, Uranus, and Neptune – are composed mostly of hydrogen, helium and methane, and have a density less than 2 g/cm^3.

Gaspra Russian resort and spa near Yalta, Crimea.

geodetic control network An ellipsoid or sphere-shaped 'net' used as a type of base-map or control to correlate coordinates on a flat image with their actual location on the spherical planetary body, to obtain accurate measurements from images.

geological activity The expression of the internal and external processes and events that affect a planetary body.

geological time The time extending from the end of the formative period of the Earth to the beginning of human history.

geological unit A body of rock (or ice, in Europa's case) that has a distinct origin and consists of dominant, unifying features that can be easily recognised and mapped.

geomorphology The study of the external structure, form, and arrangement of rocks in relation to the development of landforms.

geyser A type of water or gas vent that intermittently erupts jets of material.

Gossamer ring The outermost component of the Jovian ring system. It actually consists of two faint, fairly uniform rings, one enclosing the other, visibly spreading from the outer boundary of the main ring and fading somewhere beyond 221,000 kilometres from the centre of Jupiter, at Thebe's orbit. The denser, enclosed gossamer ring extends radially in from Amalthea's orbit at 181,000 kilometres from Jupiter's centre, while the fainter ring is situated interior to Thebe's orbit at 222,000 kilometres. As Amalthea's orbit is crossed, the ring's brightness drops to one-fifth, while near Thebe's path the drop is by a factor of three. Both rings have cross-sections that are crudely rectangular. A striking feature of both rings is that each is banded with top and lower edges brighter than the centres.

graben A long, relatively depressed crustal unit or block bounded by tectonic faults along its sides; a trough.

gravity The acceleration produced by the mutual attraction of two masses, and of magnitude inversely proportional to the square of the distance between the two centres of mass.

gravity gradient Refers to the difference in the acceleration of gravity from one point to another.

great circle The line of intersection of the surface of a sphere and any plane which passes through the centre of the sphere.

groove In the case of Ganymede grooves refer to depressions which occur in the bright terrain, often in parallel sets.

ground truth The on-site gathering of data at a particular location, for the purpose of calibrating and interpreting remotely acquired data.

Halo The innermost component of the Jovian ring system is a toroidal halo extending radially from about 92,000 kilometres to about 122,500 kilometres from the planet's centre. Its brightness decreases with the height of the equatorial plane and decreases as the planet is approached. The halo is actually a nimbus of fine particles that 'blooms' vertically at the main ring's inner boundary and continues towards the planet and vertically.

heterosphere The upper level of an atmosphere in which there is gross compositional variation, with chemical stratification.

high Sun When the Sun is directly overhead, it does not throw shadows on the surface. When planetary images are obtained in high Sun conditions they are examined for albedo variation rather than morphological information (as can be done near the terminator); sites imaged under both high and low Sun conditions can be compared to link albedo and morphological features.

Himalia [hih-MAL-yuh] A small moonlet of Jupiter. It was discovered by Charles Perrine in 1904.

homopause The level in the atmosphere which marks the transition from the homosphere to the heterosphere.

homosphere The lower region of an atmosphere in which there is no gross compositional variation.

hummocky Uneven; describing a terrain abounding in irregular knolls, mounds, or other small elevations.

ice-rock Water-ice (frozen H_2O) is an important geological material on many of the outer planet satellites. Under various conditions ice can have viscosities and flow properties similar to different types of molten rock, and therefore may also result in the formation of similar geological structures. Ice that is at near freezing temperatures (273 K) and low pressure (< 1 kilobar) – known as 'Ice-I' – has different physical properties than ice that is at lower temperatures and higher pressures. (e.g. Ice-II is stable at 200 K and 6 kilobar, while Ice-VI is stable at 200 K and 15 kilobar). These 'phases' of ice have different densities and flow properties that may be expressed through tectonic and volcanic processes on icy satellites. When these ices contain small amounts of 'impurities' (e.g. ammonia) they are known as 'clathrates', and have a lower melting temperature than pure water-ice – which may be a factor in volcanic processes on these satellites.

ice-volcanic melt The semi-fluid or fluid material associated with ice volcanism, and like molten rock, can have a wide variety of viscosities and other flow properties.

ice volcanism The eruption of molten ice or gas-driven solid fragments onto the surface of a planetary body.

Ida In Greek mythology, the mountain on Crete where Zeus spent his childhood.

igneous rock Rock solidified from a molten state.

impact crater A crater formed on a surface by the collision of a projectile.

impactor An object which strikes the surface of a celestial body

infrared Electromagnetic radiation with wavelengths longer than those perceived by the eye (visible light) but shorter than radio waves; objects between room temperature and 1,000 K emit infrared radiation.

insolation The energy in sunlight.

International Astronomical Union The governing body which – among other things – assigns names to features on planetary bodies.

Io [Eye-oh] One of the large moons of Jupiter. It was discovered by Galileo Galilei in 1610. In Greek mythology, Io was a young woman, the daughter of Inachus, who was seduced by Zeus and transformed into a cow to protect her from his jealous wife, Hera, but Hera recognised her and sent a gadfly to torment her. Io, maddened by the fly, wandered throughout the Mediterranean region.

ionosphere That region in the upper atmosphere which is sufficiently ionised by ultraviolet solar radiation for the concentration of free electrons to significantly affect the propagation of radio waves.

Jovian Of or relating to the planet Jupiter.

JPL The Jet Propulsion Laboratory in Pasadena, California.

Jupiter The largest and most massive of the planets was named Zeus by the Greeks and Jupiter by the Romans; he was the most important deity in both pantheons. It is the fifth in order from the Sun.

Kelvin The fundamental unit of temperature. It is not calibrated in terms of the freezing and boiling points of water, but in terms of energy itself. The Kelvin scale's zero point is defined to be the lowest possible temperature, also referred to as absolute zero.

kinetic energy The energy of motion; equivalent to one-half an object's mass multiplied by its velocity squared.

latitude The angular distance north or south from the equator.

leading hemisphere The hemisphere that faces forward, into the direction of motion of a satellite that keeps the same face towards the planet.

Leda [LEE-duh] A small moonlet of Jupiter. It was discovered by Charles Kowal in 1974. Leda was seduced by Zeus in the form of a swan, she was the mother of Pollux and Helen.
limb The outer edge of the apparent disk of a celestial body.
lineament A linear topographic feature, such as a fault line, aligned volcanoes, or straight stream course.
longitude The angular distance east or west from the prime meridian.
Lysithea [ly-SITH-ee-uh] A small moonlet of Jupiter discovered by Seth Nicholson in 1938.

mafic An igneous rock rich in magnesium and iron minerals.
magma Molten rock material (liquids and gases).
magnetic The property of a material to attract iron, cobalt, or nickel.
magnetosphere The volume around a magnetic body in which charged particles are subject to the bodies magnetic field rather than the magnetic field of another body, such as the Sun or Jupiter.
magnetotail The portion of a planetary magnetosphere pulled downstream by the solar wind.
main ring The brightest of the Jovian rings reaches from the halo's outer boundary at 122,500 kilometres from the planet's centre across 6,440 kilometres to 128,940 kilometres just interior to Adrastea's orbit at 128,980 kilometres. At its outer edge the main ring takes nearly 1,000 kilometres to develop its full brightness. The ring's brightness noticeably decreases around 127,850 kilometres in the vicinity of Metis. The precise location and nature of the main ring's outer periphery might shift slightly from image to image.
mantle The main bulk of a planet between the crust and the core; on Earth, the mantle ranges from about 40 to 2,900 kilometres below the surface.
mare A low-lying lunar plain of dark lava.
mass wasting The downslope movement of rock, regolith, and/or soil under the influence of gravity.
massif A massive topographical feature, commonly formed of rocks more rigid than those of its surroundings, which is displaced as a block.
mesosphere The region in the atmosphere above the stratosphere in which the temperature decreases with increasing altitude.
meteorite A stony or metallic object from interplanetary space that impacts a planetary surface.
Metis This Jovian moon is embedded within the main ring, discovered by Steve Synnott, a member of the Voyager team, in 1979. Its longest dimension is about 60 kilometres. It was named for the first wife of Zeus; he swallowed her when she became pregnant; Athena was subsequently born from the forehead of Zeus.
micron A unit of distance equivalent to one-millionth (10^{-6}) of a metre; a micrometer.
mid-ocean ridge A continuous mountain range with a central valley, located on the ocean floor where two tectonic plates move away from each other allowing molten rock from the Earth's interior to move towards the surface.
Moons of Jupiter Sixteen moons have been discovered since 1610. In addition to Io, Europa, Ganymede, and Callisto (the four largest Galilean moons), Jupiter has two groups of small satellites. Four orbit closer than Io, and eight orbit well beyond Callisto. The IAU decided to assign the outer moons with prograde orbits names ending in 'a' and those orbiting in a retrograde manner names ending with 'e', so they are: Leda, Himalia, Lysithea and Elara, and Ananke, Carme, Pasiphae and Sinope. They are almost certainly captured asteroids.
morphology The the external structure, form, and arrangement of rocks in relation to the development of landforms.

NASA The National Aeronautics and Space Administration.
nephelometer An instrument that uses measurements of scattered light from a laser beam to detect and study cloud particles.
nucleus The frozen core of a comet which contains almost the entire cometary mass and is located in the comet's head.

occultation A period of time when the view to one celestial body is blocked by the body of another.
OPNAV Galileo SSI images were taken to support optical naviation. Such images typically consisted of the limb of one main body (Jupiter or a satellite) and three to four stars.
opposition A configuration in which a celestial body is in the opposite direction (180 degrees) from the Sun as seen from another body located along the line between them (e.g. when Mars, Earth and the Sun are located along a straight line, Mars is in opposition as seen from Earth).
orbit The path followed by an object in space as it goes around another object.
orbital period The time required for an object to make a complete revolution along its orbit.
orbital resonance A relationship in which the orbital period of one body is related to that of another by a simple integer fraction.
organic compounds Complex chemical compounds that contain carbon.
OWLT One-way light-time is the time a signal takes to travel from a spacecraft to Earth, or from the Earth to a spacecraft.

pahoehoe A basaltic lava with a smooth, undulating surface.
palimpsest A roughly circular albedo spot on an icy satellite that marks the site of a crater and its rim deposit. Little, if any, of the topographic structure exists, but visual distinction from adjacent crust remains.
particles and fields instruments A complement of instruments on a spacecraft designed to provide data on the structure and dynamical variations of a magnetosphere.
Pasiphae [pah-SIF-ah-ee] A small moonlet of Jupiter, discovered by P Melotte in 1908. Named for the wife of Minos; mother of the Minotaur.
perijove The closest point to Jupiter in an orbit about that planet.
perihelion The point in the path of a planet, asteroid, comet, or other body that is closest to the Sun.
phase angle The angle between the Sun, an object, and an observer. Zero degrees of phase means that the Sun is behind the observer. A low phase angle provides high Sun illumination, similar to that at noon (with the Sun directly overhead). Such illumination emphasises the brightness contrasts of light and dark areas. High phase corresponds to sunset or sunrise. Such illumination emphasises the topography of the terrain.
Phobos In Greek mythology, a son of Ares (Mars) who, with brother Deimos, was a constant companion to his father.
phosphene PH_3.
photometry The accurate quantitative measurement of the amount of light received from an object or area.
pixel The contraction of 'picture element'. It is the area on the ground represented by each digital number in a digitised image; an individual element in a detector.
planetesimals Primordial bodies of intermediate size that accreted into planets or asteroids.
plasma A highly ionised gas, consisting of almost equal numbers of free electrons and positive ions.
plasma sheet A low-energy plasma, largely concentrated within a few planetary radii of the

equatorial plane, distributed throughout the magnetosphere throughout which concentrated electric currents flow.

plate tectonics A geological model in which the Earth's lithosphere (crust and uppermost mantle) is divided into a number of more-or-less rigid segments which move in relation to one another.

plateau Any comparatively flat area of great extent or elevation.

polarimetry The measurement and study of the polarisation of light reflected off of a surface.

polarisation The suppression of the vibration of light waves in a certain direction.

radiation Energy that is emitted in the form of electromagnetic waves.

raw data The original data we receive from the spacecraft before it has been processed or corrected.

reddening The phenomenon of the trailing hemisphere of a planetary body being darker at shorter wavelengths ('redder') than the leading hemisphere. This effect may be due to magnetospheric bombardment acting preferentially on the trailing hemisphere and impact gardening on the leading hemisphere. Of the Galilean satellites, Europa displays this effect most prominently, and Ganymede to a lesser extent.

refraction The bending of light due to a change in its velocity as it passes the boundary between two materials.

refractive index The ratio of the velocity of light in a vacuum to the velocity of light with in a material.

regolith The layer of pulverised rocky or icy debris and dust made by meteoritic impact that forms the uppermost surface of planets, satellites and asteroids.

relative age The age of one thing, such as a geological unit, in comparison to another. Relative ages are usually determined by cross-cutting relationships and the number of impact craters on planetary surfaces.

relief Variation in elevation.

resolution Ability to distinguish visual detail, usually expressed in terms of the size of the smallest features that can be distinguished.

resurfacing Creation of a new surface on a planetary body by volcanic or tectonic processes.

rift valley A valley developed at a divergence zone or other area of extension.

rotation The turning or spinning of a body about an axis running through it.

satellite Any object, man-made or natural, that orbits another body.

satellite wake A region created in front of the Galilean satellites as the charged particles that corotate with the Jovian magnetosphere sweep past the satellites.

saturation The condition in which a surface becomes completely covered with craters, such that the addition of new craters does not increase the overall crater density.

scarp A line of cliffs produced by faulting or erosion; a topographic boundary; a relatively straight, cliff-like face or slope of considerable linear extent, breaking the general continuity of the land by separating surfaces lying at different levels.

SCET Spacecraft-event-time, is the time of an event as experienced by the spacecraft (as opposed to ERT).

seafloor spreading The movement of two oceanic plates away from each other, resulting in the construction of a mid-oceanic ridge.

shearing The motion resulting from stresses that cause or tend to cause contiguous parts of a body to slide relatively to each other.

shield volcano A broad volcanic cone with gentle slopes constructed of successive non-viscous, mostly basaltic, lava flows.

Shoemaker Levy 9 The comet that broke up and fell into Jupiter in June 1994.
silicates A group of minerals constituting about 95 percent of the Earth's crust, and containing silicon and oxygen combined with one or more other elements.
Sinope [sah-NOH-pee] A small moonlet of Jupiter, discovered by Seth Nicholson in 1914. It was named after the daughter of the river gods Asopus and Merope; she was abducted by Apollo.
slumping A landslide that results from the downward sliding of rock debris as a single mass, usually with a backward rotation relative to the slope along which the movement takes place.
solar conjunction A period of time during which the Sun is in or near the spacecraft–Earth communications path, thus corrupting the communications signals.
South Atlantic Anomaly Located off the coast of Argentina, south of the Brazilian border, this is an 'anomalous' region where the innermost of the Earth's radiation belts, a torus around equatorial latitudes at an altitude of 1,000–5,000 kilometres, dips several hundred kilometres towards the surface. A spacecraft passing through this zone is exposed to a more intense dose in a few minutes than it accumulates during the rest of the orbit.
spectrometer An instrument that measures the wavelength of electromagnetic radiation.
stratigraphy The layering of rock or ice strata, from which information on succession, age relations, and origin can be deduced.
stratosphere An upper portion of the atmosphere, above the troposphere and below the mesosphere where the temperature increases with height. Convection does not take place in the stratosphere.
subduction zone A place on the surface of the Earth where two plates move towards each other, and the oceanic plate plunges beneath the other tectonic plate.
sublimation When a substance changes directly from a solid state to a gaseous state without becoming liquid.
Sub-jovian The point on a moon of Jupiter directly facing the planet.
sub-solar The point on a body that is directly beneath the Sun. The Sun's rays will hit this point at 90 degrees to the surface.
sulcus A complex area of subparallel furrows and ridges.

tectonic Relating to the structural deformation, especially folding and faulting of the crust of a moon or planet, the forces involved in or producing such deformation, and the resulting forms.
tectonism The processes of faulting, folding or other deformation of the lithosphere of a planetary body, often the result of large-scale internal movements below the lithosphere.
terminator The line of sunrise or sunset on a planet or its satellite. At dawn and dusk when the Sun is lowest in the sky (low Sun), topographic features cast their longest shadows. This reveals information about the size and shape of the objects casting the shadows. Therefore, features near the terminator are imaged in order to obtain morphological information.
terrain The surface features of an area of land.
terrestrial The class of planets that are similar to the Earth in density and composition (i.e. Mercury, Venus and Mars) and also, to some extent, to the large satellites of the outer planets.
texture The general appearance or character of a rock (e.g. the size, shape and arrangement of its constituent elements).
Thebe This small moonlet of Jupiter, discovered by Steve Synnott, a member of the Voyager team, in 1979. It was named after a nymph abducted by Zeus, and also the namesake of the Greek city of Thebes. It orbits near the outer periphery of the outer gossamer ring. Galileo

images show the satellite to have bright spots near the rim of one large crater. Thebe's longest dimension is 116 kilometres.

thermosphere The region of the upper atmosphere which is distinguished by a progressive increase in temperature with altitude. It lies above the mesosphere and includes the exosphere and the ionosphere.

Titan In Greek mythology, Titans were the firstborn children of Uranus (the sky) and Gaea (the Earth). The ruler of the Titans was Cronos, whose Roman name is Saturn.

topography The general configuration, shape and form of the surface of a planet.

torus In context with the Galileo mission, the doughnut-shaped cloud of plasma surrounding Jupiter near Io's orbit, believed to be supplied by the volcanic eruptions on Io.

trailing hemisphere The hemisphere that faces backwards, away from the direction of motion of a satellite that keeps the same face towards the planet.

tropopause The level in an atmosphere which marks the transition from the troposphere to the stratosphere.

troposphere The lowermost part of an atmosphere in which the temperature decrease with increasing height. The troposphere is dominated by convection cells.

trough A long linear depression.

Van Allen belts The Earth's magnetosphere contains a pair of toroidal radiation belts. The inner belt range from 1,000 to 5,000 kilometres above the equator, and contains electrons and protons which have escaped from the atmosphere as a consequence of interactions with cosmic rays. The outer belt (13,000 to 20,000 kilometres) comprises particles originating from the Sun.

vent An opening or fissure in a planet's surface through which volcanic material erupts.

viscosity The measure or property of a material to be resistant to flow; the internal friction of a material. Materials with a high viscosity are more resistant to flow, while those with a low viscosity are more fluid.

viscous relaxation The process whereby topographic features become subdued over time due to the flow of the surrounding geological material.

volcanic rock Rock formed by eruption onto a planet's surface.

volcanism The eruption of molten material or gas-driven solid fragments onto the surface of a planetary body. On icy moons the phenomenon is sometimes called cryovolcanism to distinguish it from silicate or sulphur volcanism.

volcano A mountain formed from the eruption of igneous matter through a source vent.

watt A unit of power, equivalent to one joule of energy per second.

wavelength The distance between successive crests or troughs in a wave.

weathering The chemical and physical alteration of materials exposed to the environment on or near the surface of a planetary body.

zeus Another name for the god, Jupiter.

Notes and References

CHAPTER 1: EARLY DAYS

1 The name 'telescope' was not coined until 1612, when it was suggested to an Italian cardinal by Ionnes Dimisiani, a Greek mathematician, because it meant 'to see at a distance'.
2 Galileo's first telescope magnified only 9 times, so he could not see very much using it. His improved instrument, with which he discovered that Jupiter possessed satellites, magnified 30 times.
3 For accounts of Galileo's discoveries see 'Galileo's Telescopic Observations', G V Coyne, and 'The Discovery by Galileo of Jupiter's Moons', E Bellone, both in '*The Three Galileos' Conference*, 1997.
4 Copernicus developed the idea in 1512, and wrote it up in a book in 1530, but this was not published until his death in 1543.
5 In fact, Galileo was found guilty of heresy by the Church for advocating Copernicus's theory, and he was placed under house arrest. When he died in 1642, he was still essentially a prisoner in his own home. It was nearly 400 years before the Church issued a formal apology.
6 'How Galileo Changed the Rules of Science', Owen Gingerich. *Sky & Telescope*, March 1993.
7 Galileo announced his discovery in March 1610 in a periodical entitled *The Starry Messenger*.
8 As Marius explained his rationale for naming the moons: "Jupiter is much blamed by the poets on account of his irregular loves. Three maidens are especially mentioned as having been clandestinely courted by Jupiter with success: Io, the daughter of Inachus; Callisto of Lycaon; Europa of Agenor; and Ganymede, the son of King Tros...".
9 *Satellites of the Solar System*, Werner Sandner. The Scientific Book Club, 1965.
10 One Astronomical Unit (AU) is defined to be the average radius of the Earth's orbit around the Sun, and it can reasonably be considered to be 150 million kilometres.
11 *Jupiter*, Garry Hunt and Patrick Moore. Mitchell Beazley, 1981.
12 *The Planet Jupiter*, Bertrand M Peek. Faber & Faber, 1958.

CHAPTER 2: RECONNAISSANCE

1 One Jovian radius (Rj) is 71,500 kilometres.
2 *Pioneer Odyssey*, Richard Fimmel, William Swindell and Eric Burgess, NASA SP-396, 1977.

3 'Pathfinder to the Rings – The Pioneer Saturn Trajectory Decision', Mark Wolverton. *QUEST* (The History of Spaceflight, University of North Dakota), 2000.
4 The Pioneer Venus mission flew in 1978. It placed a satellite in orbit around Venus and successfully directed four probes into its atmosphere.
5 *Journey into Space – The First Thirty Years of Space Exploration*, Bruce Murray Norton Co., 1989.

CHAPTER 3: GALILEO'S ORDEAL

1 *The Space Shuttle – Roles, Missions and Accomplishments*, David Harland. Wiley–Praxis, 1998.
2 *Journey into Space – The First Thirty Years of Space Exploration*, Bruce Murray. Norton Co., 1989.
3 'Galileo: Mission To Jupiter', J R Casani, IAF-81-204, IAF (International Astronautics Federation) September 1981.
4 Asteroids are assigned numbers as well as names, and it is traditional to cite both when discussing them.
5 D V Byrnes, Lou D'Amario and R E Diehl. AAS 87-420, *AAS/AIAA Astrodynamics Conference*, 1987.

CHAPTER 4: AN EXPLORING MACHINE

1 In contrast to the 3-tonne Galileo spacecraft, the Pioneers were 260 kg and the Voyagers were 815 kg.
2 'The Near-Infrared Mapping Spectrometer Experiment on Galileo', R W Carlson, P R Weissman, W D Smythe and J C Mahoney. *Space Science Reviews*, 60, pp. 457–502, 1992.
3 'Near-Infrared Mapping Spectrometer for Investigation of Jupiter and its Satellites', I Aptaker. SPIE [Society of Photo-optical Instrumentation Engineers] Vol. 331, *Instrumentation in Astronomy IV*, 182, 1982; SPIE Vol. 366, *Modern Utilisation of IR Technology VIII*, 96, 1982; SPIE Vol. 395, *Advanced IR Sensor Technology*, 132, 1983; SPIE Vol. 834, *Imaging Spectroscopy II*, 196, 1987.
4 'Design and Test of the Near-Infrared Mapping Spectrometer (NIMS) Focal Plane for the Galileo Jupiter Orbiter Mission', G Bailey. SPIE Vol. 197, *Modern Utilisation of Infrared Technology*, 210, 1979.
5 Methane, ammonia, and water vapour act as 'greenhouse' gases, and trap energy within the Jovian atmosphere.

CHAPTER 5: THE LONG HAUL

1 In fact, Voyager 2 had passed Neptune only two months earlier, in August 1989.
2 When Pioneer Venus Orbiter finally dipped into the atmosphere and burned up in October 1992, its 14 years of data had far exceeded expectation.
3 A software error prompted Galileo's solid-state imaging system to 'take' 452 additional images immediately after closest approach to Venus, but these were not recorded on tape.
4 The Venusian jetstream flows east to west because the planet rotates in a retrograde manner.
5 In 1984, David Allen at the Anglo-Australian Observatory in Australia made the fortuitous discovery that there are 'windows' through the Venusian atmosphere at 1.7 and

2.3 microns. These wavelengths provided the first means of directly observing the 'middle' and 'lower' regions of the atmosphere, below the top-level sulphuric acid cloud deck. See *Venus Revealed*, David Grinspoon. Addison-Wesley, 1997.

6 *Galileo NIMS Imagery of the Surface of Venus*, Robert Carlson, K H Baines, M Giraud, L W Kamp, P Drossart, T Encrenaz and F W Taylor. JPL Technical Report.

7 *Exploring the Moon – The Apollo Expeditions*, David Harland. Springer–Praxis, 1999.

8 The average cow produces 600 litres of methane per day.

9 *Life on Other Worlds and How to Find It*, Stuart Clark. Springer–Praxis, 2000.

10 'Earth as a Planet – Atmosphere and Oceans', Timothy Dowling, in *Encyclopedia of the Solar System*, Paul Weissman, Lucy-Ann McFadden and Torrence Johnson (Eds). Academic Press, 1999.

11 Carl Sagan *et al. Nature*, October 1993.

12 Stony meteorites are classified as either chondrites or achondrites depending upon whether they contain small spherules (chondrules). Chondrites are classified as 'carbonaceous' if they contain carbon and hydrated minerals, and as 'ordinary' if they do not. The first observation to link meteorites compositionally to a particular asteroid was linking the basaltic achondrites to 4 Vesta. The first asteroidal link for low-iron type chondrites was 3628 Boznemcova, reported in Science in 1993 by Richard Binzel of MIT (see *Astronomy Now*, p. 12, February 1994). Carbonaceous chondrites comprise barely 2 percent of recovered meteorites.

13 *Asteroids – Their Nature and Utilisation*, Charles Kowal. Wiley–Praxis, 1996.

14 *Galileo NIMS Thermal Observations of Asteroid 951 Gaspra*, P R Weissman, Robert Carlson, William Smythe, L C Byrne, A C Ocampo, L Kamp, H H Kieffer, Laurence Soderblom, F P Fanale, J C Granahan and T B McCord. JPL Technical Report.

15 One of the mysteries concerning asteroids and meteorites was that the most common type of meteorites are the ordinary chondrites and the most common type of asteroid likely to have produced them are 'stony'. It was discovered that the solar wind sputtering that darkens and reddens the asteroidal regolith can cause the spectra of powdered ordinary chondrites to closely resemble those of 'S'-type asteroids. This appeared to resolve the long-standing conundrum. See: 'Space Weathering in the Asteroid Belt', B Hapke. LPSC, 2000; and also: 'S-Type Asteroids, Ordinary Chondrites, and Space Weathering: The Evidence from Galileo's Fly-bys of Gaspra and Ida', Clark Chapman. *Meteoritics & Planetary Science*, 1996.

16 'Cratering on Gaspra', Clark Chapman, Joseph Veverka, Michael Belton, Gerhard Neukum and David Morrison. *Icarus*, 120, p. 231, 1996.

17 'Reanalysis of the Galileo SSI/NIMS-Derived Spectra of Asteroid 951 Gaspra', M S Kelley, J C Granahan, P A Abell, M J Gaffey and A Prudente. *LPSC, (Lunar and Planetary Sciences Conference)*, 2000. This study confirmed that Gaspra is an olivine-rich asteroid with a high olivine-to-pyroxene abundance ratio. It identified two substantial olivine-rich surface units differing only slightly in their olivine-to-pyroxene abundance ratios.

18 *Introduction to the Ulysses Encounter with Jupiter*, Edward Smith, Edgar Page and Klaus-Peter Wenzel. JPL Technical Report; and in *J. Geophys. Res.*, 1992.

19 'New Voyager 1 Hot Spot Identification and the Heat Flow on Io', Alfred McEwen, N R Isbell, K E Edwards and John Pearl. *Bulletin of the American Astronomical Society*, 24, p. 935, 1992.

20 Referred to in 'News Update', *Astronomy Now*, December 1992.

21 'Galileo Imaging Observations of Lunar Maria and Related Deposits', Ron Greeley *et al. J. Geophys. Res.*, 1993.

22 'Navigation of the Galileo Spacecraft', Lou D'Amario. '*The Three Galileos*' *Conference*, 1997.

23 'Galileo's Telecommunications Using the Low-Gain Spacecraft Antenna', Joe Statman and Les Deutsch. '*The Three Galileos' Conference*, 1997.

24 It had shut down the gyroscopic inertial reference system on the scan platform.
25 'Cratering on Ida', Clark Chapman, Eileen Ryan, William Merline, Gerhard Neukum, Roland Wagner, Peter Thomas, Joseph Veverka, and Robert Sullivan. *Icarus*, 120, p. 77, 1996.
26 *Galileo NIMS Thermal Observations of Asteroid 243 Ida and 1993(243)1*, P R Weissman, Robert Carlson, Marcia Segura, William Smythe, Dennis Matson, Torrence Johnson, H H Kieffer, Laurence Soderblom, F P Fanale, J C Granahan and T B McCord. JPL Technical Report.
27 See, for example, '444 Gyptis: A Double Asteroid?', Mark Kidger, Richard Casas and Javier Sanchez. *Astronomy Now*, August 1990.
28 *Radar Investigations of Asteroids*, Steven Ostro. JPL Technical Report.
29 The nucleus of Halley's Comet, which had been closely inspected by the Giotto probe in 1986, was dark due to a tarry 'skin' of hydrocarbons.

CHAPTER 6: TARGET IN SIGHT

1 The RTG booms could be adjusted to control the balance of the spinning section as mass (that is, propellant) was consumed, and also to eliminate wobble by holding the moment of inertia on the centreline.
2 Navigational (OPNAV) images were simpler to transmit because the planet did not fill the field of view and most of the image was black sky.
3 Galileo has a small amount of solid-state memory, but it had been installed as 'working space' for the main computer rather than for mass storage.
4 Unless stated otherwise, times are UT/SCET (spacecraft event time).
5 'Discovery of Currently Active Extraterrestrial Volcanism', Linda Morabito, S P Synnott, P N Kupferman, S A Collins. *Science*, 204, p. 972, 1979.
6 'Melting of Io by Tidal Dissipation', Stanton Peale, Patrick Cassen and Ray Reynolds. *Science*, 203, p. 892, 1979.
7 *Manoeuvre Design for Galileo Jupiter Approach and Orbital Operations*, Michael Wilson, Christopher Potts, Robert Mase, Allen Halsell and Dennis Byrnes. JPL Technical Report.
8 *Final Galileo Propulsion System In-Flight Characterisation*, Todd Barber, Friedrich Krug and Klaus-Peter Renner, AIAA 97-2946, 1997.
9 'Volcanic Resurfacing of Io – Post-Repair HST Imaging', John Spencer, Alfred McEwen, Melissa McGrath, P Sartoretti, D B Nash, K S Noll and D Gilmore. *Icarus*, 127, p. 221, 1997.
10 The probe's timeline was defined in terms of its entry interface, which was at 22:04 UT/SCET.
11 The L-Band antenna was mounted on the despun part of the spacecraft opposite the scan platform, so that it would be able to be held facing the probe during its penetration of the Jovian atmosphere.
12 The Huygens probe which is to be deployed by the Cassini spacecraft is not meant for Saturn, but for its mysterious cloud-enshrouded moon, Titan.
13 Mars Observer had been lost while preparing its engine to enter orbit around Mars in 1993.

CHAPTER 7: ATMOSPHERIC PROBE

1 Interestingly, helium gained its name because it was first identified in early spectroscopic observations of the Sun (Helios) by Norman Lockyer in England in 1866.
2 'Chemistry and Clouds of Jupiter's Atmosphere – A Galileo Perspective', Sushil Atreya, M H Wong, Tobias Owen, H B Niemann and P R Mahaffy. '*The Three Galileos' Conference*, 1997.

3 'On the Origins of Jupiter's Atmosphere and the Volatiles on the Medicean Moons', Tobias Owen, Sushil Atreya, M H Wong, P R Mahaffy and N B Niemann. '*The Three Galileos' Conference*, 1997.
4 *Nature*, 18 November 1999.
5 *Origin and Evolution of Planetary Atmospheres*, Alan Boss, G E Morfill and W M Tsharmutter. University of Arizona Press, 1989.
6 'Galileo Probe Measurements of the Deep Zonal Winds of Jupiter', David Atkinson. '*The Three Galileos' Conference*, 1997.
7 It has recently been discovered that terrestrial thunderstorms send lightning upwards, discharging to the ionosphere, so such lightning may well take place in Jupiter too. Interestingly, the terrestrial ionospheric discharges are an order of magnitude more powerful than the discharges to the ground.

CHAPTER 8: THE CAPTURE ORBIT

1 At 715,000 kilometres, Galileo's perijove was twice as far from Jupiter's centre as the Moon is from the Earth's, but giant Jupiter's diameter is ten times that of the Earth.

CHAPTER 9: TECTONIC GANYMEDE

1 'The Geology of Ganymede', Gene Shoemaker, B K Lucchitta, J B Plescia, S W Squyres and Don Wilhelms, in *Satellites of Jupiter*. University of Arizona Press, 1982.
2 David Senske, James Head, Robert Pappalardo, G Collins, Ron Greeley, K Magee, Gerhard Neukum and Clark Chapman. *LPSC*, 1997.
3 'Topography on Satellite Surfaces and the Shape of Asteroids, Torrence Johnson and T R McGetchin. *Icarus*, 18, p. 612, 1973.
4 'Pedestal Craters on Ganymede', V M Horner and Ron Greeley. *Icarus*, 51, p. 549, 1982.
5 The regio were named for Galileo Galilei, Simon Marius, Seth Nicholson, Edward Barnard and Charles Perrine.
6 'Evolution of Planetary Lithospheres: Evidence from Multi-Ringed Structures on Ganymede and Callisto', William McKinnon and H J Melosh. *Icarus*, 44, p. 454, 1980.
7 'Constraints on the Expansion of Ganymede and the Thickness of the Lithosphere', Matthew Golombek. *J. Geophys. Res.*, 87 (supplement), p. 77, 1982.
8 Water has the interesting property that its solid phase is less dense than its liquid phase, so it expands when it freezes (which happens to be why pipes burst in winter).
9 *NIMS Science Objectives and Observational Plans for Ganymede during the Galileo Tour*, J Hui, William Smythe, Robert Carlson, W M Calvin and H H Kieffer. JPL Technical Report.
10 L Prockter, James Head, David Senske, Gerhard Neukum, R Wagner, U Wolf and Ron Greeley. *LPSC*, 1997.
11 C M Weitz, James Head, Robert Pappalardo, Gerhard Neukum, B Giese, J Oberst, A Cook, B Schreiner, Ron Greeley, P Helfenstein and Clark Chapman. *LPSC*, 1997.
12 'Bombardment History of the Jovian System', Gerhard Neukum. '*The Three Galileos' Conference*, 1997.
13 *Science*, 18 October 1996.
14 The results were not announced until 23 October 1996.
15 'CCD Spectra of the Galilean Satellites – Molecular Oxygen on Ganymede', John Spencer, W M Calvin and M J Person. *J. Geophys. Res.*, 1995.
16 *Possibilities for an Atmosphere of Ganymede*, Claudia Alexander, Torrence Johnson, S J Bolton, William Smythe, John Spencer, W Ip and Robert Carlson. JPL Technical Report.

17 *A Model for the Ganymede Atmosphere*, Claudia Alexander, John Spencer, Robert Carlson, William Smythe, L Frank, B Patterson and S J Bolton. JPL Technical Report.
18 *Modelling Sublimation from Ganymede*, Claudia Alexander, W Ip and S J Bolton. JPL Technical Report.
19 'Gravitational Constraints on the Internal Structure of Ganymede', John Anderson, Ed Lau, W L Sjogren, G Schubert and W B Moore, *Nature*, 384, p. 541, 1996.
20 Strictly speaking, this would be eutectic melting.
21 'Flooding of Smooth Terrain on Ganymede by Low-Viscosity Aqueous Lavas – Direct Evidence from VGR-GLL Stereo Synergism', Paul Schenk, D Gwynn, William McKinnon and J M Moore. *LPSC*, p. 2037, 2000.
22 B Lucchitta. *Icarus*, 1980.
23 Paul Schenk and J Moore. *J. Geophys. Res.*, 1995.
24 James Head *et al.*, *LPSC*, 1998.
25 'Estimate of Areal Coverage of Bright Terrain on Ganymede', Paul Schenk, S Sobieszczyk. *Bulletin of the Astronomical Society of America*, 1999.
26 'A Global Database of Grooves and Dark Terrain on Ganymede, Enabling Quantitative Assessment of Terrain Features', Geoffrey Collins, James Head and Robert Pappalardo. *LPSC*, p.1034. 2000.

CHAPTER 10: BATTERED CALLISTO

1 'Origin of the Valhalla Ring Structure – Alternative Models', W Hale, James Head and E M Parmentier, in *Conference on Multi-Ring Basins*. Lunar and Planetary Institute, Houston, 1980.
2 T B McCord *et al. J. Geophys. Res.*, 103, p. 8603, 1998.
3 'Possible Exogenic and Impact Origins for CO_2 on the Surface of Callisto', C A Hibbitts, T B McCord, G B Hansen and J E Klemaszewski. *LPSC*, 1999.
4 'Distribution of CO_2 and SO_2 on the Surface of Callisto', C A Hibbitts, T B McCord and D B Hansen. *LPSC*, 2000.
5 '*NIMS – Callisto Science Objectives and Observational Plans*, Marcia Segura, William Smythe, Robert Carlson, J Sunshine and T B McCord. JPL Technical Report.
6 K C Bender, K S Homan, Ron Greeley, Clark Chapman, J Moore, C Pilcher, W J Merline, James Head, Michael Belton and Torrence Johnson, *LPSC*, 1997.
7 Clark Chapman, W J Merline, B Bierhaus, J Keller, S Brooks, Alfred McEwen, Randy Tufts, J Moore, Michael Carr, Ron Greeley, K C Bender, R Sullivan, James Head, Robert Pappalardo, Michael Belton, Gerhard Neukum, R Wagner and C Pilcher. *LPSC*, 1997.
8 'Cratering Timescales for the Galilean Satellites', Gene Shoemaker and R F Wolfe, in *Satellites of Jupiter*. University of Arizona Press, 1982.
9 'Bombardment History of the Jovian System', Gerhard Neukum. '*The Three Galileos' Conference*, 1997.
10 'Gravitational Evidence for an Undifferentiated Callisto', John Anderson, Ed Lau, W L Sjogren, G Schubert and W B Moore. *Nature*, 1997.
11 'Model Assessment and Refinement of Multi-Ring Structures on Callisto from Galileo SSI Data Analysis', J E Klemaszewski and Ron Greeley, *LPSC*, p. 2064, 2000.

CHAPTER 11: EUROPAN ENIGMA

1 Geologists prefer to base their initial studies of planetary surfaces on purely observational aspects such as the albedo, colour and texture, so as to distinguish between objective

mapping and subjective theorising concerning the active processes.

2 'The Geology of Europa', B K Lucchitta and Laurence Soderblom, in *Satellites of Jupiter* published by the University of Arizona Press, 1982.
3 *NIMS – Science Observations Designs For Europa*, A Ocampo, Dennis Matson, William Smythe, Robert Carlson and J Alonso. JPL Technical Report.
4 In addition to the Earth (known to have life), Mars (where there may be fossil evidence of ancient, extinct life), and Saturn's moon Titan (in many respects a frozen 'proto-Earth', that is widely regarded as being a likely future source of life).
5 Robert Pappalardo, James Head, Ron Greeley, R Sullivan, Michael Carr and Randy Tufts. *LPSC*, 1979.
6 'Photoclinometric Analysis of Resurfaced Regions on Europa', C Thomas and L Wilson. *LPSC*, 2000.
7 Ganymede's field was itself one-thousandth the strength of the Earth's magnetic field.
8 A variety of salts exist, not just 'common' salt.
9 Icebergs, of course, float with 90 percent of their bulk below the water-line.
10 'A Sea Ice Analog for the Surface of Europa', Robert Pappalardo and Max Coon. *LPSC*, 1996.
11 'Order from Chaos: Determining Regional Ice Lithosphere Thickness Variations on Europa using Isostatic Modelling of Chaos Regions', Steven Kadel and Ron Greeley. *LPSC*, 2000.
12 Gene Shoemaker, '*Europa Ocean Conference*', San Juan Capistrano, November 1997.
13 'Bombardment History of the Jovian System', Gerhard Neukum. '*The Three Galileos' Conference*, 1997.
14 These E11 and E12 observations were published in *Science* in March 1999.
15 Paul Schenk and William McKinnon. *Icarus*, 79, p. 75, 1989.
16 There had been nine 'safings' in the first few years of the interplanetary cruise, in response to spurious resets from debris within the spin-bearing assembly. After this 'wearing in' period, however, the bearing had operated without interruption. Evidently, the state of the electrical connectors in the bearing assembly were deteriorating in the Jovian environment.
17 This was written up in *Science* in late May 1999.
18 Recall that Europa's rotation is synchronised with its orbit, so diurnal rotation matches orbital motion.
19 'Formation of Cycloidal Features on Europa', Greg Hoppa, Randy Tufts, Richard Greenberg and Paul Geissler. *Science*, 285, p. 1899, 1999.
20 'Thrace Macula, Europa: Characteristics of the Southern Margin and Relations to Background Plains and Libya Linea', Brian Kortz, James Head and Robert Pappalardo. *LPSC*, 2000.
21 L Wilson *et al. J. Geophys. Res.*, p. 9263, 1997.
22 'Salts on Europa's Surface Detected by Galileo's Near Infrared Mapping Spectrometer', Thomas McCord *et al. Science*, 280, p. 1242, 1998.
23 'Behaviour of Hydrated Sulphates and Carbonates under Europa Surface Conditions', Thomas McCord *et al. LPSC*, 2000.
24 It was later realised that this hydrated spectral signature was likely to be hydrated sulphuric acid. See the paper by Robert Carlson, *Science*, 286, pp. 97–99, 1999.
25 'Strike-Slip Duplexing on Jupiter's Icy Moon Europa', L M Prokter, Robert Pappalardo and James Head. *J. Geophys. Res.*, 105, p. 9483, 2000.
26 The preliminary analysis of the data did not indicate any glint which would suggest the presence of hexagonal ice on the surface of Europa at the observed site.
27 Reported in *Science* on 1 October 1999.
28 The closest Europa fly-by had been on orbit 12, when it had passed 200 kilometres above the icy surface.
29 For a thorough review of the geological evidence for a subsurface ocean, see 'Does Europa

have a subsurface ocean? Evaluation of the geological evidence', Robert Pappalardo *et al.*, *J. Geophys. Res.*, 104, p. 24,015, October 1999.
30 See *2010 – Odyssey Two* by Arthur C Clarke, 1982, for example.
31 Published in *Science* on 3 June 1999.
32 Bruce Jakosky and Everett Shock. *J. Geophys. Res.*, 25 August 1998.
33 *Nature*, 27 January 2000.

CHAPTER 12: JUPITER FROM ORBIT

1 The Great Red Spot was recorded in drawings in 1879, and reports of red spots date back to the seventeenth Century.
2 'The Dynamics of Jupiter's Atmosphere from the Galileo Orbiter Imaging System', Don Banfield, M Bell, Peter Gierasch, E Ustinov, Michael Belton, Andrew Ingersoll and Ashwin Vasavada. '*The Three Galileos' Conference*, 1997.
3 'Galileo Infrared Observations of Jupiter', T Encrenaz, P Drossart, M Roos, E Lettouch, Robert Carlson, K Baines, Glenn Orton, T Martin, Fredric Taylor, P Irwin, '*The Three Galileos' Conference*, 1997.
4 Robert Carlson *et al. Science*, 274, p. 385, 1996.
5 Glenn Orton, John Spencer, Larry Travis, T Z Martin and L K Kamppari. *Science*, 274, p. 389, 1996.
6 A similar layer blankets Saturn's moon Titan and prevents us from seeing Titan's surface. Although thinner than Titan's, the Jovian haze is unexpectedly substantial.
7 *Jupiter – The Giant Planet*, Reta Beebe. Smithsonian Institute Press, 1994.
8 'Jupiter's Cloud Structure from Galileo Imaging Data', Don Banfield, Peter Gierasch, M Bell, E Ustinov, Andrew Ingersoll, Ashwin Vasavada, R A West and Michael Belton. *Icarus*, 1998.
9 'Observation of Moist Convection in Jupiter's Atmosphere', Peter Gierasch, Andrew Ingersoll, Shawn Ewald, Don Banfield, Paul Helfenstein, Amy Simon-Miller, Ashwin Vasavada, Herbert Breneman and David Senske. *Nature*, 10 February 2000.
10 Andrew Ingersoll, Ashwin Vasavada, David Senske, Herbert Breneman, William Borucki, Blaine Little and Clifford Anger. *Icarus*, December 1999.
11 Andrew Ingersoll, Peter Gierasch, Don Banfield, Ashwin Vasavada, *Nature*, 10 February 2000.
12 'Supersonic Winds in Jupiter's aurarae', Daniel Rego, Nicholas Achilleus, Tom Stallard, Steve Miller, Renee Prange, Michelle Dougherty and Robert Joseph. *Nature*, 13 May 1999.

CHAPTER 13: MOONLETS AND RINGS

1 In order of range from Jupiter, the four are Leda (Charles Kowal, 1974), Himalia (Charles Perrine, 1904), Lysithea (Seth Nicholson, 1938) and Elara (Charles Perrine, 1905).
2 Harald Krüger *et al. Nature*, 10 June 1999.
3 Interestingly, the dust is derived from surfaces which are predominantly icy.

CHAPTER 14: FIERY IO

1 'Melting of Io by Tidal Dissipation', Stanton Peale, Patrick Cassen and Ray Reynolds. *Science*, 203, p. 892, 1979.

2 'The Jupiter System through the Eyes of Voyager 1' and 'The Galilean Satellites and Jupiter – Voyager 2 Imaging Science Results', both by Brad Smith *et al. Science*, 204, p. 951, and 206, p. 927, 1979.
3 *The Lunar Sourcebook – A User's Guide to the Moon*, Grant Heiken, David Vaniman and Bevan French (eds). Cambridge University Press, 1991.
4 'Heat Flow from Io', Dennis Matson, G A Ransford and Torrence Johnson. *J. Geophys. Res.*, 86, p. 1664, 1981.
5 'Volcanic Eruptions On Io – Implications for Surface Evolution and Mass Loss', Torrence Johnson and Laurence Soderblom, in *Satellites of Jupiter*. University of Arizona Press, 1982.
6 'Two Classes of Volcanic Plume on Io', Alfred McEwen and Laurence Soderblom. *Icarus*, 55, p. 191, 1983.
7 'Volcanic Features on Io', Michael Carr, Hal Masursky, Robert Strom and Richard Terrile. *Nature*, 280, p. 729, 1979.
8 'The Surface of Io: Geological Units, Morphology and Tectonics', Gerry Schaber. *Icarus*, 43, p. 302, 1980.
9 'The Geology of Io', Gerry Schaber, in *Satellites of Jupiter*. University of Arizona Press, 1982.
10 'Volcanic Resurfacing Rates and Implications for Volatiles on Io', Torrence Johnson, A F Cook, Carl Sagan and Laurence Soderblom. *Nature*, 280, p. 746, 1979.
11 The tallest mountain currently measured is Boosaule Mons, which is 16 kilometres high. 'Formation Models of Ionian Mountains', Elizabeth Turtle, W L Jaeger, Laszlo Keszthelyi and Alfred McEwen. *LPSC*, p. 1960, 2000.
12 'Erosional Scarps on Io', Jack McCauley, Brad Smith and Laurence Soderblom. *Nature*, 280, p. 736, 1979.
13 'Dynamics and Thermodynamics of Volcanic Eruptions: Implications for the Plumes on Io', Susan Kieffer, in *Satellites of Jupiter*. University of Arizona Press, 1982.
14 'Initial Galileo Gravity Results and the Internal Structure of Io', John Anderson, W L Sjogren and Gerald Schubert. *Science*, 272, p. 709, 1996.
15 'Io on The Eve of the Galileo Mission', John Spencer and N M Schneider. *Annual Review of Earth and Planetary Sciences*, 24, p. 125, 1996.
16 'Physics and Chemistry of Sulphur Lakes On Io', J I Lunine and D J Stevenson. *Icarus*, 64, p. 345, 1985.
17 'Silicate Volcanism on Io', Michael Carr. *J. Geophys. Res.*, 91, p. 3521, 1986.
18 'Volcanic Eruptions on Io – Heat Flow, Resurfacing and Lava Composition', Diana Blaney, Torrence Johnson, Dennis Matson and Glenn Veeder. *Icarus*, 113, p. 220, 1995.
19 'Thermal Emission from Lava Flows on Io', Robert Howell. *Icarus*, 127, p. 394, 1997.
20 'Volcanic Resurfacing of Io – Post-Repair HST Imaging', John Spencer, Alfred McEwen, Melissa McGrath, P Sartoretti, D B Nash, K S Noll and D Gilmore. *Icarus*, 127, p. 221, 1997.
21 'Sulphur Flows of Ra Patera', D Pieri, S M Baloga, R M Nelson and Carl Sagan. *Icarus*, 60, p. 685, 1984.
22 'Dynamic Geophysics of Io', Alfred McEwen, J Lunine and Michael Carr, in '*Time-Variable Phenomena in the Jovian System*. NASA SP-494, 1989.
23 'Observations of Industrial Sulphur Flows – Implications for Io', Ron Greeley et al. *Icarus*, 84, p. 374, 1990.
24 'Phase Transformations and the Spectral Reflectance of Solid Sulphur – Can Metastable Sulphur Allotropes Exist on Io?', J Moses and D Nash. *Icarus*, 89, p. 277, 1991.
25 'Galileo's First Images of Jupiter and the Galilean Satellites', Michael Belton *et al. Science*, 274, p. 377, 1996.
26 'Geology and Topography of Ra Patera, Io, in the Voyager Era – Prelude to Eruption', Paul Schenk, Alfred McEwen, Ashley Davies, Trevor Davenport, Kevin Jones and Brian Fessler. *Geophys. Res. Lett.*, 24, p. 2467, 1997.

27 'Road log from American Falls to Split Butte', R Greeley and J S King, in *Volcanism of the Eastern Snake River Plain, Idaho: A Comparative Planetary Geology Guidebook.* NASA CR 15554621, p. 295, 1977.
28 'The Snake River Plain, Idaho – Representative of a New Category of Volcanism', Ron Greeley, *J. Geophys. Res.*, 1982.
29 'Active Volcanism on Io as seen by Galileo SSI', Alfred McEwen, Laszlo Keszthelyi, Paul Geissler, Damon Simonelli, Michael Carr, Torrence Johnson, Kenneth Klaasen, Herbert Breneman, Todd Jones, James Kaufman, Kari Magee, David Senske and Gerald Schubert. *Icarus*, 135, p. 181, 1998.
30 'Hot Spots on Io – Initial Results from Galileo's Near Infrared Mapping Spectrometer', Rosaly Lopes-Gautier, Ashley Davies, Robert Carlson, William Smythe, L Kamp, Laurence Soderblom, F E Leader and R Mehlmen. *Geophys. Res. Lett.*, 24, p. 2439, 1997.
31 'Myriads of Small Hot Eruptions on Io', Diana Blaney, D L Matson, Torrence Johnson, Glenn Veeder and Ashley Davies. *LPSC*, 2000.
32 'The Pele Plume – Observations by the HST', John Spencer, P Sartoretti, G E Ballaster, Alfred McEwen, J T Clarke and Melissa McGrath. *Geophys. Res. Lett.*, 24, p. 2471, 1997.
33 Such an observer would have a magnificent view, but would promptly succumb to the radiation.
34 'A History of High Temperature Io Volcanism – February 1995 to May 1997', John Spencer, J Stansberry, C Dumas, D Vakil, R Pregler and M Hicks. *Geophys. Res. Lett.*, 24, p. 2451, 1997.
35 'Io Meteorology – How Atmospheric Pressure is Controlled Locally by Volcanoes and Sulphur Frosts', Andrew Ingersoll. *Icarus*, 81, p. 298, 1989.
36 See *Sky & Telescope*, August 1993.
37 'Active Volcanism on Io as seen by Galileo SSI', Alfred McEwen, Laszlo Keszthelyi, Paul Geissler, Damon Simonelli, Torrence Johnson, Kenneth Klaasen, Herbert Breneman, Todd Jones, James Kaufman, Kari Magee, David Senske, Michael Belton and Gerald Schubert. *Icarus*, 135, p. 181, 1998.
38 'The 1997 Mutual Event Occultations of Io', Robert Howell, John Spencer and J A Stansberry, in *Io During the Galileo Era*. Lowell Observatory, Flagstaff.
39 'High-Temperature Silicate Volcanism on Jupiter's Moon Io', Alfred McEwen, Laszlo Keszthelyi, John Spencer, Gerry Schubert, D L Matson, Rosaly Lopes-Gautier, Kenneth Klaasen, Torrence Johnson, James Head, Paul Geissler, S Gagents, Ashley Davies, Michael Carr, Herbert Breneman and Michael Belton. *Science*, 281, p. 87, 1998.
40 'Earth as a planet: surface and interior', David Pieri and Adam Dziewonski, in *Encyclopedia of the Solar System*. Academic Press, 1999.
41 'Stealth Plumes on Io', Torrence Johnson, Dennis Matson, Diana Blaney, Glenn Veeder and Ashley Davies. *Geophys. Res. Lett.*, 22, p. 3293, 1995.
42 *The Geology of Io*, Gerry Schaber, in *Satellites of Jupiter*. University of Arizona Press, 1982.
43 'New Results on Io's Colour and Composition', Paul Geissler *et al. LPSC*, 2000.
44 'Galileo imaging of atmospheric emissions from Io', Paul Geissler, Alfred McEwen, W Ip, Michael Belton, Torrence Johnson, W H Smythe and Andrew Ingersoll. *Science*, 6 August 1999.
45 *EUV Imaging Spectrography of the Io Plasma Torus*, Bill Sandel, Floyd Herbert, Lyle Broadfoot and A J Dessler. Division for Planetary Sciences of the American Astronomical Society (October 1999).
46 'Galileo Europa Mission Tour Design', Julia Bell and Jennie Johannesen. *AIAA*, August 1997.
47 'Two Classes of Volcanic Plumes On Io', Alfred McEwen and Laurence Soderblom. *Icarus*, 55, p. 191, 1983.
48 'Active Volcanoes on Io as seen by Galileo SSI', Alfred McEwen, Laszlo Keszthelyi, Paul Geissler, Damon Simonelli, Michael Carr, Torrence Johnson, Kenneth Klaasen, Herbert

Breneman, Todd Jones, James Kaufman, Kari Magee, David Senske and Gerald Schubert. *Icarus*, 135, p. 181, 1998.
49 'Modelling Io's Heat Flow – Constraints from Galileo's PPR', J A Rathburn and John Spencer. *LPSC*, 2000.
50 The ultraviolet spectrometer was not to be used during the Io fly-by. It had been switched off because it had been suffering a grating problem for two months. This was to protect the instrument from additional radiation damage. The engineers' hope was that the instrument's damaged electronic components would anneal, restoring the instrument's functionality. Annealing is the process by which heat is applied to a cooling material in order to relieve stresses, change properties, improve machinability, or — in this case — for realignment of atoms in a distorted crystal.
51 For a description of the threat posed by a resurgent calderas such as Yellowstone, see *Apocalypse – A Natural History of Global Disasters*, Bill McGuire. Cassell Co., 1999.
52 'Characteristics of Calderas on Io – Surface Morphology, Sizes and Distribution', Jani Radebaugh, Laszlo Keszthelyi and Alfred McEwen. *LPSC*, 2000.
53 'Eruption Evolution of Major Volcanoes on Io – Galileo Takes a Close Look', A G Davies, Laszlo Keszthelyi, Rosaly Lopes-Gautier, William Smythe, L Kamp and Robert Carlson. *LPSC*, p.1754, 2000.
54 'A Close-Up View of Io in the Infrared – NIMS Results from the Galileo Fly-Bys', Rosaly Lopes-Gautier, William Smythe, Robert Carlson, Ashley Davies, S Doute, Paul Geissler, L W Kamp, S W Kieffer, F E Leader, Alfred McEwen, R Mehlman and Laurence Soderblom. *LPSC*, p.1767, 2000.
55 'The Thermal Structure of Loki seen in Galileo's NIMS Data from the I24 Orbit', William Smythe, Rosaly Lopes-Gautier, L Kamp, Ashley Davies, Robert Carlson, *LPSC*, 2000.
56 Interestingly, the image of Pele's caldera obtained by Galileo in October 1999 includes only about 1 percent of the hot area known to be on the volcano. This indicates that 99 percent of the activity at Pele is in a region that was not imaged in this fly-by.
57 Melissa McGrath, *American Geophysical Union Fall Meeting 1999* (presented by F Bagenal).
58 'Eruption Conditions of Pele Volcano on Io Inferred from Chemistry of its Volcanic Plume', M Yu Zoltov and B Fegley. *LPSC*, 2000.
59 'Lava Channels on Io – Latest Galileo Imaging Results', D A Williams and Ron Greeley. *LPSC*, 2000.
60 Most of the SSI imagery during the Io fly-by in October 1999 was taken using 'fast camera' mode, in which the camera itself preprocessed the image to average the brightness in adjacent parts of the picture prior to storing the image on tape. However, it turned out that radiation had caused the process to become desynchronised, and this type of imagery was seriously degraded.
61 Laszlo Keszthelyi *et al. EOS*, F624, 1999.
62 'Origin of Mountains on Io by Thrust Faulting and Large-Scale Mass Movements', P M Schenk and M H Bulmer. *Science*, 279, p. 1514, 1998.
63 Landslides in Valles Marineris, and around Olympus Mons on Mars are also claimed to be the largest in the Solar System, as indeed are some submarine landslides on Earth.
64 J Radebaugh *et al. LPSC*, p. 1983, 2000.
65 E P Turtle *et al. LPSC*, p. 1948, 2000.
66 'Formation Models of Ionian Mountains', Elizabeth Turtle, W L Jaeger, Laszlo Keszthelyi and Alfred McEwen. *LPSC*, p. 1960, 2000.
67 'Observations of Ionian Mountains', E P Turtle, Laszlo Keszthelyi, Alfred McEwen, M Milazzo and D P Simonelli. *LPSC*, p.1948, 2000.
68 J M Moore *et al. LPSC*, p. 1531, 2000.
69 'Degradation and Deformation of Scarps and Slopes on Io – New Results', J M Moore, R J Sullivan, Robert Pappalardo and E P Turtle. *LPSC*, 2000.
70 'Lava Channels on Io – Latest Galileo Imaging Results', D A Williams and Ron Greeley. *LPSC*, 2000.

71 Laszlo Keszthelyi *et al.* *LPSC*, 1999.
72 'Revisiting the Hypothesis of a Mushy Global Magma Ocean on Io', Laszlo Keszthelyi, Alfred McEwen and Jeffrey Taylor. *Icarus*, 141, p. 415, 1999.
73 'Does Io Have a Mushy Magma Ocean?', Laszlo Keszthelyi, Alfred McEwen and Jeffrey Taylor. *LPSC*, 1999.
74 'Io Spectra from Hubble – Volatile Distribution, Composition, and Processes', Jeffrey Kargel, John Spencer, Laurence Soderblom, T Becker and G Bennett. *LPSC*, 2000.
75 Komatiite basalt has not been extruded onto the Earth's surface since the 'Archaean' era which drew to a close 2.5 billion years ago.
76 'Komatiites from the Commondale Greenstone Belt, South Africa: A Potential Analog to Ionian Ultramafics?', D A Williams, A H Wilson and Ron Greeley. *LPSC*, 1999.
77 'New Results on Io's Colour and Composition', Paul Geissler *et al.* *LPSC*, 2000.
78 'A Close-Up View of Io in the Infrared – NIMS Results from the Galileo Fly-Bys', Rosaly Lopes-Gautier, William Smythe, Robert Carlson, A G Davies, S Doute, Paul Geissler, L W Kamp, S W Kieffer, F E Leader, Alfred McEwen, R Mehlman and Laurence Soderblom. *LPSC*, p.1767, 2000.

CHAPTER 15: PASSING THE TORCH

1 Italian-born Giovanni Domenico Cassini was appointed to run the Paris Observatory in 1671. Upon adopting French citizenship two years later he converted his name to Jean-Dominque Cassini.
2 *Satellites of the Solar System*, Werner Sandner. The Scientific Book Club, 1965.
3 'Galileo's Legacy to Cassini', Dennis Matson and J P Lebreton. '*The Three Galileos' Conference*, 1997.
4 Peter Smith, Mark Lemmon, Ralph Lorenz, John Caldwell, Larry Sromovsky, and Michael Allison (see STScI-PR94-55, December 1994).

Further reading and websites

FURTHER READING

Since each chapter has been appended with notes and sources, this section represents only a few suggestions for books, articles and websites which are well worth seeking out. To provide a sense of time, the further reading list is presented in chronological order.

The Planet Jupiter, Bertrand Peek. Faber & Faber, 1958
Satellites of the Solar System. Werner Sandner, Scientific Book Club, 1965
Eyes on the Universe – A History of the Telescope, Isaac Asimov. Quartet, 1975
Asimov's Biographical Encyclopedia of Science and Technology, Isaac Asimov. Pan, 1975.
Pioneer Odyssey, Richard Fimmel, William Swindell and Eric Burges. NASA SP–396, 1977.
Planets and Moons, William Kaufmann. Freeman Co., 1978.
Voyage to Jupiter, David Morrison and Jane Samz. NASA SP–439, 1980.
Planetary Encounters – The Future of Unmanned Spaceflight, Robert Powers. Warner Books, 1980.
The Solar System and its Strange Objects, Brian Skinner (ed.). Readings from *American Scientist*.
Jupiter, Garry Hunt and Patrick Moore. Mitchell Beasley in conjunction with the Royal Astronomical Society, 1981.
Galileo: The Mission to Jupiter, Clayne Yeates and Theodore Clarke. *Astronomy*, February 1982.
Satellites of Jupiter, David Morrison (ed.). University of Arizona Press, Tucson, 1982
The Planets, Heather Couper and Nigel Henbest. Pan, 1985.
Far Travelers – The Exploring Machines, Oran Nicks. NASA SP–480, 1985.
Planetary Landscapes, Ron Greeley. Allen & Unwin, 1987
Rings – Discoveries from Galileo to Voyager, James Elliot and Richard Kerr. MIT Press, 1987.
Solar System Log, Andrew Wilson. *Jane's*, 1987.
The Cambridge Photographic Atlas of the Planets, Geoffrey Briggs, and Fredric Taylor. Cambridge University Press, 1988.
Journey into Space - The First Three Decades of Space Exploration, Bruce Murray. Norton Co., 1989.
Time-Variable Phenomena in the Jovian System, Michael Belton, Robert West and Jurgen Rahe. NASA SP–494, 1989.
Volcanoes, Robert Decker and Barbara Decker. Freeman Co., 1989.
The Atlas of the Solar System, Patrick Moore and Garry Hunt. Artists House in conjunction with the Royal Astronomical Society, 1990

Exploring the Planets, Kenneth Hamblin and Eric Christiansen. Macmillan Co., 1990.
The New Solar System, Kelly Beatty and Andrew Chaikin (eds). Sky Publishing Corporation, 1990.
The Lunar Sourcebook: A User's Guide to the Moon, Grant Heiken, David Vaniman and Bevan French (eds). Cambridge University Press, 1991.
Wanderers in Space - Exploration and Discovery of the Solar System, Kenneth Lang and Charles Whitney. Cambridge University Press, 1991.
Satellites of the Outer Planets, D Rothery. Clarendon Press, Oxford, 1992.
Moons and Planets, William Hartmann. Wadsworth Co., 1993.
Astronomers Await the Crash of '94, Steve Miller. *Astronomy Now*, January and June 1994.
Awaiting the Crash, Kelly Beatty and David Levy. *Sky & Telescope*, January and July 1994.
Observing Jupiter at Impact Time, Clark Chapman. *Sky & Telescope*, July 1994.
Jupiter – The Giant Planet, Reta Beebe. Smithsonian Institute Press, 1994.
The Giant Planet Jupiter, John Rogers. Cambridge University Press, 1995.
Impact Jupiter – The Crash of Comet Shoemaker Levy 9, David Levy. Plenum Press, 1995.
The Great Comet Crash – The Impact of Comet Shoemaker Levy 9 on Jupiter, John Spencer and Jacqueline Mitton (eds). Cambridge University Press, 1995.
Pale Blue Dot, Carl Sagan. Headline Books, 1995.
Asteroids: Their Nature and Utilisation, Charles Kowal. Wiley–Praxis, 1996.
Volcanoes of the Solar System, Charles Frankel. Cambridge University Press, 1996.
Planetary Volcanism – A Study of Volcanic Activity in the Solar System, Peter Cattermole. Wiley–Praxis, 1996.
'*The Three Galileos' Conference*, Cesare Barbieri, Jürgen Rahe, Torrence Johnson, Anita Sohus (eds). Kluwer Academic Publishers, 1997.
The NASA Atlas of the Solar System, Ronald Greeley and Raymond Batson. Cambridge University Press, 1997.
The Space Shuttle – Roles, Missions and Accomplishments, David Harland. Wiley-Praxis, 1998.
Planetary Astronomy – From Ancient Times to the Third Millennium, Ronald Schorn. Texas A&M University Press, 1998.
Our Worlds - The Magnetism and Thrill of Planetary Exploration, Alan Stern. Cambridge University Press, 1999.
Exploring the Moon – The Apollo Expeditions, David Harland. Springer-Praxis, 1999.
Journey Beyond Selene, Jeffrey Kluger. Little Brown Co., 1999.
Encyclopedia of the Solar System, Paul Weissman, Lucy-Ann McFadden and Torrence Johnson (eds). Academic Press, 1999.
Life on Other Worlds and How to Find It, Stuart Clark. Springer–Praxis, 2000.

WEBSITES

A number of websites are also worth checking regularly for updates on ongoing planetary missions. Over its more than 40 years of existence, NASA's spacecraft have taken hundreds of thousands of pictures of Solar System objects. Much of this image resource is available only to the science community through the National Space Science Data Center, but many of the more recent images are on-line in various forms.

The National Space Science Data Center located at the Goddard Space Flight Center can be browsed at:

<http://nssdc.gsfc.nasa.gov/astro/astro_home.html>.

'Welcome to the Planets' is an educational CD-ROM which includes 190 images acquired over approximately 20 years of planetary exploration. An on-line version with a selection of images is available at:
<http://pds.jpl.nasa.gov/planets>.

The 'Planetary Photojournal' is available at:
<http://photojournal.jpl.nasa.gov>.

The 'Planetary Image Atlas', which can be browsed using keywords, is a comprehensive archive designed to provide access to all of the unprocessed data collected on NASA missions. It is available at:
<http://www-pdsimage.jpl.nasa.gov/PDS/ public/Atlas/Atlas.html>.

The 'Planetary Data System' archives and distributes digital data from NASA missions, astronomical observations and laboratory measurements. It is available at:
<http://pds.jpl.nasa.gov>.

Images intended for Press Releases are available at:
<http://galileo.jpl.nasa.gov/images/images.html>

The Ames Research Centre's site for the Galileo atmospheric probe is:
<http://ccf.arc.nasa.gov/galileo_probe>.

For reflections on the life of Gene Shoemaker, I recommend:
<http://wwwflag.wr.usgs.gov/USGSFlag/Space/Shoemaker>.

For a comprehensive listing of Shoemaker Levy 9 imagery, consult:
<http://www.jpl.nasa.gov/sl9/images.html>.

To keep up to date on scientific observations of Jupiter's satellites, consult the International Jupiter Watch site at:
<http://www.lowell.edu/users/ijw>.

Several NASA Centres have on-line archives of Technical Reports, which can be searched by keyword. The JPL archive is available at:
<http://jpltrs.jpl.nasa.gov>
and the Ames archive is at:
<http://atrs.arc.nasa.gov>.

To follow the progess of the Cassini mission, the issues of the 'Cassini-Huygens Journal' available at:
<http://www.jpl.nasa.gov/cassini/MoreInfo/newslets>.

This is just a partial listing, of course, but each site contains links to others, and browsers are sure to find additional extremely useful resources.

Index

Note that a citation in bold is a reference to an image.